Lecture Notes in Computer Science   8196

Commenced Publication in 1973
Founding and Former Series Editors:
Gerhard Goos, Juris Hartmanis, and Jan van Leeuwen

T0241151

FoLLI Publications on Logic, Language and Information

Subline of Lectures Notes in Computer Science

Davide Grossi   Olivier Roy   Huaxin Huang (Eds.)

Logic, Rationality, and Interaction

4th International Workshop, LORI 2013
Hangzhou, China, October 9-12, 2013
Proceedings

 Springer

Volume Editors

Davide Grossi
University of Liverpool
Department of Computer Science
Ashton Building, Ashton Street
Liverpool L69 3BX, UK
E-mail: d.grossi@liverpool.ac.uk

Olivier Roy
Ludwig-Maximilians-Universität München
Fakultät für Philosophie
Geschwister-Scholl-Platz 1
80539 München, Germany
E-mail: olivier.roy@lrz.uni-muenchen.de

Huaxin Huang
Zhejiang University
Center for the Study of Language and Cognition
Xixi Campus, Tianmushan Road 148
Hangzhou 310028, China
E-mail: rw211@zju.edu.cn

ISSN 0302-9743   e-ISSN 1611-3349
ISBN 978-3-642-40947-9   e-ISBN 978-3-642-40948-6
DOI 10.1007/978-3-642-40948-6
Springer Heidelberg New York Dordrecht London

Library of Congress Control Number: 2013947929

CR Subject Classification (1998): F.4, G.2, I.2.6, F.3, I.2.3

LNCS Sublibrary: SL 1 – Theoretical Computer Science and General Issues

Typesetting: Camera-ready by author, data conversion by Scientific Publishing Services, Chennai, India

Printed on acid-free paper

Springer is part of Springer Science+Business Media (www.springer.com)

Preface

This volume collects the papers presented at LORI-4, the Fourth International Workshop on Logic, Rationality and Interaction, held in Hangzhou, P.R. China, during October 9–12 and hosted by the Center for the Study of Language and Cognition (CSLC) of Zhejiang University.

The workshop received 42 submissions and the final program consisted of 23 full papers and 10 short papers presented at a dedicated poster session. Each paper was selected on the basis of 2 to 4 reviews. The topics covered in this program well represent the span and depth that has meanwhile become a trademark of the LORI workshop series, where logic is interfaced with disciplines as diverse as game theory and decision theory, philosophy and epistemology, linguistics, computer science and artificial intelligence. The technical program of the workshop was further enriched with invited addresses by Giuseppe Dari-Mattiacci, Valentin Goranko, Hannes Leitgeb, Beishui Liao, Christian List, Sonja Smets and Dongmo Zhang.

The LORI series was kickstarted with a first event (LORI-1) hosted in August 2007 by Beijing Normal University in Beijing. That event was a great success providing an effective platform for Chinese and non-Chinese logicians to meet and exchange research ideas. The wish to perpetuate such a platform led to two later editions: LORI-2, hosted by South-West University in Chongqing, and LORI-3, hosted by Sun Yet-sen University in Guangzhou. A history of the series can be accessed at a glance through the dedicated web portal www.golori.org.

As Organization and Program Committee chairs we would like to thank the PC members and all the external reviewers for a truly outstanding job under extremely tight time constraints. The program is greatly indebted to their contribution. Our activity has been further widely supported by the indefatigable work of the LORI Standing Committee: Fenrong Liu and Johan van Benthem. We would also like to acknowledge the use of EasyChair, which has been a fantastic tool for both organizing the reviewing process and creating these proceedings. The final thanks should go to our colleagues from CSLC, for the hard 'ground work' they put into making this workshop happen.

July 2013

Davide Grossi
Olivier Roy
Huaxin Huang

Organization

Program Committee

Natasha Alechina	University of Nottingham, UK
Albert Anglberger	Ludwig Maximilians University, Germany
Alexandru Baltag	University of Amsterdam, The Netherlands
Ulle Endriss	University of Amsterdam, The Netherlands
Nina Gierasimczuk	University of Amsterdam, The Netherlands
Davide Grossi	University of Liverpool, UK
Jiahong Guo	Beijing Normal University, China
Wesley Holliday	University of California at Berkeley, USA
Tomohiro Hoshi	Stanford University, USA
Fangzhen Lin	Hong Kong University of Science and Technology, China
Fenrong Liu	Tsinghua University, China
Yongmei Liu	Sun Yat-sen University, China
Guo Meiyun	South-West University, China
Eric Pacuit	University of Maryland, USA
Henry Prakken	University of Utrecht and University of Groningen, The Netherlands
Ram Ramanujam	Institute of Mathematical Sciences, India
Antonino Rotolo	University of Bologna, Italy
Olivier Roy	Ludwig Maximilians University, Germany
Jeremy Seligman	University of Auckland, New Zealand
Kaile Su	Tsinghua University, China and Griffith University, Australia
Hans van Ditmarsch	LORIA, France
Jan van Eijck	CWI, The Netherlands
Wenfang Wang	National Yang Ming University, China
Yanjing Wang	Peking University, China
Minghui Xiong	Sun Yat-sen University, China
Tomoyuki Yamada	Hokkaido University, Japan
Thomas Ågotnes	University of Bergen, Norway

Additional Reviewers

Benda, Thomas	Bulling, Nils
Berwanger, Dietmar	Feldbacher, Christian
Bex, Floris	Isaac, Alistair

Ju, Fengkui

Paul, Soumya

Simon, Sunil Easaw

Slavkovik, Marija

Steinert-Threlkeld, Shane

Suresh, S.P.

Wang, Linton

Wen, Xuefeng

Shen, Yuping

Wáng, Yì N.

Xiong, Zuojun

Table of Contents

Full Papers

Short Papers

Boolean Games with Epistemic Goals

Thomas Ågotnes[1], Paul Harrenstein[2],
Wiebe van der Hoek[3], and Michael Wooldridge[2]

[1] University of Bergen, Norway
thomas.agotnes@infomedia.uib.no
[2] University of Oxford, UK
{mjw,paulh}@cs.ox.ac.uk
[3] University of Liverpool, UK
wiebe@csc.liv.ac.uk

Abstract. We introduce and formally study games in which the goals of players relate to the epistemic states of players in the game. For example, one player might have a goal that another player knows a certain proposition, while another player might have as a goal that a certain player does not know some proposition. The formal model we use to study epistemic games is a variation of the increasingly popular *Boolean games* model in which each player controls a number of Boolean variables, but has limited ability to see the truth values of the overall set of formulae that hold in the game. Each player in an *epistemic Boolean game* has a goal, defined as a formula of modal epistemic logic. Using such a language for goals allows us to explicitly and compactly represent desirable epistemic states. After motivating and formally defining epistemic Boolean games as a concise representation of epistemic Kripke structures, we investigate their complexity and study their properties.

1 Introduction

In our everyday lives, we all quite naturally have goals and aspirations that relate to the epistemic states (knowledge and belief) of other agents. You want your children to know that you love them; you want your boss to know you work hard; the politician wants you to know he is honest; and so on. The formal analysis of such epistemic states is of course a well-established research topic in artificial intelligence, with modal logic and Kripke semantics being the pre-eminent tools of choice in such work [1]. Our aim in this paper is to begin to extend this research to the *game theoretic* aspects of systems in which the motivations of players relate to the epistemic states of others. Doing so, the following question seems very relevant: *If all players act rationally to bring about their goals, then what epistemic states will result in equilibrium?* We will refer to games in which the goals and preferences of players relate to the epistemic states of other players as epistemic games. In the present paper, we shall restrict our attention to knowledge, leaving the study of belief to future work.

As the basis for our study, we adapt the increasingly popular game theoretic model of *Boolean games* [2–5]. More precisely, we introduce *epistemic Boolean*

D. Grossi, O. Roy, and H. Huang (Eds.): LORI 2013, LNCS 8196, pp. 1–14, 2013.
© Springer-Verlag Berlin Heidelberg 2013

games (EBGs). In such a game, (as in regular Boolean games), each player i is associated with a set of Boolean variables Φ_i, which are under his control in the sense that he can assign Boolean values to the variables Φ_i in any way that he chooses. That is, the strategies available to a player i in an EBG are the possible Boolean assignments that can be made to the variables Φ_i. The outcome of a Boolean game is a valuation for the overall set of Boolean variables Φ, which will be composed of the individual assignments made by the players i in the game to their variables Φ_i. A player i is not assumed to have perfect information of the game. This we formally capture by *visibility sets* Θ_i, that is, sets of propositional formulae. If a formula φ is in Θ_i—in which case we say that "i can see φ"—it means that player i can distinguish states in which φ holds from ones in which this is not the case. A limit case is where every player can see every atomic formula, in which case the game is one of perfect information.

As in regular Boolean games, each player i is assumed to have a goal that he desires to be achieved. In conventional Boolean games, player i's goal γ_i is represented as a formula of propositional logic. In our present work, however, the goal is assumed to be represented as a formula of modal multi-agent epistemic logic [1]. Thus, player i might have the goal that another player j comes to know something ($\gamma_i = K_j\varphi$), while player j might have a goal that another player k does not know something ($\gamma_j = \neg K_k\psi$). As in conventional Boolean games, the ability of a player i to influence whether his goal is achieved lies within the variables Φ_i under his control; but our EBGs bring a new twist to this story, since the visibility sets of each player will have a part in determining whether a players's goal is achieved. The underlying EBG model we use (with sets of controlled variables and visibility sets) derives in part from the work of van der Hoek *et al.* on epistemic logics of propositional control [6], which in turn derives from logics of propositional control [7].

The remainder of this paper is structured as follows. First, in the following section we introduce EBGs. We then formalise Nash equilibria for our games, and investigate the complexity of decision problems relating to Nash equilibria. Sections 4 and 5 concern the conciseness of the representation of the strategic and epistemic situations that EBGs provide with respect to Kripke structures and regular Boolean games, respectively. We conclude in section 6 with a brief discussion of related work and some issues for future research.

2   Epistemic Boolean Games

We adapt the basic model of Boolean games (see, e.g., [2–5]) to model partial information.

Epistemic Logic. Let $\mathbb{B} = \{\top, \bot\}$ be the set of Boolean truth values, with "$\top$" being truth and "$\bot$" being falsity. Throughout the paper, we will use Φ to denote a fixed, finite, non-empty set of Boolean variables, with typical members $p, q, \ldots$ etc. A *valuation* is a total function $v : \Phi \to \mathbb{B}$, assigning truth or falsity to every Boolean variable. Let $\mathcal{V}$ denote the set of all valuations (over Φ). In the interests of brevity, we sometimes use a binary representation of valuations; for example, if $\Phi = \{p, q\}$, then the valuation 01 would be the one making p false and q true, assuming a natural order on the propositional variables.

We make use of a *multi-agent epistemic modal logic*, which we will refer to as $\mathcal{EL}$ [1]. The language of $\mathcal{EL}$ is that of the well-known multi-modal logic S5$_n$. The language $\mathcal{EL}$ extends classical propositional logic with a collection of indexed unary modal operators K_i, where the intended interpretation of a formula $K_i\varphi$ is that "agent i knows φ." Formally, given a set Φ of propositional variables the syntax of $\mathcal{EL}(\Phi)$ is given by the following grammar:

$$\varphi ::= p \mid \neg\varphi \mid \varphi \vee \varphi \mid K_i\varphi,$$

where p is a Boolean variable in Φ and i is an agent. We assume that the remaining classical connectives—"$\wedge$" (conjunction), "$\rightarrow$" (material implication), "$\leftrightarrow$" (material bi-implication)—are defined in the standard way. The propositional fragment of $\mathcal{EL}$ over Φ—i.e., the set of formulae without occurrences of the epistemic operators K_i—we denote by $\mathcal{L}(\Phi)$.

The semantics for the epistemic logic S5$_n$, as formulated in the language $\mathcal{EL}(\Phi)$, is defined with respect to tuples $K = (W, R_1, \ldots, R_n, \pi)$, also referred to as S5$_n$ Kripke structures. Here, W is a non-empty set of *(possible) worlds* and, for each agent i, $R_i \subseteq W \times W$ is an equivalence relation over W. Finally, $\pi : W \times \Phi \rightarrow \mathbb{B}$ is a *valuation* function, indicating the truth value of every Boolean variable in every world. For the purposes of this paper we assume W to be finite. The formulae of $\mathcal{EL}(\Phi)$ are interpreted with respect to *pointed structures*, i.e., pairs of the form (K, w) as follows [1, pp.18–19]: An atom p is true in (K, w) iff $\pi(w, p) = \top$; the clauses for negation and disjunction are standard, and for knowledge, we have

$$(K, w) \models_\mathcal{K} K_i\varphi \quad \textit{iff} \quad \text{for all worlds } w' \text{ with } R_i(w, w'): (K, w') \models_\mathcal{K} \varphi.$$

Epistemic Boolean Games. The standard framework of Boolean games involves a set of agents or players, each of which has a goal formulated as a formula γ_i of classical propositional logic and a set of Boolean variables she controls, in the sense that he has the unique ability to set their value. Each player strives to satisfy her goal by appropriately setting the values of the variables she controls. Every such setting can be analysed as a strategic game in which the strategies of the players are given by the ways they can assign values to the variables they control. A profile of strategies then determines a unique truth value assignment or valuation to all propositional variables. These are taken as the outcomes of the game. Moreover, each player strictly prefers states that satisfy her goal to ones that do not and is indifferent otherwise. The difficulty is that a player's goal γ_i may contain variables controlled by other players $j \neq i$, who will also be trying to choose values for their variables in Φ_j so as to get their goals γ_j satisfied, the satisfaction of which may in turn dependend on the variables Φ_i.

The setting we consider in this paper is similar, except that each player's goal is represented as a *formula γ_i of the epistemic language $\mathcal{EL}(\Phi)$*. In order to evaluate the players' epistemic goals, each player is endowed with a *visibility set Θ_i* consisting of formulae in the propositional fragment $\mathcal{L}(\Phi)$ of $\mathcal{EL}(\Phi)$. The idea is that player i can completely and correctly perceive truth values of the formulae in Θ_i, i.e., player i can only distinguish outcomes that differ with respect to the truth values they assign to some of the formulae contained in Θ_i. A formula of the form $K_i\varphi$ then holds in a particular outcome if φ holds in all outcomes i cannot distinguish from that outcome.

Formally, we define an *epistemic Boolean game* for $\mathcal{EL}(\Phi)$ (hereafter simply "game") as a tuple

$$G = (N, \Phi, \Phi_1, \ldots, \Phi_n, \gamma_1, \ldots, \gamma_n, \Theta_1, \ldots, \Theta_n), \text{ where:}$$

- $N = \{1, \ldots, n\}$ is a set of agents (also called the *players* of the game);
- $\Phi = \{p, q, \ldots\}$ is a finite set of Boolean variables;
- $\Phi_i \subseteq \Phi$ is the set of Boolean variables under the unique control of $i \in N$;
- γ_i is an $\mathcal{EL}(\Phi)$ formula representing the goal of player $i \in N$; and
- $\Theta_i \subseteq \mathcal{L}(\Phi)$ is a finite *visibility set* for player $i \in N$ consisting of propositional formulae, with the intended interpretation that player i is able to correctly observe the truth values of the propositional formulae in Θ_i.

As usual in Boolean games, we will require that $\Phi_i \cap \Phi_j = \emptyset$ for $i \neq j$, and that $\Phi_1 \cup \cdots \cup \Phi_n = \Phi$ (i.e., the sets $\Phi_1, \ldots, \Phi_n$ form a partition of Φ).

A *choice* for player $i \in N$ is a function $v_i : \Phi_i \to \mathbb{B}$, i.e., an allocation of truth or falsity to all the variables under i's control. Let $\mathcal{V}_i$ denote the set of choices for player i. The intuitive interpretation we give to $\mathcal{V}_i$ is that it defines the *actions* or *strategies* available to player i. An *outcome* is a collection of choices, one for each player. Formally, an outcome for a game is a tuple $\vec{v} = (v_1, \ldots, v_n) \in \mathcal{V}_1 \times \cdots \times \mathcal{V}_n$. An outcome uniquely defines an overall valuation for the variables in Φ and for this reason we often treat outcomes for games as if they were valuations, for example, writing $\vec{v}(p)$ to denote the value of variable $p \in \Phi$ under the assignment corresponding to outcome $\vec{v}$. We will also equivocate the set of valuations and the set of outcomes writing $\mathcal{V}$ for $\mathcal{V}_1 \times \cdots \times \mathcal{V}_n$.

To model partial information, for every visibility set Θ_i, we define an equivalence relation $\sim_{\Theta_i}$ over outcomes as follows such that for all $\vec{v}, \vec{v}' \in \mathcal{V}$:

$$\vec{v} \sim_{\Theta_i} \vec{v}' \quad \text{iff} \quad \text{for all } \varphi \in \Theta_i: \ \vec{v} \models \varphi \text{ iff } \vec{v}' \models \varphi.$$

Thus, $\vec{v} \sim_{\Theta_i} \vec{v}'$ if and only if $\vec{v}$ and $\vec{v}'$ agree on the truth value of the formulae in Θ_i. In the interests of readability, where there is no possibility of confusion, we will write $\sim_i$ instead of $\sim_{\Theta_i}$.

One natural constraint to consider in our setting would be that $\Theta_i \subseteq \Phi$, i.e., that player i can only perceive the value of atomic propositions or propositional variables. Such games we will refer to as *atomic games*. Most of the examples in this paper pertain to this setting and we will see that they have quite a specific structure. It would also be natural to require that $\Phi_i \subseteq \Theta_i$, i.e., a player can see his own choice with respect to the variables he controls. However, we will generally not place this as a requirement on games. Furthermore, if G is such that for every player $i \in N$ we have $\Phi \subseteq \Theta_i$, we say that G is a game of *perfect information*: in a game of perfect information, every player can see every propositional variable.

We interpret formulae of $\mathcal{EL}(\Phi)$ with respect to *pointed games*, i.e., pairs $(G, \vec{v})$ consisting of a game $G = (N, \Phi, \Phi_1, \ldots, \Phi_n, \gamma_1, \ldots, \gamma_n, \Theta_1, \ldots, \Theta_n)$ and an outcome $\vec{v}$, as follows (again, we only give the epistemic clause)

$$(G, \vec{v}) \models_{\mathcal{EL}} K_i \varphi \quad \textit{iff} \quad \text{for all valuations } \vec{v}_2 \text{ with } \vec{v} \sim_{\Theta_i} \vec{v}_2 \colon (G, \vec{v}_2) \models_{\mathcal{EL}} \varphi.$$

Observe that this semantics depend on neither $\Phi_1, \ldots, \Phi_n$ nor $\gamma_1, \ldots, \gamma_n$.

We assume that a player i strictly prefers all those outcomes that satisfy its goal γ_i over all those that do not, but is indifferent between outcomes that satisfy its goal, and is indifferent between outcomes that do not satisfy its goal. We define for each player i a *utility function* $u_i \colon \mathcal{V}_1 \times \cdots \times \mathcal{V}_n \to \{0, 1\}$ over outcomes representing these preferences as follows.

$$u_i(\vec{v}) = \begin{cases} 1 & \text{if } (G, \vec{v}) \models \gamma_i, \text{ and} \\ 0 & \text{otherwise.} \end{cases}$$

Example 1. Father (player 1) and Mother player (player 2) reason about picking up their child Baby (player 3) from nursery: either father does this (p), or mother (q). Assume $\Theta_1 = \Phi_1 = \{p\}, \Theta_2 = \Phi_2 = \{q\}$, and $\Theta_3 = \emptyset$. Mother wants to know that Baby is being picked up, either by father or by herself. Father wants to know that Mother knows, but also wishes to avoid both parents showing up at nursery. Baby just wants to be picked up. Formally, the goals are:

$$\gamma_1 = K_1(K_2(p \vee q) \wedge \neg(p \wedge q))$$
$$\gamma_2 = K_2(p \vee q) \text{ and } \gamma_3 = p \vee q$$

The situation (call the game G_1) is depicted in Figure 1(a). Now suppose that Mother decides to pick up Baby, but Father does not, i.e., $\neg p \wedge q$. This obviously fulfils the goal γ_2, and even $K_2(p \vee q) \wedge \neg(p \wedge q)$. Note, however, that Father is still unhappy with this outcome. Indeed, it is easy to see that $\neg \gamma_1$ is true in this outcome. Note that each parent can bring about $p \vee q$, but not the fact that the spouse knows this.

We are now in a position to apply the well-known notion of (pure strategy) Nash equilibrium [8] to EBGs. Formally, an outcome $(v_1, \ldots, v_i, \ldots, v_n)$ is a *Nash equilibrium* if there is no player $i \in N$ and choice $v_i' \in \mathcal{V}_i$ for i such that

$$u_i(v_1, \ldots, v_i', \ldots, v_n) > u_i(v_1, \ldots, v_i, \ldots, v_n).$$

Thus, an outcome is a Nash equilibrium if no player can unilaterally deviate to obtain a better outcome for themselves, under the assumption that every other player stays with their choice. The Nash equilibria of game G will be denoted by $\mathcal{N}(G)$. Observe that, in general, we have that $\vec{v}$ is a Nash equilibrium in G if and only if for all players i,

$$(G, \vec{v}) \not\models \gamma_i \quad \text{implies} \quad (G, (v_1, \ldots, v_{i-1}, v_i', v_{i+1}, \ldots, v_n)) \not\models \gamma_i \text{ for all } v_i' \in \mathcal{V}_i.$$

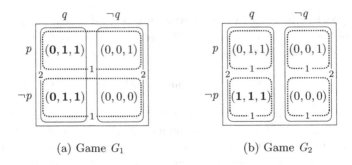

Fig. 1. The games G_1 and G_2 as in Example 1. The utilities to the players of each outcome are indicated by vectors (x, y, z), where x represents father's utility, y mother's, and z baby's. The outcomes father and mother cannot distinguish are indicated by solid and dotted boxes, respectively and the Nash equilibria are in bold face.

Example 1 (continued). Recall the nursery Example 1. We noted that the outcome where $\neg p \wedge q$ satisfies $K_2(p \vee q) \wedge \neg(p \wedge q)$. This is indeed a Nash equilibrium, even though Father's goal γ_1 is not satisfied (Father does not *know* that Mother will go). It is a Nash equilibrium, because, given this outcome, father on his own cannot fulfil his goal: the epistemic goal $K_1 K_2(p \vee q)$ cannot be satisfied in our model. Note that, since Baby cannot influence any proposition, every outcome has the property that it cannot unilaterally deviate and improve its outcome, i.e., it can never prevent an outcome from being a Nash equilibrium.

Note now that knowledge is power: if we change the game to G_2, in which $\Theta_1 = \{p, q\}$ (but keep the rest as in G_1), then Father will be informed as to whether mother satisfies her own goal by making q true, and hence $\neg p \wedge q$ is a Nash equilibrium for this game in which both parents and the child are happy.

3   Computational Complexity

In this section we analyse the computational complexity of a number of natural problems relating to EBGs. First, the MODEL CHECKING problem for $\mathcal{EL}$ is as follows (cf. [10]):

MODEL CHECKING:
Given: Pointed game $(G, \vec{v})$ and $\mathcal{EL}$ formula φ.
Question: Is it the case that $(G, \vec{v}) \models_{\mathcal{EL}} \varphi$?

We leave out the proof of the following due to lack of space.

Proposition 2. *The* MODEL CHECKING *problem for $\mathcal{EL}$ is PSPACE-complete.*

As an aside, we remark that this result may at first seem surprising, given that the problem of model checking formulae of S5$_n$ over Kripke structures may be solved in polynomial time [1, pp. 63–64]. There is no contradiction,

```
Input:    Game G and outcome v⃗ for G
Output: "yes" if v⃗ ∈ N(G), "no" otherwise.
 1  for i := 1 to n do
 2    if (G, (v₁, ..., vᵢ, ..., vₙ)) ⊭_{EL} γᵢ then
 3      for each v′ᵢ ∈ Vᵢ do
 4        if (G, (v₁, ..., v′ᵢ, ..., vₙ)) ⊨_{EL} γᵢ then
 5          return "no"
 6        end if
 7      end for
 8    end if
 9  end for
10  return "yes"
```

Algorithm 1. Algorithm for MEMBERSHIP

however: we are interpreting our $\mathcal{EL}$ formulae with respect to games, which can be understood as *compact* representations of Kripke structures. The polynomial time computational results for model checking over Kripke structures assumes that each state is explicitly listed in the input to the problem, which we do not assume. Thus, model checking $S5_n$ over Kripke structures assumes an input that in the worst case is exponentially larger than our game representation. As a general rule of thumb in complexity analysis, the more compact a representation is, the higher will be the complexity of the decision problems associated with these structures (see, e.g., [11, pp. 492–495]).

Two natural decision problems suggest themselves relating to the Nash equilibria of EBGs (cf. [12, pp. 8–9]). The first asks whether a given outcome $\vec{v}$ is a Nash equilibrium of a game G. We call this problem MEMBERSHIP:

MEMBERSHIP:
Instance: Game G and outcome $\vec{v}$ for G.
Question: Is it the case that $\vec{v} \in \mathcal{N}(G)$?

We first prove the upper bound for the MEMBERSHIP problem; we postpone the lower bound for the moment.

Proposition 3. MEMBERSHIP *for* EBG*s is in PSPACE.*

Proof. First observe that from Proposition 2, the model-checking problem for $\mathcal{EL}$ is PSPACE-complete. Let G be the game and let $\vec{v} = (v_1, \ldots, v_i, \ldots, v_n)$ be the outcome given in the problem instance. Then, Algorithm 1. decides the problem in PSPACE. The loop on lines (1)–(9) checks whether any player has a beneficial deviation: if it finds such a beneficial deviation, the algorithm returns "no" (line 5), indicating $\vec{v} \notin \mathcal{N}(G)$. We claim that the overall algorithm operates in PSPACE. To see this, observe that for each player $i \in N$ we carry out a single PSPACE check (line 2, Proposition 2) followed by a loop (lines 3–7) that iterates through all elements of $\mathcal{V}_i$, in each case carrying out a PSPACE check

(Proposition 2 again). The loop on lines 3–7 is in PSPACE as a consequence of the fact that

$$\text{PSPACE} = \text{PSPACE}^{\text{PSPACE}}.$$

Since PSPACE is closed under sequential PSPACE operations, the outer loop operates in PSPACE. We conclude that the overall algorithm operates in PSPACE.

The second problem simply asks whether there exist *any* Nash equilibria for a given EBG G. We call this problem NON-EMPTINESS:

NON-EMPTINESS:
Instance: Game G.
Question: Is it the case that $\mathcal{N}(G) \neq \emptyset$?

The proofs of the following propositions are left out due to lack of space.

Proposition 4. NON-EMPTINESS *for* EBGs *is PSPACE-complete.*

Given this, we can prove the lower bound for the MEMBERSHIP problem:

Proposition 5. MEMBERSHIP *for* EBGs *is PSPACE-hard.*

4   Epistemic Boolean Games *versus* Kripke Structures

The semantics we have given to $\mathcal{EL}$ in terms of epistemic Boolean games is very close to Kripke semantics, with outcomes for games essentially playing the role of possible worlds. Let us make this idea both explicit and precise. We argue that Boolean games of incomplete information can represent the same situations modelled by $S5_n$ Kripke structures and, moreover, that they sometimes can do so significantly more concisely.

We adapt the standard definition of *bisimulation* for modal logic [13, pp.64–67] to games and Kripke structures. Given a game G for $\mathcal{EL}(\Phi)$ and a Kripke structure K for $\mathcal{EL}(\Phi')$, for any $\Psi \subseteq \Phi \cap \Phi'$, we say that G and K are Ψ-*bisimilar* if there exists a relation

$$\mathcal{Z} \subseteq \mathcal{V} \times W$$

such that for all outcomes $\vec{v}$ and all possible worlds w:

- if $\mathcal{Z}(\vec{v}, w)$, then for all $p \in \Psi$, we have $\vec{v}(p) = \pi(w, p)$;
- if $\mathcal{Z}(\vec{v}, w)$ and for some $i \in N$ and $\vec{v}'$ we have $\vec{v} \sim_i \vec{v}'$, then there is some w' such that $\mathcal{Z}(\vec{v}', w')$ and $R_i(w, w')$; and
- if $\mathcal{Z}(\vec{v}, w)$ and for some $i \in N$ and w' we have $R_i(w, w')$, then there is some $\vec{v}'$ such that $\mathcal{Z}(\vec{v}', w')$ and $\vec{v} \sim_i \vec{v}'$.

We write $(G, \vec{v}) \cong_\Psi (K, w)$ to mean that there is a Ψ-bisimulation relation $\mathcal{Z}$ between G and K such that $\mathcal{Z}(\vec{v}, w)$. If $\Phi = \Phi' = \Psi$ we omit the subscript in $\cong_\Psi$. The key point about bisimulations is the following, readily established result.

Proposition 6. *Let* $\Psi \subseteq \Phi$. *If* $(G, \vec{v}) \cong_\Psi (K, w)$, *then for all formulae* $\varphi \in \mathcal{EL}(\Psi)$ *we have:*

$$(G, \vec{v}) \models_{\mathcal{EL}} \varphi \quad \textit{iff} \quad (K, w) \models_K \varphi.$$

We easily show that for every game G there exists a bisimilar Kripke structure K_G. Given an EBG $G = (N, \Phi, \Phi_1, \ldots, \Phi_n, \gamma_1, \ldots, \gamma_n, \Theta_1, \ldots, \Theta_n)$, we define the *Kripke structure* $K^G = (W^G, R_1^G, \ldots, R_n^G, \pi^G)$ *induced by* G as follows.

- $W^G = \mathcal{V}$, i.e., W^G is the set of outcomes for G;
- $R_i^G(\vec{v}, \vec{v}')$ iff $\vec{v} \sim_i \vec{v}'$; and
- for all $\vec{v}$ and $p \in \Phi$, we have $\pi^G(\vec{v}, p) = \vec{v}(p)$.

Then it is immediate by construction that we have:

Proposition 7. *Let G be a game, $\vec{v}$ an outcome, and K^G be the Kripke structure induced by G. Then, $(G, \vec{v}) \cong_\Phi (K^G, \vec{v})$.*

Notice that the construction of the set of worlds in K^G makes the exponential blow up of size in moving from games G to Kripke structures explicit. The previous result essentially tells us that for every game G there is an "equivalent" Kripke structure K, where equivalence is measured in terms of bisimulation.

Likewise, for every finite Kripke structure $K = (W, R_1, \ldots, R_n, \pi)$ for $\mathcal{EL}(\Phi)$, a Φ-bisimilar EBG G^K exists. This EBG, however, is defined on an extended language, which also involves propositional variables for every possible world $w \in W$. Assuming without loss of generality that $W \cap \Phi = \emptyset$, we let $\Psi = \Phi \cup W$. Moreover, define for every $w \in W$ a formula $\chi(w)$ as follows.

$$\chi(w) = w \wedge \bigwedge_{w' \neq w} \neg w' \wedge \bigwedge_{\pi(w,p)=\top} p \wedge \bigwedge_{\pi(w,p)=\bot} \neg p$$

Let for each possible world w the outcome $\vec{v}_w : \Psi \to \mathbb{B}$ such that for all $p \in \Psi$,

$$\vec{v}_w(p) = \begin{cases} \top & \text{if either } \pi(w, p) = \top \text{ or } p = w, \\ \bot & \text{otherwise.} \end{cases}$$

Note that the formula $\chi(w)$ characterises outcome $\vec{v}_w$, in the sense that

$$(G, \vec{v}_w) \models \chi(w') \quad \text{if and only if} \quad w = w'.$$

Given $K = (W, R_1, \ldots, R_n, \pi)$, we define for each agent i and each world w,

$$\theta_i^K(w) = \bigvee_{R_i(w,w')} \chi(w').$$

Intuitively, $\theta_i^K(w)$ characterises (in $\mathcal{L}(\Psi)$) the set of worlds that player i cannot distinguish from w in K along with their interpretation of the propositional variables. That is, if $\theta_i^K(w)$ is in player i's visibility set, i can distinguish worlds that are in $\{w' \in W : R_i(w, w')\}$ from those that are not. Moreover, as K is an $S5_n$ model, each R_i is an equivalence relation. Hence, $R_i(w, w')$ implies $\theta_i^K(w) = \theta_i^K(w')$. The game $G^K = (N, \Phi^K, \Phi_1^K, \ldots, \Phi_n^K, \gamma_1^K, \ldots, \gamma_n^K, \Theta_1^K, \ldots, \Theta_n^K)$ induced by K is now defined such that $\Phi^K = \Phi \cup W$ and for each agent i,

$$\Theta_i^K = \{\theta_i^K : w \in W\}.$$

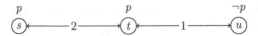

Fig. 2. S5 Kripke structure with no bisimilar atomic game models

The choices for $\Phi_1^K, \ldots, \Phi_n^K$ and $\gamma_1^K, \ldots, \gamma_n^K$ are arbitrary. Observe, however, that the game G^K does *not* in general represent the situation much more concisely than the Kripke structure K.

Now define the relation $\mathcal{Z}$ such that for all $\vec{v} \in V$ and all $w \in W$,

$$\mathcal{Z}(\vec{v}, w) \quad \text{if and only if} \quad \vec{v} = \vec{v}_w.$$

We find that $\mathcal{Z}$ is witness to the fact that G^K and K are Φ-bisimilar, giving us the following result (proof left out due to lack of space).

Proposition 8. *Let K be a Kripke structure, w a world, and let G^K be the game induced by K. Then, $(G^K, \vec{v}_w) \cong_\Phi (K, w)$.*

However, if we restrict ourselves to atomic games, i.e., games with visibility sets Θ_i that consist of propositional variables, Proposition 8 (or the converse of Proposition 6) does not hold.

Proposition 9. *There are Kripke structures K over N, Φ for which there is no atomic game G over N, Φ such that $G \cong K$.*

Proof. Consider the $S5_n$ Kripke structure K with $N = \{1, 2\}$ and $\Phi = \{p\}$ as depicted in Figure 2. (Both R_1 and R_2 are equivalence relations, but for clarity we have not drawn the reflexive arrows). Note that $K_1 p \vee K_1 \neg p$ means that 1 knows the value of p. In K, this is true in s, but not in t or u. Hence we have $K, s \models_K K_1 p \wedge \neg K_2(K_1 p \vee K_1 \neg p)$. However, we claim that in any atomic game G and any valuation $\vec{v}$ we have

$$(G, \vec{v}) \models_{\mathcal{EL}} K_1 p \to K_2(K_1 p \vee K_1 \neg p)$$

To see the latter, suppose that $(G, \vec{v}) \models_{\mathcal{EL}} K_1 p$. This means not only that $\vec{v}(p) = \top$, but also that $p \in \Theta_1$. To appreciate the latter, observe that $p \notin \Theta_1$ would imply that there is some $\vec{u}$, which only differs from $\vec{v}$ in that it assigns $\bot$ to p rather than $\top$, and, since $\vec{v}$ and $\vec{u}$ agree on all other variables (in particular those in Θ_1), for which $\vec{v} \sim_i \vec{u}$. But this in turn implies that $(G, \vec{v}) \models_{\mathcal{EL}} \neg K_i p$, a contradiction. But since $p \in \Theta_1$, it is obvious that $G \models_{\mathcal{EL}} K_1 p \vee K_1 \neg p$, and hence we have $(G, \vec{v}) \models_{\mathcal{EL}} K_2(K_1 p \vee K_1 \neg p)$. So there is no $(G, \vec{v})$ that satisfies $K_1 p \wedge \neg K_2(K_1 p \vee K_1 \neg p)$, and hence, by Proposition 2, there is no $(G, \vec{v})$ for which $(G, \vec{v}) \cong (K, s)$.

The upshot of this discussion is that atomic games induce Kripke structures with a very particular structure. Specifically, they induce Kripke structures in which: (i) the set of worlds corresponds exactly to the set of possible valuations;

(*ii*) the accessibility relations R_i are derived solely from the visibility sets Θ_i that consist of propositional variables; and (*iii*) which variables agents can see, which variables they can manipulate, and what goals they have, is all common knowledge. Issues like this (in the closely related model of *interpreted systems* for knowledge [1, pp.103–114]) were studied in depth by Lomuscio [14].

5   Epistemic Boolean Games *versus* Boolean Games

What distinguishes epistemic Boolean games from regular Boolean games is the language in which the players' goals are phrased, viz., the language of $\mathcal{EL}$ and raw propositional logic, respectively. Otherwise, the strategic structure of epistemic Boolean games is very similar to that of regular Boolean games. Rather, in this section, we argue that for every epistemic Boolean game a corresponding regular Boolean game can be found that is in an important sense strategically equivalent to it. However, we also show that, in the case of atomic games, the regular Boolean game may be exponentially larger than the epistemic Boolean game it corresponds to.

Given a set Φ of propositional variables the set of $2^{|\Phi|}$ outcomes are determined by the set of valuations. As the only role of the visibility sets in EBGs is to enable the evaluation of the epistemic formulae in $\mathcal{EL}(\Phi)$, it can now easily be seen that for every epistemic Boolean game there is a regular Boolean game with the same *strategic* properties. (The claim in the opposite direction is trivial.)

Let $G = (N, \Phi, \Phi_1, \ldots, \Phi_n, \gamma_1, \ldots, \gamma_n, \Theta_1, \ldots, \Theta_n)$ be an epistemic Boolean game. Associate with each outcome $\vec{v}$ a *propositional formula* $\chi(\vec{v})$ in much the same way as in the previous section.

$$\chi(\vec{v}) = \bigwedge_{\vec{v}(p)=\top} p \wedge \bigwedge_{\vec{v}(p)=\bot} \neg p.$$

Now consider the regular Boolean game $G' = (N, \Phi, \Phi_1, \ldots, \Phi_n, \gamma_1', \ldots, \gamma_n')$, where for each player i,

$$\gamma_i' = \bigvee_{(G,\vec{v})\models_{\mathcal{EL}}\gamma_i} \chi(\vec{v}).$$

We then have that for players i and all outcomes $\vec{v}$,

$$(G, \vec{v}) \models_{\mathcal{EL}} \gamma_i \quad \text{if and only if} \quad (G, \vec{v}) \models_{\mathcal{EL}} \gamma_i'.$$

Hence, G and G' agree on the players, strategies, and outcomes. Moreover, each goal γ_i induces the same preferences over the outcomes in G as γ_i' does in G'. Accordingly, G and G' could justifiably be said to be strategically equivalent.

It should be observed though, that the size of γ_i', as defined above, is exponential in the size of γ_i. This raises the obvious question of whether we are in fact gaining anything by using the epistemic language: can we find an equivalence preserving translation τ from epistemic formulae to propositional formulae such that $\tau(\varphi)$ is guaranteed to be of size polynomial in the size of φ? If the answer was "yes", then this would indicate that our epistemic language was somewhat redundant, in terms of raw expressive power and succinctness. In fact, there is a compelling complexity theoretic argument that the epistemic language is *exponentially more succinct* than the propositional language on our game structures:

Proposition 10. *If $P \neq PSPACE$, then the epistemic language is exponentially more succinct than the propositional language over game structures.*

Proof. Recall that model checking for the epistemic language over our game structures is PSPACE-complete, while the model checking problem for the propositional language is in P. Now, suppose there exists a polynomial time computable equivalence preserving translation τ from the epistemic language to the propositional language. The existence of such a translation would imply PSPACE=P: to solve a PSPACE hard model checking problem for the epistemic language, we could apply the polynomial time translation τ and apply the polynomial time model checking algorithm for the resulting formula, yielding a polynomial time decision procedure for the PSPACE-hard model checking problem. We conclude that no such translation can exist unless P=PSPACE.

Note that it is considered *highly* unlikely that PSPACE=P.

Thus, the epistemic language provides concrete benefits with respect to succinctness. We might also note that the use of epistemic modalities of course provides benefits with respect to naturalness of expression and the readability of formulae when compared to the use of raw propositional logic.

6   Related Work and Conclusions

In this paper, we formally defined and investigated games in which the goals of players relate to the epistemic states of other players. In such a game, a player will be strategically motivated to act in such a way as to bring about states of knowledge—or indeed ignorance—in other players. We formally defined epistemic Boolean games, an extension to the now well-established Boolean games model in which players have goals represented not as propositional logic formulae, but as formulae of modal epistemic logic. We then investigated the computational complexity of questions relating to Nash equilibria in such games.

Our research is closely related to several other papers that have appeared in the literature. In [15] we also considered Boolean games with visibility sets, but the setting in that paper is very different: goals are propositional rather than epistemic, and the focus is rather on identifying *verifiable* Nash equilibria; equilibria that players know to be equilibria.

Van Otterloo *et al.* introduced *knowledge condition games* [16]. Knowledge condition games are extensive form games, in which state sets are explicitly listed. The basic question considered in the work of van Otterloo *et al.* is when sets of players in extensive form games are able to act in such a way as to bring about a state of knowledge; the basic result relating to this problem is that the problem is Σ_2^p-complete. Because the problem relates to strategic ability, van

Otterloo also considered the use of the strategic ability logic ATL [17] and its epistemic variant ATEL [18]. The work differs from the present paper in that we consider a Boolean games framework, with visibility sets implicitly defining epistemic accessibility relations; we also allow for players to have goals explicitly represented as formulae of epistemic logic.

Ågotnes and van Ditmarsch also considered closely related issues [19]. They introduced *public announcement games*. The foundation for their study was the growing body of work on *dynamic epistemic logic* (DEL) [20]. In dynamic epistemic logic, agents are allowed to make announcements, which may be simple objective statements about the state of the world (e.g., "p is true") or may be more complex announcements involving statements about the knowledge or ignorance of agents (e.g., "I don't know p"). Ågotnes and van Ditmarsch investigated how players in a game could use such announcements to bring about states of knowledge. The main difference is that the announcements considered by Ågotnes and van Ditmarsch are much richer than the mechanisms available to players in our games for bringing about epistemic states; and the semantics are correspondingly much more technically involved than in our setting. Our setting offers a much more *compact* representation for epistemic games than that of Ågotnes and van Ditmarsch, and one that is much closer to computational models. Additionally, we are able to prove results relating to (e.g.) the complexity of Nash equilibria that were not considered in [19].

Finally, and perhaps the closest to the present paper is the work of Grant *et al.* on Boolean games in which players have (possibly incorrect) beliefs about the value of certain variables, and where an external principal is able to make announcements about these variables in order to influence the beliefs of the players within the game, and hence the rational choices that they subsequently make [21]. The model of belief studied by Grant *et al* is much more restricted than the model of knowledge we use in the present paper (every player simply believes every proposition is either true or false).

Several avenues suggest themselves for future work. First, we might look at richer computational models than simply setting variables to be true or false. For example, we might model games using a practical system specification language such as REACTIVE MODULES [22], and, correspondingly, allow for richer strategies as in the Alternating-time Temporal Logic [17]. This would bring the setting of epistemic games much closer to practical systems, and would make it possible to model practical protocols and systems. Second, we might consider the use of mixed (probabilistic) strategies. If players are permitted to use randomised strategies, then the set $\mathcal{N}(G)$ can be replaced by a probability distribution over outcomes. In this case, instead of asking whether, for example, player i knows φ in all outcomes, we can ask what the probability is that player i comes to know φ, assuming that players play Nash equilibrium strategies.

Third, the epistemic concept we have worked with in this paper is *knowledge*, in the sense of S5 modal logic [1]. It would also be interesting to consider the use of *belief*, e.g., in the sense of the modal logic KD45. Finally, it looks worthwhile to see whether the exponential blow-up when moving from games to Kripke models, can be avoided by adopting a slightly different perspective on Kripke models, as proposed in [9], viewing them as compositions of smaller agent models.

Acknowledgements. Harrenstein and Wooldridge were supported by the ERC under Advanced Investigator Grant 291528 ("RACE").

References

1. Fagin, R., Halpern, J.Y., Moses, Y., Vardi, M.Y.: Reasoning About Knowledge. The MIT Press (1995)
2. Harrenstein, P., van der Hoek, W., Meyer, J.J., Witteveen, C.: Boolean games. In: van Benthem, J. (ed.) Proc. TARK VIII, pp. 287–298 (2001)
3. Bonzon, E., Lagasquie, M., Lang, J., Zanuttini, B.: Boolean games revisited. In: Proc. ECAI 2006 (2006)
4. Dunne, P.E., Kraus, S., van der Hoek, W., Wooldridge, M.: Cooperative boolean games. In: Proc. AAMAS 2008 (2008)
5. Endriss, U., Kraus, S., Lang, J., Wooldridge, M.: Designing incentives for boolean games. In: Proc. AAMAS 2011 (2011)
6. van der Hoek, W., Troquard, N., Wooldridge, M.: Knowledge and control. In: Proc. AAMAS 2011 (2011)
7. van der Hoek, W., Wooldridge, M.: On the logic of cooperation and propositional control. Artiifcial Intelligence 164, 81–119 (2005)
8. Osborne, M.J., Rubinstein, A.: A Course in Game Theory. MIT Press (1994)
9. van Eijck, J., Sietsma, F., Wang, Y.: Composing Models. Journal of Applied Non-Classical Logic 21(3-4), 397–425 (2011)
10. Clarke, E.M., Grumberg, O., Peled, D.A.: Model Checking. MIT Press (2000)
11. Papadimitriou, C.H.: Computational Complexity. Addison-Wesley (1994)
12. Chalkiadakis, G., Elkind, E., Wooldridge, M.: Computational Aspects of Cooperative Game Theory. Morgan-Claypool (2011)
13. Blackburn, P., de Rijke, M., Venema, Y.: Modal Logic. Cambridge University Press (2001)
14. Lomuscio, A.: Knowledge Sharing among Ideal Agents. PhD thesis, School of Computer Science, University of Birmingham, Birmingham, UK (1999)
15. Ågotnes, T., Harrenstein, P., van der Hoek, W., Wooldridge, M.: Veri able equilibria in boolean games. In: Proc. IJCAI 2013 (2013)
16. van Otterloo, S., van der Hoek, W., Wooldridge, M.: Knowledge condition games. Journal of Logic, Language, and Information 15, 425–452 (2006)
17. Alur, R., Henzinger, T.A., Kupferman, O.: Alternating-time temporal logic. Journal of the ACM 49, 672–713 (2002)
18. van der Hoek, W., Wooldridge, M.: Time, knowledge, and cooperation: Alternating-time temporal epistemic logic and its applications. Studia Logica 75, 125–157 (2003)
19. Ågotnes, T., van Ditmarsch, H.: What will they say? – public announcement games. Synthese 179, 57–85 (2011)
20. van Ditmarsch, H., van der Hoek, W., Kooi, B.: Dynamic Epistemic Logic. Springer (2007)
21. Grant, J., Kraus, S., Wooldridge, M., Zuckerman, I.: Manipulating boolean games through communication. In: Proc. IJCAI 2011 (2011)
22. Alur, R., Henzinger, T.A.: Reactive modules. Formal Methods in System Design 15, 7–48 (1999)

Minimal Preference Change

Natasha Alechina[1], Fenrong Liu[2], and Brian Logan[1]

[1] School of Computer Science, University of Nottingham, UK
[2] Department of Philosophy, Tsinghua University, Beijing, China
{nza,bsl}@cs.nott.ac.uk, fenrong@tsinghua.edu.cn

Abstract. We propose a novel approach to preference change. We treat a set of preferences as a special kind of theory, and define minimal change contraction and revision operations in the spirit of minimal change as advocated by the Alchourron, Gardenfors, and Makinson (AGM) theory of belief revision. We characterise minimal contraction of preference sets by a set of postulates and prove a representation theorem. We also give a linear time algorithm which implements minimal contraction by a single preference. We also define minimal contraction by a set of preferences, and for a significant special case state postulates, prove a representation theorem, and provide an efficient algorithm implementing minimal contraction by a set of preferences.

1 Introduction

Preference plays a crucial role in agents' reasoning and their intelligent interaction with other agents. In this paper we consider the problem of *preference change*: the contraction and revision of an agent's set of preferences by a single preference and by a set of preferences. We are motivated by an analogy between preference change and the Alchourrón, Gärdenfors, and Makinson (AGM) theory of belief revision. Both contraction and revision require maintaining consistency of the agent's set of preferences. When a new preference is inconsistent with an agent's existing preferences, a rational agent should remove as few preferences from its set of preferences as possible to restore consistency. We are interested in *efficient* algorithms (at most polynomial in the size of the agent's preference set) for minimal contraction and revision that may be employed by feasible, resource-bounded reasoners.

The contribution of this paper is as follows. We define minimal preference contraction and show how to define revision in terms of contraction. Next, we give postulates for rational minimal preference contraction and prove a representation theorem. As far as we know, this is the first representation theorem for minimal change preference contraction in the literature. We then give a linear time minimal preference contraction algorithm. We also investigate the problem of contracting by a *set* of preferences rather than by a single preference, or minimal iterative preference contraction. We define minimal iterative preference set contraction and state postulates characterising an important special case in which contraction is by an uncoupled set of preferences. We also give a polynomial time algorithm to compute minimal contraction by a set for that case.

D. Grossi, O. Roy, and H. Huang (Eds.): LORI 2013, LNCS 8196, pp. 15–26, 2013.

2 Formal Preliminaries

We assume an agent's preferences are given by a binary relation over some finite set of alternatives $\mathcal{A}$. An agent's preference state is represented by a *preference set* consisting of preference sentences (or simply preferences) which are atomic statements of the form $A < B$ (B is preferred to A), $A \equiv B$ (A and B are equally preferred) or $A \# B$ (A and B are incomparable). In using these three basic relations, we essentially follow [1]. [1] In addition, the agent's preference set may contain a special sentence $\bot$, which is used to indicate a problem (derivability of an inconsistency).

We assume that preference sets are not necessarily complete, in the sense that they may include no sentences expressing a relation between A and B, for $A, B \in \mathcal{A}$. We take this to be a natural feature of resource-bounded agents. We do assume that the agents are rational, i.e., they don't accept $A < B$ and $B < A$ or $A \# B$ at the same time, they can complete their preference sets using transitivity of $<$ and $\equiv$ and symmetry of $\#$, etc.

The agent's *rational reasoning rules* or integrity constraints in the sense of [7] are given below. Rule 1 states that $\#$ is symmetric, rules 2-4 state that $\equiv$ is an equivalence relation, rule 5 states that $<$ is transitive, and the rest of the rules state that at most one of $\#, \equiv, <, >$ can hold between two alternatives[2].

1. $A \# B \ \Rightarrow \ B \# A$
2. $A \equiv A$
3. $A \equiv B \ \Rightarrow \ B \equiv A$
4. $A \equiv B, B \equiv C \ \Rightarrow \ A \equiv C$
5. $A < B, B < C \ \Rightarrow \ A < C$
6. $A < B, B < A \ \Rightarrow \ \bot$
7. $A \equiv B, A < B \ \Rightarrow \ \bot$
8. $A \equiv B, A \# B \ \Rightarrow \ \bot$
9. $A \# B, A < B \ \Rightarrow \ \bot$

We denote by $Cn(S)$ the closure of a set S under the rules above. Formally, $Cn(S)$ is the set of preferences which contains S, $A \equiv A$ for every $A \in \mathcal{A}$, and in addition for every rule $p_1, \ldots, p_n \Rightarrow p$ above, if $p_1, \ldots, p_n \in Cn(S)$, then $p \in Cn(S)$. A set of preferences S is *deductively closed* iff $S = Cn(S)$.

Sometimes we will use the notation $S \vdash p$ to say that p can be derived from S and the reasoning rules above by application of the following inference rule (where $n \leq 2$):

---

[1] One may start with a different initial setting, for instance, taking as $A \leq B$ (B is at least as good as A, cf. [8]) as the primitive relation, then define other relations and explore the similar questions. We leave this possibility for another occasion.

[2] Note that we do not have the following rules:

$A \equiv B, B < C \ \Rightarrow \ A < C$
$A \equiv B, B \# C \ \Rightarrow \ A \# C$

the agent may have a preference regarding B and C, and consider A and B indistinguishable, but may not have a preference regarding A and C.

$$\frac{p_1,\ldots,p_n \quad p_1,\ldots,p_n \Rightarrow p}{p}$$

Clearly for any p, $\vdash p$ (p is derivable from an empty set) if, and only if, p is of the form $A \equiv A$. Note that we do not assume any logical connectives or any other inference rules.

In what follows, we always assume that the agent's set of preferences S is deductively closed. The set of preferences is consistent if and only if it does not contain $\bot$.

3   Preference Revision

Clearly, if an agent acquires a new preference, its preference set may become inconsistent. For example, if the agent used to prefer B to A ($A < B$) and C to B ($B < C$) and has decided that it prefers A to C, its preference set is inconsistent since it contains both $A < C$ by transitivity from the old preferences and $C < A$ (the new preference). In order to incorporate the new preference and have a consistent preference set, the agent needs to remove some of the old preferences. We are interested in *minimal* preference revision, namely removing as few sentences as possible to restore consistency. As in AGM belief revision, we define revision in terms of contraction by a preference sentence.

Revision of a preference set S by a preference p is defined as adding p to S if the result is consistent and deductively closing the resulting set, otherwise first contracting S by p^{-S}, denoted $S - p^{-S}$, where p^{-S} is the *S-complement* of p. For a consistent set of preferences S and a preference p such that $S \cup \{p\}$ is inconsistent, the S-complement of p is defined as follows:

- $A \equiv B^{-S} = S \cap \{A < B, B < A, A \# B\}$
- $A < B^{-S} = S \cap \{A \equiv B, B < A, A \# B\}$
- $A \# B^{-S} = S \cap \{A \equiv B, A < B, B < A\}$

Contracting S by the S-complement of p makes p consistent with the result, and we can add p to the resulting set and close it under consequence. Revision $S * p$ of S by p is thus defined as $Cn(S - p^{-S} \cup \{p\})$. This is essentially the Levi identity [6, 12] $S * p = (S - \neg p) + p$.

3.1   Minimal Contraction

Definition 1. *(Minimal contraction) Given a preference set S and a preference p, such that $\nvdash p$, the result of a minimal contraction of S by p is a set $S - p$ such that:*

(1) $S - p \subseteq S$
(2) $S - p \nvdash p$
(3) for any other set S' satisfying (1) and (2), $|S'| \leq |S - p|$.

The removal of minimal number of preferences is similar to Hansson's definition of revising to the most similar preference relation (where the distance between preference relations is defined as the symmetrical difference between the sets of pairs of alternatives in the two relations) [9].

3.2 Minimal Contraction Postulates

Before we can state the postulates characterising minimal contraction, we need to introduce the following abbreviations. By $A_S^<$ we will denote $\{C : A < C \in S\}$. By $A_S^>$ we will denote $\{C : C < A \in S\}$. By $A_S^\equiv$ we will denote $\{C : A \equiv C \in S\} \setminus \{A\}$. The *cost* $c_S(p)$ of $p \in S$ (intuitively, the number of preferences a contraction by p has to remove from S) is defined as follows:

- $c_S(A < B) = |A_S^< \cap B_S^>| + 1$
- $c_S(A \equiv B) = 2 * |A_S^\equiv|$
- $c_S(A\#B) = 2$

The following postulates characterise minimal contraction. For readability, we will omit subscript S when it is unambiguous.

C-Closure $S - p = Cn(S - p)$
C-Inclusion $S - p \subseteq S$
C-Vacuity If $p \notin S$, $S - p = S$
C-Success If p is not of the form $A \equiv A$, then $p \notin S - p$
C-Equivalence If $Cn(p_1) = Cn(p_2)$, then $S - p_1 = S - p_2$
C-Minimality If $p \in S$, then $|S - p| = |S| - c_S(p)$

The postulates of C-Closure, C-Inclusion, C-Vacuity, C-Success and C-Equivalence are standard postulates for contraction of beliefs. Recovery ($S \subseteq Cn((S - p) \cup p)$) does not hold, but this postulate has always been considered controversial [13]. The C-Minimality postulates characterise specifically minimal contraction of preferences, because for preferences it is possible to predict the cardinality of the resulting set.

Theorem 1. *The result of any minimal contraction satisfies the minimal preference contraction postulates above, and every contraction satisfying these postulates is a minimal preference contraction.*

Proof. For the case when $p \notin S$, clearly the minimal contraction is S itself, and all the postulates hold for $S - p = S$ trivially.

Let us consider the case when $p \in S$. We show first that every minimal contraction satisfies the postulates. C-Inclusion holds by Definition 1, and C-Vacuity trivially since $p \in S$. To show that C-Closure holds, assume by contradiction that $S - p$ is a minimal contraction and it is not deductively closed. Since $S - p \nvdash p$ (by Definition 1 (2)) and $S - p$ is not deductively closed, then there must be a consequence q of $S - p$ such that $q \notin S - p$. Since $S - p \nvdash p$ and $S \vdash q$, it follows that $(S - p) \cup \{q\} \nvdash p$. Since $S - p \subseteq S$ (by Definition 1 (1)), $S \vdash q$, and since S is deductively closed, $q \in S$. Hence there is a set $S' = (S - p) \cup \{q\}$ such that conditions (1) and (2) of Definition 1 hold for S', and its cardinality is greater than that of $S - p$. Hence $S - p$ is not a minimal contraction because it violates condition (3): a contradiction. C-Success holds for all p which are not derivable from an empty preference set because there is always a subset of S which does not derive p (in the worst case, $\emptyset$). C-Equivalence holds rather trivially because the only cases when two syntactically different preferences have the same set of consequences are: $Cn(A \equiv B) = Cn(B \equiv A)$ and $Cn(A\#B) = Cn(B\#A)$;

due to symmetry rules, any successful contraction by one of $A \equiv B$, $B \equiv A$ has to get rid of both of them, similarly for $A\#B$, $B\#A$. Now let us consider the minimality postulates. We need to prove that any minimal contraction removes exactly $|S| - |S - p|$ sentences for each of the cases. In particular, we need to prove that:

- a minimal contraction by $A \equiv B$ removes exactly $2 * |A^{\equiv}|$ preferences;
- a minimal contraction by $A < B$ removes exactly $|A^< \cap B^>| + 1$ preferences;
- a minimal contraction by $A\#B$ removes exactly 2 preferences.

Let us consider the easiest case first. If $A\#B \in S$ and we want to remove it and make sure that $S \not\vdash A\#B$, we need to remove $A\#B$ itself, and $B\#A$ (note that since $A\#B \in S$ and S is deductively closed, $B\#A \in S$). Clearly if one of those preference is left in S then it would be possible to derive $A\#B$. So both $A\#B$ and $B\#A$ *have to be removed*. On the other hand, from the inspection of the reasoning rules, there is no other way to derive $A\#B$. So these two preferences are the *only* ones which have to be removed. Hence any contraction satisfying (2) will remove these 2 sentences, and any contraction satisfying (3) will only remove these 2 sentences.

Now consider the case of $A < B \in S$. In order to contract by $A < B$, we need to remove $A < B$ itself from S. However $A < B$ may still be derivable, namely using the transitivity rules. The number of possible derivations of $A < B$ using the rule $A < C, C < B \Rightarrow A < B$ is exactly $|A^< \cap B^>|$. We need to 'destroy' each such derivation, and in order to do this we need to remove *at least one* of the premises in each derivation, namely either $A < C$ or $C < B$. So any contraction satisfying (1) and (2) needs to remove at least $|A^< \cap B^>| + 1$ preferences (1 is for $A < B$ itself). Conversely, if one of the preferences for each possible derivation is removed, then $A < B$ is no longer derivable, so the operation already satisfies (1) and (2). (Note that if $A < C$ for $C \in A^< \cap B^>$ is itself derivable, one premise in the derivation of $A < C$ is $A < D$ where $D < C$ since $C < B, D < C$, so $D \in A^< \cap B^>$, so $A < D$ will be removed and hence $A < C$ is not re-derivable.) Hence, in order to satisfy (3), the operation should not remove anything else. Hence any minimal contraction removes exactly $|A^< \cap B^>| + 1$ preferences.

In the case when $A \equiv B \in S$, any contraction operation needs to remove $A \equiv B$ and $B \equiv A$. However after this $A \equiv B$ may still be derivable by transitivity, using $A \equiv C, C \equiv B \Rightarrow A \equiv B$. The number of such derivations is the number of elements in $A^{\equiv} \setminus \{B\}$ (we are only considering uses of transitivity rule where C is different from both A and B). If for some of those derivations, both premises are left in S, then $A \equiv B$ can be re-derived. So any contraction satisfying (1) and (2) needs to remove at least one of the premises, either $A \equiv C$ or $C \equiv B$. Note that in order to properly remove $A \equiv C$, we also need to remove $C \equiv A$, otherwise $A \equiv C$ will be rederivable by symmetry. This means that any contraction needs to remove at least $2 * |A^{\equiv}|$ preferences: $A \equiv B, B \equiv A$, and $2 * (|A^{\equiv} \setminus \{B\}|)$. To show that this number of removed preferences is sufficient, and hence that no minimal contraction needs to remove more, we exhibit a concrete contraction which satisfies (1) and (2) and removes only $2 * |A^{\equiv}|$ preferences. Namely, consider a contraction which removes A from its equivalence class in S: it removes all $A \equiv C, C \equiv A$ for $C \in A^{\equiv}$. In the resulting set, A is not connected by $\equiv$ to any other alternative, hence $A \equiv B$ is not derivable.

The other direction: if an operation satisfies the postulates, it is a minimal contraction. Clearly, since the operation satisfies C-Closure, C-Inclusion and C-Success, it satisfies conditions (1)-(2) of Definition 1. To show that it satisfies (3), we need to prove that there is no set of strictly larger cardinality than $S - p$ which still satisfies (1)-(2), in other words that every successful contraction has to remove at least as many preferences as is stated in C-Minimality postulates. The argument is exactly as above.   □

3.3   Minimal Contraction Algorithm

We give an algorithm for the case when p is derivable from S, and $\nvdash p$.

The algorithm for computing $S - p$ is given by cases (see Algorithm 1).

Algorithm 1. Minimal preference contraction algorithm

---

procedure MINIMAL-CONTRACTION(S,p)
    case $p \notin S$
        return
    case $p == A < B$
        $A^< := \{C \mid A < C\}$
        $B^> := \{C \mid C < B\}$
        for each $C \in A^< \cap B^>$ **do**
            $S := S \setminus \{A < C\}$
        end for
        $S := S \setminus \{A < B\}$
    case $p == A \equiv B$
        $A^\equiv := \{C \mid A \equiv C, C \neq A\}$
        for each $C \in A^\equiv$ **do**
            $S := S \setminus \{A \equiv C, C \equiv A\}$
        end for
    case $p == A \# B$
        $S := S \setminus \{A \# B, B \# A\}$

Theorem 2. *The minimal preference contraction algorithm computes a minimal preference contraction.*

Proof. We show that the result of applying the algorithm to a preference set S and $p \in S$, p not of the form $A \equiv A$, always satisfies the conditions in Definition 1. Condition (1) holds because the algorithm only removes sentences from S. Condition (2) holds because the algorithm removes a premise from every possible derivation of p. Condition (3) holds because the algorithm result satisfies the minimal contraction postulates hence it is a minimal contraction by Theorem 1.   □

Theorem 3. *The time complexity of the algorithm for minimal contraction is in $O(|\mathcal{A}|)$.*

Proof. We assume that we can order the alternatives in some order (e.g., lexicographic order) and for each relation $(<, \equiv, \#)$ we can recover the ordered set of alternatives to which an alternative A is related in constant time (e.g., a hash table for each relation/position mapping from alternatives to sets (lists) of alternatives).

Then we can determine in constant time whether $p \notin S$ (recall that S is deductively closed).

For the $A < B$ case, the maximum size of $A^<$ and $B^>$ is bounded by $|\mathcal{A}|$, since A and B can be related to at most $|\mathcal{A}| - 1$ alternatives by $<$. Computing the set of alternatives $C \in A^< \cap B^>$ is also linear in $|\mathcal{A}|$ (to be precise it requires at most $2|\mathcal{A}|$) and the number of such alternatives C is again bounded by $|\mathcal{A}|$. Removing the preferences $A < C$ for $C \in A^< \cap B^>$ requires at most $|\mathcal{A}|$ operations (if the set of preferences is implemented as, e.g., a linked list) and replacing the new set in the map is constant time. For the $A \equiv B$ case, replacing the entry for A in the $\equiv$ map is a constant time operation. For the $A \# B$ case, we need to remove a single entry from the set of preferences for A in the $\#$ map. This requires at most $|\mathcal{A}|$ steps.   □

4   Minimal Set Contraction

In this section we turn to the problem of contracting by a *set* of preferences, which is similar to the problem of iterated belief revision [3–5]. As in the case of single preferences, we concentrate on the contraction rather than revision by a set of preferences, since 'minimal change' has a more intuitive and straightforward interpretation in the case of contraction.

We define a minimal contraction of a preference set S by a set of preference sentences X as follows:

Definition 2. *(Minimal contraction by a set) An operation $-$ is a minimal contraction of S by a set X if it satisfies the following properties:*

1. $S - X \subseteq S$
2. *if $p \in X$ and $\not\vdash p$, then $S - X \not\vdash p$*
3. *for every other set S' which satisfies properties (1)-(2) above, $|S'| \leq |S - X|$.*

A minimal revision of a preference set S by a set of preferences S' can be defined analogously to Hansson's *consolidation* [10]: first compute $Cn(S \cup S')$, then minimally contract by contradictions. Note that contracting S' by all sentences X inconsistent with S may not be enough to make $\perp$ underivable from $S \cup (S' \setminus X)$.

A natural question to ask is whether a minimal contraction of S by p_1 followed by a minimal contraction of $S - p_1$ by p_2 is a minimal contraction of S by $\{p_1, p_2\}$. The answer is negative. Consider the following example:

- $S = \{A < B, A < C, C < B\} \cup \{A \equiv A : A \in \mathcal{A}\}$
- $p_1 = A < B$
- $p_2 = C < B$

A minimal contraction of S by $A < B$ computed by Algorithm 1 is $S - A < B = \{C < B\} \cup \{A \equiv A : A \in \mathcal{A}\}$. It removes two preferences, $A < B$ itself and $A < C$.

A minimal contraction of this set by $C < B$ removes $C < B$. The set $(S - p_1) - p_2$ is $\{A \equiv A : A \in \mathcal{A}\}$ which is the result of removing three preferences from S. However, it is possible to make $A < B$ and $C < B$ underivable from S by removing just two preferences: $A < B$ and $C < B$. Recall that Algorithm 1 makes a particular choice in contraction by $A < B$: it removes sentences of the form $A < C$ where $C \in A^< \cap B^>$. It could have just as well removed sentences of the form $C < B$; for a single step contraction is does not matter which choice is made, since the number of removed sentences would be the same in each case. However for the iterated case, we need to look ahead to decide which choice to make. The problem of computing a minimal contraction by a set is, of course, decidable, but may require considering exponentially many (in $|X|$) choices. The problem only arises when several preferences in X share alternatives, as $A < B$ and $C < B$ above.

We can however characterise minimal set contraction in an important special case, when X has a specific form (which we call uncoupled) defined below. This special case covers, for example, contraction by a set of (some other) agent's preferences when that agent has a linearly ordered preference set. Given a set X, we will denote by $X(<)$ all elements of X which are of the form $A < B$, by $X(\equiv)$ all elements of X which are of the form $A \equiv B$ (we assume X does not contain tautologies $A \equiv A$), and by $X(\#)$ all elements of X of the form $A\#B$. The set X can therefore be represented as a set of disjoint sets $X_1, \ldots, X_k$.

Definition 3. *(Uncoupled set of preferences)*

– *the subset $X(<)$ of X is partitioned into subsets $X(A, <)$ of the form*

$$\{A < A_1, A_1 < A_2, A < A_2, \ldots, A < A_n\}$$

(where all A_i are linearly ordered between A and A_n) and $X(B, >)$ of the form

$$\{B_1 < B, B_1 < B_2, B_2 < B, \ldots, B_n < B\}$$

(where all B_i are linearly ordered between B_1 and B) and no alternative occurs in two different partitions of $X(<)$
– *the subset $X(\equiv)$ of X is partitioned into subsets $X(A, \equiv)$ of the form*

$$\{A \equiv A_1, A_1 \equiv A_2, \ldots, A_{n-1} \equiv A_n\}$$

(or an equivalent way of stating that $A, A_1, \ldots, A_n$ form an equivalence class) and no alternative occurs in two different partitions of $X(\equiv)$
– *$X(\#)$ is partitioned into two parts, $X(\#)_1$ which contains $A\#B$ such that $B\#A \notin X$, and $X(\#)_2$ which contains $A\#B, B\#A$ such that $A\#B, B\#A \in X$.*

Note that the last condition is not a restriction on X, just a notational convenience for the postulates below.

4.1   Minimal Set Contraction Postulates

It is possible to provide a representation theorem and an efficient algorithm for the case of contraction of S by an uncoupled set of preferences X. Essentially, minimal contraction by an uncoupled set X can be reduced to to an unordered set of minimal contractions by single sentences, where those single sentences correspond to partitions of X.

The following postulates characterise a minimal contraction of S by an uncoupled $X \subseteq S$.

CX-Closure $S - X = Cn(S - X)$
CX-Inclusion $S - X \subseteq S$
CX-Vacuity If $X \cap S = \emptyset$, $S - X = S$
CX-Success If $p \in X$ is not of the form $A \equiv A$, then $p \notin S - X$
CX-Minimality If $X \subseteq S$, and X is uncoupled, then $|S - X| = |S| - \Sigma_i c_s(X_i)$ where
  the costs of contracting by each X_i are defined as follows:
  - $c_S(X(A, <)) = c_S(A < A_n)$
  - $c_S(X(B, >)) = c_S(B_1 < B)$
  - $c_S(X(A, \equiv)) = (|A_{\overline{X}}^{\equiv}| + 1) * |A_{\overline{S}}^{\equiv}|$, where by $A_{\overline{X}}^{\equiv}$ we denote the set of alternatives occurring in $X(A, \equiv)$
  - $c_S(X(\#)) = 2 * |X(\#)_1| + |X(\#)_2|)$

Theorem 4. *The result of any minimal set contraction by an uncoupled set of preferences satisfies the minimal set contraction postulates above, and every contraction by an uncoupled set satisfying these postulates is a minimal set contraction.*

Proof. Let us prove that every minimal set contraction satisfies the postulates. The proof for CX-Closure, CX-Inclusion, CX-Vacuity, CX-Success is very similar to Theorem 1. For CX-Minimality, observe that since the partitions X_i do not share alternatives, the sets of sentences which have to be removed to contract by each X_i are disjoint. Note that

- for each $X(A, <)$, it is sufficient and necessary to remove $\{A < C : C \in A^< \cap A_n^>\}$ to make all sentences in $X(A, <)$ underivable
- for each $X(B, >)$, it is sufficient necessary to remove $\{B < C : C \in B_1^< \cap B^>\}$ to make all sentences in $X(B, >)$ underivable
- for each $X(A, \equiv)$, it is sufficient and necessary to remove connections between alternatives occurring in $X(A, \equiv)$ and other members of the equivalence class of A in S, so assuming that $|A^{\equiv}| = m$ and $X(A, \equiv)$ contains occurrences of alternatives $A, A_1, \ldots, A_n$, then we need to remove $(n + 1) * m$ sentences ($2 * m$ for removing sentences connecting A to the equivalence set, $2 * (m - 1)$ for removing sentences connecting $A_1, \ldots, 2 * (m - n)$ for removing sentences connecting A_n).
- for the whole of $X(\#)$, we need to remove $2 * |X(\#)_1|$ and $|X(\#)_2|$.

For the other direction, assume an operation satisfies the postulates for minimal set contraction. Then it clearly satisfies (1) and (2) of Definition 2. It also satisfies (3), since any other contraction by X has to remove at least as many preferences.   □

A postulate corresponding to C-Equivalence: if $Cn(X_1) = Cn(X_2)$, then $S - X_1 = S - X_2$ does not hold. For example, let $X_1 = \{A < B, B < C, A < C\}$ $X_2 = \{A < B, B < C\}$, and $S = \{A < C, A \equiv A, B \equiv B, C \equiv C\}$. Clearly $Cn(X_1) = Cn(X_2)$. However, $S - X_1 = S \setminus \{A < C\}$ and $S - X_2 = S$.

4.2 Minimal Set Contraction Algorithm

We can also give a concrete polynomial time algorithm for contraction by an uncoupled set of preferences.

Algorithm 2. Minimal preference set contraction algorithm

---

procedure MINIMAL-SET-CONTRACTION(S, X)
   for each $X_i \subseteq X$ **do**
      case $X_i == X(A, <)$
         $A^< := \{C \mid A < C\}$
         $A_n^> := \{C \mid C < A_n\}$
         for each $C \in A^< \cap A_n^>$ **do**
            $S := S \setminus \{A < C\}$
         end for
         $S := S \setminus \{A < A_n\}$
      case $X_i == X(B, >)$
         $B_1^< := \{C \mid B_1 < C\}$
         $B^> := \{C \mid C < B\}$
         for each $C \in B_1^< \cap B^>$ **do**
            $S := S \setminus \{C < B\}$
         end for
         $S := S \setminus \{B_1 < B\}$
      case $X_i == X(A, \equiv)$
         $A^\equiv := \{C \mid A \equiv C\}$
         $A^{X,\equiv} := \{D \mid D \text{ occurs in } X(A, \equiv)\}$
         for each $D \in A^{X,\equiv} \cup \{A\}$ **do**
            for each $C \in A^\equiv \setminus \{D\}$ **do**
               $S := S \setminus \{D \equiv C, C \equiv D\}$
            end for
         end for
      case $X_i == X(\#)$
         for each $A \# B \in X(\#)$ **do**
            $S := S \setminus \{A \# B, B \# A\}$
         end for
   end for

The algorithm contracts by each $X_i \subseteq X$ in turn; since X_i are disjoint, in the worst case there are $|X|$ members of the partition. Each contraction by X_i is linear in $|\mathcal{A}|$, by an argument similar to the proof of Theorem 3. This means that the time complexity of Algorithm 2 is $O(|X| \times |\mathcal{A}|)$.

5 Related Work

In this section we compare our results to those works which are most closely related to our own, focusing on the main ideas rather than providing a full-fledged comparison, and highlighting some future research directions.

[9] describes four types of preference change: contraction and revision of preference relations, and addition and subtraction of alternatives. We do not consider changes in alternatives in our framework, thus we compare with the first two kinds. Hansson defines contraction in terms of revision with the intuition that "to contract your state of preference by α means to open it up for the possibility that $\neg\alpha$" and gives postulates for this operation. To define a *minimal* preference revision operator, Hansson introduces a measure of similarity between preference relations. This involves a calculation of the symmetric difference between two sets X and Y ($X \Delta Y$), which is equal to $(X \backslash Y) \cup (Y \backslash X)$. The result of the preference change is a preference relation that has as small a distance from the original relation as possible. This idea inspired our notion of minimal contraction. Since Hansson considers a full logical language with negations, disjunctions etc. of preferences, the complexity of his operations is clearly much higher than ours. [7] discuss logical constraints on preference — formal requirements that a preference state has to satisfy. These are called *reasoning rules* in our framework. A further distinction between logical constraints, input constraints that come with a specific input, and priorities has been made in the same discussion, and various ways of formalizing those aspects in logical models are proposed. In our work, we consider reasoning involving merely logical constraints. It would be interesting to study how to modify our algorithms to incorporate other kinds of constraints.

There exists considerable work on iterated belief revision, see for example [3–5, 11]. We consider set contraction in this paper. However our focus is a special case of contraction by an uncoupled set of preferences. [2] point out potential connections between this area and preference aggregation as they study revision of a total preorder in the context of iterated belief revision. We would like to extend our result to preference aggregation in the future.

6   Conclusion

In this paper, we introduce a simple setting of preference change where it is possible to define minimal preference contraction (revision). We propose rationality postulates and an efficient algorithm for that setting. Then we study contraction by a set of preferences and provide a characterisation and an efficient algorithm for the case where the set of preferences is uncoupled. Finally, we compare our work with some related work and highlight some directions for future research.

Acknowledgement. The authors thank the three anonymous referees for their useful comments. Fenrong Liu is supported by the Project of National Social Science Foundation of China (NO.13AZX018) and Tsinghua University Project (NO. 2012WHYX003).

References

1. Andréka, H., Ryan, M., Schobbens, P.-Y.: Operators and laws for combining preference relations. Journal of Logic and Computation 12(1), 13–53 (2002)
2. Booth, R., Meyer, T.: How to revise a total preorder. Journal of Philosophical Logic 40, 193–238 (2011)

3. Boutilier, C.: Iterated revision and minimal change of conditional beliefs. Journal of Philosophical Logic 25(3), 263–305 (1996)
4. Darwiche, A., Pearl, J.: On the logic of iterated belief revision. Artificial Intelligence 89, 1–29 (1997)
5. Delgrande, J.P., Dubois, D., Lang, J.: Iterated revision as prioritized merging. In: Doherty, P., Mylopoulos, J., Welty, C.A. (eds.) Proceedings, Tenth International Conference on Principles of Knowledge Representation and Reasoning, Lake District of the United Kingdom, June 2-5, pp. 210–220. AAAI Press (2006)
6. Gärdenfors, P.: Knowledge In Flux: Modeling the Dynamics of Epistemic States. MIT Press (1988)
7. Grüne-Yanoff, T., Hansson, S.O.: From belief revision to preference change. In: Preference Change: Approaches from Philosophy, Economics and Psychology, pp. 159–184. Springer (2009)
8. Guo, M., Seligman, J.: Making choices in social situations. In: Logic and Interactive Rationality: Yearbook 2011, pp. 176–202. ILLC, University of Amsterdam (2012)
9. Hansson, S.O.: Changes in preference. Theory and Decision 38, 1–28 (1995)
10. Hansson, S.O.: Semi-revision (invited paper). Journal of Applied Non-Classical Logics 7(2) (1997)
11. Hansson, S.O.: Multiple and iterated contraction reduced to single-step single-sentence contraction. Synthese 173(2), 153–177 (2010)
12. Levi, I.: Subjunctives, dispositions and chances. Synthese 34, 423–455 (1977)
13. Makinson, D.: On the status of the postulate of recovery in the logic of theory change. Journal of Philosophical Logic 16, 383–394 (1987)

The Topology of Belief, Belief Revision and Defeasible Knowledge

Alexandru Baltag[1], Nick Bezhanishvili[2,*], Aybüke Özgün[1], and Sonja Smets[1,**]

[1] University of Amsterdam, The Netherlands
[2] Utrecht University, The Netherlands

Abstract. We present a new topological semantics for doxastic logic, in which the belief modality is interpreted as the closure of the interior operator. We show that this semantics is the most general (extensional) semantics validating Stalnaker's epistemic-doxastic axioms [22] for "strong belief", understood as *subjective certainty*. We prove two completeness results, and we also give a topological semantics for update (dynamic conditioning), i.e. the operation of revising with "hard information" (modeled by restricting the topology to a subspace). Using this, we show that our setting fits well with the defeasibility analysis of knowledge [18]: topological knowledge coincides with undefeated true belief. Finally, we compare our semantics to the older topological interpretation of belief in terms of Cantor derivative [23].

1 Introduction

Ever since Edmund Gettier published his famous counterexamples [14], formal epistemologists have been concerned with understanding the relation between belief and knowledge, and in particular with finding the conditions that distinguish an item of belief (no matter how true and justified) from an item of knowledge. This question can be approached from two sides: 1) we start with the weakest notion of true justified (or justifiable) belief and add conditions in order to argue that they establish a "good" (e.g. factive, correctly-justified, unrevisable, coherent, stable, truth-sensitive) notion of knowledge; or 2) we start from a chosen notion of knowledge and weaken it to obtain a "good" (e.g. consistent, introspective, possibly false) notion of belief. Most research in formal epistemology follows the first approach. In particular, the standard *topological semantics for knowledge* (in terms of the interior operator) can be included within this first approach, as based on a notion of knowledge as "correctly justified belief": according to the interior semantics, a proposition (set of possible worlds) P is known if there exists some "true evidence" (i.e. an open set A containing the real

* Supported by the Dutch NWO grant 639.032.918 and the Rustaveli Science Foundation of Georgia grant FR/489/5-105/11.
** S. Smets contribution to this paper has received funding from the ERC under the European Community's 7th Framework Programme/ERC Grant agreement no. 283963.

D. Grossi, O. Roy, and H. Huang (Eds.): LORI 2013, LNCS 8196, pp. 27–40, 2013.

world s) that entails P (i.e. $A \subseteq P$). Another example of the first approach is the so-called *defeasibility analysis of knowledge* proposed by Lehrer and Paxson [18], Klein [16] and other authors: knowledge is defined as *undefeated (justified) true belief*; i.e. true belief that cannot be defeated by any new (true) evidence.

While most research in formal epistemology follows the first approach, the second approach has to date received much less attention from formal logicians. This is rather surprising, since such a "knowledge-first" approach has been persuasively defended by one of the most influential contemporary epistemologists (Williamson [26]). The only formal account following this second approach that we are aware of (prior to our own work) is the one given by Stalnaker [22], using a relational semantics for knowledge, based on Kripke models in which the accessibility relation is a directed preorder. In this setting, Stalnaker argues that the "true" logic of knowledge is the modal logic $S4.2$ and that belief can be defined as the *epistemic possibility of knowledge*. In other words, *believing p is equivalent to "not knowing that you don't know"* p:

$$Bp = \neg K \neg Kp.$$

Stalnaker justifies this identity from first principles based on a particular notion of belief, namely *belief as "subjective certainty"*. Stalnaker refers to this concept as "strong belief", but we prefer to call it *full belief*[1]. What is important about this type of belief is that it is *subjectively indistinguishable from knowledge*: an agent "fully believes" p iff in fact she "believes that she knows" p.

The resulting conception of knowledge is clearly different from Williamson's (who rejects the KK principle[2]), but it is closely related to the above-mentioned *defeasibility analysis* [17]. Indeed, Stalnaker proceeds to formalize AGM belief revision, based on a special case of the above semantics, in which the accessibility relation is assumed to be a weakly connected preorder, and (conditional) beliefs are defined by minimization. This validates the AGM principles for belief revision. Stalnaker shows that in this special case his notion of knowledge coincides with (a simplified and idealized version of) Lehrer's concept of *undefeated (justified) true belief*: i.e. true belief that cannot be defeated by (revising with) any new (true) evidence. However, this special case supports a stronger logic of knowledge (the system $S4.3$). Since Stalnaker defends the weaker $S4.2$ as the "true" logic of knowledge, he is lead to argue *against* the defeasibility theory.

---

[1] We adopt this terminology both because we want to avoid the clash with the very different notion of strong belief (due to Battigalli and Siniscalchi [2]) that is standard in epistemic game theory, and because we think that the intuitions behind Stalnaker's notion are very similar to the ones behind Van Fraassen's probabilistic concept of full belief [13].

[2] In formal epistemology, the "KK principle" is one of the names given to the axiom of Positive Introspection $K\varphi \to KK\varphi$, also known as the modal axiom 4. The well-known modal system $S4$ consists of this Positive Introspection axiom 4 together with the axiom T of Factivity, or Truthfulness ($K\varphi \to \varphi$), as well as Kripke's axiom K ($K(\varphi \to \psi) \to (K\varphi \to K\psi)$) and the rules of Modus Ponens and Necessitation. Stalnaker's system $S4.2$ includes the system $S4$, and hence contains the KK principle, contradicting Williamson's conception.

In this paper, we aim to generalize Stalnaker's formalization, making it independent from the concept of plausibility order and from relational semantics. In fact, we are looking for *the most general extensional semantics for "full belief"* (in the above-mentioned sense). By an "extensional" semantics we mean here any semantics that assigns the same meaning to sentences having the same extension. Essentially, an extensional semantics takes the meaning of a sentence to be given by a "U.C.L.A. proposition" in the sense of Anderson-Belnap-Dunn[3]: a set of possible worlds (intuitively thought of as the set of worlds at which the proposition is true). We prove that the most general extensional semantics validating Stalnaker's axioms is a *topological* one, that extends the standard topological interpretation of knowledge (as *interior* operator) with a new topological semantics for belief, given by the *closure of the interior* operator (with respect to an extremally disconnected topology). We compare our new semantics with the older topological interpretation of belief in terms of Cantor derivative, giving several arguments in favor of our semantics.

We prove that the logic of knowledge and belief with respect to our semantics is completely axiomatized by Stalnaker's epistemic-doxastic principles. Furthermore, we show that the complete logic of knowledge in this setting is indeed the system $S4.2$, while the complete logic of belief is the standard system $KD45$. We formalize the action of learning (conditioning with) new "hard" (true) information P as a *topological update* operator, using the *relativization* of the original topology to (the subspace corresponding to) the set P. This allows us to model belief revision of a more general type than the one axiomatized by the AGM theory. We show that, in this generalized setting, *Stalnaker's objections to the defeasibility theory of knowledge do not apply*: when interpreted (as interior) over topological spaces, Stalnaker's notion of knowledge (having $S4.2$ as its complete logic) *coincides with undefeated (justified) true belief.*

2 Background: Topological Interpretation of Knowledge

2.1 Topological Preliminaries

For the basic definitions of general topology we refer to [11] or any other textbook in General Topology. Here we just recall that a *topological space* is a pair (X, τ), where X is a non-empty set and τ is a *topology* on X, i.e. a family $\tau \subseteq \mathcal{P}(X)$ containing X and $\emptyset$ and closed under finite intersections and arbitrary unions. Elements of τ are called *open sets*. Complements of open sets are called *closed sets*. An open set containing $x \in X$ is called an *open neighbourhood* of x. The *interior* $\mathrm{Int}(A)$ of a set $A \subseteq X$ is the largest open set contained in A. The *closure* $\mathrm{Cl}(A)$ of A is the least closed set containing A. It is easy to check that $\mathrm{Int}(A) = X \setminus \mathrm{Cl}(X \setminus A)$.

---

[3] Dunn [10] explains this name as follows: 'The name honors the university that has had both R. Carnap and R. Montague in its faculty, since in modern times they (together with others, e.g. S. Kripke and R. Stalnaker) have been proponents of this construction. But the idea actually originates with Boole, who suggested thinking of propositions as "sets of cases" (...).'

An *alternative definition of topological spaces* (due to Kuratowski) takes the closure operation (or equivalently, the interior) as the basic notion. Stated in terms of interior, a topological space is a pair (X, Int), where X is a non-empty set and $\mathrm{Int} : \mathcal{P}(X) \to \mathcal{P}(X)$ is an operation satisfying the (dual of the) so-called *Kuratowski axioms*: $\mathrm{Int}(X) = X$, $\mathrm{Int}(A) \subseteq A$, $\mathrm{Int}(\mathrm{Int}(A)) = \mathrm{Int}(A)$, $\mathrm{Int}(A \cap B) = \mathrm{Int}(A) \cap \mathrm{Int}(B)$. In this setting, the family of open sets is defined by putting $\tau = \{A \subseteq X : \mathrm{Int}(A) = A\}$. It is easy to see that this is a topology. Indeed, the two definitions of topological spaces are equivalent.

2.2   The Interior Semantics for Modal Logic

We start by recalling the standard topological semantics of modal (epistemic) logic, originating in the work of Tarski and McKinsey [19]. We consider the standard unimodal language $\mathcal{L}_K$ with a countable set of propositional letters Prop, and a modal operator K. Formulas are defined as usual by

$$\varphi ::= \bot \mid p \mid \neg\varphi \mid \varphi \wedge \varphi \mid K\varphi,$$

where $p \in$ Prop. Abbreviations for the connectives $\vee$, $\to$ and $\leftrightarrow$ are standard. The possibility operator $\langle K \rangle \varphi$ is defined as $\langle K \rangle \varphi := \neg K \neg \varphi$.

Definition 1. *A **topological model** $\mathcal{M} = (X, \tau, \nu)$ is a tuple where (X, τ) is a topological space and ν is a valuation, i.e., a map $\nu : $ Prop $\to \mathcal{P}(X)$. We let Cl and Int denote the closure and interior operators, respectively. The **topological semantics for modal formulas** is defined by the following inductive definition, where $\mathcal{M} = (X, \tau, \nu)$ is a topological model and $p \in$ Prop:*

$$\begin{array}{ll} [\![\bot]\!]^{\mathcal{M}} = \emptyset, & [\![p]\!]^{\mathcal{M}} = \nu(p) \\ [\![\varphi \wedge \psi]\!]^{\mathcal{M}} = [\![\varphi]\!]^{\mathcal{M}} \cap [\![\psi]\!]^{\mathcal{M}} & [\![\neg\varphi]\!]^{\mathcal{M}} = X \setminus [\![\varphi]\!]^{\mathcal{M}} \\ [\![K\varphi]\!]^{\mathcal{M}} = \mathrm{Int}[\![\varphi]\!]^{\mathcal{M}} \end{array}$$

As $\langle K \rangle \varphi$ is equivalent to $\neg K \neg \varphi$, it is easy to see that $[\![\langle K \rangle \varphi]\!]^{\mathcal{M}} = \mathrm{Cl}[\![\varphi]\!]^{\mathcal{M}}$. We skip the index $\mathcal{M}$ if it is clear from the context. Truth, validity, soundness and completeness wrt topological semantics are defined as usual.

Proposition 1. *(see e.g. [3], [20] and [7]) The modal logic S4 is sound and complete wrt all topological spaces.*

2.3   Epistemic Interpretation: Open Sets as Pieces of Evidence

The original reason for interpreting interior as knowledge was that the Kuratowski axioms match exactly the $S4$ axioms, and in particular the principles

$$(T) \quad Kp \to p$$

of Truthfulness of Knowledge ("factivity") and

$$(KK) \quad Kp \to KKp$$

of Positive Introspection of Knowledge (known as axiom 4 in modal logic).

Philosophically, one of the best arguments in favor of the topological semantics is negative: namely, the fact that it does *not* validate the principle

$$\neg Kp \to K\neg Kp$$

This principle, known as (5) or Negative Introspection, is rejected by essentially all philosophers. One of its undesirable consequences is that it makes it impossible for a rational agent to have wrong beliefs about her knowledge: *she always knows whatever she believes that she knows*. This is known in the literature as Voorbraak's paradox [25]: it contradicts the day-to-day experience of encountering agents who believe they know things that they do not actually know[4].

But, even beyond the issue of negative introspection, the topological semantics can arguably give us a deeper insight into the nature of knowledge and its evidential basis than the usual Kripke semantics. From an extensional point of view, the properties U that are directly observable by an agent naturally form an *open basis* for a topology: closure under finite intersections captures an agent's ability to combine finitely many pieces of evidence into a single piece[5]. A proposition P is true at world w if $w \in P$. If an open U is included in a set P, then we can say that *proposition P is entailed (supported, justified) by evidence U*. Open neighbourhoods U of the actual world w play the role of *sound (correct, truthful) evidence*. The actual world w is in the interior of P iff there exists such a sound piece of evidence U that supports P. So *the agent "knows" P if she has a correct justification for P* (based on a sound piece of evidence supporting P). Moreover, open sets will then correspond to properties that are in principle *verifiable* by the agent: whenever they are true they can be known. Dually, closed sets will correspond to *falsifiable* properties. See Vickers [24] and Kelly [15] for more on this interpretation and its connections to Epistemology, Logic and Learning Theory.

So the knowledge-as-interior conception can be seen as an implementation of one of the most widespread intuitive responses to Gettier's challenge: knowledge is "correctly justified belief" (rather than being simply true justified belief). To qualify as knowledge, not only the content of one's belief has to be truthful, but its evidential justification has to be sound.

2.4   Extensions and Improvements

The interior-based semantics for knowledge has been extended to multiple agents [4], to common knowledge [1,6], to logics of learning ("topo-logic", see [20]), to topological versions of dynamic-epistemic logic [27]. See [3] for a comprehensive survey of the field.

---

[4] This common experience can be considered the starting point of all epistemological reflection, and historically played such a role, see e.g. in Platonic dialogues.

[5] But see van Benthem and Pacuit [5] for a more general logical account of evidence-management which relaxes this assumption: by using instead a neighbourhood semantics, this account can deal with agents who have not yet managed to combine all their pieces of evidence.

But there are two other topologically-based logics that are of particular interest in this paper. The first is an alternative semantics for modal logic, in terms of Cantor's derivative operation, which has been proposed as a semantics for belief. We will give a critical presentation of this alternative in Section 6. The second is a logic that *strengthens* S4, namely:

$$S4.2 = S4 + (\langle K \rangle K p \rightarrow K \langle K \rangle p).$$

By $L + \varphi$ we denote the smallest normal modal logic containing L and φ.

Recall that a topological space (X, τ) is called *extremally disconnected* if the closure of every open subset of X is open.

Proposition 2. *[3, p. 253]* S4.2 *is sound and complete wrt all extremally disconnected topological spaces.*

We also recall that a topological space (X, τ) is called an *Alexandroff space* if the intersection of open sets of X is open. It is well known that Alexandroff spaces correspond to reflexive and transitive Kripke frames, see e.g., [3], [20] or [7]. Moreover, the evaluation of modal formulas in an Alexandroff space coincides with their evaluation in the corresponding Kripke frame.

A Kripke frame (X, R) is called *directed*[6] if

$$(\forall x, y, z)(xRy \wedge xRz) \rightarrow (\exists u)(yRu \wedge zRu).$$

It is well known, see e.g., [8] or [9] that S4.2 is sound and complete wrt reflexive, transitive and directed Kripke frames.

We give a few examples of extremally disconnected spaces. Alexandroff spaces corresponding to reflexive, transitive and directed Kripke frames are extremally disconnected. Another classical example of an extremally disconnected space is the Stone-Čech compactification $\beta(\mathbb{N})$ of the set of natural numbers with a discrete topology. Also it is well known that topological spaces that are Stone-dual to complete Boolean algebras are extremally disconnected [21].

3   The Topology of Full Belief

3.1   Stalnaker's Epistemic-Doxastic Axioms

In his paper [22], Stalnaker proposes a very interesting analysis of the relationship between knowledge and (justified or justifiable) belief. This is based on a conception of belief as "subjective certainty": from the point of the agent in question, her belief is subjectively indistinguishable from her knowledge. In this paper, we will refer to Stalnaker's notion as "full belief".

The *bimodal language* $\mathcal{L}_{KB}$ of knowledge and (full) belief is given recursively:

$$\varphi ::= \bot \mid p \mid \neg\varphi \mid \varphi \wedge \varphi \mid K\varphi \mid B\varphi,$$

where $p \in \mathsf{Prop}$. We will also consider two unimodal fragments of this language $\mathcal{L}_K$ (having K as its only modality) and $\mathcal{L}_B$ (having only B).

---

[6] Directedness is also called *confluence* or the *Church-Rosser property*.

Stalnaker's epistemic-doxastic axioms for the logic KB are given in the Table below.

	Stalnaker's Epistemic-Doxastic Axioms	
(K)	$K(\varphi \to \psi) \to (K\varphi \to K\psi)$	Knowledge is additive
(T)	$K\varphi \to \varphi$	Knowledge implies truth
(KK)	$K\varphi \to KK\varphi$	Positive introspection for K
(CB)	$B\varphi \to \neg B\neg\varphi$	Consistency of belief
(PI)	$B\varphi \to KB\varphi$	(Strong) positive introspection of B
(NI)	$\neg B\varphi \to K\neg B\varphi$	(Strong) negative introspection of B
(KB)	$K\varphi \to B\varphi$	Knowledge implies Belief
(FB)	$B\varphi \to BK\varphi$	Full Belief
	Inference Rules	
(MP)	From φ and $\varphi \to \psi$ infer ψ.	Modus Ponens
(K-Nec)	From φ infer $K\varphi$.	Necessitation

We will refer to this axiomatic system as KB. The axioms seem very natural and uncontroversial: the first three are the $S4$ axioms for knowledge; (CB) captures the consistency of beliefs, and in the context of the other axioms will be equivalent to the modal axiom (D) for beliefs: $\neg B\bot$; (PI) and (NI) capture strong versions of introspection of beliefs: the agent knows what she believes and what not; (KB) means that agents believe what they know; and finally, (FB) captures the essence of "full belief" as subjective certainty (the agent believes that she knows all the things that she believes). Finally, the rules of Modus Ponens and Necessitation seem uncontroversial (for implicit knowledge, if not for explicit knowledge) and are accepted by a majority of authors (and in particular, they are implicitly used by Stalnaker).

The above axioms imply that *belief can be defined in terms of knowledge*:

Proposition 3. *(Stalnaker) The equivalence*

$$B\varphi \leftrightarrow \neg K\neg K\varphi$$

is provable in the system KB. Moreover, all the axioms of the standard system $KD45$ for belief are provable in the system KB, and in particular: Kripke's axiom for belief $(B(\varphi \to \psi) \to (B\varphi \to B\psi))$; the so-called axiom (D) $(\neg B\bot)$; axiom 4 (positive introspection) for belief $(B\varphi \to BB\varphi)$; the axiom 5 (negative introspection) for belief $(\neg B\varphi \to B\neg B\varphi)$.

Finally, the formula $\langle K \rangle K\varphi \to K\langle K \rangle\varphi$ is also provable in KB: i.e. all the axioms of the system $S4.2$ hold for knowledge in the system KB.

3.2   Our Topological Semantics for Full Belief

Definition 2. *An **extensional (and compositional) semantics** for the language $\mathcal{L}_{KB}$ of knowledge and full belief is a triplet (X, B, K), where X is a set of possible worlds, and $B : \mathcal{P}(X) \to \mathcal{P}(X)$ and $K : \mathcal{P}(X) \to \mathcal{P}(X)$ are unary operations on (sub)sets of worlds.*

Any extensional semantics (X, B, K), *together with a valuation* $\nu : Prop \rightarrow \mathcal{P}(X)$, *gives us an* **extensional model** $M = (X, B, K, \nu)$, *in which we can interpret the formulas* φ *of* $\mathcal{L}_{KB}$ *in the obvious way: the clauses for propositional connectives are the same as in the topological semantics above, and in rest we put*

$$[\![K\varphi]\!]^{\mathcal{M}} = K[\![\varphi]\!]^{\mathcal{M}} \qquad [\![B\varphi]\!]^{\mathcal{M}} = B[\![\varphi]\!]^{\mathcal{M}}.$$

As usual, a formula is valid in an extensional semantics (X, B, K) *if it is true at all worlds of all models* $M = (X, B, K, \nu)$ *based on it. An inference rule is valid if it preserves validity of formulas.*

A special case of extensional semantics for the language $\mathcal{L}_{KB}$ is *our proposed topological semantics*:

Definition 3. *A* **topological semantics for the language** $\mathcal{L}_{KB}$ *is an extensional semantics* (X, K^{τ}, B^{τ}), *where* (X, τ) *is a topological space,* $K^{\tau} = Int^{\tau}$ *is the interior operator with respect to the topology* τ, *and* $B^{\tau} = Cl^{\tau}(Int^{\tau})$ *is the closure of the interior with respect to* τ.

Proposition 4. *A topological space validates all the axioms and rules of the system* $KD45$ *for belief (with the semantics given above) iff it is extremally disconnected.*

The proof of this result is rather long and intricate, and is left for a future journal publication.

Proposition 5. *A topological space validates all the axioms and rules of Stalnaker's system* KB *(with the semantics given above) iff it is extremally disconnected.*

Proof. It is easy to check that extremally disconnected spaces validate all the the axioms of KB. The other direction follows from Propositions 3 and 4.

Now we can give a Topological Representation Theorem for extensional models of KB:

Theorem 1. *An extensional semantics* (X, B, K) *validates all the axioms and rules of Stalnaker's system* KB *iff it is a topological semantics given by an extremally disconnected topology* τ *on* X *(such that* $K = K^{\tau} = Int^{\tau}$ *and* $B = B^{\tau} = Cl^{\tau}(Int^{\tau})$).

Proof. One direction is proved in the previous Proposition, so let us look at the other direction. Suppose an extensional semantics (X, B, K) validates the axioms and rules of KB. Then the validity of the $S4$ axioms implies that K satisfies the Kuratowski conditions for topological interior, and so it gives rise to a topology τ in which $K = Int^{\tau}$. By the Proposition in the previous section, the KB axioms imply that $B = \neg K \neg K$, i.e. $B = \neg Int^{\tau} \neg Int^{\tau} = Cl^{\tau} Int^{\tau}$. Hence, (X, B, K) is a topological semantics, in the sense above, for a topology τ. By the previous Proposition, the validity of KB implies that τ is extremally disconnected.

This last result shows that Stalnaker's axioms are just an alternative axiomatization of extremally disconnected topological spaces, in which both interior and the closure of interior are taken as primitive operations. The conclusion is that *our topological semantics is indeed the most general (extensional compositional) semantics validating Stalnaker's axioms.*

3.3 Completeness Results

Let us first look at the bimodal logic KB of knowledge and (full) belief.

Theorem 2. *The (sound and) complete logic of knowledge and belief on extremally disconnected spaces is given by Stalnaker's system KB.*

Proof. This follows trivially from our Topological Representation Theorem for extensional models of KB (Theorem 1 in Section 3.2).

Next, we look at the unimodal fragment $\mathcal{L}_K$ having K as the only modality. In fact, this language has exactly the same expressivity as KB (since the belief operator can be eliminated via the identity $B\varphi = \neg K\neg K\varphi$). Moreover, we already know (by Proposition 2 in section 2.3) that *the sound and complete logic of knowledge on extremally disconnected spaces is $S4.2$.*

Further, we look at the unimodal fragment $\mathcal{L}_K$ having B as the only modality: this logic is less expressive than the bimodal language KB, since knowledge is not reducible to belief.

Theorem 3. *The complete logic of belief on extremally disconnected spaces is $KD45$.*

This result, though unsurprising, is technically the hardest result in this paper. The proof is long and intricate, and is left for a future journal publication.

4   From Updates to Defeasible Knowledge

Conditioning (with respect to some qualitative plausibility order or to a probability measure) is the most widespread way to model the learning of "hard" information[7]. The prior plausibility/probability assignment (encoding the agent's original beliefs before the learning) is changed to a new such assignment, obtained from the first one by conditioning with the new information P. In the qualitative case, this means just restricting the original order to P-worlds; while in the probabilistic case, restriction has to be followed by re-normalization (to ensure that the probabilities newly assigned to the remaining worlds add up to 1). In Dynamic Epistemic Logic, one makes also a distinction between *simple ("static") conditioning* and *dynamic conditioning* (also known as "update").

---

[7] This term is used to denote information that comes with an inherent warranty of veracity, e.g. because of originating from an infallibly truthful source.

The first essentially corresponds to conditional beliefs: the change is made only locally, affecting only one occurrence of the belief operator $B\varphi$ (which is thus locally replaced by conditional belief $B^P\varphi$) or of the probability measure (which is locally replaced by conditional probability). In contrast, an update is a global change, at the level of the whole model (thus recursively affecting the meaning of all occurrences of doxastic operators). In this paper, due to space restrictions, we only investigate the natural topological analogue of *dynamic conditioning*.

Topological Updates. As recognized already in [27] (among others), the natural topological analogue of dynamic conditioning (update) is the operation of taking the *restriction (or "relativization") of a topology* τ on X to a subset $P \subseteq X$. What we obtain in this way is a *subspace* of the original topological space. Given a topological space (X, τ) and a set $P \subseteq X$, a space (P, τ_P) is called a *subspace* of (X, τ) if $\tau_P = \{U \cap P : U \in \tau\}$. It is well-known that the closure and interior operators in the relativized topology (P, τ_P), denoted by Cl_{τ_P} and Int_{τ_P} respectively, satisfy the following equations for every $A \subseteq P$:

$$\mathrm{Cl}_{\tau_P}(A) = \mathrm{Cl}(A) \cap P, \qquad\qquad \mathrm{Int}_{\tau_P}(A) = \mathrm{Int}((X \setminus P) \cup A) \cap P.$$

Update Modalities. The *dynamic language* $\mathcal{L}_{KB!}$ is obtained by extending $\mathcal{L}_{KB}$ with *(existential) dynamic update modalities* $\langle!\varphi\rangle\psi$, meaning that: φ is true and after the agent learns this, ψ becomes true. The corresponding universal modality is defined by putting $[!\varphi]\psi := \neg\langle!\varphi\rangle\neg\psi$.

Definition 4. *(Semantics of Updates)* Let $\mathcal{M} = (X, \tau, \nu)$ be a *topological model. Given a formula φ we will denote by $\mathcal{M}_\varphi$ the **relativized model***

$$\mathcal{M}_\varphi = (\llbracket\varphi\rrbracket, \tau_{\llbracket\varphi\rrbracket}, \nu_{\llbracket\varphi\rrbracket}),$$

*where $\llbracket\varphi\rrbracket = \llbracket\varphi\rrbracket^{\mathcal{M}}$ is the interpretation of φ in $\mathcal{M}$, $\tau_{\llbracket\varphi\rrbracket}$ is the relativized topology and $\nu_P(p) = \nu(p) \cap P$, for each $p \in \mathsf{Prop}$. The **semantics of** $\mathcal{L}_{KB!}$ is obtained by extending the semantics of $\mathcal{L}_{KB}$ with the following clause:*

$$\llbracket\langle!\varphi\rangle\psi\rrbracket^{\mathcal{M}} = \llbracket\psi\rrbracket^{\mathcal{M}_\varphi}.$$

Connection to the Defeasibility Theory of Knowledge. As promised in the Introduction, we show now that in the generalized Belief Revision Theory given by our topological semantics for conditional beliefs, *topological knowledge coincides with the one given by the defeasibility analysis*:

Theorem 4. *Let $\mathcal{M} = (X, \tau, \nu)$ be a topological model. The following are equivalent, for all worlds $x \in X$ and atomic sentences[8] p:*

1. $x \in \llbracket Kp \rrbracket^{\mathcal{M}}$;
2. $x \in \llbracket[!\theta]Bp\rrbracket^{\mathcal{M}}$ for every formula θ;
3. $x \in \llbracket Bp\rrbracket^{\mathcal{M}_\theta}$ for every formula θ such that $x \in \llbracket\theta\rrbracket^{\mathcal{M}}$.

---

[8] The restriction to atomic sentences in the other clauses is necessary because of the so-called Moore sentences: these are epistemic formulas which change their truth value after being learnt.

Proof. The equivalence between (2) and (3) follows immediately from the semantics of dynamic update modalities, so we only prove the equivalence between (1) and (3). For this, let us first put $A := [\![p]\!]^{\mathcal{M}}$, $P := X \setminus \text{Int}(A)$, which gives us $X \setminus P = \text{Int}(A)$. Then by the above equations we have $\text{Int}_{\tau_P}(A) = \text{Int}((X \setminus P) \cup A) \cap P = \text{Int}(\text{Int}(A) \cup A) \cap P = \text{Int}(\text{Int}(A)) \cap P = \text{Int}(A) \cap (X \setminus \text{Int}(A)) = \emptyset$.

To show (1) $\Rightarrow$ (3): assume that $x \in [\![Kp]\!]^{\mathcal{M}}$, so $x \in \text{Int}([\![p]\!]^{\mathcal{M}})$. Let θ be any formula s.t. $x \in [\![\theta]\!]^{\mathcal{M}}$. Note that $\text{Int}([\![p]\!]^{\mathcal{M}}) \subseteq \text{Int}((X \setminus [\![\theta]\!]^{\mathcal{M}}) \cup [\![p]\!]^{\mathcal{M}})$, since $[\![p]\!]^{\mathcal{M}} \subseteq (X \setminus [\![\theta]\!]^{\mathcal{M}}) \cup [\![p]\!]^{\mathcal{M}}$. Then, since $x \in [\![\theta]\!]^{\mathcal{M}}$ and $[\![p]\!]^{\mathcal{M}_\theta} = [\![p]\!]^{\mathcal{M}} \cap [\![\theta]\!]^{\mathcal{M}}$, we have $x \in \text{Int}(X \setminus [\![\theta]\!]^{\mathcal{M}} \cup [\![p]\!]^{\mathcal{M}}) \cap [\![\theta]\!]^{\mathcal{M}} = \text{Int}((X \setminus [\![\theta]\!]^{\mathcal{M}} \cup [\![p]\!]^{\mathcal{M}}) \cap X) \cap [\![\theta]\!]^{\mathcal{M}} = \text{Int}((X \setminus [\![\theta]\!]^{\mathcal{M}} \cup [\![p]\!]^{\mathcal{M}}) \cap (X \setminus [\![\theta]\!]^{\mathcal{M}} \cup [\![\theta]\!]^{\mathcal{M}})) \cap [\![\theta]\!]^{\mathcal{M}} = \text{Int}(X \setminus [\![\theta]\!]^{\mathcal{M}} \cup ([\![p]\!]^{\mathcal{M}} \cap [\![\theta]\!]^{\mathcal{M}})) \cap [\![\theta]\!]^{\mathcal{M}} = \text{Int}_{\tau_{[\![\theta]\!]\mathcal{M}}}([\![p]\!]^{\mathcal{M}_\theta}) \subseteq \text{Cl}_{\tau_{[\![\theta]\!]\mathcal{M}}}(\text{Int}_{\tau_{[\![\theta]\!]\mathcal{M}}}([\![p]\!]^{\mathcal{M}_\theta})) = [\![Bp]\!]^{\mathcal{M}_\theta}$.

For (3) $\Rightarrow$ (1): assume that (3) holds but (1) fails, i.e $x \notin [\![Kp]\!]^{\mathcal{M}}$, and hence $x \in [\![\neg Kp]\!]^{\mathcal{M}}$. By applying (3) to the formula $\theta := \neg Kp$, we obtain that $x \in [\![Bp]\!]^{\mathcal{M}_\theta}$. But since $[\![\theta]\!]^{\mathcal{M}} = [\![\neg Kp]\!]^{\mathcal{M}} = X \setminus \text{Int}(A) = P$, we have $x \in [\![Bp]\!]^{\mathcal{M}_\theta} = \text{Cl}_{\tau_P}(\text{Int}_{\tau_P}(A)) = \text{Cl}_{\tau_P}(\emptyset) = \emptyset$. Contradiction!

5 Comparison with Related Work

We compare now our topological interpretation of belief with a different (and older) topological semantics that has been proposed for doxastic logic.

Cantor's Derivative and Its Dual. Let (X, τ) be a topological space. We recall that a point x is called a *limit point* (limit points are also called *accumulation points*) of a set $A \subseteq X$ if for each open neighbourhood U of x we have $(U \setminus \{x\}) \cap A \neq \emptyset$. Let $d(A)$ denote the set of all limit points of A. This set is called the *derived set* and d is called the *derived set operator*. For each $A \subseteq X$ we let $t(A) = X \setminus d(X \setminus A)$. We call t the *co-derived set operator*. Also recall that there is a close connection between the derived and co-derived set operators and the closure and interior operators. In particular, for each $A \subseteq X$ we have $\text{Cl}(A) = A \cup d(A)$ and $\text{Int}(A) = A \cap t(A)$. Unlike the closure operator there may exist elements of A that are not its limit points. In other words, in general $A \not\subseteq d(A)$. Also note that for each $x \in X$ we have $x \notin d(x)$, where $d(x)$ is a shorthand for $d(\{x\})$.

Definition 5. *Let $\mathcal{M} = (X, \tau, \nu)$ be a topological model. The **co-derived set semantics for** $\mathcal{L}_{KB}$ is obtained by extending the standard topological semantics for $\mathcal{L}_K$ (interpreting K as interior) with the following clause:*

$$[\![B\varphi]\!]^{\mathcal{M}} = t([\![\varphi]\!]^{\mathcal{M}})$$

This immediately gives us that $[\![\langle B \rangle \varphi]\!]^{\mathcal{M}} = d([\![\varphi]\!]^{\mathcal{M}})$. We again skip the index $\mathcal{M}$ if it is clear from the context. See [3], [20] and [7] for an overview of the results on the co-derived set semantics. Here we only mention the completeness results for the unimodal language $\mathcal{L}_B$ with the co-derived set semantics: the complete logic of belief over all topological spaces is $wK4 = K + ((p \wedge Bp) \rightarrow BBp)$ [12], while the doxastic logic $KD45$ is complete wrt so-called *DSO-spaces*. Here, a *DSO-space* is a topological space (X, τ) satisfying the following conditions:

the T_D-separation axiom[9]; for every $A \subseteq X$ the set $d(A)$ is open; and (X, τ) is dense-in-itself, i.e., $d(X) = X$. See [23] for more details.

Criticism and Comparison with Our Conception. Steinsvold [23] was the first to propose the co-derived set operator as a semantics for belief. However, this interpretation has a major disadvantage: *it entails (not just the possibility, but) the necessity of error*. To explain: all authors agree that one of the main characteristics of belief is the *possibility of error*: it is possible that some of the agent's beliefs are false. In other words, any good semantics for belief should allow for models and worlds at which some beliefs are false. However, we claim that, according to the co-derived semantics *the existence of false beliefs is a necessary fact* (holding for all possible agents at all possible worlds in all possible models!).

Indeed, as we pointed out above, for each $x \in X$ we always have $x \notin d(x)$. So $x \in B(X \setminus \{x\})$. Thus, at a point x the agent believes $X \setminus \{x\}$, which is false (since $x \notin X \setminus \{x\}$). This means that in *any topological model and any world in this model, there is at least one false belief*. Hence, the co-derived set interpretation implies the "necessity of error": *the actual world is always dis-believed*.

We think this consequence is an intuitively undesirable property. It generally prevents any act of learning (updating with) the actual world. Indeed, the main problem of Formal Learning Theory (learning the true world, or the correct possibility, from a given set of possibilities) becomes automatically unattainable. Similarly, the physicist's dream of finding a true "theory of everything" is declared impossible by fiat, *as a matter of logic*. More importantly, even if necessity of error might seem realistic within a Lewisian "large-world interpretation" of possible-world semantics (in which each world must really come with a full description of all the myriad of ontic facts of the world), this property seems completely unrealistic when we adopt the more down-to-earth "small-world" models that are common in Computer Science, Game theory and other applications. In these fields, the "worlds" in any usable model come only with the description of the facts that are *relevant* for the problem at hand: e.g. in a scenario involving the throwing of a fair coin, the relevant fact is the upper face of the coin. A model for this scenario will involve typically only two possible worlds: Head and Tail. Requiring that the agent must always have a false belief means in this context that the agent can never find out which of the coin's faces is the upper one: an obviously absurd conclusion!

There is another objection, maybe even more decisive, against the co-derived set semantics, namely that it can be easily "Gettierized". As mentioned above, we have $\mathrm{Int}(A) = A \cap t(A)$, which means that in *the co-derived set interpretation, knowledge is exactly the same as true (justified) belief*. So this semantics is easily vulnerable to all the well-known Gettier-type counterexamples!

Finally, here is an argument of a more technical nature. As mentioned above, the co-derived set semantics validates the $KD45$ axioms only on DSO-spaces, while our semantics validates them on extremally disconnected spaces. So the

---

[9] Recall that the T_D separation axiom states that every point is the intersection of a closed and open set. This condition is equivalent to $d(d(A)) \subseteq d(A)$, see e.g., [11].

following result shows that *our topological interpretation "works" on a larger class of models than the co-derived set semantics*:

Proposition 6. *Every DSO-space is extremally disconnected.*

Proof. Let (X, τ) be a *DSO*-space and $U \in \tau$. Recall that for any $A \subseteq X$, $\mathrm{Cl}(A) = d(A) \cup A$. So $\mathrm{Cl}(U) = d(U) \cup U$. Since (X, τ) is a *DSO*-space, $d(U)$ is an open subset of X. Thus, since U is open as well, $d(U) \cup U = \mathrm{Cl}(U)$ is open.

6   Conclusions and Future Work

In this paper, we proposed a new topological semantics for belief and argued that it is the "correct" one, at least as far as full belief (understood as subjective certainty) is concerned: it is *the "canonical" (most general) semantics for (Stalnaker's axioms for) full belief*. Moreover, our proposal comes with an *independent motivation* and has an *intrinsic philosophical and intuitive value*. Topologically, a point is in the interior of a set P iff it can be sharply distinguished (separated) from all non-P points (by an open set); similarly, a point is in the closure of P iff it is "very close" to P, i.e. it cannot be sharply distinguished from all P points. Thus, an agent knows P if she can sharply distinguish the actual world from all non-P-worlds. Hence, according to our semantics for full belief, an agent (fully) believes P if she cannot sharply distinguish the actual world w from the worlds in which she has knowledge of P. In this sense, one can say that *belief is topologically "very close" to knowledge*: indeed, the agent cannot sharply distinguish it from knowledge. We thus think that our topological semantics perfectly captures the essence of full belief as "subjective certainty".

From a philosophical perspective, the main importance of our paper is that *it connects three different epistemological conceptions* that were proposed as responses to Gettier's challenge: Stalnaker's epistemic definition of full belief (in the spirit of the "knowledge-first" approach), the "knowledge as correctly-justified-belief" approach (underlying the topological semantics of knowledge) and the defeasibility analysis of knowledge. Indeed (as shown in Section 4), in our semantics knowledge is undefeated-true-belief. To show this, we needed the topological analogue of dynamic conditioning (update), as it was already defined in [27].

In on-going work, we also explore the corresponding "static" conditioning, by giving a topological semantics for *conditional belief*; we investigate the properties of the resulting topological belief revision, showing that it satisfies only the AGM Postulates 1-6 (but not postulates 7 and 8). In the same work, we give a complete axiomatization of the logic of conditional beliefs (with the topological semantics), as well as a complete axiomatization of the corresponding dynamic logic (obtained by adding dynamic update operators, as in Section 4). We plan to present these results (as well as the proofs that are missing from the current paper) in a future journal publication.

References

1. Barwise, J.: Three views of common knowledge. In: Proceedings of the Second Conference on Theoretical Aspects of Reasoning about Knolwedge, pp. 365–379 (1988)
2. Battigalli, P., Siniscalchi, M.: Strong belief and forward induction reasoning. Journal of Economic Theory 105, 356–391 (2002)
3. van Benthem, J., Bezhanishvili, G.: Modal logics of space. In: Handbook of Spatial Logics, pp. 217–298. Springer, Dordrecht (2007)
4. van Benthem, J., Bezhanishvilli, G., ten Cate, B., Sarenac, D.: Modal logics for products of topologies. Studia Logica 84(3), 369–392 (2005)
5. van Benthem, J., Pacuit, E.: Dynamic logics of evidence-based beliefs. Studia Logica 99(1), 61–92 (2011)
6. van Benthem, J., Sarenac, D.: The geometry of knowledge. Aspects of Universal Logic, 17 (2005)
7. Bezhanishvili, N., van der Hoek, W.: Structures for epistemic logic. In: Baltag, A., Smets, S. (eds.) Logical and Informational Dynamics. A volume in honour of Johan van Benthem. Trends in Logic. Springer (to appear)
8. Blackburn, P., de Rijke, M., Venema, Y.: Modal Logic. Cambridge University Press (2001)
9. Chagrov, A., Zakharyaschev, M.: Modal Logic. The Clarendon Press (1997)
10. Dunn, J.M.: Generalized ortho negation. In: Wansing, H. (ed.) Negation: A Notion in Focus (Perspectives in Analytical Philosophy, Bd 7). Perspectives in analytical philosophy, pp. 3–26. de Gruyter (1996)
11. Engelking, R.: General topology, 2nd edn., vol. 6. Heldermann Verlag, Berlin (1989)
12. Esakia, L.: Weak transitivity-restitution. Study in logic 8, 244–254 (2001) (in Russian)
13. van Fraassen, B.: Fine-grained opinion, probability, and the logic of full belief. Journal of Philosophical Logic 24, 349–377 (1995)
14. Gettier, E.: Is justified true belief knowledge? Analysis 23, 121–123 (1963)
15. Kelly, K.: The Logic of Reliable Inquiry. Oxford University Press (1996)
16. Klein, P.: A proposed definition of propositional knowledge. Journal of Philosophy 68, 471–482 (1971)
17. Lehrer, K.: Theory of Knowledge. Westview Press (2000)
18. Lehrer, K., Paxson Jr., T.: Knowledge: Undefeated justified true belief. Journal of Philosophy 66, 225–237 (1969)
19. McKinsey, J.C.C., Tarski, A.: The algebra of topology. Ann. of Math (2) 45, 141–191 (1944)
20. Parikh, R., Moss, L., Steinsvold, C.: Topology and epistemic logic. In: Handbook of Spatial Logics, pp. 299–341. Springer, Dordrecht (2007)
21. Sikorski, R.: Boolean Algebras. Springer, Heidelberg (1964)
22. Stalnaker, R.: On logics of knowledge and belief. Philosophical Studies 128(1), 169–199 (2006)
23. Steinsvold, C.: Topological models of belief logics. PhD thesis, City University of New York, New York, USA (2006)
24. Vickers, S.: Topology via logic. Cambridge Tracts in Theoretical Computer Science, vol. 5. Cambridge University Press, Cambridge (1989)
25. Voorbraak, F.P.J.M.: As Far as I Know. PhD thesis, Utrecht University (1993)
26. Williamson, T.: Knowledge and its Limits. Oxford Univ. Press (2000)
27. Zvesper, J.: Playing with Information. PhD thesis, University of Amsterdam ILLC PhD Thesis (2010)

Plan Recognition, Indefinites, and the Semantics-Pragmatics Boundary

Hsiang-Yun Chen

Institute of European and American Studies, Academia Sinica
hsiangyun.chen@utexas.edu

Abstract. Planning and plan recognition are arguably essential to all rational, co-operative activities, and linguistic communication is no exception. Recently, Lewis (2012) argues that recognizing the importance of plan helps settle a debate regarding the semantics and pragmatics of indefinites. More specifically, Lewis argues against the dynamic approach (e.g. Kamp (1981), Heim (1982), Groenendijk and Stokhof (1991), Kamp and Reyle (1993), and Asher and Lascarides (2003)), according to which indefinites are subject to a *semantic* "Novelty" condition; instead, she offers a neo-Gricean account and analyzes Novelty as a pragmatic, cancelable implicature. I argue that Lewis' analysis is inadequate. Her pragmatic picture not only rests on dubious assumptions concerning plan recognition, but offers no real explanation of the alleged counterexamples against the dynamic theories. Moreover, I provide evidence that supports a more semantic analysis of the Novelty condition.

1 Introduction

According to the traditional *static* approach, the semantic content of a sentence is its truth-conditions, and the semantic content of sub-sentential expressions is their contributions to the truth-conditions of the sentence in which they are embedded. Following Russell (1905), indefinite expressions—expressions of the form "a F"— are semantically equivalent to existential quantification. Such is a widely endorsed picture. In actual practice, however, it is not difficult to see indefinites play a dual function: they not only assert existence, but introduce an element that can figure in subsequent discourse. For example, take

(1) John wrote a paper on quantifier domain restriction, and he has submitted it to a prestigious journal for publication.

(2) Every man who has a daughter adores her.

We have no problem understanding the pronouns as anaphoric, but semantic theories of indefinites which follow Russell have few resources to explain how indefinites can license anaphoric pronouns beyond their syntactic binding scope.

By contrast, *dynamic* semantics takes the meaning of a discourse (i.e., a single or multiple sentences) as its *context change potential* (CCP); the meaning of sub-sentential expressions is consequently defined as their contributions to the overall

D. Grossi, O. Roy, and H. Huang (Eds.): LORI 2013, LNCS 8196, pp. 41–53, 2013.

CCP. This is not to say that truth-conditions are not important; yet in order to fully capture what goes on in linguistic communication, one needs to keep track of something more. In particular, one needs to keep a record of "things being talked about," or the "objects under discussion" in a conversation. To be sure, the discourse interlocutors' task in understanding what is being said in the course of a conversation consists in (a) cataloguing *discourse referents* as well as (b) altering the information associated with them as the discourse unfolds; only the latter is strictly speaking truth-evaluable.

Regarding the aforementioned dual function of definites, the dynamic theorists offer a non-quantificational analysis: the CCP of indefinites is the introduction of a new "discourse referent" in Discourse Representation Structures (DRS) as in Kamp's (1981) and Kamp and Ryle's (1993) Discourse Representation Theory (DRT), or the addition of a "file card" in Heim's (1982, 1983) File Change Semantics. The existential quantification traditionally associated with indefinites is implicit: the quantificational force is construed as part of the verification condition of a DRS, or the satisfaction condition of updating a file with an utterance that contains an indefinite. By taking the "Novelty Condition" as the defining characteristic of indefinites, together with a formalism that allows for a wider binding scope, dynamic theories can successfully account for sentences like (1) and (2).

Lewis (2012) claims that the dynamic approach cannot be right, since Novelty is definitely not semantic. She argues on the basis of some interesting data concerning what she calls the "summary" uses of indefinites that Novelty must be analyzed as a pragmatic, cancelable implicature. Moreover, by appealing to the notion of plan and plan recognition, Lewis offers a neo-Gricean account that purportedly explains cases in which the behavior of indefinites does conform to Novelty, and those that do not, respectively.

This paper argues that Lewis' analysis is deeply problematic and offers empirical evidence that a more semantic construal of Novelty should not be easily dismissed. Section 2 reviews Lewis' argument that Novelty cannot be a semantic feature of indefinites and her own pragmatic picture. Section 3 argues that the criticism of dynamic theories results from a confusion. Moreover, I demonstrate that Lewis' pragmatic proposal not only rests on problematic bases, but falls short of explaining the very examples that she herself brings to salience and a wider range of linguistic phenomena involving indefinites. Section 4 discusses some general lessons from the dialectic.

2   A Pragmatic Account of Indefinites

Lewis holds that indefinites have their traditional Russellian semantic content, i.e. they are simply existential quantifiers. The static semantic account, however, needs to be supplanted with a broadly Gricean story to explain how indefinites are capable of (a) introducing a new "object under discussion" into the conversation and (b) licensing anaphora beyond their standard binding scope. Lewis argues that not only can these two features, which she calls *Novelty* and *Licensing*, respectively, be accounted for pragmatically, but that a pragmatic picture is empirically superior to the competing dynamic semantic view.

The critical evidence for the pragmatic view comes from the following:[1]

(3) a. A student walked into Sue's office and asked her about his exam.
     b. Finally, a student needed her help!

(4) a. I went to see *Star Trek* on Sunday.
     b. That's pretty much all I did all weekend: I saw a movie.

(5) a. We have this nail here.
     b. Unfortunately, now we have a nail and no hammer.

(6) a. I went out to dinner with the woman from the bar last night.
     b. Can you believe it— a woman went out to dinner with me!

Another example of this kind is found in Gundel, Hedberg and Zacharski (1993):

(7) a. Dr. Smith told me that exercise helps.
     b. Since I heard it from a doctor, I'm inclined to believe it.[2]

In all these examples, the indefinite expressions in (b)— "a student," "a movie," "a nail," "a woman," and "a doctor"— do not pick out a new object in the discourse. Rather, their use is justified by an object previously mentioned in (a).[3] So, Lewis argues that (3) through (7) are not *introductory* uses; rather they exemplify the *summary* uses of indefinites. More importantly, if Novelty is a semantic feature of indefinites, it must be conventional, systematic, and cannot be overridden. That summary uses of indefinites are felicitous and robust thus argues strongly against treating Novelty as semantic.[4]

Lewis then contends that a broadly Gricean pragmatic analysis of Novelty is preferable, as it will allow for both the introductory and summary uses. So long as Novelty is treated as an implicature, that it is sometimes cancelable poses no problem. In fact, cancelability is often viewed as an indicator that the phenomenon in question is pragmatic rather than semantic. Now, the real challenge is how to provide a plausible and coherent pragmatic story. According to Lewis, the key lies in recognizing planning as fundamental in conversation and communication.

Humans are essentially planning creatures. We are intelligent actors that inhabit complex, dynamic environments, which we manipulate in complex ways. One of the important ways that we connect to and affect our environments, including other agents, is through language. From this perspective, "a well-run conversation is just like any other co-operative, rational activities."[5]

---

[1] These examples are from Lewis (2012) examples (6), (7), (8), and (9), p.318.

[2] Gundel, Hedberg, Zacharski (1993), p296, example (49).

[3] The "antecedent" of the indefinites may be an indefinite (as in (3) and (7)), a proper name (as in (4)), a demonstrative (as in (5)), or a definite (as in (6)).

[4] Szabo (2000, 2003) also makes the same argument.

[5] Lewis (2012), p.322.

"A successful conversation [also] requires a coherent series of plans: not just what to talk about or how to answer a question under discussion, but also how an object under discussion relates to a question under discussion."[6] In other words, interlocutors "do not make random, disconnected utterances."

Planning and plan recognition are clearly closely related to intending and intention recognition. Lewis maintains that thinking of plan recognition as central is compatible with and extends Gricean pragmatics, since it emphasizes, besides 'what a speaker wants the interlocutors to believe (or understand, or presume)," "how the speaker wants to fit her contribution into the overall conversation." Moreover, the plan recognition framework provides a natural explanation of the fundamental inter-relatedness between Grice's maxim of relation (i.e. be relevant) and the maxim of of manner (i.e. be perspicuous). Lewis acknowledges that a complete plan for a conversation is oftentimes not pre-determined,[7] nevertheless, *local discourse plans* should be recognizable, as they are the driving forces of particular utterances. A local plan is recognizable partly because it connects to the overall discourse plan in a transparent way. Put differently, "recognizable, perspicuous plans go hand in hand with relevant utterances."

How does this explain Novelty and the introductory uses of indefinites? Consider (8):

(8)   a. A woman walked in.

      b. She looked gloomy.

The derivation goes like this. Semantically speaking, a sentence with an indefinite is simply a general, existential claim. By assumption, participants of a conversation are co-operative so that they only make relevant contributions to the conversation. So, the existential claim made in (8a) must be relevant to the conversational context and the overall discourse plan. If the speaker had wanted to talk about a woman already under discussion, she had less misleading ways to do so: pronouns, definite descriptions, or names would all be more appropriate. Hence, using the indefinite "a woman" is indicative of a plan to covey information about a new woman under discussion. Furthermore, the use of an indefinite is frequently a marker of a plan to say something further about its referent, which accounts for the anaphoric pronoun in (8b).

To sum up, Lewis thinks that Novelty and Licensing should be analyzed pragmatically. So long as the file-card metaphor is construed in pragmatic terms, she has no objection against it. Tracking a conversation, or updating a conversational context, is a pragmatic process that involves plan recognition. The conversational context that interlocutors must keep track of is, at the very least, a stack of file-cards, or a collection of the objects under discussion. From the speaker's point of view, the use of an indefinite is a perspicuous way to signal that a new object is being introduced into the conversation; the addressee grasps

---

[6] ibid. p.323.

[7] ibid. "[A] complete plan for a typical conversation is not decided upon beforehand, but the sort of plans we will be concerned with are speakers' short-term plans, which we can call *local* plans."

the speaker's communicative intention and understands the speaker's utterance as relevant to the overall discourse, resulting in the addition of a new card. Planning and plan recognition are not *ad hoc*; they are general reasoning mechanisms that are independently motivated. Taking them seriously as the underlying principles governing discourses makes explicit the coordination necessary for communication.

3   Adjudicating between Semantics and Pragmatics

While I agree that planning and plan recognition are crucial in rational, cooperative activities, and that linguistic communication is arguably no exception, I think both Lewis' objection against the dynamic semantic theories and her own pragmatic account do not stand close examination. First, the criticism of the dynamic theories is a misreading. Once the thesis of the dynamic approach is properly understood, the so-called summary uses constitute no counter examples. Moreover, the positive proposal lacks explanatory power. Not only does it hinge on dubious assumptions, it does not adequately account for the specific examples that Lewis herself brings to spotlight, nor the data involving indefinite expressions in general.

3.1   Is There Anything Wrong with the Dynamic Theories?

To begin, as the dynamic theories conceive it, Novelty is not a matter of reference, or objects in the model, but a constraint on the construction of the semantic representation of the utterance containing an indefinite. In Heim's FCS or Kamp's DRT, the CCP of an indefinite is the introduction of a new file-card to the file or a new discourse referent to the DRS, where a file or a DRS is a theoretical, representational construct mediating between language and the world. Novelty simply leaves open whether distinct cards or discourse referents are mapped to the same or different objects in the model.

Discourse reference and genuine reference are two distinct notions in dynamic theories. Here are some quotes from Heim (1983):

> "[D]iscourse referents behave in ways which it wouldn't make any sense to attribute to real referents: not only are there discourse referents for NPs that have no referents, but moreover, discourse referents may suddenly go out of existence, depending on certain properties of the utterance."
> "[I]t is quite conceivable for there to be a file card that fails to describe a referent, or for two different file cards to happen to describe the same thing, or for file cards to be introduced into and be removed from the file, depending on what is getting uttered."[8]

In fact, with this distinction firmly in place, Heim discusses an example that bears much on the present discussion:

---

[8] The quotes are from Heim (1983), p.166 and p.168, respectively.

(9)   John came, and so did Mary. *One of them* bought a cake.[9]

"On of them" is an indefinite noun phrase. Without doubt, its referent, be it John or Mary, has already been mentioned in the first part of (9). This is not a violation of the Novelty condition, however. The prediction about "one of them" is simply that "its discourse referent must be new and must be distinct from the discourse referents of "John" and "Mary" in particular. There is no prediction about the reference of these three NPs, and we may consistently hold any assumption we please about those. In particular, we may assume that NPs with *discourse reference sometimes happen to coincide in reference*(my italics), and that [(9)], being a case of this kind, involves three discourse referents, but only two referents."[10]

As is clear from this example, a new discourse referent, or file-card, does not entail a new individual.[11] Judging from this light, examples of the summary uses are no challenges to the Novelty condition as the dynamic theories depict it.

Of course, what is interesting about (3) through (7) is that the file-cards must have the same genuine reference. Lewis briefly considers a potential response from the proponent of a dynamic semantic account that explores the *merging* of file-cards: file-cards may be merged when conversation participants realize that what were being treated as distinct objects under discussion are in fact satisfied by the same object in the world. She then criticizes that merging is *ad hoc* and unsatisfactory as it "saves a technical notion of Novelty" by sacrificing the the significance and explanatory power of the file-card metaphor. While I am not convinced that the merging process is ever needed, I am sympathetic to the concern of how contentful the Novelty constraint really is.

Still, treating the summary uses as a decisive evidence against a semantic account strikes me as a hasty conclusion. First, note that in the examples of summary uses (i.e.(3)–(7)), expressions such as "finally," "can you believe it," and "since" play an important role. The minimal pair (10) and (11) provides a vivid illustration:

(10)   a.   A student walked into Sue's office and asked her about his exam.

  b.   A student needed her help!

(11)   a.   A student walked into Sue's office and asked her about his exam.

  b.   Finally, a student needed her help!

---

[9] Ibid. p.165.

[10] Ibid. p.166.

[11] To be fair, Lewis is not completely unaware of this. She notes that "[i]t is important to note that Novelty is not a matter of reference or denotation; no one claims that the object in the world that actually satisfies the indefinite description has to be new to the conversation. Novelty is the claim that, roughly, a speaker is talking about something that is novel for the purposes of the conversation."(Lewis (2012), p.316.) Nevertheless, it seems to me that her characterization of Novelty is unsatisfactory. Novelty has nothing to do with "the purpose of the conversation"; it is just a property of indefinite expressions.

In the absence of "finally," the discourse in (10) allows for various interpretations; but in (11), there is no such flexibility. It is no longer ambiguous whether the two occurrences of "a student" pick out the same individual. The summary uses become natural only when there is a discourse particle that signals the discourse structure by marking the rhetoric relation between sentence (a) and (b). This is exactly what the dynamic theories predict.

Second, the effects these structural markers contribute to do not seem to be cancelable. Take

(12)   a. A student walked into Sue's office and asked her about his exam.

b. Finally, a student needed her help!

c. #But he is not the same student as the first one./ #But they are not the same students.

What happens in (12) is that once the second occurrence of "a student" is interpreted as an instance of the summary use, that bit of information cannot be overridden no matter how the conversation further develops.

The contrast between (10), (11) and (12) is evidence that indefinite expressions (e.g. "a student") and discourse particles (e.g. "finally") must interact in such a way that systematically constrains how the discourse can be interpreted. One the one hand, the two occurrences of "a student" need not pick out the same individual in the model in (10), though they must so in (11). On the other hand, the use of subsequent anaphora is highly regulated: while the speaker in (10) may carry on with the information that she is really talking about two distinct students, she cannot do so once the sentence that contains "finally" appears in the discourse, as (12) demonstrates. If the interplay between indefinites and markers of the rhetoric relations is confined at the pragmatic level, however, it makes no sense why the summary uses cannot be retracted.

In short, Lewis' objection to the dynamic accounts is misguided. She fails to recognize the status of file-cards as theoretical, representational entities, so there is actually no objection to the dynamic approach. Even if examples of the summary uses raise a question of the exact content of the Novelty constraint, they are no knock-down arguments against a semantic treatment of indefinites. As a matter of fact, considerations of the interaction between indefinites and other parts of the discourse, particularly those signaling the conversational structure, favor such a treatment.

3.2   What Is Wrong with the Pragmatic Story?

The pragmatic account of Novelty that Lewis proposes does not stand close scrutiny. Besides lacking crucial details regrading the nature of plans, her theory suffers from obvious counterexamples and does not even explain the data she herself raises to salience.

One fundamental difficulty with the kind of account Lewis proposes concerns the speaker's explicit denial of any discourse plan. Take

(13) a. I do not have any plan in telling you the following.

   b. A student walked into Sue's office and asked her about his exam.

   c. Finally, a student needed her help!

Despite the speaker's straightforward confession that she has no plan for the conversation, the addressee would engage in some plan recognition: the speaker's utterance of the indefinite "a student" in (4b) introduces into the conversational context a new file-card no matter what. This raises the question of the nature of plans that the addressee is supposed to be able to recognize.

At one point, Lewis states that "speakers use and participants recognize maximally strategic plans."[12] Taken as an unrestricted, empirical claim, this is plainly false as conversations are oftentimes random and extemporaneous. Lewis' claim is more realistically viewed as an idealization or the goal of conversations. But what are maximally strategic plans and what makes them recognizable? One would expect that an account that rests on the centrality of plans to address these fundamental questions. Yet Lewis says surprisingly little on either, and what she does say raises more worries.

In the AI literature, planning is typically understood as "the process of formulating a program of action to achieve some specific goal."[13] Given some initial conditions and the specification of a specific goal, the planning agent (or system) produce a series of actions whose execution will achieve that goal. I am not sure if this is the picture Lewis has in mind, for she wants to "remain neutral" on the nature of plan. She does, however, assert that her focus is on the speaker's short-term, or *local*, plans, which may be thought of as elements or sub-plans of an overall plan. Crucially, local plans should be recognizable, as they are the type of plans that "drive particular utterances."[14]

It strikes me that there is a puzzle regarding the connection between local plans and the overall discourse plan. On the one hand, Lewis admits that "a complete plan for a typical conversation is not decided upon beforehand." Yet according to her, a well-run conversation must be one where the local plans are maximally relevant and perspicuous with respect to the discourse plan. But if a complete plan is not established in the first place, it is unclear how local plans—the pragmatic import of sub-sentential, sub-discourse elements—can ever be judged as relevant and perspicuous. On the other hand, the problem that discourses such as (4) bring out is even more telling. In the sheer absence of an overall discourse plan, what maximally strategic local plans can there be? To maintain the idea that conversation participants recognize maximally strategic local plans, Lewis would have to admit that these plans must, in general, be autonomous. But then it makes little sense to talk about local plans coming together and being relevant and perspicuous for the purpose of a conversation.

---

[12] Lewis (2012), p.329.

[13] Pollack (1992), p.3.

[14] The recognition of local plans allows "the participants to track the discourse, i.e. know what to expect will likely be a topic of conversation, an object under discussion, or a question being addressed."(Lewis (2012). p.329.)

Once again, the relation between local plans and the entire discourse becomes a mystery. Furthermore, as the denial of a complete discourse plan can be easily generalized, it is not helpful to counter the challenge by restricting the analysis to task-oriented dialogues. So doing seriously reduces the significance of the theory and leaves the real problem unresolved.

Whether or not (4) is deviant, the general points it illustrates are transparent. A theory of linguistic understanding and communication that builds upon planning and plan recognition is faced with two inter-connected tasks of coordination. First, it must explain what makes possible the coordination between the speaker and the addressee. It must allow for the possible gap between the speaker's possibly nonexistent plan, incomplete plan, multiple plans and what the address is able to recognize. Second, it must explain what makes up a discourse plan. If its composition involves fine-grained levels of sub-plans, it must account for the contributions these sub-elements make to the global plan, and how this bears on the speaker's production and the addressee's understanding. There can be no dodging a precise explication of the nature of plans and what makes them recognizable. Lewis' proposal fails to adequately address these metaphysical and epistemological issues.

My second objection concerns the pragmatic account's inability to successfully account for the relevant linguistic phenomena, including both the summary uses and the introductory uses.

First, consider the following:

(14)　a. A student walked into Sue's office and asked her about his exam.

　　　 b. Finally, a student needed her help!

(15)　a. A student walked into Sue's office and asked her about his exam.

　　　 b. ?Finally, some/at least one student needed her help!

(16)　a. A student walked into Sue's office and asked her about his exam.

　　　 b. #Finally, he/the student/John needed her help!

Replacing the second occurrence of "a student" in (b) of the above discourses with other truth-conditionally equivalent phrases results in, if not infelicity, at least some difference in acceptability. But why is this the case? The question has two faces: (a) substituting "a" with other indefinite expressions like "some" and "at least one"; (b) substituting "a student" with a definite expression— the pronoun "he," the definite description "the student," or a proper name. Even if type (a) substitution is marginally acceptable, type (b) substitution appears much worse.[15] However, it is not clear how the pragmatic analysis of indefinites can coherently explain these phenomena without being self-defeating. Here is a quote from Lewis:

---

[15] If (16) is to make sense at all, it seems to me, the discourse as a whole means something quite different, and the speaker must assume her addressee to have some *familiarity* with the said individual.

In the summary uses the existential, general meaning of the indefinite is emphasized. [(14b)] is appropriate to utter in a context in which Sue had been waiting and hoping for *some student or other* to need her help—she isnt happy or relieved because that particular student came to her office in need of help, but that some student at all needed her help. The speaker has a special reason to use an existential claim, since it *expresses something a definite expression cannot* (italics mine). If we replaced *a* with *the* in the summary uses, they would each convey something different, if they made sense at all. Since there is this special reason to use an indefinite and only an indefinite, we have reason to believe novelty wont be implicated.[16]

This remark is curious. For one thing, Lewis recognizes that what indefinites do in their summary uses is something definites cannot. For another, while the introductory uses are meant to pick out an individual, Lewis says here that the summary uses are *not* supposed to pick out any specific individual, though such uses are justified by such a person. She too emphasizes that it is the purely existential, general meaning that underlies the summary uses.

This admission seems to me to suggest strongly that the summary uses of indefinites, just like the introductory uses, introduce some new file-cards into the conversational context. However, the file-cards so triggered are special: they denote a concept or a category rather than individual instances thereof. This way, the summary uses of an indefinite does something more than the introductory uses in that they contributes to a conversation, besides an individual (who is precisely the individual picked out by the introductory use in the same conversation), a reference to the *kind* of which the individual is an instance. Hence, one is committed to discourse referents of different *types* in the representations: one for the particular instances and one for the general kind. This, of course, is not a challenge to but a confirmation of a *semantic* account of Novelty.

Further problems for the pragmatic approach concerns the introductory uses are abundant. Take the embedding of indefinites in negation:

(17)   a. Bill didn't see a woman. # She was walking her dog.[17]

b. Bill didn't see a woman who was walking her dog.

The occurrence of "a woman" is embedded in negation in both (17a) and (17b). Hence, the pragmatic account ought to predict no addition of a file-card, and *a fortiori* no later anaphoric expression on the presumably non-existent object under discussion. Anaphora is nevertheless permitted in (17b).

Consider also the contrast between (18) and (19):

(18)   a. A woman walked in.

b. She looked gloomy.

---

[16] Lewis (2012), p.322.

[17] The intended reading here is not one where "a woman" receives a wide-scope, *de re* interpretation.

(19)   a. It is not the case that not every woman didn't walk in.

    b. # She looked gloomy.

Since Lewis equates content to truth-conditions, (18a) and (19a) have exactly the same existential entailing content. But the use of anaphora in (19b) apparently is infelicitous. What is truth-conditionally equivalent to an indefinite (e.g. "a F G" and "not every F note G") does not possess the matching Licensing capacity. This discrepancy cannot be explained away by claiming that a file-card is only introduced via linguistic acts that contain an explicit device of existential quantification. That response begs the question; it is not an argument nor an explanation of the phenomena, but merely a restatement of the view that indefinites, but not other truth-conditionally equivalent expressions, provide the optimal, most perspicuous way of signaling a new object under discussion.

In addition, it is not transparent what answers the pragmatic account can supply regarding the ensuing minimal pair:[18]

(20)   a. A wolf might come in.

    b. It would will eat you first.

(21)   a. A wolf might come in.

    b. # It will eat you first.

The kind of analysis that Lewis advocates faces a general difficulty. The introductory uses of indefinites are supposedly the default, but subsequent use of anaphora is not ubiquitous. Various particles in the discourse can give rise to a control effect— negation and modals, for example, often constitute barriers to back-referencing the object previously mentioned. Yet such control effects and the lack thereof cannot be sufficiently justified by pragmatics. By contrast, a semantic account along the lines of dynamic theories offers a straightforward and more plausible explanation. The difference in the availability of subsequent anaphoric reference is analyzed in terms of a well-defined accessibility constraint, with no need to appeal to any equivocal notion of plan recognition.

4   Concluding Remarks

Let me conclude with two general morals from the foregoing discussion. First, whether one takes the dynamic or the static stance, the right analysis must make recourse to a two-stage process. The dynamic theories have the two-level analysis built in in its own nature. The Novelty condition associated with indefinites applies at the level of the construction of the representation, that is, an indefinite invariably adds a new file-card to the representation of the discourse; it is a separate issue if more than one files-cards are mapped to the same object in the

---

[18] Such phenomena is referred to as *modal subordination* in the literature. See, for example, Roberts (1987, 1989), Frank and Kamp (1997), and Asher and Pogodalla (2010).

model. In this sense, cases involving the summary uses of indefinites are no counterexamples to a semantic construal of Novelty. What those examples do show, however, is that there are clear further constraints on the verification or satisfaction conditions of the dynamic discourse representations. Recent developments in the dynamic approach confront precisely this: how various linguistic expressions, for instance, those that signal rhetoric relations and discourse structures, affect the question under discussion and the at-issueness to which conversation participants are remarkably sensitive.[19]

The static approach that Lewis defends and supplements with a neo-Gricean pragmatic account invokes two distinct steps in the analysis as well. She needs to first accounts for the systematicity of Novelty, and then explain how it is cancelable. However, given the many difficulties that I manifest in the previous sections, Lewis fails to make a strong case.

Second, one sees that what Lewis calls local plans get recognized regardless of the speaker's intention or plan for the entire discourse. This strikes me as evidence that local plans are nothing but the semantic content. Given the close tie between plan and intention, a case in point is Bratman's (1984) distinction between "the intention to A" and "intentionally A." According to Bratman, when an agent intentionally A , she intends something, but she may not specifically intend to A. A is an intended action when it is an agent's intention to carry out A; by contrast, when one A intentionally, A may be an unintended consequence, or side effect, of one's intended action.

In the case of linguistic communication, a speaker's use of an indefinite is indicative of some communicative intention, but whatever that is, it need not be identical to the addition of a file-card. In other words, one should distinguish between "introducing a file-card intentionally" and "the intention to introduce a file-card." Since a file-card may be introduced by the use of an indefinite whether or not the speaker has a plan for the very introduction, the mechanism of file-card addition must operate in a way independent of planning and plan recognition. The best explanation is that the so-called local plan associated with a speaker's use of an expression is simply its semantic content. Uttering an indefinite triggers the introduction of a new file-card; what the speaker plans or whether she has a plan is beside the point.

To conclude, there is no knock-down argument for the pragmatic interpretation of Novelty, and a semantic characterization should not be easily dismissed. Once a wider range of linguistic phenomena involving indefinites is taken into consideration, it seems that it is the semantic approach that makes more sense.

References

Asher, N., Lascarides, A.: Logics of Conversation. Cambridge University Press (2003)

Asher, N., Pogodalla, S.: A Montagovian treatment of modal subordination. In: Proceedings of SALT, vol. 20, pp. 1–15 (2010)

---

[19] See, for example, Asher and Lascarides (2003) and Beaver et al. (2010).

Beaver I., D., Roberts, C., Simons, M., Tonhauser, J.: What projects and why. In: Li, N., Lutz, D. (eds.) Semantics and Linguistic Theory (SALT), vol. 20, pp. 309–327. CLC Publicatication, Ithaca (2010)

Frank, A., Kamp, H.: On Context Dependence in Modal Constructions. In: Proceedings of SALT VII. CLC Publications and Cornell University (1997)

Groenendijk, J., Stokhof, M.: Dynamic Predicate Logic. Linguistics and Philosophy 14, 39–100 (1991)

Gundel, J.K., Hedberg, N., Zacharski, R.: Cognitive Status and the Form of Referring Expressions in Discourse. Language 69(2), 274–307 (1993)

Heim, I.: The semantics of definite and indefinite noun phrases PhD thesis University of Massachusetts at Amherst (1982)

Heim, I.: File change semantics and the familiarity theory of deifniteness, pp. 164–189. De Gruyter, Berlin (1983)

Kamp, H., Reyle, U.: From Discourse to Logic. Kluwer, Dordrecht (1993)

Kamp, H.: A theory of truth and semantic representation. In: Janssen, T.M.V., Groenendijk, J.A.G., Stokhof, M.B.J. (eds.) Formal Methods in the Study of Language, Amsterdam. Mathematical Centre Tracts, vol. 135, pp. 277–322 (1981)

Lewis, K.: Discourse dynamics, pragmatics, and indefinites. Philosophical Studies 158, 313–342 (2012)

Pollack, M.E.: The Use of Plans. Artiifcial Intelligence 57(1), 43–68 (1992)

Roberts, C.: Modal Subordination, Anaphora, and Distributivity PhD thesis University of Massachusetts at Amherst (1987)

Roberts, C.: Modal Subordination and pronominal anaphora in discourse. Linguistics and Philosophy 12, 683–721 (1989)

Russell, B.: On Denoting. Mind 14, 479–493 (1905)

Szabo, Z.G.: Descriptions and Uniqueness. Philosophical Studies 101(1), 29–57 (2000)

Szabo, Z.G.: Definite descriptions without uniqueness: A reply to Abbott. Philosophical Studies 114(3), 279–291 (2003)

A Semantic Model for Interrogatives
Based on Generalized Quantifiers and Bilattices

Ka-Fat Chow

The Hong Kong Polytechnic University
kfzhouy@yahoo.com

Abstract. In this paper, I will develop a semantic model for interrogatives, an important sentence type expressing a special aspect of uncertainty. The model is based on the notions of generalized quantifiers and bilattices, and is used to model several aspects of interrogative semantics, including resolvedness conditions, answerhood, exhaustivity and interrogative inferences. It will be shown that the semantic model satisfies a number of adequacy criteria.

1   Introduction

Interrogatives are an important sentence type in natural language expressing a special aspect of uncertainty. Yet the study on interrogatives in logic and formal semantics has been a difficult task because classical logical and formal semantics are basically truth conditional while there is not an intuitive and uncontroversial notion of truth values for interrogatives. While the topic of interrogatives seems to be a linguistic one, some scholars (such as [2], [6]) have studied this topic from the perspectives of logic and formal semantics and have identified a number of aspects for study. The aspects studied in this paper, i.e. resolvedness, exhaustivity, answerhood, interrogative inferences, are the standard ones in the studies on interrogatives. A good summary of these aspects can be found in [6].

According to [11], an adequate semantic model for interrogatives should satisfy the following adequacy criteria: 1. material adequacy – semantic notions of answerhood, entailment and equivalence should be definable under the model; 2. formal adequacy – the semantic notions should be interpretable as set-theoretic relations / operations; 3. empirical adequacy – the semantic notions should correspond to native speaker intuitions. What this criterion in fact meant according to [11] is that certain inferential relations that are intuitively correct should be provable under the model.

In this paper, I will propose a semantic model that satisfies the aforesaid adequacy criteria. This model combines elements from a framework developed by [7, 8] that is based on generalized quantifiers (GQs) and a framework developed by [10, 11] that is based on bilattices. In Section 2, I will review the basic ideas proposed in [7, 8] and [10, 11], and point out some of their merits and demerits, and the need for enhancing and combining the frameworks. In Section 3, I will introduce the enhanced model and discuss the formal semantics of various types of interrogatives. In Section 4, I will

D. Grossi, O. Roy, and H. Huang (Eds.): LORI 2013, LNCS 8196, pp. 54–67, 2013.
© Springer-Verlag Berlin Heidelberg 2013

discuss the issue of interrogative inferences. In Section 5, I will conclude the paper by discussing how the adequacy criteria are satisfied by my model.

2   Two Previous Models on Interrogatives

According to [5], the semantic frameworks for interrogatives may be classified into two broad approaches – the Categorial Approach and the Propositional Approach. Under the Categorial Approach, an interrogative is seen as an incomplete object, i.e. a function, which requires a constituent answer for completion. Since different constituent answers correspond to different semantic types, this approach does not assume a uniform type for interrogatives.

The semantic framework developed by [7, 8] is an example of the Categorial Approach. This framework is based on the Generalized Quantifier Theory (GQT)[1] and views WH-words as a special type of GQs, i.e. interrogative quantifiers (IQs). In a nutshell, a GQ can be seen as a second-order predicate with ordinary sets as arguments. The semantics of a GQ can be delineated by its truth condition expressed as a set-theoretic relation. For example, *every* can be seen as a GQ with 2 sets as arguments satisfying the truth condition $\|every(A, B)\| = t \Leftrightarrow A \subseteq B$ [2]. Different from ordinary GQs, IQs have an additional argument corresponding to the answer to the interrogative. The purpose of this answer argument is to make the interrogaive become a proposition. For example, under this framework, the truth condition of *which* can be represented by $\|which(A, B, X)\| = t \Leftrightarrow A \cap B = X$ [3], where X is the answer argument. This truth condition says that 'X' is the answer to the interrogative "Which 'A' is 'B'?" iff $A \cap B = X$ [4].

Contrary to the Categorial Approach, the Propositional Approach assumes that interrogatives are of one uniform type and the semantic type of interrogatives is to be analysed in terms of propositions.

The semantic framework developed by [10, 11] is an example of the Propositional Approach. This framework adopts the language of First Order Logic augmented by the symbol "?" for forming questions. Semantically, this framework adopts a bilattice model. According to [1], a bilattice is an algebraic structure composed of two

---

[1] This paper adopts the standard notation as used in [12] for denoting quantified statements. Using this notation, a quantified statement such as "Every boy sang" is represented by *every*(BOY, SING), where *every* (in italics) is a quantifier with BOY and SING as arguments. Here the sets BOY and SING are semantic denotations of "boy" and "sang", respectively.

[2] In this paper, I use $\|s\|$ to denote the truth value of a proposition / question s and "t" to denote the truth value "true".

[3] Strictly speaking, according to [7, 8], the truth condition of *which* should involve a context set because the interpretation of *which* is dependent on context. For simplicity, I have ignored the context-dependent effect of *which* in this paper.

[4] In this paper, I use 'S' to denote the natural language word / phrase corresponding to the set S. I also use 's' to denote the natural language declarative / interrogative sentence corresponding to the proposition / question s.

complete lattices ordered separately but sharing a common negation operator "¬", such that "¬" reverses the order in one constituent lattice but preserves the order in the other. Now the framework developed by [10, 11] assumes a uniform type for both declarative and interrogative sentences. The denotation of declaratives and interrogatives are thus both truth values. However, to distinguish declaratives and interrogatives, they distinguish 2 subsets of truth values. For declaratives, there are 3 truth values: t ("known to be true"), f ("known to be false") and uk ("unknown whether true or false"). For interrogatives, they borrow the concept of "resolvedness" from [4] and assume 2 truth values: r ("resolved") and ur ("unresolved"). These 5 truth values thus form a bilattice composed of a declarative lattice and an interrogative lattice. The declarative lattice is ordered by $f \leq uk \leq t$ and the interrogative lattice ordered by $ur \leq r$. Obviously, "¬" reverses the order in the declarative lattice because if p_1 and p_2 are propositions and $\| p_1 \| \leq \| p_2 \|$, then $\| \neg p_2 \| \leq \| \neg p_1 \|$. As for the negation of interrogatives, the discussion will be postponed to Section 4.

Under this framework, the semantics of interrogatives is expressed by the resolvedness conditions which relate the two groups of truth values. For illustration, consider the polar interrogative "Did Mary kiss John?". The formal representation and resolvedness condition of this interrogative is $\| ?(KISS(m, j)) \| = r$ iff $\| KISS(m, j) \| \in \{t, f\}$, which means this polar interrogative is resolved iff it is known whether Mary kissed John.

Compared with the GQT framework developed by [7, 8], the bilattice framework developed by [10, 11] has some merits. Since their framework has a clear definition for truth values of both declaratives and interrogatives, it is straightforward to define entailment and equivalence relations between interrogatives and is thus convenient to study the issue of interrogative inferences under this model. On the other hand, interrogatives are interpreted as functions under the GQT framework. Although we may define entailment as set inclusion (note that functions can be seen as sets), this definition is only applicable to objects of the same category. Since interrogatives may belong to different categories, it is not clear how to come up with an appropriate definition for the general entailment relation between interrogatives.

Nevertheless, the bilattice framework developed by [10, 11] can only deal with a very small set of WH-words because it uses only one operator "?x" for WH-interrogatives. This is in sharp contrast with the GQT framework developed by [7, 8] which has defined a whole range of IQs for different WH-words. Thus, in comparison with the bilattice framework, the GQT framework has greater expressive power. It is also an attractive model because WH-words do share certain characteristics with ordinary GQs. In fact, in the GQT literature, WH-words are sometimes seen as a subtype of quantifiers. Moreover, it is also found that IQs possess certain properties that are thoroughly studied in GQT, such as conservativity, monotonicity, intersectivity, etc.

Given the merits and demerits of the aforesaid two frameworks, I will propose a novel semantic model for interrogatives that combines the merits and avoids the demerits of the two frameworks. Moreover, this semantic model will also deal with certain phenomena that are not dealt with in the two frameworks, such as non-exhaustive interrogatives and certain types of interrogative inferences.

3   A Novel Semantic Model for Interrogatives

3.1   Strongly Exhaustive IQs

The novel semantic model will adopt a bilattice structure that distinguishes 5 truth values as described in Section 2. Apart from this, we need some additional definitions. Under a 2-valued universe, with respect to every concept we have two complementary sets, e.g. X and ¬X. But under a 3-valued universe, we need 3 notions: X_t, X_f and X_{uk}, with the following definitions (in what follows, U denotes the universe or domain of discourse):

$$X_t = \{x \in U: \|x \in X\| = t\} \tag{1}$$

$$X_f = \{x \in U: \|x \in X\| = f\} \tag{2}$$

$$X_{uk} = \{x \in U: \|x \in X\| = uk\} \tag{3}$$

Thus, X_t, X_f and X_{uk} are sets containing elements that are known to belong to X, known not to belong to X and unknown whether to belong to X, respectively.

We now consider WH-interrogatives. Following [7, 8], I will treat these interrogatives as quantified statements containing IQs. But contrary to [7, 8], I do not employ the notion of "answer arguments" and will treat IQs in the same way as ordinary GQs. For instance, since in everyday use, "which" is used with a noun phrase and a verb phrase, such as in "Which boy sang?", *which* will be treated as an IQ with 2 arguments, just like the ordinary GQ *every*. Thus, the WH-interrogative "Which boy sang?" will be expressed as *which*(BOY, SING). In this way, IQs are similar to ordinary GQs as they function as second-order predicates with ordinary sets as arguments. Moreover, just like ordinary GQs, the semantics of IQs will be delineated by their truth conditions (or more precisely, resolvedness conditions) expressed as set-theoretic relations.

How can we derive the resolvedness condition of an IQ like *which*? Before answering this question, I need to introduce the notion of "exhaustivity", which is concerned with what constitutes an acceptable answer to a certain interrogative. From the literature, we can identify two most important types of exhaustivity: strong exhaustivity and non-exhaustivity. While non-exhaustivity only requires the answer to contain some true and no false information requested by the interrogative, strong exhaustivity requires the answer to contain all and only (i.e. exactly) the true information. In other words, strongly exhaustive answers differ from non-exhaustive ones in that the former are unique while the latter are not. For example, if it is known that John and Bill are exactly the boys who sang, then "John and Bill and no other boys" would be a strongly exhaustive answer to the interrogative "Which boy sang?". Since strong exhaustivity is easier to handle and is assumed by the most important theories on interrogatives, including [7, 8] and [10, 11], the simplest IQs are interpreted as strongly exhaustive IQs in this paper.

Under the strongly exhaustive interpretation, we know the answer to the interrogative "Which boy sang?" iff for every element x of U, we know whether x is a boy who sang. In other words, there is no element x such that we do not know whether x is a boy who sang. Thus, the resolvedness condition of "Which boy sang?" can be written as $\|which(BOY, SING)\| = r \Leftrightarrow (BOY \cap SING)_{uk} = \varnothing$. This condition reflects the following intuition: $(BOY \cap SING)_{uk}$ represents the area of uncertainty with respect to the interrogative "Which boy sang?". If this area is empty, then the uncertainty does not exist and the interrogative is thus resolved.

The resolvedness conditions of *which* and some other commonly used strongly exhaustive IQs can be generalized as follows[5]:

$$\|which(A, B)\| = r \Leftrightarrow (A \cap B)_{uk} = \varnothing \tag{4}$$

$$\|(all\ except\ which)(A, B)\| = r \Leftrightarrow (A - B)_{uk} = \varnothing \tag{5}$$

$$\|who(B)\| = r \Leftrightarrow (PERSON \cap B)_{uk} = \varnothing \tag{6}$$

$$\|(everybody\ except\ who)(B)\| = r \Leftrightarrow (PERSON - B)_{uk} = \varnothing \tag{7}$$

Note that the right-hand side of the above all have the form $S_{uk} = \varnothing$ for an appropriate set S.

The semantics of IQs is more complicated than other GQs in that one does not only need to study the resolvedness conditions but also the resolved answers of IQs. For a typical WH-interrogative, the resolved answer may take two forms. The short form appears as a noun phrase. This form is called the constituent answer (CA). The full form appears as a complete sentence. This form is called the sentential answer (SA).

For strongly exhaustive IQs, it is easy to specify the semantic denotations of their CAs, as the form they take is closely related to their resolvedness conditions. For example, provided that $\|which(BOY, SING)\| = r$, we have $(BOY \cap SING)_{uk} = \varnothing$, and the semantic denotation of the CA to "Which boy sang?" is then $(BOY \cap SING)_t$, i.e. all those entities who are known to be boys who sang. We can generalize the above: let q be a strongly exhaustive question whose resolvedness condition has the form $S_{uk} = \varnothing$, then the semantic denotation of the CA to 'q' is S_t.

Moreover, since SA is just the result of writing a CA in the form of a complete sentence, we can express the semantic denotation of an SA by making use of this relation as follows: let q be a strongly exhaustive question whose resolvedness condition has the form $S_{uk} = \varnothing$, then the semantic denotation of the SA to 'q' is the proposition $S = S_t$. Note that this proposition can often be re-expressed in the standard form as appears in the GQT literature by using the truth conditions of GQs. For illustration, suppose in a universe all entities who are known to be boys who sang are John and Bill,

---

[5] Due to limited space, only the resolvedness conditions of a handful of IQs are given in this paper. The resolvedness conditions of other IQs may be derived in a similar fashion, although one needs to define some additional notions or domains for some IQs, such as a "possession" predicate for *whose*, a spatial domain for *where*, etc.

i.e. $(BOY \cap SING)_t = \{j, b\}$, and suppose $(BOY \cap SING)_{uk} = \varnothing$. Then the semantic denotation of the SA to "Which boy sang?" is the proposition $BOY \cap SING = \{j, b\}$, which can be re-expressed as *(no ... except {j, b})*$(BOY, SING)$[6]. This expression corresponds to the natural language sentence "No boy except John and Bill sang".

An advantage of modeling WH-words as quantifiers is that we can derive the resolvedness conditions of WH-interrogatives involving predicates with an arity of 2 or higher by applying certain established operations in GQT. Under GQT, a sentence containing a higher-arity predicate can be viewed as containing a polyadic quantifier. There is an important subtype of polyadic quantifiers, called iterated quantifiers, whose truth conditions can be derived by using an operation called "iteration". For example, consider the sentence "Every boy loves every girl" which contains a binary predicate "loves". The truth condition of this sentence can be expressed as:

$$\| every(BOY, \{x: every(GIRL, \{y: LOVE(x, y)\})\}) \| = t \Leftrightarrow BOY \subseteq \{x: GIRL \subseteq \{y: LOVE(x, y)\}\} \tag{8}$$

The above formula says that "Every boy loves every girl" is true iff every boy x is such that for every girl y, x loves y.

A WH-interrogative containing a higher-arity predicate can be treated in a similar fashion. For illustration, consider the interrogative "Which girl does every boy love?". According to the literature, this interrogative has at least 2 different readings: an individual reading and a pair-list reading. In this paper, I will only consider the individual reading, which can be paraphrased as "Which girl is such that every boy loves her?"[7]. Using iteration, one can easily derive the resolvedness condition of this reading as:

$$\| which(GIRL, \{y: every(BOY, \{x: LOVE(x, y)\})\}) \| = r \Leftrightarrow (GIRL \cap \{y: BOY \subseteq \{x: LOVE(x, y)\}\})_{uk} = \varnothing \tag{9}$$

The above formula says that "Which girl does every boy love?" is resolved iff there is no entity y such that it is not known whether y is a girl and is loved by every boy.

3.2   Non-exhaustive IQs

Apart from "strongly exhaustive" interrogatives requesting complete information concerning a subject matter, there are also "non-exhaustive" interrogatives which request only partial information. [3] listed certain markers in natural languages and pointed out that interrogatives with these markers have inherent exhaustivity. For instance, "for example" is a marker of non-exhaustivity as exemplified by the interrogative "Which boy sang, for example?". In this paper, WH-phrase "which ... for example" will be expressed as a non-exhaustive IQ *(at least which)*. A non-exhaustive WH-interrogative such as "Which boy sang, for example?" is resolved in two mutually exclusive

---

[6] According to the standard GQT literature, the truth condition of the GQ "*(no ... except C)*(A, B)" where C is a non-empty set of individuals manifested as (conjoined) proper names is $\| (no ... except C)(A, B) \| = t \Leftrightarrow A \cap B = C$.

[7] To handle the pair-list reading properly, we need more notions which are definable under the semantic model developed in this paper.

situations: (1) at least one member of U is known to belong to BOY $\cap$ SING; (2) all members of U are known not to belong to BOY $\cap$ SING. Thus, the resolvedness condition can be written as $\|(at\ least\ which)(\text{BOY}, \text{SING})\| = r \Leftrightarrow (\text{BOY} \cap \text{SING})_t \neq \emptyset$ $\vee (\text{BOY} \cap \text{SING})_f = U$. Note that situations (1) and (2) are represented by the two disjuncts on the right-hand side of this resolvedness condition. Generalizing the above discussion, the resolvedness conditions of two non-exhaustive IQs are given below:

$$\|(at\ least\ which)(A, B)\| = r \Leftrightarrow (A \cap B)_t \neq \emptyset \vee (A \cap B)_f = U \tag{10}$$

$$\|(at\ least\ who)(B)\| = r \Leftrightarrow (\text{PERSON} \cap B)_t \neq \emptyset \vee (\text{PERSON} \cap B)_f = U \tag{11}$$

Next I derive the semantic denotation of the CA to the non-exhaustive interrogative "Which boy sang, for example?". Since the CA to a non-exhaustive interrogative is not unique, I will provide the set of the semantic denotations of all possible CAs, called the CA set, as follows:

$$\text{CA set} = \begin{cases} \{X: X \subseteq (\text{BOY} \cap \text{SING})_t \wedge X \neq & \text{if } (\text{BOY} \cap \text{SING})_t \neq \emptyset \\ \emptyset\}, & \\ \{\emptyset\}, & \text{if } (\text{BOY} \cap \text{SING})_f = U \end{cases} \tag{12}$$

The above piecewise-defined function provides the CA set under two mutually exclusive situations. If $(\text{BOY} \cap \text{SING})_f = U$, no boy sang and so the unique CA should be "none of them", represented by a set consisting of $\emptyset$ as the unique member. If $(\text{BOY} \cap \text{SING})_t \neq \emptyset$, then every non-empty subset of $(\text{BOY} \cap \text{SING})_t$, i.e. any set X satisfying $X \subseteq (\text{BOY} \cap \text{SING})_t \wedge X \neq \emptyset$, is the semantic denotation of an acceptable CA. So all these Xs are collected into a set, and the CA can be represented by any member of this set. For illustration, suppose $(\text{BOY} \cap \text{SING})_t = \{j, b\}$, then the CA set is $\{\{j\}, \{b\}, \{j, b\}\}$, i.e. any one of "John", "Bill" and "John and Bill" is an acceptable CA to the non-exhaustive interrogative "Which boy sang, for example?".

Similar to CA, the SA to "Which boy sang, for example?" is also not unique and may be represented by a set of propositions, called the SA set, as shown below:

$$\text{SA set} = \begin{cases} \{X \subseteq \text{BOY} \cap \text{SING}: X \subseteq (\text{BOY} \cap & \text{if } (\text{BOY} \cap \text{SING})_t \neq \emptyset \\ \text{SING})_t \wedge X \neq \emptyset\}, & \\ \{\text{BOY} \cap \text{SING} = \emptyset\}, & \text{if } (\text{BOY} \cap \text{SING})_f = U \end{cases} \tag{13}$$

For illustration, suppose $(\text{BOY} \cap \text{SING})_t = \{j, b\}$, then the SA set is $\{\{j\} \subseteq \text{BOY} \cap \text{SING}, \{b\} \subseteq \text{BOY} \cap \text{SING}, \{j, b\} \subseteq \text{BOY} \cap \text{SING}\}$, i.e. any one of the sentences "John sang", "Bill sang" and "John and Bill sang" is an acceptable SA to the non-exhaustive interrogative "Which boy sang, for example?".

It is not hard to generalize (12) and (13) to a general non-exhaustive question q whose resolvedness condition has the form $S_t \neq \emptyset \vee S_f = U$ for an appropriate set S. All we need to do is replace the set BOY $\cap$ SING in (12) and (13) by S.

3.3   Polar Interrogatives

In this subsection I will discuss polar interrogatives. I propose that a polar interrogative be represented by *whether*(p) where *whether* is a Boolean operator asking for the truth value of p, where 'p' is the declarative associated with the polar interrogative. In this respect, *whether* is similar to the unary Boolean operator "¬". While the latter may be manifested as "It is not the case that", the former may be manifested as "Is it the case that". For example, since the declarative associated with the polar interrogative "Does John love Mary?" is "John loves Mary", the formal representation of this polar interrogative is *whether*(LOVE(j, m)).

Since a polar interrogative is resolved iff its associated declarative is known to be true or false, we can easily write down the resolvedness conditions for polar interrogatives:

$$\| whether(p) \| = r \Leftrightarrow \| p \| \neq uk \tag{14}$$

I next determine the semantic denotations of the CA and SA to a polar interrogative. For a polar question *whether*(p), the semantic denotation of its CA can be easily written down as $\| p \|$. In English, $\| p \|$ can be represented by particular words, such as "yes" (corresponding to $\| p \| = t$) and "no" (corresponding to $\| p \| = f$). As for the semantic denotation of the SA to a polar interrogative, it can be expressed as

$$\text{Semantic denotation of SA} = \begin{cases} p, & \text{if } \| p \| = t \\ \neg p, & \text{if } \| p \| = f \end{cases} \tag{15}$$

4   Interrogative Inferences

To study interrogative inferences, we need to define entailment and equivalence relations involving questions. Under the present framework, it is straightforward to define these notions. First, we define the notion of entailments: let $S = \{s_1, \dots s_n\}$ be a set of questions / propositions (called the premises) and q a question (called the consequence), then S entails q (denoted $S \Rightarrow q$) iff in every model, if $\| s_1 \| \in \{t, r\}$ and $\dots \| s_n \| \in \{t, r\}$, then $\| q \| = r$.

Next we define the notion of equivalence: let q_1 and q_2 be questions, then q_1 is equivalent to q_2 (denoted $q_1 \Leftrightarrow q_2$) iff in every model, $\| q_1 \| = r$ if and only if $\| q_2 \| = r$.

4.1   Interrogative Entailments

Based on the resolvedness conditions of IQs and the above definitions, we can derive valid inferential patterns of IQs. We first consider some basic entailments:

$$which(A, B) \Rightarrow (at\ least\ which)(A, B) \tag{16}$$

$$which(A, B) \Rightarrow whether(some(A, B)) \tag{17}$$

These two entailments are in accord with our intuition. For example, if we know the answer to the strongly exhaustive interrogative "Which boy sang?", we automatically know an answer to the non-exhaustive interrogative "Which boy sang, for example?" as well as the answer to the polar interrogative "Did any boy sing?".

To prove (16) and (17), we first assume that $\|which(A, B)\| = t$. By (4), this is true iff $(A \cap B)_{uk} = \emptyset$, i.e. for all $x \in U$, $\|x \in A \cap B\|$ is either equal to t or f. From this we can deduce that either there is an x such that $\|x \in A \cap B\| = t$, or for all x, $\|x \in A \cap B\| = f$, which is equivalent to the following two propositions:

$$(A \cap B)_t \neq \emptyset \vee (A \cap B)_f = U \tag{18}$$

$$\|A \cap B \neq \emptyset\| = t \vee \|A \cap B \neq \emptyset\| = f \tag{19}$$

From (18) we can then deduce $\|(at\ least\ which)(A, B)\| = r$ by (10) and thus complete the proof of (16). From (19) we can then deduce $\|A \cap B \neq \emptyset\| \neq uk$. By (14), we have $\|whether(some(A, B))\| = r$ and thus complete the proof of (17) since $A \cap B \neq \emptyset$ is the truth condition of $some(A, B)$, according to the GQT literature.

Apart from inferences with only one premise, we may also consider interrogative inferences with more than one premise, such as the following:

$$\{which(C, B), which(C, A), A \subseteq C\} \Rightarrow which(A, B) \tag{20}$$

Note that (20) is a generalization of a result in [5]. An instance of this inference schema is that the two questions "Which child does Mary teach?" and "Which child is a boy?" collectively entail the question "Which boy does Mary teach?" (on the understanding that boys are children).

To prove (20), we first write down the resolvedness conditions of the first two premises:

$$(C \cap B)_{uk} = \emptyset, (C \cap A)_{uk} = \emptyset \tag{21}$$

We then observe that $(C \cap B) \cap (C \cap A) = A \cap B$, given the third premise $A \subseteq C$. Next we need to apply the following result:

$$\text{For any sets A and B, } (A \cap B)_{uk} \subseteq A_{uk} \cup B_{uk}. \tag{22}$$

(22) can be proved as follows: let $x \in (A \cap B)_{uk}$, then $\|x \in A \cap B\| = uk$. This implies that $\|x \in A\| = uk$ or $\|x \in B\| = uk$. But this is equivalent to $x \in A_{uk}$ or $x \in B_{uk}$, i..e. $x \in A_{uk} \cup B_{uk}$.

Combining the above results, we have $(A \cap B)_{uk} \subseteq (C \cap B)_{uk} \cup (C \cap A)_{uk}$. By (21) we have $(A \cap B)_{uk} \subseteq \emptyset \cup \emptyset$, i.e. $(A \cap B)_{uk} = \emptyset$. The consequence of (20) thus obtains.

Monotonicity inferences constitute a special subtype of entailments. Monotonicity is concerned with truth preservation of a quantified statement when the arguments of the statement are replaced by their supersets / subsets. Here are the definitions of increasing and decreasing monotonicities: let $Q(X_1, \ldots X_n)$ be a GQ with n arguments, then Q is increasing in the ith argument $(1 \leq i \leq n)$ iff for all $X_1, \ldots X_i, X_i', \ldots X_n$ such that $X_i \subseteq X_i'$, $Q(X_1, \ldots X_i, \ldots X_n) \Rightarrow Q(X_1, \ldots X_i', \ldots X_n)$. Q is decreasing in the ith

argument iff for all $X_1, \ldots X_i, X_i', \ldots X_n$ such that $X_i \supseteq X_i'$, $Q(X_1, \ldots X_i, \ldots X_n) \Rightarrow Q(X_1, \ldots X_i', \ldots X_n)$. Q is called monotonic in the ith argument iff it is either increasing or decreasing in the ith argument. Otherwise, it is called non-monotonic in the ith argument.

By treating WH-words as quantifiers, we may also talk about the monotonicities of WH-words. But the basic results turn out to be negative. First, according to the definition of entailments, "Which boy sang?" does not entail "Which boy sang Auld Lang Syne?". The point is that even if you know exactly who sang, you may still not know exactly who sang Auld Lang Syne, because the latter interrogative requires more information than the former. In fact, we can show that[8]

Proposition 1   All strongly exhaustive IQs studied in this paper are non-monotonic in all of their arguments.

Here I will only prove *which* is not decreasing. The remaining part of the proof and the proofs for other strongly exhaustive IQs are similar. I construct a counterexample model. Let $U = A_t = \{a, b, c\}$, $A_f = A_{uk} = \varnothing$, $B_t = \{b, c\}$, $B_f = \{a\}$, $B'_t = \{b\}$, $B'_f = \{a\}$, $B'_{uk} = \{c\}$. It is obvious that this model satisfies $B \supseteq B'$ and $(A \cap B)_{uk} = \varnothing$. Thus, according to (4), $\| which(A, B) \| = r$. But since $(A \cap B')_{uk} \neq \varnothing$, we have $\| which(A, B') \| = ur$. The above fact shows that *which* is not decreasing.

I next consider the non-exhaustive IQs. According to (10) and (11), the resolvedness condition of each of these IQs is composed of two disjuncts. Due to this complexity, it turns out that all these IQs are in general non-monotonic in all of their arguments. For example, from the fact that all boys are children, we cannot deduce the following entailment:

$$(at\ least\ which)(\text{BOY}, \text{SING}) \Rightarrow (at\ least\ which)(\text{CHILD}, \text{SING}) \qquad (23)$$

because it may be the case that the children in question consist of boys and girls, and it is known that no boy sang, while it is not known whether there was any girl who sang. In this case, the premise is resolved, but the consequence is not. The invalidity of (23) is mainly due to the fact that the resolvedness condition of the premise of (23) is composed of two disjuncts: "either at least one boy is known to have sung, or it is known that no boy sang". If we now discard the second disjunct, then the resulting premise entails that at least one child is known to have sung. Thus in this case, (23) is valid.

The above discussion shows that the non-exhaustive IQs are in general non-monotonic, but may become increasing in certain specific cases. In fact, we have:

Proposition 2   Within the domain $\{<A, B>: (A \cap B)_f \neq U\}$, (*at least which*) is increasing in both of its arguments, whereas within the domain $\{B: (\text{PERSON} \cap B)_f \neq U\}$, (*at least who*) is increasing in its unique argument.

---

[8]   As a matter of fact, [7] contended that all IQs are decreasing. But his conclusion is based on his special definitions for monotonicities of IQs, which look very different from the usual definitions for monotonicities as used in the GQT literature. I thus do not adopt his definitions and obtain a different conclusion.

Here I will only prove that within the domain $\{<A, B>: (A \cap B)_f \neq U\}$, (*at least which*) is increasing in A. The remaining part of the proof and the proof for (*at least who*) is similar. Suppose $\|(at\ least\ which)(A, B)\| = r$ and $A \subseteq A'$. According to (10), within the given domain, $\|(at\ least\ which)(A, B)\| = r$ iff $(A \cap B)_t \neq \emptyset$. So we must have $(A' \cap B)_t \neq \emptyset$. This implies that $\|(at\ least\ which)(A', B)\| = r$, thus showing that (*at least which*) is increasing in A.

4.2   Interrogative Equivalences

Based on the definition of equivalences and the resolvedness conditions of IQs, we can easily derive (and prove) simple equivalences between IQs such as the following:

$$who(B) \Leftrightarrow which(PERSON, B) \tag{24}$$

$$which(A, \neg B) \Leftrightarrow (all\ except\ which)(A, B) \tag{25}$$

These two equivalences are in accord with our intuition that "Who sang?" has the same meaning as "Which person sang?"[9], whereas "Which boy did not sing?" has the same meaning as "All except which boy sang?"

The equivalence in (25) involves "inner negation", i.e. negation on an argument of the IQ. We now consider the notion of "outer negation", i.e. negation of a question, and see if we can derive any equivalence[10]. Our first problem is whether we can make a proper definition for the negation of a question which should conform to the requirement of the negation operator "$\neg$" in the definition of bilattices set out in Section 2, i.e. "$\neg$" should preserve the order in the interrogative lattice. One way to achieve this is to define "$\neg$" such that for any question q,

$$\|\neg q\| = r \Leftrightarrow \|q\| = r \tag{26}$$

For example, [10, 11] stipulated that "$\neg$" has null effect on questions, which is equivalent to defining $\neg q = q$. But this definition runs counter to our intuition about negation. Therefore, I will try to provide alternative definitions for negated questions which satisfy (26). I will consider WH-questions and polar questions in turn.

First consider a strongly exhaustive WH-question q asking for S, where S is a certain set. Its outer negation $\neg q$ can be defined as another strongly exhaustive WH-question asking for $\neg S$. Now the resolvedness condition of q is $S_{uk} = \emptyset$. When $S_{uk} = \emptyset$, we have $S = S_t$ and $\neg S = S_f$. Thus, S_t and S_f are the semantic denotations of the resolved CAs to 'q' and '$\neg q$', respectively. In contrast, when $S_{uk} \neq \emptyset$, we cannot determine S and $\neg S$ because we do not know whether the elements in S_{uk} belong to S or $\neg S$. Thus, $S_{uk} = \emptyset$ is both a resolvedness condition of q and $\neg q$, and so (26) is satisfied.

---

[9] This is true only if we ignore the context-dependent effect of *which*.

[10] The formal definitions of inner negation and outer negation can be found in [12].

However, under the above definition, the outer negation of a strongly exhaustive WH-question often results in an unnatural question. For example, the outer negation of "Which boy sang" is something like the following:

$$\text{Which individual was not a boy who sang?} \tag{27}$$

Note that the above is a rather strange way to form an interrogative. It is completely different from the natural interrogative "Which boy did not sing?". While the latter asks for $\text{BOY} \cap \neg\text{SING}$, the former asks for $\neg(\text{BOY} \cap \text{SING})$. Since the outer negation of a strongly exhaustive WH-question is unnatural, no sensible equivalence can be derived for this type of questions[11].

Next consider a polar question $whether(p)$. Its outer negation can be defined as the polar question $whether(\neg p)$[12]. Now the resolvedness condition of $whether(p)$ is $\|p\| \neq$ uk, which is equivalent to $\|p\| \in \{t, f\}$. This last statement is true iff $\|\neg p\| \in \{t, f\}$, which is equivalent to $\|\neg p\| \neq$ uk. Thus we have the following equivalence:

$$whether(p) \Leftrightarrow whether(\neg p) \tag{28}$$

and (26) is satisfied.

Under the above definition, the outer negation of a polar interrogative is its negative counterpart. For example, the outer negation of "Does John love Mary?" is "Doesn't John love Mary?", provided that this negative polar interrogative is not read as a rhetorical question.

According to [9], we can derive logical equivalences by combining the inner negation and outer negation of different GQs. For example, [9] proposed the following valid inference schema:

$$Q_1(A, \{x: Q_2(B, \{y: P(x, y)\})\}) \Leftrightarrow (Q_1\neg)(A, \{x: \neg Q_2(B, \{y: P(x, y)\})\}) \tag{29}$$

where $Q_1\neg$ represents the inner negation of Q_1, whereas $\neg Q_2$ represents the outer negation of Q_2. If we now substitute a suitable IQ for Q_1 and an ordinary GQ for Q_2 in (29), we will obtain an equivalence relation involving both IQs and ordinary GQs, such as the following:

$$which(A, \{x: some(B, \{y: P(x, y)\})\}) \Leftrightarrow (all \; except \; which)(A, \{x: no(B, \{y: P(x, y)\})\}) \tag{30}$$

In the above, I have made use of the fact that (all except which) is the inner negation of which whereas no is the outer negation of some. The above schema may be exemplified by a concrete example:

---

[11] In principle, it is also possible to define outer negation for non-exhaustive WH-questions. But the result is even more bizarre. Given limited space, I will not discuss this issue in this paper.

[12] Although "$\neg$" appears inside the argument position of whether, $whether(\neg p)$ should be seen as the outer negation of $whether(p)$, because $whether(\neg p)$ satisfies the definition of outer negation.

Which boy has got some prize? $\Leftrightarrow$ All except which boy has got no prize?   (31)

Note that the above is a sensible equivalence.

4.3   Answerhood

According to [10, 11], there is a special entailment relation between an interrogative and the SA to that interrogative. Let p be a proposition and q a question. Then we have the following[13]:

Proposition 3   If 'p' is an SA to 'q', then $p \Rightarrow q$.

To prove this, I have to consider several cases. When q is a polar question in the form *whether*(s), then p is either s or $\neg$s. Let $\|p\| = t$, which entails $\|s\| = t$ or $\|s\| = f$. In either case, $\|s\| \neq uk$. Then by (14), we have $\|q\| = r$. When q is a strongly exhaustive WH-question whose resolvedness condition has the form $S_{uk} = \varnothing$, p has the form $S = S_t$. Now let $\|p\| = t$, i.e. $\|S = S_t\| = t$. According to the definition of S_t, we have for all x, $\|x \in S_t\|$ is either equal to t or f. This is equivalent to $(S_t)_{uk} = \varnothing$. But since $S = S_t$, we have $S_{uk} = \varnothing$. Thus the resolvedness condition of q is satisfied, and so $\|q\| = r$. When q is a non-exhaustive WH-question whose resolvedness condition has the form $S_t \neq \varnothing \vee S_f = U$, p has the form $X \subseteq S$ (where $X \subseteq S_t \wedge X \neq \varnothing$) or $S = \varnothing$. Now let $\|p\| = t$, i.e. either $\|X \subseteq S\| = t$ or $\|S = \varnothing\| = t$. In the former case, we have $S_t \neq \varnothing$. In the latter case, we have $S_f = U$. In either case, the resolvedness condition of q is satisfied, and so we have $\|q\| = r$.

Proposition 3 shows that $p \Rightarrow q$ is a necessary condition for 'p' is an SA to 'q'. In other words, we can show that 'p' is not an SA to 'q' by showing that $p \#\Rightarrow q$. For instance, we can show that "John sang" is not a resolved SA to "Which boy sang?" by proving that "John sang" (assuming that "John" is a boy) does not entail "Which boy sang?". To prove this, we may construct a counterexample model. Let $U = BOY_t = \{j, b\}$, $BOY_f = BOY_{uk} = \varnothing$, $SING_t = \{j\}$, $SING_f = \varnothing$, $SING_{uk} = \{b\}$. With respect to this model, on the one hand, we find that $\|j \in SING\| = t$, i.e. "John sang" is true. On the other hand, we also find that $\|b \in BOY \cap SING\| = uk$, which shows that $(BOY \cap SING)_{uk} \neq \varnothing$. According to (4), we have $\|which(BOY, SING)\| \neq r$. Thus, "John sang" does not entail "Which boy sang?", and so the former is not a resolved SA to the latter.

5   Conclusion

In Section 1, I have mentioned 3 adequacy criteria which an adequate framework for interrogatives should satisfy. It is now time to see if the semantic model developed in this paper satisfies these criteria.

---

[13] The central idea of the following proposition is from [11]. But the proof is my own and is presented in terms of the definitions and results in this paper.

It is clear that the semantic model is materially adequate, as the notions of answer-hood, entailments and equivalences are all definable under the model. Not only have I provided the resolvedness conditions for various types of interrogatives, but I have also provided explicit expressions for the semantic denotations of the CAs and SAs corresponding to these interrogatives.

The model is also formally adequate, as the semantic notions are interpretable as set-theoretic relations / operations. This point is particularly obvious as the resolved-ness conditions and the semantic denotations of the CAs and SAs corresponding to various types of interrogatives are all expressed as set-theoretic relations or operations.

Finally, the model is also empirically adequate in that certain inferential relations that are intuitively correct are provable under the model. These include the interroga-tive entailments and equivalences recorded in Section 4 as well as Propositions 1 and 2 concerning the monotonicities of IQs.

References

1. Arieli, O., Avron, A.: Reasoning with Logical Bilattices. Journal of Logic, Language and Information 5(1), 25–63 (1996)
2. Belnap Jr., N.D., Steel Jr., T.B.: The Logic of Questions and Answers. Yale University Press, New Haven (1976)
3. Beck, S., Rullmann, H.: A Flexible Approach to Exhaustivity in Questions. Natural Lan-guage Semantics 7, 249–298 (1999)
4. Ginzburg, J.: Resolving Questions I and II. Linguistics and Philosophy 18, 459–527, 567–609 (1995)
5. Groenendijk, J., Stokhof, M.: Type-shifting rules and the semantics of interrogatives. In: Portner, P., Partee, B.H. (eds.) Formal Semantics: the Essential Readings, pp. 421–456. Blackwell, Oxford (1989)
6. Groenendijk, J., Stokhof, M.: Questions. In: van Benthem, J., ter Meulen, A. (eds.) Hand-book of Logic and Language, 2nd edn., pp. 1059–1132. Elsevier Science, Amsterdam (2011)
7. Gutierrez-Rexach, J.: Questions and Generalized Quantifiers. In: Szabolcsi, A. (ed.) Ways of Scope Taking, pp. 409–452. Kluwer Academic Publishers, Dordrecht (1997)
8. Gutierrez-Rexach, J.: Interrogatives and Polyadic Quantification. In: Scott, N. (ed.) Pro-ceedings of the International Conference on Questions, pp. 1–14. University of Liverpool (1999)
9. Keenan, E.L.: Excursions in Natural Logic. In: Casadio, C., et al. (eds.) Language and Grammar: Studies in Mathematical Linguistics and Natural Language, pp. 31–52. CSLI, Stanford (2003)
10. Nelken, R., Francez, N.: The Algebraic Semantics of Interrogative NPs. Grammars 3, 259–273 (2000)
11. Nelken, R., Francez, N.: Bilattices and the Semantics of Natural Language Questions. Lin-guistics and Philosophy 25, 37–64 (2002)
12. Peters, S., Westerståhl, D.: Quantifiers in Language and Logic. Clarendon Press, Oxford (2006)

A Two-Tiered Formalization of Social Influence

Zoé Christoff[1] and Jens Ulrik Hansen[2]

[1] Institute for Logic, Language and Computation, University of Amsterdam
[2] Department of Philosophy, Lund University

Abstract. We propose a new dynamic hybrid logic to reason about social networks and their dynamics building on the work of "Logic in the Community" by Seligman, Liu and Girard. Our framework distinguishes between the purely private sphere of agents, namely their mental states, and the public sphere of their observable behavior, i.e., what they seem to believe. We then show how such a distinction allows our framework to model many social phenomena, by presenting the case of pluralistic ignorance as an example and discussing some of its dynamic properties.

In recent years, information dynamics and belief formation in groups of interacting agents have been widely studied within the field of *logic* [1]. The topic has also been extensively studied for agents situated in a network structure within the field of *social network analysis* [2, 3]. However, there has been very little information flow between these two fields. A recent exception is the work on influence in a community and peer pressure effects by Patrick Girard, Fenrong Liu, and Jeremy Seligman [4–8]. This paper attempts to continue building the bridge between the two fields.

A possible explanation for the lack of interaction between logic and social network analysis is their very distinct paradigmatic cases of inspiration. In social network research, the inspiration mainly comes from diffusion phenomena such as the spreading of diseases. In the logic tradition, rational agents are taken to be equipped with unlimited higher-order reasoning powers aiming for the truth. The work of Girard, Liu, and Seligman goes in both directions. On the one hand, their initial work [4–6] presents an extremely simple model of how knowledge, belief and preferences change under influence within social networks. On the other hand, their latest work [7, 8] aims at fully describing information dynamics in networks of agents with unlimited higher-order reasoning power.

Our goal is to design a framework to model real-life social phenomena and their corresponding information dynamics. As we will show, the setting of [6] cannot model situations involving a discrepancy between what the agents actually believe and what they seem to believe. Yet, we claim that the very possibility of such a discrepancy is an important feature of many social phenomena. However, we do not want to turn to much more complex frameworks such as [7] either. Therefore, we will build on the setting of [6] to design a framework which remains relatively simple but is capable of capturing more complex social phenomena.

In the next section, we briefly recall the "one-layer" framework of Seligman, Girard and Liu and we explain why it cannot model some particular social

D. Grossi, O. Roy, and H. Huang (Eds.): LORI 2013, LNCS 8196, pp. 68–81, 2013.

phenomena. In Section 2, we give an example of such a phenomenon, known from social psychology as *pluralistic ignorance* – a situation where all individuals of a group believe that their private attitudes differ from the ones of the rest of the group despite the fact that everyone in the group acts identically. This example illustrates the need for a more fine-grained definition of social influence which we then offer, by distinguishing what agents *privately believe* from how they *publicly behave*. In Section 3, we introduce a new general hybrid logic to reason about network dynamics taking into account these two different layers. Finally, in Section 4, we model the case of pluralistic ignorance and characterize some of its dynamic properties within this new framework.

1 The Network Logic of Girard, Liu, and Seligman

In [6], a hybrid logic in the original Facebook logic style of [4] is designed to model belief change induced by social influence in a community. The social network structure is represented by a set of agents and an irreflexive and symmetric relation (as in a real Facebook friendship) between them. A modal operator F quantifies over friends (or accessible agents): F reads "all of my friends" and its dual, $\langle F \rangle$, "some of my friends". Some hybrid logic machinery is also used: *nominals*, to refer to the agents, and operators $@_i$, to switch the evaluation point to the unique agent named by i. Each agent is always in one of the three following doxastic states, relatively to a given proposition φ: either she believes that φ ($B\varphi$), or she believes that $\neg\varphi$ ($B\neg\varphi$), or she is undecided about φ: ($U\varphi$ – an abbreviation of $\neg B\varphi \wedge \neg B\neg\varphi$). Sentences are interpreted indexically at an agent: if p means "I am blonde", BFp reads "I believe that all my friends are blonde" and FBp reads "each of my friends believes that s/he is blonde".

This static framework is combined with an influence operator to represent how belief repartition changes in a community, according to the following *peer pressure principle*: every agent tends to align her belief with the ones of her friends. The notions of Strong Influence and Weak Influence are defined, corresponding respectively to the belief changing operators of revision and contraction in the tradition of [9]. An agent is *strongly influenced* (*SI*) to believe φ when *all* of her friends (and at least one) believe that φ:

$$SI\varphi := FB\varphi \wedge \langle F \rangle B\varphi$$

An agent under strong influence with φ will come to believe φ too (assuming that revision is successful) whatever her initial attitude towards φ. An agent is already *weakly influenced* (*WI*) with φ when *some* of her friends believe that φ and none of her friends believe that $\neg\varphi$:

$$WI\varphi := F\neg B\neg\varphi \wedge \langle F \rangle B\varphi$$

Under weak influence, if the agent was undecided or if she already believed that φ, nothing changes; but if she believed that $\neg\varphi$, she will drop her belief and become undecided.

This simple framework makes it unproblematic to identify the stability and stabilization conditions of social-doxastic configurations, both of which can be characterized directly in the language of friendship and belief. However, this simplicity is pricey: even though this is not explicitly mentioned as such, it relies on an extremely strong assumption: *agents' belief states are influenced directly by their friends' belief states*. Thus, either all agents have direct access to their friends' beliefs (as mind-readers would), or their observed behavior always reflects their private beliefs, i.e., there is no difference between what they *seem to believe* and what they *actually believe*. This *transparency* assumption (all agents always automatically know what their friends believe) trivially rules out the modeling of situations where agents act in a way which does *not* reflect their mental states.[1]

Similar issues arise for preferences. Indeed, even if we agree that you are influenced in the very similar way described in [5], if you end up wearing a hat rather than none, it is probably not directly because all of your friends privately prefer to wear one too, but because they *act as if* they did. They could all be pretending because they all observe that everybody else is wearing a hat, and everyone could be following a trend that nobody actually likes. This is a crucial component of social science if we think for instance about real life cases where agents are enforcing a norm which they individually do not agree with. It is precisely because they do not have access to each other's preferences and beliefs that a collective behavior can result which goes against the opinions of most or even all agents, considered individually.[2] In the next section we will consider a class of similar situations called *pluralistic ignorance* and develop a two-layer notion of social influence to represent the distinction between what an agent privately believes and what beliefs she publicly expresses.

2   Pluralistic Ignorance and a Two-Layer Definition of Influence

The term "pluralistic ignorance" originates in the social and behavioral sciences in the work Allport and Katz [11]. It can be roughly defined as a situation where each individual of a group believes that her private attitude towards a proposition or norm differs from the rest of the group members', even though everyone in the group acts identically. For instance, after a difficult lecture which none of the students understood, it can happen that none of them asks any question even though the teacher explicitly requested them to do so in case they did not understand the material. There are numerous examples of pluralistic ignorance in the social and psychological literature such as, in addition to this classroom

---

[1] Such an additional layer is also necessary for cases involving higher-order beliefs, since the complexity of such cases usually arises precisely from the fact that there might be a difference between what agent a believes that agent b believes and what agent b actually believes. However, we will not pursue the issue of higher-order beliefs any further in this paper.

[2] See for instance [10] on this issue.

example, drinking among college students, attitudes towards racial segregation, and many more.[3]

Even though different definitions have been given in the literature [16–18, 11, 19], we will follow [19] and define pluralistic ignorance as *a collective discrepancy between the agents' private attitudes and their public behavior*, namely a situation where all the individuals of a group have the same private attitude towards a proposition φ (say a belief in φ), but publicly "display" a conflicting attitude towards φ (say a belief in $\neg\varphi$).

From a dynamic perspective, pluralistic ignorance is often reported as being both a *robust* and *fragile* phenomenon. It is robust in the sense that, if nothing changes in the environment, the phenomenon might persist over a long period of time – the college students might keep obeying an unwanted drinking norm for generations. On the other hand, it is fragile in the sense that if just one agent announces her private belief, it may be enough to dissolve the phenomenon – if just one student of the classroom example starts to ask questions about the difficult lecture the rest of the students might soon follow. The two-layer definition of social influence which we develop below will allow us to explain how pluralistic ignorance may dissolve in a community by cascading effects and thus allow us to illustrate both its robustness and its fragility. Moreover, in Section 4 we will show formal results about these dynamic properties of pluralistic ignorance.

To reflect the fact that agents do *not* have access to what the others privately believe, we introduce a distinction between *private belief*, which we name "inner belief" (I_B) and *public (or observable) behavior*, which we name "expressed belief" (E_B). We define *two* undecidedness or "unbelief" notions accordingly:

$$U_{IB}\varphi := \neg I_B\varphi \wedge \neg I_B\neg\varphi \qquad\qquad (inner \text{ unbelief})$$

$$U_{EB}\varphi := \neg E_B\varphi \wedge \neg E_B\neg\varphi \qquad\qquad (expressed \text{ unbelief})$$

To define our new influence operator, we make the following simplifying assumption: from the subjective perspective of each agent, what matters (what influences her) is what she herself privately believes and what the others seem to believe. This reflects the fact that influence occurs (at least in good part) at the behavioral (observable, visible, displayed) level. We now redefine strong and weak influence accordingly: *2-layer strong influence* (SI^2) with respect to φ is the situation where all (and some) of my friends express the belief that φ.

$$SI^2 := FE_B\varphi \wedge \langle F\rangle E_B\varphi$$

In this case, whatever my own initial (inner and expressed) state, I end up expressing the belief that φ ($E_B\varphi$). Similarly, *2-layer weak influence* (WI^2) with

---

[3] An extensive study of the classroom phenomenon was done by Miller and McFarland [12]. In a study of college students, Prentice and Miller [13] found that most students believed that the average student was much more comfortable with alcohol norms than they themselves were. Fields and Schuman [14] conducted a similar study, which showed that on issues of racial and civil liberties most people perceived others to be more conservative than they actually were. O'Gorman and Garry [15] found a similar tendency among whites to overestimate other whites' support for racial segregation.

respect to φ is the situation where some of my friends express the belief that φ and none of them expresses the belief that $\neg\varphi$.

$$WI^2 := \langle F\rangle E_B\varphi \wedge F\neg E_B\neg\varphi$$

As a result, I will express the belief that φ ($E_B\varphi$) if I was initially privately undecided about φ (U_{IB}) or if I already privately believed that φ ($I_B\varphi$), and I will act as if I was indifferent ($U_{EB}\varphi$) if I initially privately believed $\neg\varphi$ ($I_B\varphi$).

According to these definitions, my reaction depends on asymmetrical information: what *I privately believe* and what *the others seem to believe*. This reflects the fundamental asymmetry between the first and third person perspectives which is needed to model pluralistic ignorance. It is symmetrical in that everybody reacts in the same way and in that everybody interprets the behavior of others in the same way; but it is asymmetric in that people don't have access to others' mental states and have a "privileged" access to their own.

	Inner state	$\langle F\rangle E_B\varphi$	$\langle F\rangle E_B\neg\varphi$	$\langle F\rangle E_U\varphi$	Type 1	Type 2	Type 3
1	$I_B\varphi$				$\leadsto E_B\varphi$	$\leadsto E_U\varphi$	$\leadsto E_B\varphi$
2	$I_B\neg\varphi$	1	1	1	$\leadsto E_B\neg\varphi$	$\leadsto E_U\varphi$	$\leadsto E_B\neg\varphi$
3	$I_U\varphi$				$\leadsto E_U\varphi$	$\leadsto E_U\varphi$	$\leadsto E_U\varphi$
4	$I_B\varphi$				$\leadsto E_B\varphi$	$\leadsto E_U\varphi$	$\leadsto E_B\varphi$
5	$I_B\neg\varphi$	1	1	0	$\leadsto E_B\neg\varphi$	$\leadsto E_U\varphi$	$\leadsto E_B\neg\varphi$
6	$I_U\varphi$				$\leadsto E_U\varphi$	$\leadsto E_U\varphi$	$\leadsto E_U\varphi$
7	$I_B\varphi$				$\leadsto E_B\varphi$	$\leadsto E_B\varphi$	$\leadsto E_B\varphi$
8	$I_B\neg\varphi$	1	0	1	$\leadsto E_U\varphi$	$\leadsto E_U\varphi$	$\leadsto E_U\varphi$
9	$I_U\varphi$				$\leadsto E_U\varphi$	$\leadsto E_U\varphi$	$\leadsto E_U\varphi$
10	$I_B\varphi$						
11	$I_B\neg\varphi$	1	0	0	$\leadsto E_B\varphi$	$\leadsto E_B\varphi$	$\leadsto E_B\varphi$
12	$I_U\varphi$						
13	$I_B\varphi$				$\leadsto E_U\varphi$	$\leadsto E_U\varphi$	$\leadsto E_U\varphi$
14	$I_B\neg\varphi$	0	1	1	$\leadsto E_B\neg\varphi$	$\leadsto E_B\neg\varphi$	$\leadsto E_B\neg\varphi$
15	$I_U\varphi$				$\leadsto E_B\neg\varphi$	$\leadsto E_B\neg\varphi$	$\leadsto E_B\neg\varphi$
16	$I_B\varphi$						
17	$I_B\neg\varphi$	0	1	0	$\leadsto E_B\neg\varphi$	$\leadsto E_B\neg\varphi$	$\leadsto E_B\neg\varphi$
18	$I_U\varphi$						
19	$I_B\varphi$				$\leadsto E_B\varphi$	$\leadsto E_B\varphi$	$\leadsto E_U\varphi$
20	$I_B\neg\varphi$	0	0	1	$\leadsto E_B\neg\varphi$	$\leadsto E_B\neg\varphi$	$\leadsto E_U\varphi$
21	$I_U\varphi$				$\leadsto E_U\varphi$	$\leadsto E_U\varphi$	$\leadsto E_U\varphi$
22	$I_B\varphi$				$\leadsto E_B\varphi$	$\leadsto E_B\varphi$	$\leadsto E_B\varphi$
23	$I_B\neg\varphi$	0	0	0	$\leadsto E_B\neg\varphi$	$\leadsto E_B\neg\varphi$	$\leadsto E_B\neg\varphi$
24	$I_U\varphi$				$\leadsto E_U\varphi$	$\leadsto E_U\varphi$	$\leadsto E_U\varphi$

Fig. 1. Influence on three different types of agents

Figure 1 lists the 24 possible situations of an individual among her friends, from her perspective, and describes her (observable) reaction. Her private attitude appears in the first column, the possible repartition of her friends' behaviors (expressed belief states) in columns 2,3,4 (in a truth table format – 1 for "true" and 0 for "false"), and her resulting behavior in one of the last three columns (depending of which type of agents we are considering). It is easy to see that our strong influence (rows 10 to 12 and 16 to 18 of the table) is still similar to the one from Seligman et al. but defined on the level of "expressed belief" instead of what was simply called "belief". However, weak influence (when not strong, rows 7 to 9 and 15 to 18) now results in a different state depending on the initial *private* belief state of the agent herself (see for instance rows 7 and 8).

There are two possible cases in which I have friends (unlike in rows 22 to 24) but I am neither strongly nor weakly influenced: whenever all of my friends express undecidedness (rows 19 to 21) and whenever some of them express the belief that φ while some express the belief that $\neg\varphi$ (rows 1 to 6). In the setting of [6], nothing happens, i.e, the agent continues to believe whatever she did before. In our setting, we have to make a choice as to what the agent expresses. The simplest one is to assume that in both these cases, agents express their true private belief (act sincerely). This corresponds to agent of type 1 in the table. However, some agents might be more inclined to follow the others, and in different ways. Types 2 and 3 in the table are examples of other possible types of agents which still comply with our definition of two-layer strong and weak influence. If I am a type 2 agent, I will be sincere (i.e., my expressed belief state will correspond to my inner belief state) whenever I face no opposition. For instance, if I privately believe that φ, I will express this belief if none of my friends expresses a belief in $\neg\varphi$. And if I am a type 3 agent, I will be sincere whenever some of my friends express support for my private belief state, I will for instance express my inner belief in φ if some of my friends express a belief in φ too. Type 1 agents are thus simply the ones that are sincere in both cases: when they get some support and when they face no opposition.

We will see in section 4 how the dynamic properties of social phenomena like pluralistic ignorance depend on the type of agents involved but let us first introduce the formal framework we will use to represent changes of the (multi-layered) state of agents in a social network.

3   A Hybrid Network Logic

In this section, we introduce a hybrid logic to reason about networks and their dynamics, which will allow us to model cases like pluralistic ignorance. We start with a static logic and then move on to give the full dynamics.

In section 2 we introduced two characteristics of each agent, namely her inner (private) belief state and her expressed belief state. Each of the two could be of three kinds. For instance, the inner belief state could be inner belief, inner non-belief, or inner undecidedness. We will generalize this idea by assuming that each agent has n different characteristics, each of which is taken from a finite set of possible values. More formally, we assume a finite set of variables/characteristics $\{V_1, V_2, ..., V_n\}$, where each variable V_l takes a value from a finite set R_l, for each $l \in \{1, ..., n\}$.

The atomic propositions of our language will then be of the form

$$V_l = r,$$

for an $l \in \{1, ..., n\}$ and an $r \in R_l$. We will refer to these as characteristic propositions and we will refer to the set of all characteristic propositions as PROP. If the proposition $V_l = r$ is true of an agent, we will read it as the agent possessing the particular characteristic r of type V_l.[4]

In addition to characteristic propositions, we will assume a countable infinite set of nominals (NOM) used as names for agents in possible networks, just as nominals are used to refer to possible states in traditional hybrid logic [20]. The syntax of our static language is then be given by:

$$\varphi \; ::= \; p \mid i \mid \neg\varphi \mid \varphi \wedge \varphi \mid F\varphi \mid G\varphi \mid @_i\varphi \, ,$$

where $p \in$ PROP and $i \in$ NOM. We will use the standard abbreviations for $\vee$, $\rightarrow$, and $\leftrightarrow$ and denote the dual operator of F by $\langle F \rangle$ and the dual of G by $\langle G \rangle$. The intuitive meaning of the F and $@_i$ operators were already discussed in Section 1. The G-operator is the global modality quantifying over all agents in the network and $G\varphi$ is read as "all agents (satisfy) φ".

We now move on to define the semantics of our language. A (network) model is a tuple $\mathcal{M} = (A, \sim, g, \nu)$, where A is a non-empty set of agents, $\sim$ is a binary relation on A representing the network structure[5], $g :$ NOM $\rightarrow A$ is a function assigning an agent to each nominal, and $\nu : A \rightarrow \mathcal{V}$ is a valuation assigning characteristics to all agents in the network. Here $\mathcal{V}$ denotes the set of all *assignments* $s : \{1, ..., n\} \rightarrow R_1 \times ... \times R_n$. Hence, an assignment assigns a value in R_l to each variable V_l and given an agent $a \in A$, $\nu(a)$ is an assignment assigning characteristics to a for all variables $V_1, ..., V_n$.

Given a $\mathcal{M} = (A, \sim, g, \nu)$, an $a \in A$ and a formula φ, we define the truth of φ at a in $\mathcal{M}$ inductively by:

$$
\begin{array}{lll}
\mathcal{M}, a \models V_l = r & \text{iff} & \nu(a)(l) = r \\
\mathcal{M}, a \models i & \text{iff} & g(i) = a \\
\mathcal{M}, a \models \neg\varphi & \text{iff} & \text{it is not the case that } \mathcal{M}, a \models \varphi \\
\mathcal{M}, a \models \varphi \wedge \psi & \text{iff} & \mathcal{M}, a \models \varphi \text{ and } \mathcal{M}, a \models \psi \\
\mathcal{M}, a \models G\varphi & \text{iff} & \text{for all } b \in A; \mathcal{M}, b \models \varphi \\
\mathcal{M}, a \models F\varphi & \text{iff} & \text{for all } b \in A; a \sim b \text{ implies } \mathcal{M}, b \models \varphi \\
\mathcal{M}, a \models @_i\varphi & \text{iff} & \mathcal{M}, g(i) \models \varphi
\end{array}
$$

Satisfiability, validity etc. are as usual. To obtain the full dynamic language, we add dynamic modalities, which, as in standard Dynamic Epistemic Logic

---

[4] Characteristic propositions are obviously a generalization of classical propositional variables.

[5] If we are talking about undirected networks, we will assume that $\sim$ is symmetric.

[21, 22], come from event models. On the syntactic level, given an event model $\mathcal{E}$ and a formula φ, we will add the construct $[\mathcal{E}]\varphi$ to our language. Event models are defined by simultaneous induction with the syntax of the language: An *event model* is a pair $\mathcal{E} = (\Phi, \mathsf{post})$ consisting of a finite set Φ of pairwise inconsistent formulas of our language and a post-condition function $\mathsf{post} : \Phi \to V$. The set Φ will be referred to as "preconditions", and given a precondition $\varphi \in \Phi$, we will call $\mathsf{post}(\varphi) \in V$ the post-condition of φ. The intuition behind this is that if an agent satisfy a $\varphi \in \Phi$ (in which case, φ is necessarily unique), then after the event $\mathcal{E}$, a will have the characteristics specified by $\mathsf{post}(\varphi)$.

As in standard Dynamics Epistemic Logic, the semantics of formulas involving event models requires a definition of product update of models with event models. Given a model $\mathcal{M} = (A, \sim, g, \nu)$ and an event model $\mathcal{E} = (\Phi, \mathsf{post})$, the product update is $\mathcal{M} \otimes \mathcal{E} = (A, \sim, g, \nu')$, where ν' is defined by:

$$\nu'(a) = \begin{cases} \mathsf{post}(\varphi) & \text{if there is a } \varphi \in \Phi \text{ such that } \mathcal{M}, a \models \varphi \\ \nu(a) & \text{otherwise} \end{cases} \tag{1}$$

Then, the semantics of a formula of the form $[\mathcal{E}]\varphi$ is given by:

$$\mathcal{M}, a \models [\mathcal{E}]\varphi \quad \text{iff} \quad \mathcal{M} \otimes \mathcal{E}, a \models \varphi$$

This way, we obtain the semantics of our full dynamic logic and satisfiability, validity etc. are extended to this in the obvious way.[6]

Given a model $\mathcal{M} = (A, \sim, g, \nu)$ and an event model $\mathcal{E} = (\Phi, \mathsf{post})$, let

$$\mathcal{M} \otimes^k \mathcal{E} := \underbrace{(...((\mathcal{M} \otimes \mathcal{E}) \otimes \mathcal{E}) \otimes ...) \otimes \mathcal{E}}_{k \text{ times}} \quad \text{for every } k \in \mathbb{N}_0.$$

An interesting question is whether the network stabilizes, that is if successive updates by $\mathcal{E}$ will result in a network model that does not change under update by $\mathcal{E}$, i.e. a fixed-point of $\mathcal{E}$. Let us formally define this.

Definition 1. *A network model* $\mathcal{M} = (A, \sim, g, \nu)$ *is said to be* stable *under the dynamics of an event model* $\mathcal{E} = (\Phi, \mathsf{post})$ *if* $\mathcal{M} = \mathcal{M} \otimes \mathcal{E}$. $\mathcal{M}$ *is said to* stabilize *under the dynamics of* $\mathcal{E}$ *if there is a* $k \in \mathbb{N}$ *such that* $\mathcal{M} \otimes^k \mathcal{E}$ *is stable.*

We can express in our language that a network is stable. Given a model $\mathcal{M} = (A, \sim, g, \nu)$, the assignment $\nu(a)$ completely describes the characteristics of a, thus the complete characteristics of a is expressed by:

$$\varphi_{\nu(a)} := \bigwedge_{l=1}^{n} V_l = \nu(a)(l).$$

---

[6] A sound and complete Hilbert-style proof system for the logic can be obtained from the authors.

Moreover, note that the set of all possible assignments $\mathcal{V}$ is finite. Thus, we can quantify over it in our language and express that a network model is stable by[7]:

$$\varphi_{stable} := \bigwedge_{s \in \mathcal{V}} \left(\varphi_s \to [\mathcal{E}]\varphi_s \right). \tag{2}$$

Then, it is easy to see that:

Lemma 1. *A network model* $\mathcal{M}$ *is stable if, and only if,*

$$\mathcal{M} \models \varphi_{stable}.$$

4   Pluralistic Ignorance Revisited

We will use the logic of the previous section to model pluralistic ignorance. We assume that everyone in a "group" is connected to everyone else through some finite number of steps (a "community" in the sense of [4]). In other words, we will work with connected network models, i.e, networks containing a unique community. Moreover, we will assume that the $\sim$ relation is symmetric in the rest of the section.

To begin with, we consider two variables V_I and V_E as corresponding respectively to inner belief and expressed belief. Moreover, we assume that $R_I = R_E = \{B\varphi, B\neg\varphi, UP\varphi\}$, such that $V_I = B\varphi$ corresponds to an inner belief in φ, for instance. Note that "$B\varphi$" is a value assigned to a variable, and as such, the "φ" here is NOT a formula of our formal language – φ will only occur as part of a value for a variable. However, we write $I_{B\varphi}$ as a short hand notation for $V_I = B\varphi$ etc..

Pluralistic ignorance, in the sense that everybody inner believes φ but expresses a belief in $\neg\varphi$, can be formalized by:

$$PI\varphi := G(I_{B\varphi} \wedge E_{B\neg\varphi}) \tag{3}$$

If $PI\varphi$ is true in a network model $\mathcal{M}$ we will say that $\mathcal{M}$ *is in a state of pluralistic ignorance*.

To investigate how social influence affects pluralistic ignorance we need to define an event model that captures the two-layer influence described in Section 2. This is fairly straightforward given the table of Figure 1. For now we assume that all agents are of type 1 mentioned in Section 2. We will return to considering other types of agents later on. For each of the 24 rows, the conjunction of the

---

[7] Another way of expressing that a network model is stable would be to follow the line of [6]. If $V_L = r$ is true of some agent and the network is stable, this means that none of the preconditions $\varphi \in \Phi$ of $\mathcal{E}$ for which $\mathsf{post}(\varphi)$ would change the value of V_l can be satisfied at the agent. Then, for every full characteristic we can write the conjunction of the negation of all preconditions that would change this characteristic. Finally, we can take the disjunction over all possible full characteristics and thereby obtain a formula for a network being stable.

first four columns will be a precondition. These 24 preconditions will clearly be pairwise inconsistent. For instance, the fourth row gives the precondition formula

$$I_B\varphi \wedge \langle F\rangle E_B\varphi \wedge \langle F\rangle E_B\neg\varphi \wedge \neg\langle F\rangle E_U\varphi.$$

The corresponding post-condition will be the assignment assigning $B\varphi$ to V_I and $B\varphi$ to V_E as specified by the first and the fifth column of the table. The resulting event model will be denoted $\mathcal{I}$.

As claimed in Section 2, pluralistic ignorance constitutes a "robust" state, or "equilibrium", in the sense that if a network is in a state of pluralistic ignorance it will stay in this state. We now formalize this in the following lemma:

Proposition 1. *A connected network model in a state of pluralistic ignorance is stable and the condition for being stable reduces to*

$$PI\varphi \to [\mathcal{I}]PI\varphi. \tag{4}$$

Proof. If a network model $\mathcal{M}$ satisfies $PI\varphi$, then clearly every agent satisfies the assignment that assigns $B\varphi$ to V_I and $B\neg\varphi$ to V_E and thus the truth of (2) reduces to the truth of (4). Now, an inspection of row 16 in the table of Figure 1 shows that all agents will keep expressing a belief in $\neg\varphi$ and keep their inner belief in φ after an update with $\mathcal{I}$. Thus, $PI\varphi$ will remain true after the update, i.e. $[\mathcal{I}]PI\varphi$ is true and the network is stable.   □

The "fragility" component of pluralistic ignorance is a little more complex. If just one agent announces her private belief this may "dissolve" the phenomenon or it may not, depending on the structure of the network. We take pluralistic ignorance (in the form of (3)) to be *dissolved* when it is true that $G(I_B\varphi \wedge E_B\varphi)$. Assume that the network model $\mathcal{M}$ is in a state of pluralistic ignorance, i.e. $\mathcal{M}$ satisfies $PI\varphi$. Now assume that some agent (maybe by mistake) suddenly expresses her true inner belief in φ. Let us refer to this agent by the nominal i. Then the following is now satisfied in $\mathcal{M}$

$$UPI\varphi := @_i(I_B\varphi \wedge E_B\varphi) \wedge G\big(\neg i \to (I_B\varphi \wedge E_B\neg\varphi)\big).$$

A model satisfying $UPI\varphi$ (where i might be replaced by another nominal) will be said to be in a state of *unstable pluralistic ignorance*.[8] How $\mathcal{M}$ will evolve under the influence event $\mathcal{I}$ depends on several factors. First, consider the case where i will keep expressing her true belief.[9] Then, if $\mathcal{M}$ is connected (and finite) it is easy to show that after a finite number of updates by the influence event $\mathcal{I}$, $\mathcal{M}$ will end up in a stable state where everyone expresses their true beliefs: By

---

[8] The reader should not be confused by the name of "unstable pluralistic ignorance", which does not refer to a particular case of pluralistic ignorance, but to a state of "almost" pluralistic ignorance, a state which minimally differs from it at the observable level, by one agent expressing her private beliefs, when the others do not.

[9] Formally, we have to make a small change to $\mathcal{I}$ to make sure that i will not change her expressed belief.

inspecting row 4 in the table of Figure 1, it follows that after one update by $\mathcal{I}$ all of i's friends will express a belief in φ and that after another update with $\mathcal{I}$ the friends of friends of i will also express their true belief. In this way, a cascade effect will spread the change throughout the network and result in a stable state where everyone expresses the same true belief.

Now, if i is only made to express her true belief for a *single* round, things get more complicated as she will, in the next round already, revert to expressing a belief in $\neg\varphi$ by the influence event $\mathcal{I}$ (as all of i's friends originally expressed a belief in $\neg\varphi$). For this reason, the network might keep "fluctuating" and never stabilize. Here is an example of the later case, where i refers to agent a^{10}:

$$
\begin{array}{c}
\mathbf{a} \\
E_B\varphi \\
I_B\varphi
\end{array}
\!\!-\!\!
\begin{array}{c}
\mathbf{b} \\
E_B\neg\varphi \\
I_B\varphi
\end{array}
\quad \underset{\mathcal{I}}{\rightsquigarrow} \quad
\begin{array}{c}
\mathbf{a} \\
E_B\neg\varphi \\
I_B\varphi
\end{array}
\!\!-\!\!
\begin{array}{c}
\mathbf{b} \\
E_B\varphi \\
I_B\varphi
\end{array}
\quad \underset{\mathcal{I}}{\rightsquigarrow} \quad
\begin{array}{c}
\mathbf{a} \\
E_B\varphi \\
I_B\varphi
\end{array}
\!\!-\!\!
\begin{array}{c}
\mathbf{b} \\
E_B\neg\varphi \\
I_B\varphi
\end{array}
\quad \underset{\mathcal{I}}{\rightsquigarrow} \quad \cdots
$$

The above example shows that an unstable state of pluralistic ignorance will not necessarily stabilize, and hence not necessarily result in a state where pluralistic ignorance is dissolved. Below, we give a characterization of the ones which do result in such a state, given our assumption that all agents are of type 1.

Proposition 2. *Let* $\mathcal{M} = (A, \sim, g, \nu)$ *be a finite, connected, symmetric network model in a state of unstable pluralistic ignorance. Then the following are equivalent:*

(i) *After a finite number of updates by the influence event* $\mathcal{I}$, $\mathcal{M}$ *will end up in a stable state where pluralistic ignorance is dissolved, i.e. there is a* $k \in \mathbb{N}$ *such that* $\mathcal{M} \otimes^k \mathcal{I} \models G(I_B\varphi \wedge E_B\varphi)$ *and* $\mathcal{M} \otimes^k \mathcal{I} = \mathcal{M} \otimes^{k+1} \mathcal{I}$.

(ii) *There is an agent that expresses her true belief in* φ *for two rounds in a row, i.e. there is an* $a \in A$ *and a* $k \in \mathbb{N}$ *such that* $\mathcal{M} \otimes^k \mathcal{I}, a \models E_B\varphi$ *and* $\mathcal{M} \otimes^{k+1} \mathcal{I}, a \models E_B\varphi$.

(iii) *There are two agents that are friends and both express their true beliefs in* φ *in the same round, i.e. there are* $a, b \in A$ *and a* $k \in \mathbb{N}$ *such that* $a \sim b$, $\mathcal{M} \otimes^k \mathcal{I}, a \models E_B\varphi$, *and* $\mathcal{M} \otimes^k \mathcal{I}, b \models E_B\varphi$.

(iv) *There are two agents that are friends and have paths of the same length to the agent named by* i, *i.e. there are agents* $a, b \in A$ *and a* $k \in \mathbb{N}$ *such that* $a \sim b$, $\mathcal{M}, a \models \langle F \rangle^k i$, *and* $\mathcal{M}, b \models \langle F \rangle^k i$.

(v) *There is a cycle in* $\mathcal{M}$ *of odd length starting at the agent named by* i, *i.e. there is a* $k \in \mathbb{N}$ *such that* $\mathcal{M} \models @_i \langle F \rangle^{2k-1} i$.

(vi) *There is a cycle in* $\mathcal{M}$ *of odd length, i.e. there is a* $k \in \mathbb{N}$ *and* $a_1, a_2, ..., a_{2k-1} \in A$ *such that* $a_1 \sim a_2, a_2 \sim a_3, ..., a_{2k-2} \sim a_{2k-1}, a_{2k-1} \sim a_1$.

The proof of this proposition is a little lengthy and will be omitted here, however, it can be obtained from the authors.

---

[10] Here we regain the same fluctuation case that was given in [6], except that it now occurs, as wanted, at the level of expressed belief instead of "belief".

By this proposition (and its proof) we can also come up with an upper bound of the number of update-steps needed for a network model in an unstable pluralistic ignorance state to dissolve, if it stabilizes. If a network model $\mathcal{M} = (A, \sim, g, \nu)$ stabilizes it follows from (iv) that there are $a, b \in A$ and a $k \in \mathbb{N}$ such that $a \sim b$ and a and b both have a path of length k to $g(i)$. Choose the smallest such k. For all $c \in A$, let $m(c)$ be the length of the shortest path to either a or b. Then, by inspecting the proof it is not hard to see that $\mathcal{M}$ stabilizes in a state where pluralistic ignorance is dissolved in at most $k + max_{c \in A}\{m(c)\}$ steps.

As mentioned in section 2, the type of agents might also influence whether unstable pluralistic ignorance will dissolve. In the above we have focused on what happens when agents are of type 1. If one wants all agents to be of another type, then one can simply change the definition of $\mathcal{I}$. First, note that agents in a state of pluralistic ignorance will always be strongly influenced and since all the three different kinds of agents react the same to strong influence, Proposition 1 remains true for all types.

Now, let us consider a network of type 3 agents (expressing their inner belief whenever they have some support for it). The lines 1, 4, 7, and 10 of Figure 1 will stay unchanged. Thus, Proposition 2 will remain true for this type of agents. The only case left to consider is therefore whether Proposition 2 holds for type 2 agents (expressing their inner belief whenever they face no opposition). We leave this as an open problem.

Another interesting case would be networks with *mixed* types of agents. Our framework can be used to model this as well. We simply add another variable V_T to keep track of the agents' types, i.e. we take $R_T = \{1, 2, 3\}$. Now, we can modify the definition of $\mathcal{I}$ such that in the lines where the agent's type affects what they will do we split each line into three new lines distinguished by the extra preconditions of the form $V_T = k$. Then we change the corresponding post-conditions accordingly. In this way, a new event model $\mathcal{I}'$ can be defined, resulting in an influence dynamics that also depends on the agents types. We will leave the details of this for future research.

Even though we have shown that pluralistic ignorance is stable, there is a sense in which the phenomenon will not continue forever. The discrepancy between one's inner beliefs and one's expressed beliefs is a conflict which might have negative consequences for the agents and as such they may very well try to resolve it. This is a well studied issue in the social and psychological literature on pluralistic ignorance. It is usually assumed [13] that the agents have three different ways in which they can act to resolve this conflict: They can either *internalize the perceived view of their peers*, i.e. change their private beliefs, *attempt to change the perceived view of their peers*, or *alienate themselves form their peers*. In our setting, the first option simply corresponds to the agents changing their inner beliefs in φ to inner beliefs in $\neg\varphi$. the only way they can try and change the opinion of others is by their expressed belief. Thus, the most natural interpretation of the second option would be that the agents will start expressing their true beliefs in φ. Finally, one interpretation of the action of alienating oneself from one's peers would be to remove friendship links to all agents that express a belief in $\neg\varphi$.

Different agents might choose different reactions to a conflict between their inner and expressed beliefs. Therefore, it would be natural to add a new variable V_A to keep track of what action an agent will chose in case of such a conflict. Moreover, it would be natural to assume that agents only try to eliminate this conflict after experiencing it for some time, i.e., for a given number of rounds. We could also capture this by adding another variable that acts as a "counter" of rounds. These new variables can then be included in the preconditions of the influence event $\mathcal{I}$. For the first two options it is obvious what the new post-conditions should be, but for the third option we need an extension of our notion of event model such that it can also change the links in a network model. We believe this can be done, but we leave the details for future research.

5   Conclusion

We developed a hybrid logic to describe networks dynamics. Obtained as a formalization and extension of the simple framework of [6] with added dynamic modalities and event models, this new setting allows for agents to have multiple (changing) characteristics.

We extended the notion of social influence from [6] to a two-layered version, distinguishing between what agents actually (privately) believe and what they express to their friends. We argued that this distinction is a component of many social phenomena, and discussed the case of pluralistic ignorance – a phenomenon widely discussed in social psychology and behavioral economics.

We then formalized pluralistic ignorance and some of its dynamic properties in our new framework. Finally, we obtained a characterization result of the network configurations for which pluralistic ignorance will dissolve into a stable state where everybody agrees into expressing what they truly believe.

Acknowledgments. We would like to thank Johan van Benthem and Fenrong Liu for suggestions and comments during the elaboration of this paper. The research of Zoé Christoff leading to these results has received funding from the European Research Council under the European Communitys Seventh Framework Programme (FP7/2007-2013)/ERC Grant agreement no. 283963. Jens Ulrik Hansen is sponsored by the Swedish Research Council (VR) through the project "Collective Competence in Deliberative Groups: On the Epistemological Foundation of Democracy".

References

1. van Benthem, J.: Logical Dynamics of Information and Interaction. Cambridge University Press, The Netherlands (2011)
2. Easley, D., Kleinberg, J.: Networks, Crowds, and Markets: Reasoning About a Highly Connected World. Cambridge University Press, New York (2010)

3. Jackson, M.O.: Social and Economic Networks. Princeton University Press (2010)
4. Seligman, J., Liu, F., Girard, P.: Logic in the community. In: Banerjee, M., Seth, A. (eds.) ICLA 2011. LNCS (LNAI), vol. 6521, pp. 178–188. Springer, Heidelberg (2011)
5. Zhen, L., Seligman, J.: A logical model of the dynamics of peer pressure. Electronic Notes in Theoretical Computer Science 278, 275–288 (2011)
6. Seligman, J., Girard, P., Liu, F.: Logical dynamics of belief change in the community. Under Submission (2013)
7. Girard, P., Seligman, J., Liu, F.: General dynamic dynamic logic. In: Bolander, T., Brauner, T., Ghilardi, S., Moss, L. (eds.) Advances in Modal Logic, vol. 9, pp. 239–260. College Publication (2012)
8. Seligman, J., Girard, P.: Facebook and the epistemic logic of friendship Extended Abstract
9. Alchourrn, C.E., Gärdenfors, P., Makinson, D.: On the logic of theory change: Partial meet contraction and revision functions. Journal of Symbolic Logic 50(2), 510–530 (1985)
10. Schelling, T.C.: Micromotives and Macrobehavior. W. W. Norton and Company (1978)
11. O'Gorman, H.J.: The discovery of pluralistic ignorance: An ironic lesson. Journal of the History of the Behavioral Science 22, 333–347 (1986)
12. Miller, D.T., McFarland, C.: Pluralistic ignorance; when similarity is interpreted as dissimilarity. Journal of Personality and Social Psychology 53, 298–305 (1987)
13. Prentice, D.A., Miller, D.T.: Pluralistic ignorance and alcohol use on campus: Some consequences of misperceiving the social norm. Journal of Personality and Social Psychology 64(2), 243–256 (1993)
14. Fields, J.M., Schuman, H.: Public beliefs about the beliefs of the public. The Public Opinion Quarterly 40(4), 427–448 (1976)
15. O'Gorman, H.J., Garry, S.L.: Pluralistic ignorance – a replication and extension. The Public Opinion Quarterly 40(4), 449–458 (1976)
16. Krech, D., Crutchfield, R.S.: Theories and Problems of Social Psychology. McGraw-Hill, New York (1948)
17. Halbesleben, J.R.B., Buckley, M.R.: Pluralistic ignorance: historical development and organizational applications. Management Decision 42(1), 126–138 (2004)
18. Miller, D.T., McFarland, C.: When social comparison goes awry: The case of pluralistic ignorance. In: Suls, J., Wills, T. (eds.) Social Comparison: Contemporary Theory and Research, pp. 287–313. Erlbaum, Hillsdale (1991)
19. Bjerring, J.C., Hansen, J.U., Pedersen, N.J.L.L.: On the rationality of pluralistic ignorance. Synthese (to appear)
20. Areces, C., ten Cate, B.: Hybrid logics. In: Blackburn, P., van Benthem, J., Wolter, F. (eds.) Handbook of Modal Logic, pp. 821–868. Elsevier, Amsterdam (2007)
21. Baltag, A., Moss, L.S., Solecki, S.: The logic of public announcements, common knowledge and private suspicious. Technical Report SEN-R9922, CWI, Amsterdam (1999)
22. van Ditmarsch, H., van der Hoek, W., Kooi, B.: Dynamic Epistemic Logic. Syntese Library, vol. 337. Springer, The Netherlands (2008)

A Unified Epistemic Analysis of Iterated Elimination Algorithms from Regret Viewpoint

Jianying Cui and Xudong Luo*

Institute of Logic and Cognition, Philosophy Department,
Sun Yat-sen University, Guangzhou, 510275, China

Abstract. In this paper, we re-explain four types of players' rationality from the viewpoint of strategy-choosing regret, and we provide a unified logic of epistemic characterization of the four iterated elimination algorithms IESD (Iterated Elimination Strictly Dominated strategy), Rationalizablity (also called iterated elimination strategies that are never best responses), IA (Iterated Admissibility) and IERS (Iterated Elimination Regret-dominated Strategy). The unified characterization extends van Benthem's work of linking game theory with epistemic logic and provides further insights into exploring the rationale of these iterated eliminating algorithms. In addition, to clarify the proof-theoretic principles assumed in players' reasoning, we also develop an axiomatic presentation for our results.

1 Introduction

It is well-known in game theory that rationality implies that every player is motivated by maximizing his own utilities. Thus, every player should be able to calculate the result of every strategy profile. So, the utility maximization is one of the customary criterions for judging whether or not a player is rational in game theory. However, the criterion of rationality is also one of the factors that lead to many differences between predicted outcomes (Nash Equilibriums [18]) and empirical observations in many games, such as in the Traveler Dilemma [5], the Centipede game [17] and so on. Actually, if a player considers every strategy of his opponents, whichever strategy he chooses, he would feel regret more or less. That is, the regret consideration can implicitly play an important role in the players' decision making in a game. In this sense, one of reasons why a rational player will not choose a strictly dominated strategy can be that there must be a better strategy for him in a game, which can mitigate his regret irrespective of how his opponents move. Moreover, it has been justified that a player feels inclined to choose a strategy min-maximizing his regret value in many games [14].

Thus, what can we conclude if we reconsider the meaning of players' rationality in a game from a strategy-choosing regret perspective? To answer this question, we construct in this paper a unified epistemic game structure for the four customary iterated elimination algorithms, i.e., IESD (Iterated Elimination Strictly Dominated strategy), Rationalizablity (also called iterated elimination strategies that are never best responses), IA (Iterated Admissibility)

---

* Corresponding author.

D. Grossi, O. Roy, and H. Huang (Eds.): LORI 2013, LNCS 8196, pp. 82–95, 2013.

and IERS (Iterated Elimination Regret-dominated Strategy). More specifically, we re-explain the four types of players' rationality associated with these algorithms from the strategy-choosing regret viewpoint and provide a unified logic epistemic characterization of these iterated elimination algorithms in Public Announcement Logic from a regret viewpoint. The unified characterization extends van Benthem's work of linking game theory with dynamic epistemic logic [21] and provides further insight into exploring the rationale of an iterated eliminating algorithm. In addition, to clarify the proof-theoretic principles assumed in players' reasoning, we develop an axiomatic presentation of our result.

The paper is organized as follows. Section 2 recalls the iterated elimination algorithms in game theory. In Sections 3 and 4, after transforming a strategic form game model into a regret game model, we offer an epistemic regret-game structure and semantic interpretations of types of players' rationality. In Section 5, we construct an axiomatic epistemic game logic and discuss the relevant properties. Section 6 discusses related work. Finally, Section 7 concludes the paper with interesting future work.

2   Preliminaries

This paper explores four types of iterated elimination algorithms in game theory. Therefore, in this section we recall these algorithms and relevant concepts and notations (See [18,14] for more details). In the paper, we focus on finite games with pure strategies (i.e., the pure strategic set and players set are both finite).

Definition 1. *A strategic form game* $G = \langle N, \{S_i\}_{i \in N}, \{u_i\}_{i \in N} \rangle$, *where*

- N *is a finite set of players in game* G,
- S_i *is the finite set of strategies of player* i, *and*
- u_i *is a function that assigns a real value to every strategy profile* $s = (s_1, \ldots, s_n)$.

We are interested in a 'one-shot' strategic game, where each player i chooses a strategy from its strategy set S_i. Let $S = S_1 \times \ldots \times S_n$ be the set of strategy profiles, and S_{-i} be the set of strategy profiles of the players other than i. When focusing on player i, we denote the strategy profile $s \in S$ by (s_i, s_{-i}) where $s_i \in S_i$ and $s_{-i} \in S_{-i}$. The following concepts are about various strategy preferences over strategies of player i corresponding to dominated strategies in game theory, which play important roles in the paper.

Definition 2. *Given strategic form game* $G = \langle N, \{S_i\}_{i \in N}, \{u_i\}_{i \in N} \rangle$, *(i) strategy* s_i *is strictly dominated by* s_i' *if* $\forall s_{-i} \in S_{-i}, u_i(s_i', s_{-i}) > u_i(s_i, s_{-i})$; *(ii) strategy* s_i *is weakly dominated by* s_i' *if* $\forall s_{-i} \in S_{-i}, u_i(s_i', s_{-i}) \geq u_i(s_i, s_{-i})$ *and* $u_i(s_i', s_{-i}') > u_i(s_i, s_{-i}')$ *for some* $s_{-i}' \in S_{-i}$; *and (iii) strategy* s_i^* *is a best response in* G *to* $s_{-i} \in S_{-i}$ *if* $\forall s_i \in S_i, u_i(s_i^*, s_{-i}) \geq u_i(s_i, s_{-i})$.

Following the notation in [8], we call a weakly dominated strategy s_i by strategy s_i' an inadmissible strategy for player i, while strategy s_i' is an admissible strategy.

And in a game without players' mixed-strategies, a pure strategy for player i is rationalizable if it is a best response to the joint strategies of his opponents [11]. In order to characterize algorithm IERS developed in [14], we need the following definition:

Definition 3. *For strategic form game* $G = \langle N, \{S_i\}_{i \in N}, \{u_i\}_{i \in N} \rangle$, *player* i's *ex post regret with respect to any pure strategy profile* (s_i, s_{-i}) *is given by:*

$$Re_i(s_i) = \max\{re_i(s_i, s_{-i}) \mid \forall s_{-i} \in S_{-i}\}, \tag{1}$$

where

$$re_i(s_i, s_{-i}) = \max\{u_i(s_i', s_{-i}) \mid \forall s_i' \in S_i\} - u_i(s_i, s_{-i}). \tag{2}$$

Intuitively, $re_i(s_i, s_{-i})$ means the regret for a player choosing s_i when his opponents choose s_{-i}, and $Re_i(s_i)$ is the maximal regret value of choosing s_i whichever strategy his opponents choose. That is, $Re_i(s_i)$ refers to player i's ex post regret associated with his strategy s_i without excluding any strategy of his opponents.

Definition 4. *Given strategic form game* $G = \langle N, \{S_i\}_{i \in N}, \{re_i\}_{i \in N} \rangle$, s_i *is regret-dominated by* s_i' *if* $Re_i(s_i') < Re_i(s_i)$. *A regret-dominated strategy of* s_i *is called regrettable for player* i; *while strategy* $s_i' \in S_i$ *is un-regretted if no strategies in* G *regret-dominate* s_i'.

3   Algorithms and Regret Game Model

In this section we will redefine the elimination process of the four algorithms IESD [18], Rationalizablity [11], IA [8] and IERS [14], as the recursive sets of player strategies. And by transforming a model of a strategic form game into a model of a strategic form regret game, we redefine the concepts regarding these algorithms.

Definition 5. *Given strategic game* $G = \langle N, \{S_i\}_{i \in N}, \{u_i\}_{i \in N} \rangle$, *let* $S(P_\alpha)$ *(* $\alpha = 1, 2, 3, 4$ *) be the set of iterated strategies with properties* P_α *recursively defined by* $S(P_\alpha) = \prod_{i \in N} S_i(P_\alpha)$, *where* $S_i(P_\alpha) = \bigcap_{m \geqslant 0} S_i^{(m)}(P_\alpha)$ *(* m *is an ordinal) with* $S_i^{(0)}(P_\alpha) = S_i$ *and* $S_i^{(0)}(NP_\alpha) = \{s_i \mid s_i \in S_i^{(0)}(P_\alpha)\}$ *is a strategy without properties* P_α *respect to* $S_i^{(0)}(P_\alpha)$. *For* $m \geqslant 1$,

$$S_i^{(m)}(P_\alpha) = S_i^{(m-1)}(P_\alpha) \setminus S_i^{(m-1)}(NP_\alpha), \tag{3}$$

where $S_i^{(m)}(NP_\alpha) = \{s_i \mid s_i \in S_i^{(m)}(P_\alpha)\}$ *is a strategy without properties* P_α *in* $G^{(m)}\}$.[1]

In Definition 5, when α is $1, 2, 3$ and 4 respectively, P_α denotes properties of the strictly un-dominated, Rationalizable, admissible and un-regrettable strategy respectively. For example, $S_i(P_1)$ is the set of strategies surviving after iterated

---

[1] $G^{(m)}$ is a sub-game of G, in which $S_i = S_i^{(m)}(P_\alpha)$ and $G^{(0)} = G$.

Table 1. The game model of G_1

S_1 \ S_2	a	b	c
A	(0,5)	(3,2)	(0,3)
B	(0,0)	(3,2)	(1,1)
C	(3,0)	(0,1)	(2,0)

elimination of strictly dominated strategies for player i. More importantly, according to Definition 5, at each of the elimination stages, all strategies without P_α of all players are simultaneously deleted. Therefore, these algorithms always yield a rectangular set of strategy profiles. In G_1 as shown in Table 1, since $Re_2(a) = Re_2(c) = 2$ and $Re_2(b) = 3$, strategy b is regret-dominated for player 2, by a and c, and there are no regret-dominated strategies for player 1 in the initial game $G^{(0)}$. Then at the first deletion round, strategy b is deleted. Since strategy b does not exist there any longer, in the sub-game $G^{(1)}$, $Re_1(A)$ and $Re_1(B)$ are both 3, while $Re_1(C)$ becomes 0. Thus, regret-dominated strategies A and B of player 1 are deleted at the second round. Continuing these steps, we finally can obtain a set of strategy profiles $S(P_4) = \{(C, a), (C, c)\}$:

$$S_1^{(0)}(P_4) = \{A, B, C\}, S_1^{(0)}(NP_4) = \emptyset; S_2^{(0)}(P_4) = \{a, b, c\}, S_2^{(0)}(NP_4) = \{b\};$$
$$S_1^{(1)}(P_4) = \{A, B, C\}, S_1^{(1)}(NP_4) = \{A, B\}; S_2^{(1)}(P_4) = \{a, c\}, S_2^{(1)}(NP_4) = \emptyset;$$
$$S_1^{(2)}(NP_4) = \emptyset, S_1^{(2)}(P_4) = \{C\} = S_1(P_4);$$
$$S_2^{(2)}(P_4) = \{a, c\} = S_2(P_4), S_2^{(2)}(NP_4) = \emptyset;$$
$$S(P_4) = \{(C, a), (C, c)\}.$$

Note that in some games the outcomes or solutions computed from algorithm IERS are inconsistent with Nash Equilibriums (NE). For instance, (B, b) is the only NE of players' pure strategies in the game, which is also the outcome of the Rationalizability algorithm, i.e., $S(P_2) = \{(B, b)\}$ as show in Table 1. However, the solutions attained by the IERS algorithm exhibit the same behavior as that observed in experiments of many famous games in real-life (e.g., the Traveler's Dilemma [4], the Centipede Game [18]), which have been proved to be problematic for the Nash Equilibrium concept.[2] Therefore, the IERS algorithm is also very worth being explored further, and it has been studied in [15,14,19,9].

Definition 6. *For strategic form game* $G = \langle N, \{S_i\}_{i \in N}, \{u_i\}_{i \in N} \rangle$, *its regret game is game* $G' = \langle N, \{S_i\}_{i \in N}, \{re_i\}_{i \in N} \rangle$, *where* $re_i(s_i, s_{-i})$ *is given by formula* (2).

In effect, re_i assigns a regret value for player i's strategy s_i given other the other players' strategies s_{-i}. Game G_1 with regret is illustrated by G_1' in Table 3.[3] Thus, we can extend the above concepts relevant to dominated strategy as follows:

---

[2] More details on the IERS algorithm and its solutions can be found in [14].

[3] For convenience of comparison, game G_1 is depicted in Table 2 again.

Table 2. The game model of G_1

S_1 \ S_2	a	b	c
A	(0,5)	(3,2)	(0,3)
B	(0,0)	(3,2)	(1,1)
C	(3,0)	(0,1)	(2,0)

Table 3. The game model of G'_1

S_1 \ S_2	a	b	c
A	(3,0)	(0,3)	(2,2)
B	(3,2)	(0,0)	(1,1)
C	(0,1)	(3,0)	(0,1)

Definition 7. *Given regret game* $G' = \langle N, \{S_i\}_{i \in N}, \{re_i\}_{i \in N} \rangle$, *(i) strategy* s'_i *is strictly R-dominated by* s_i *if* $\forall s_{-i} \in S_{-i}, re_i(s_i, s_{-i}) < re_i(s'_i, s_{-i})$; *(ii) strategy* s'_i *is weakly R-dominated by* s_i *if* $\forall s_{-i} \in S_{-i}, re_i(s_i, s_{-i}) \leq re_i(s'_i, s_{-i})$ *and* $re_i(s_i, s'_{-i}) < re_i(s'_i, s'_{-i})$ *for some* $s'_{-i} \in S_{-i}$; *and (iii) strategy* s_i *is a best R-response in* G *to* $s_{-i} \in S_{-i}$ *if* $\forall s'_i \in S_i, re_i(s_i, s_{-i}) \leq re_i(s'_i, s_{-i})$.

Since according to the above definition a larger utility always means a smaller regret value, a strictly dominated strategy (or weakly dominated) amounts to a strictly R-dominated strategy (or weakly R-dominated), similar to a best response for a player. That is:

Fact1. Strategy s_i is a strictly (or weakly) dominated for player i in game G if and only if s_i is a strictly (or weakly) R-dominated for him in regret-game G' over G.

Fact2. Strategy s_i^* for player i is a best response to some strategy profile s_{-i} of his opponents in game G if and only if s_i^* is a best R-response for him given his opponents' joint strategies in regret-game G' over G.

4   An Epistemic Model for a Game with Regret

In the section, we first will extend the language in Public Announcement Logic (PAL) by adding new atomic propositions,[4] which are used to express concepts and properties in game theory. Then we provide semantic interpretations for various concepts of players' epistemic rationality based on our epistemic game-regret model.

Definition 8. *Let* G' *be a regret game regarding game* G. *Then* Θ *is a collection of the following atomic propositions set, and* $\Theta = S_i \cup PER_i \cup GP$, *where*

- $S_i = \{s_i \mid i \in N\}$ *is the set of strategies for player* i. *The intended interpretation of* s_i *is that player* i *chooses strategy* s_i.
- GP *is the set of some specific concepts in game theory. For example, its elements* $Ra_i^{(\alpha)}$ *means player* i *is* α-*type epistemically rational* ($\alpha = 1, 2, 3, 4$), *symbol* $Br_i^{(\gamma)}$ *means the* γ-*type best response of player* i ($\gamma = 1, 2$), *and* GS *is read it is a Regret-Game Solution, NE is a Nash Equilibrium.*

---

[4] Because of space limitations, we assume that the reader is familiar with PAL and omit an introduction of PAL. If not, the readers can consult [16,3].

- PRE_i denotes the strategy preference set of player i. Its elements are the atomic propositions of the form $s_i \succcurlyeq s'_i$ and $s_i \succcurlyeq' s'_i$, which encode two types of player i's weak preferences over his strategies, while $s_i \succ s'_i$ and $s_i \succ' s'_i$ stand for two types of his strict preferences over his strategies.

We call PAL including these special atomic propositions PAL-G. Thus, the language of PAL-G over an infinite set of primitive propositions Γ ($\Gamma = P \cup \Theta$, and P is a general atomic propositions set) is defined as follows:

$$\varphi ::= p \mid \neg\varphi \mid \varphi_1 \wedge \varphi_2 \mid K_i\varphi \mid C_N\varphi \mid [\varphi!]\psi$$

where p ranges over Γ. $[\varphi!]\psi$ means that ψ is true after a truthful public announcement of φ, and $\langle\varphi!\rangle\psi$ means that φ is true and ψ is true after φ is announced, $\hat{K}_i\varphi$ is read as i considers φ possible, where $\langle\varphi!\rangle\psi$ and $\hat{K}_i\varphi$ are for the dual $\neg[\varphi!]\neg\psi$ and $\neg K_i\neg\varphi$, respectively. $C_N\varphi$ means φ is common knowledge among set of players N.

Definition 9. Given $G' = \langle N, \{S_i\}_{i\in N}, \{re_i\}_{i\in N}\rangle$, an epistemic (Kripke) model over G' is a tuple of $M_{G'} = \langle W, \{\sim_i\}_{i\in N}, \{f_i\}_{i\in N}, V\rangle$, where

- $W(\neq \emptyset)$ consists of all players' pure strategy profiles;
- $\sim_i$ is an epistemic accessibility relation for player i, which is defined as the equivalence relation of agreement of profiles in the i'th coordinate;
- $f_i : W \to S_i$ is a pure strategic function, which satisfies the following property : if $w \sim_i v$ then $f_i(w) = f_i(v)$; and
- $V : \Gamma \to 2^W$ assingns atomic propositions to the worlds in which they are true.

The property of function f_i, i.e., if $w \sim_i v$ then $f_i(w) = f_i(v)$, tends to state players' intuition in a strategic form game, i.e., if i chooses strategy s_i then he knows that he chooses s_i. For convenience, $R_i(w) = \{v \mid w \sim_i v, w, v \in W\}$, i.e., the set of worlds that player i believes possible in world w, and $\|s_i\| = \{w \in W \mid f_i(w) = s_i\}$, i.e., the set of the worlds where player i chooses strategy s_i. The interpretation of formulae in pointed epistemic model $M_{G'}, w$ is defined as follows:[5]

Definition 10. Given an epistemic game structure $M_{G'}$ over game G':

- $M_{G'}, w \vDash p \Leftrightarrow w \in V(p)$, where $p \in P$;
- $M_{G'}, w \vDash s_i \Leftrightarrow w \in \|s_i\|$;
- $M_{G'}, w \vDash (s_i \succcurlyeq s'_i) \Leftrightarrow re_i(s_i, f_{-i}(w)) \leq re_i(s'_i, f_{-i}(w))$;
- $M_{G'}, w \vDash (s_i \succcurlyeq' s'_i) \Leftrightarrow \exists v \in \|s'_i\|, re_i(s_i, f_{-i}(w)) \leq re_i(s'_i, f_{-i}(v))$;
- $M_{G'}, w \vDash (s_i \succ s'_i) \Leftrightarrow re_i(s_i, f_{-i}(w)) < re_i(s'_i, f_{-i}(w))$;
- $M_{G'}, w \vDash (s_i \succ' s'_i) \Leftrightarrow \forall v \in \|s'_i\|, re_i(s_i, f_{-i}(w)) < re_i(s'_i, f_{-i}(v))$;
- $M_{G'}, w \vDash Br_i^{(1)} \Leftrightarrow \bigwedge_{s_i \neq f_i(w)}(f_i(w) \succcurlyeq a)$;
- $M_{G'}, w \vDash Br_i^{(2)} \Leftrightarrow Re_i(f_i(w)) = \min\{Re_i(s_i)|\forall s_i \in S_i\}$;
- $M_{G'}, w \vDash Ra_i^{(1)} \Leftrightarrow (M_{G'}, w) \vDash \bigwedge_{s_i \neq f_i(w)}(\hat{K}_i(f_i(w) \succcurlyeq s_i))$;

---

[5] The truth definition of formulas $\neg\varphi$, $\varphi_1 \wedge \varphi_2$ and $C_N\varphi$ are the same as in classic epistemic logic, see [6] for more details.

- $M_{G'}, w \vDash Ra_i^{(2)} \Leftrightarrow (M_{G'}, w) \vDash \hat{K}_i((\bigwedge_{s_i \neq f_i(w)}(f_i(w) \succcurlyeq s_i)));$
- $M_{G'}, w \vDash Ra_i^{(3)} \Leftrightarrow (M_{G'}, w) \vDash \bigwedge_{s_i \neq f_i(w)}(K_i((f_i(w) \succcurlyeq s_i) \vee \hat{K}_i(f_i(w) \succ s_i)));$
- $M_{G'}, w \vDash Ra_i^{(4)} \Leftrightarrow M_{G'}, w \vDash \bigwedge_{s_i \neq f_i(w)}(K_i(f_i(w) \succcurlyeq' s_i));$
- $M_{G'}, w \vDash NE \Leftrightarrow M_{G'}, w \vDash \bigwedge_{i \in N} Br_i^{(1)};$
- $M_{G'}, w \vDash GS \Leftrightarrow M_{G'}, w \vDash \bigwedge_{i \in N} Br_i^{(2)};$
- $M_{G'}, w \vDash K_i\varphi \Leftrightarrow \forall v \in R_i(w), M_{G'}, v \vDash \varphi;$ and
- $M_{G'}, w \vDash [\varphi!]\psi \Leftrightarrow M_{G'}, w \vDash \varphi$ implies $M_{G'}\mid_\varphi, w \vDash \psi$ where $M_{G'}\mid_\varphi$ is a sub-model of $M_{G'}$ where φ is true.

In the above definition, except that atomic proposition s_i holds in world w if player i's strategy is the current strategy $fi(w)$ (i.e., $fi(w) = si$) and the truths of atomic proposition p, formulas $\neg\varphi$ and $\varphi_1 \wedge \varphi_2$ in a pointed epistemic game model $(M_{G'}, w)$ are the same as the interpretations in PAL.[6] Now we focus on the interpretations of players' preferences and the properties of their rationality. There is a difference between the IERS algorithm and other algorithms. In fact, IERS is an algorithm of min-maximizing players' regret, instead of simply minimizing players' regret. So, to characterize the four algorithms in a unified epistemic framework, we need to offer a semantic interpretation of the specific strategy preference of players regarding the IERS algorithm, i.e., $s_i \succ' s_i'$ and $s_i \succcurlyeq' s_i'$. Here, $s_i \succ' s_i'$ says that s_i is strictly better than s_i' for player i in the current world w if *whichever strategies his opponents choose*, the regret raising from s_i' is larger than the regret of his current strategy; while $s_i \succ s_i'$ in world w means that player i prefers the current strategy s_i to s_i' if *given his opponents' current action* $f_{-i}(w)$, the regret deriving from s_i' is larger than the regret of s_i. In the same way, we can interpret preference $s_i \succcurlyeq' s_i'$ and $s_i \succcurlyeq s_i'$.

Furthermore, we can interpret the current strategy that is a best response, denoted as $Br_i^{(1)}$, of player i, as the strategy such that the strategy regret associated with the current strategy $f_i(w)$ is minimal among his other strategies' given his opponents' strategies $f_{-i}(w)$. The semantic interpretation is intuitive since it is true that minimizing a player's strategy regret is maximizing his utilities. However, we cannot explain $Br_i^{(2)}$ associated with the IERS algorithm in a similar way, i.e., we cannot define the truth of $Br_i^{(2)}$ in world w as $\bigwedge_{s_i \neq f_i(w)}(f_i(w) \succcurlyeq' s_i)$. The reason is that the algorithm is to compute a strategy that can minimize the maximum regret without excluding any strategy of his opponents, rather than finding a strategy minimizing regret given the opponents' strategies. So, we have to use function Re_i to achieve it, i.e., $Br_i^{(2)}$ holds in world w if $Re_i(f_i(w))$ of the current strategy is minimal among the rest strategies for player i.

Nevertheless, we can again obtain the intuition by adding epistemic ingredients into the definition of rationality $Ra_i^{(4)}$: player i is $Ra_i^{(4)}$-type epistemically rational in world w if for all of his strategies, he knows that his current strategy is at least as good as his other strategies, or if for all of his strategies, he

---

[6] More details about the interpretation of the formulas can be found in [6]. Because of the page limit, we omit them to focus on interpretations for the players' strategy preferences and their rationalities, which play generic important roles in the paper.

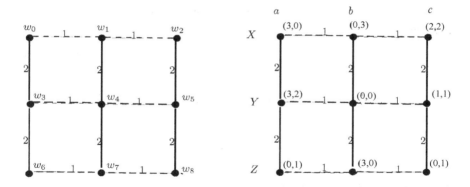

Fig. 1. The epistemic regret-game model $M_{G'_1}$ (left hand-side) corresponding to game G_1, where dashed lines for epistemic accessibility of player 1, and solid lines for player 2. For convenience, regret game G'_1 with players' utilities is also illustrated in the right hand-side.

knows that his currently chosen strategy is the strategy that can minimize his regret.[7] Meanwhile, player i is $Ra_i^{(1)}$-type epistemically rational in world w if for an arbitrary strategy s_i of him (not the current strategy $f_i(w)$), i thinks it is possible that his current strategy is at least as good as s_i; player i is $Ra_i^{(2)}$-type epistemically rational in world w if i considers it possible that his current strategy is at least as good as all of his other strategies, and $Ra_i^{(3)}$ holds in world w if for an arbitrary strategy s_i of him (except the current strategy $f_i(w)$), either player i considers it possible that his current strategy is strictly better than his other strategies or player i knows that his current strategy is at least as good as his other strategies.[8] Thus, it is easy to verify that in an epistemic regret game model $M_{G'}$, $Ra_i^{(\alpha)}(\alpha = 1, 2, 3, 4)$ fails exactly at the rows or the columns corresponding to R-dominated (or regret-dominated) strategies of players in a regret-game model G'. For example, in $M_{G'_1}$ as shown in Figure 1, $Ra_2^{(4)}$ fails in worlds $(A, b), (B, b)$ and (C, b).

5   Axiom System of PAL-G

In this section, we provide an axiom system to prove some interesting properties regarding a game, such as introspections of players' rationality, strength relationships among these different rationalities, and so on. More importantly, a characterization theorem is presented to express the relation between PAL and the above algorithms, which establishes a generic result that links true

---

[7] Again it indicates that it is necessary to add epistemic ingredients into definitions of rationality.

[8] The semantic explanations of $Ra_i^{(\alpha)}(\alpha = 1, 2, 3)$ are inspired by the definitions of rationality provided in [21] and [10] from the players' utility perspective, respectively.

common knowledge of players' rationality with outcomes obtained by using these algorithms.

Axioms: *Let* $G = \langle N, \{S_i\}_{i \in N}, \{re_i\}_{i \in N} \rangle$ *be a regret game, logic PAL-G system over G' with the following additional axioms:* $\forall i \in N$ *and* $\forall s_i, s_i' \in S_i$:

(**PA′**)   All principles of Public Announcement Logic;

(**RA1**) $\bigvee_{s_i' \neq s_i} K_i(s_i' \succ s_i) \wedge s_i \leftrightarrow \neg Ra_i^{(1)}$;

(**RA2**) $K_i(\bigvee_{s_i' \neq s_i}(s_i' \succ s_i)) \wedge s_i \leftrightarrow \neg Ra_i^{(2)}$;

(**RA3**) $\bigvee_{s_i' \neq s_i}(\hat{K}_i(s_i' \succ s_i) \wedge K_i(s_i' \succcurlyeq s_i)) \wedge s_i \leftrightarrow \neg Ra_i^{(3)}$;

(**RA4**) $\bigvee_{s_i' \neq s_i} \hat{K}_i(s_i' \succ' s_i) \wedge s_i \leftrightarrow \neg Ra_i^{(4)}$;

(**ST1**) $\bigvee_{s_i \in S_i} s_i$;

(**ST2**) $\neg(s_i \wedge s_i')$;

(**ST3**) $(s_i \succcurlyeq s_i') \vee (s_i' \succcurlyeq s_i)$;

(**ST3′**) $(s_i \succcurlyeq' s_i') \vee (s_i' \succcurlyeq' s_i)$;

(**ST4**) $(s_i \succ s_i') \leftrightarrow (s_i \succcurlyeq s_i') \wedge \neg(s_i' \succcurlyeq s_i)$;

(**ST4′**) $(s_i \succ' s_i') \leftrightarrow ((s_i \succcurlyeq' s_i') \wedge \neg(s_i' \succcurlyeq' s_i))$;

(**ST5**) $s_i \to K_i s_i$;

(**ST6**) $\neg s_i \to K_i \neg s_i$;

(**PP′**) $K_i(s_i \succcurlyeq' s_i') \to (\hat{K}_i(s_i \succcurlyeq s_i'))$.

The intuitive meanings of these axioms are as follows. Axioms **RA1-RA4** are the properties of different rationalities. For instance, **RA3** says that a player with $Ra_i^{(3)}$-type of rationality should not choose a weakly R-dominated strategy, and **RA4** means that a $Ra_i^{(4)}$-type of player i must not choose a regrettable strategy. While **ST1-ST6** plus axioms **ST3′** and **ST4′** offer some principles for strategy choosing during a game. In fact, axioms **ST1** and **ST2** together imply that each player i chooses exactly one strategy; **ST3**, **ST3′**, **ST4** and **ST4′** mean that the ordering of strategies is complete and that the corresponding strict ordering is defined as usual; **ST5** and **ST6** states that player i is aware of his own choice. Finally, axiom **PP′** indicates a strength relationship between preference $\succcurlyeq$ and preference $\succcurlyeq'$, when the preferences are intertwined players' epistemic ability.

We write $\vdash \varphi$ if φ is a theorem of logic PAL-G. The following theorems reveal the relationships between the different rationalities and all of the rationalities are epistemically introspective.[9]

Theorem 1. $\forall i \in N, \forall \alpha \in \{1, 2, 3, 4\}, \forall \beta \in \{2, 3, 4\}$ *the following hold:*

(*1a*) $\vdash_{PAL-G} Ra_i^{(\alpha)} \leftrightarrow K_i Ra_i^{(\alpha)}$; *and*

(*1b*) $\vdash_{PAL-G} Ra_i^{(\beta)} \to Ra_i^{(1)}$.

---

[9] Because of page limit, the proofs of all the theorems in the paper are put at http://logic.sysu.edu.cn/faculty/cuijianying/en/.

It is clear that $Ra_i^{(1)}$ is the weakest rationality among the rationalities discussed in the paper. However, we cannot determine a relationship between $Ra_i^{(\beta)}$ ($\beta = 2, 3, 4$). For example, in the epistemic regret game model of $M_{G_1'}$ depicted in Figure 1, proposition $Ra_2^{(3)}$ holds in the whole model, $Ra_2^{(4)}$ holds in the worlds of (A, a), (B, a), (C, a), (A, c), (B, c), and (C, c), but $Ra_2^{(2)}$ is true in the worlds of (A, a), (B, a), (C, a), (A, b), (B, b), and (C, b). While $Ra_1^{(2)}$ and $Ra_1^{(4)}$ are both true in the model, $Ra_1^{(3)}$ is false in worlds (B, a), (B, b), and (B, c).

Theorem 2. *Logic PAL-G is sound with respect to the class of models $M_{G'}$.*

According to [21], for any model M one can keep announcing φ, retaining only those worlds in which φ holds. This yields a sequence of nested decreasing sets, which must stop in finite models, denoted by $\sharp(\varphi, M)$ [21]:

Definition 11. *For any model M and formula φ, the announcement limit $\sharp(\varphi, M)$ is the first sub-model in the repeated announcement sequence where announcing φ has no further effect. If $\sharp(\varphi, M)$ is non-empty, we have a model where φ has become common knowledge. We call such statements self-fulfilling in the given model, and all others self-refuting.*

Pursuing the above idea, we see that public announcement $[!\varphi]$ of true propositions φ yields *the information* that changes the current model irrevocably, discarding worlds that fail to satisfy φ. Meanwhile, the agents' interactive knowledge about φ will be increased as the model changes. In addition, dominated strategies are deleted because in game theory players' *rationality* is postulated as high-order interactive information among players. Therefore, after re-defining the above rationalities (see Definition 10), in the following we will show that *an announcement limit* has close connections with the equilibriums or game solutions found by using many iterated elimination algorithms, when we refer to the rationalities as announcement assertions (see Theorem 4 later). As an announcement rule in PAL, players publicly announce assertions, which must be the statements that they know are true. Theorems 1 and 3 (in the following) enables us to successively remove the worlds in which $Ra^{(\alpha)}$ ($Ra^{(\alpha)} = \cap_{i \in N} Ra_i^{(\alpha)}, \alpha = 1, 2, 3, 4$) does not hold in model $M_{G'}$ after repeatedly announcing some rationality.

Theorem 3. *Every finite epistemic regret-game model has worlds in which $Ra^{(\alpha)}$ is true, where $\alpha = 1, 2, 3, 4$.*

In Figure 2, the left-most model is the epistemic regret-game model $M_{G_1'}$ shown in Figure 1. The other models are obtained by public announcements of $Ra^{(4)}$ successively for two times. So, in the last sub-model, we have:

$$M_{G_2'}, (C, a) \vDash [Ra^{(4)}!][Ra^{(4)}!]C_N(GS).$$

This indicates that if the players iteratively simultaneously announce that they are $Ra_i^{(4)}$-type of rationality, the process of regret-dominated strategies elimination leads them to the solutions that are commonly known to be a game solution

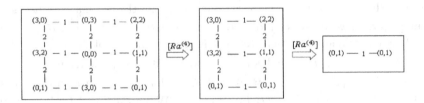

Fig. 2. The public announcement of $Ra^{(4)}$

(GS). Similarly, announcing different types of rationality can be described as the iterated elimination procedure of the four algorithms.

Theorem 4. (*Characterization Theorem*) *Let G be a game, $M_{G'}$ be an epistemic regret-game model associated with G, for all w in $M_{G'}$, if $\sharp(Ra^{(\alpha)}, M_{G'})$, ($\alpha = 1, 2, 3, 4$) is stable by repeated announcements of $Ra_i^{(\alpha)}$ in $M_{G'}$, then:*

$$w \in \sharp(Ra^{(\alpha)}, M_{G'}) \Leftrightarrow f(w) \in S(P_\alpha) \ .$$

6   Related Work

There is a large amount of literature on the algorithms of iterated elimination in the field of logic[12,21,22], computer science [1,13] and game theory [7,8,15]. In particular, [21] and [1] are closely related to our research.

In [21], van Benthem characterizes two different algorithms of IESD and Rationalizability by analyzing the two types of rationality in PAL: (*i*) *the weak rationality (WR)*, which is used to characterize the classic algorithm IESD; and (*ii*) *the strong rationality (SR)*, which is relevant to the Rationalizability algorithm of Pearce [11]. Our work in this paper further extends the work of [21] to characterize the four types of rationality, and furthermore lays epistemic foundations for all of the algorithms IESD, Rationalizability, AI and IERS in our epistemic regret-game model. Thus, the properties of these different types of rationality can be discussed in our unified framework. By contrast, the rationalities defined by us cannot be analyzed in [21]'s epistemic model for lack of semantic interpretations of some concepts related to agents' regret in that work.

In [1], Apt et al. offer a simple generalization of the above public announcement approach, which has been used to study the effect of the IESD and Rationalizability algorithms, to cover arbitrary strategic games and many optimality notions [1]. To provide a characterization of various iterated elimination algorithms based on the concept of a public announcement, they make a distinction between 'global properties' and 'local properties' to an optimal strategy for a player. They write: "...to assess the optimality of s_i globally, player i must consider all of his strategies that occur in his strategy set in the initial game". For example, the strategy s_i of player i is not strictly dominated in G by any

Table 4. The regret game model of G_2'

S_1 \ S_2	a	b	c
A	(3,3)	(0,0)	(0,2)
B	(0,2)	(2,1)	(0,0)
C	(1,1)	(0,2)	(2,0)

Table 5. The sub-game model of G_2'

S_1 \ S_2	b	c
B	(2,1)	(0,0)
C	(0,2)	(2,0)

strategy from his strategy set in the initial game, i.e., $\neg \exists s_i' \in H_i, s_i' \succ_G s_i$.[10] Therefore, although it seems that the achievements in [1] imply some of our results (as we mentioned in the proof of Theorem 4), because the property of regret-dominated strategy cannot be defined as a "global property", their observations cannot cover the epistemic analysis for the IERS algorithm. So, actually the conclusion in their theorem 7 in [1] is derived by our characterization theorem. In this sense, our work is also an extension to the work of [1].

Tables 4 and 5 illustrate the reason why the property of regret-dominated strategy cannot be defined as a "global property". In $M_{G_2'}$, since strategy a is regret-dominated by b and c for player 2 and A is a regret-dominated for player 1, we can delete a and A at the same time in the first elimination round. Thus, in the sub-game model $M_{G_2'^{(1)}}$, because of $H_1 = \{A, B, C\}$, according to the definition of "global optimal strategy" in [1], neither B nor C are globally optimal strategies for player 1. Therefore, they should be deleted simultaneously in the second round, and then there are no strategies for player 1 to take in the next sub-game. It is an impossible situation in a game. Therefore, for the IERS algorithm we cannot distinguish "global optimal" and "local optimal". Still, both [21] and [1] explore the relationship between common knowledge of rationality and the results from iterated elimination algorithms based on Fix-point Logic [20], while ours is just based on PAL.

7 Conclusion and Further Research

In this paper, we have constructed a unified epistemic regret-game model to provide a dynamic epistemic characterization of four kinds of iterated elimination algorithms. In the future, we will try to generalize the characterization result in the paper even more, providing characterizations for any iterative algorithm that as certain basic properties. Meanwhile, there are some directions of future research relevant to our work in the paper. For example:

– Comparing, combining, and reducing methods. Comparing methods like IESD and IERS, we can see that one may be better than another depending on the structure of a given game. It is interesting to investigate what happens

---

[10] In [1], G can be read as a set of strategy profiles of all players in a sub-game concerning an initial game, and H_i is a set of strategies for player i in the initial game model.

if a variety of such methods are used. One possibility is that one method may simulate another by means of translating the given game systematically into one with changed outcome values. Moreover, there are games where both methods make sense intuitively. We will start with sequential combinations of solution methods, and the eventual goal would be an algebra of solution methods.

– Linking up with limit behavior in learning theory. In our unified structure, we have only considered the cases where games are solved through iterated soft updates with regret statements. However, many other scenarios have the same features, including infinite sequences where the approximation behavior itself is the focus of interest. Particularly, it is interesting to connect our setting with the learning-theoretic scenarios and extended temporal update logics suggested by the results of [2] and [23].

Acknowledgements. We are grateful to Johan van Benthem for valuable suggestions, to Olivier Roy, who suggested to compare our work with some results in [1], and to Henry Prakken for polishing up English of the paper. We also appreciate the anonymous reviewers for their helpful comments. This paper is supported partially by National Natural Science Foundation of China (No. 61173019), Major Projects the Ministry of Education of China (No.10JZD0006), Social Science Foundation of China (No. 12CZX056), Humanity and Social Science Youth Foundation of Ministry of Education (No. 11YJC72040001), Philosophy and Social Science Youth Projects of Guangdong Province (No. GD11YZX03), and Bairen Plan of Sun Yat-sen University.

References

1. Apt, K.R., Zvesper, J.A.: Common beliefs and public announcements in strategic games with arbitrary strategy sets. arXiv preprint, arXiv:0710.3536 (2007)
2. Baltag, A., Gierasimczuk, N., Smets, S.: Belief revision as a truth-tracking process. In: Proceedings of the 12th Conference on Theoretical Aspects of Rationality and Knowledge, pp. 187–190 (2011)
3. Baltag, A., Moss, L.S., Solecki, S.: The logic of public announcements, common knowledge and private suspicious. Technical Report SEN-R9922, CWI, Amsterdam University (1999)
4. Basu, K.: The traveler's dilemma: Paradoxes of rationality in game theory. American Economic Review 84(2), 391–395 (1994)
5. Basu, K.: The traveler's dilemma. Journal of the American Statistical Association 46, 55–67 (2007)
6. Blackburn, P., van Benthem, J., Wolter, F.: Handbook of Modal Logic. Elsevier Science Inc. (2007)
7. Bonanno, G.: A syntactic approach to rationality in games with ordinal payoffs, vol. 3. Amsterdam University Press, Amsterdam (2008)
8. Brandenburger, A., Friedenberg, A., Keisler, H.J.: Admissibility in games. Econometrica 76(2), 307–352 (2008)
9. Cui, J., Luo, X., Sim, K.M.: A new epistemic logic model of regret games. In: Wang, M. (ed.) KSEM 2013. LNCS (LNAI), vol. 8041, pp. 372–386. Springer, Heidelberg (2013)

10. Cui, J., Tang, X.: A method for solving Nash equilibria of games based on public announcement logic. Science China (Information Science) 53(7), 1358–1368 (2010)
11. David, P.: Rationalizable strategic behavior and the problem of perfection. Econometrica 52(4), 1029–1050 (1984)
12. van Ditmarsch, H., Lang, J., Ju, S. (eds.): LORI 2011. LNCS, vol. 6953. Springer, Heidelberg (2011)
13. Halpern, J.Y., Pass, R.: A logical characterization of iterated admissibility. In: Proceedings of the 12th Conference on Theoretical Aspects of Rationality and Knowledge, pp. 146–155 (2009)
14. Halpern, J.Y., Pass, R.: Iterated regret minimization: A new solution concept. Games and Economic Behavior 74(1), 184–207 (2012)
15. Ludovic, R., Schlag, K.H.: Implementation in minimax regret equilibrium. Games and Economic Behavior 71(2), 527–533 (2011)
16. Plaza, J.A.: Logics of public communications. In: Proceedings of the 4th International Symposiumon Methodologies for Intelligent Systems, pp. 201–216 (1989)
17. Rosenthal, R.: Games of perfect information, predatory pricing, and the chain store. Journal of Economic Theory 25(1), 92–100 (1981)
18. Rubinstein, A.: A Course in Game Theory. The MIT Press, Cambridge (1994)
19. Stoye, J.: Axioms for minimax regret choice correspondences. Journal of Economic Theory 146(6), 2226–2251 (2011)
20. Tarski, A.: A lattice-theoretical fixpoint theorem and its applications. Pacific Journal of Mathematics 5(2), 285–309 (1955)
21. van Benthem, J.: Rational dynamics and epistemic logic in games. International Game Theory Review 9(1), 13–45 (2007)
22. van Benthem, J.: Logical Dynamcis of Information. Cambridge University Press (2011)
23. van Benthem, J., Pacuit, E., Roy, O.: Toward a theory of play: A logical perspective on games and interaction. Games 2(1), 52–86 (2011)

Listen to Me!
Public Announcements to Agents That Pay Attention — or Not

Hans van Ditmarsch[1], Andreas Herzig[2],
Emiliano Lorini[2,3], and François Schwarzentruber[4]

[1] LORIA, CNRS – Université de Lorraine, France
[2] University of Toulouse, CNRS, IRIT, France
[3] IAST, Toulouse, France
[4] ENS Cachan (Brittany Extension), IRISA, France

Abstract. In public announcement logic it is assumed that all agents pay attention (listen to/observe) to the announcement. Weaker observational conditions can be modelled in event (action) model logic. In this work, we propose a version of public announcement logic wherein it is encoded in the states of the epistemic model which agents pay attention to the announcement. This logic is called attention-based announcement logic, abbreviated ABAL. We give an axiomatization and prove that complexity of satisfiability is the same as that of public announcement logic, and therefore lower than that of action model logic [2]. We exploit our logic to formalize the concept of joint attention that has been widely discussed in the philosophical and cognitive science literature. Finally, we extend our logic by integrating attention change.

1 Introduction

In public announcement logic it is assumed that announcements are perceived by all agents: it models the consequences of each of the agents incorporating a new formula into the set of beliefs. The argument of the dynamic modal operator in public announcement logic is therefore called an *announcement*. Once the government has announced a new election, they cannot be held liable when you forget to vote on election day. You *were supposed to know*.

In this work we take one step back from that point of view. When an announcement is made, it may well be that some agents were not paying attention and therefore did not hear it. Also, there may be uncertainty among the agents about who is paying attention and who not, and therefore, who heard the message and who not. Contrarily to action model logic, in our modelling it is not an aspect of the description of the action to which subset of all agents the announcement is made, but this is now an aspect of the state in which the announcement is executed.

Additional to the usual set of propositional variables we add designated variables for each agent, that express that the agent is paying attention. A given

D. Grossi, O. Roy, and H. Huang (Eds.): LORI 2013, LNCS 8196, pp. 96–109, 2013.

state of a Kripke model therefore contains information about which agents are paying attention and which agents are not paying attention. This determines the meaning of what we call *attention-based announcements*. A special case is that of introspective agents that know whether they are paying attention. We axiomatize our attention-based announcement logic ABAL, including a version with introspection for beliefs and attention.

An announcement by an outside observer that is public for a subset of all agents is modelled in [9,6] as a *private announcement* to that subset of agents. The agents' attention configuration behind such announcements can be modelled in our logic by a particular formula built from our attention variables. Our logic generalizes Gerbrandy's because the 'attention level' of a given agent can vary between the states. Our logic can in turn be mapped to action model logic: each configuration of attention corresponds to a particular class of action models.

We show that the complexity of satisfiability in our logic remains in the same range as that of public announcement logic, viz. PSPACE. This contrasts with the higher complexity of action model logic. As the action models corresponding to attention-based announcements can be quite large, we consider that this is indeed a valuable result.

In the ABAL we can formalize a concept that has been widely discussed in the philosophical and in the cognitive science literature, namely *joint attention* [15,13,7]. This concept has been shown to be crucial for explaining the genesis of common belief in a group of agents.

Finally, we add other dynamics to our logic, namely change of attention. This is an elementary further addition to the logical framework and this logic also has a complete axiomatization.

2 Attention-Based Announcement Logic ABAL

Let AGT be a finite set of agents, let ATM be a (disjoint) countable set of propositional variables, and let $H = \{h_a \mid a \in AGT\}$ be a disjoint set of propositional variables. A proposition h_a (for 'a is *hearing* what is being said' or more simply 'a is listening') expresses that agent a is paying attention and so will hear public announcements.

Definition 1 (Language). *The* language $\mathcal{L}$ *of attention-based announcement logic* ABAL *is defined as follows, where* $p \in ATM$ *and* $a \in AGT$.

$$\mathcal{L} \ni \ \varphi \ ::= \ p \ \mid \ h_a \ \mid \ \neg\varphi \ \mid \ (\varphi \wedge \varphi) \ \mid \ \mathbf{B}_a\varphi \ \mid \ [\varphi]\varphi$$

We abbreviate $\bigwedge_{a \in A} h_a$ *by* h_A.

We write q to denote a variable that is either $p \in ATM$ or $h_a \in H$. Other propositional connectives, and the dual modalities, are defined as usual. Formula $\mathbf{B}_a\varphi$ is read as 'agent a believes that φ is true', and formula $[\varphi]\psi$ as 'after the public announcement of φ, ψ holds'.

Definition 2 (Epistemic attention model). *An epistemic attention model is a triple $M = (S, R, V)$ with S a non-empty set, R a function assigning to each agent an accessibility relation R_a and V a function assigning to each propositional variable $q \in ATM \cup H$ the subset $V(q) \subseteq S$ where the variable is true.*

Definition 3 (Attention introspection). *Given an epistemic attention model $M = (S, R, V)$, the model satisfies the property of attention introspection if for all $s, t \in S$, if $(s, t) \in R_a$, then $s \in V(h_a)$ iff $t \in V(h_a)$.*

When attention introspection holds, an agent knows whether she is paying attention.

Boolean constructions as well as operators of belief $\mathbf{B}_a$ are interpreted in the standard way. The truth condition for attention-based announcements $[\varphi]$ is different from that of (world eliminating) truthful public announcement [14] and also different from that of (arrow eliminating) public announcement [9], although it comes closer to the latter in spirit: it is also arrow eliminating.

Definition 4 (Semantics of attention-based announcements)

$$M, s \models [\varphi]\psi \text{ iff } M^\varphi, (s, 0) \models \psi$$

where $M^\varphi = (S', R', V')$ is defined as follows.

- $S' = S \times \{0, 1\}$
- *for each agent a, $((s, i), (t, j)) \in R'_a$ if and only if $(s, t) \in R_a$ and:*
  1. $i = 0$, $j = 0$, $(M, s) \models h_a$ and $(M, t) \models \varphi$, or
  2. $i = 0$, $j = 1$, and $(M, s) \not\models h_a$, or
  3. $i = 1$, $j = 1$.
- *for each $p \in ATM$, $(s, 0) \in V'(p)$ iff $s \in V(p)$ and $(s, 1) \in V'(p)$ iff $s \in V(p)$.*

The model M^φ is the extended disjoint union of the (arrows to) φ restriction of M, called $M|\varphi$, and M itself, plus — that is the extension — a number of additional accessibility pairs between states for those agents that are not attentive. Roughly speaking, $M^\varphi = M|\varphi \oplus M$ plus some edges. After the announcement of φ, the agents that are attentive only consider possible the 0-copies of the states of the original model M in which φ is true. In contrast, the agents that are not attentive only consider possible 1-copies of the states of the original model M. This construction of the updated model M^φ ensures that attentive agents learn φ while inattentive agents don't learn anything.

Example 1. Ann (a) and Bill (b) have lunch in the cafeteria and each consider the possibility of snowfall this afternoon (p) — a regular occurrence in Nancy, many times of the year. In fact Bill has seen the weather report and knows whether it will snow, while Ann does not. However, Bill never knows whether Ann is paying attention. Ann knows that Bill is attentive. Both agents know whether they are attentive. This situation is depicted as model M in Fig. 1 (there is no particular actual state: any of the four may do.) Now Cath comes along and says she just read the weather report: it will snow. This results in the model transition depicted in Figure 1, where Cath's announcement is modelled as an announcement by an outsider. Any of the four points in $M|p$ can be the actual state of the resulting model, but not any of the M copy on the right hand side. If Ann and Bill pay attention and p is true, as on the bottom-left side of $M|p$, Bill remains uncertain if Ann now knows that p, as he considers it possible that she was not paying attention (top-left of $M|p$), in which case she would have remained uncertain about p.

Proposition 1 (Preservation of attention introspection). *If M satisfies attention introspection then M^φ satisfies attention introspection.*

Although we consider Prop. 1 a valuable result, it takes somewhat away from the glamour when we realize that models with empty accessibility relations also satisfy attention introspection. For example, suppose that agent a is paying attention in the actual state s (h_a is true) and also in the (uniquely) accessible state t, and where s is also considered possible. Attention introspection is satisfied. After the announcement with attentive agents $\neg h_a$, the agent a no longer considers state t possible but also no longer considers the actual state possible. Because the agent was paying attention, she has come to believe that she is not paying attention; but at the price of still also believing that she is paying attention — where the latter remains in fact the truth.

In this paper we focus on two classes of models: K_n, where $n = |AGT|$, that is, multiple agents and no special properties of the accessibility relations, and $\mathsf{K45}_n^h$, that is, multiple agents with transitive and Euclidean accessibility relations, and with attention introspection as well. (The classes $\mathsf{S5}_n^h$ and $\mathsf{KD45}_n^h$ are unsuitable: they are not closed under announcements because the property of seriality (D) may be lost after an announcement.) The set of valid $\mathcal{L}$-formulas on the class of models K_n is called ABAL, and the set of valid $\mathcal{L}$ formulas on the class of models $\mathsf{K45}_n^h$ is called ABAL^{intro}.

3   Relation with Action Models

Every attention-based announcement is definable as an action model. Whether an announcement φ is heard in a given state depends on the value of h_a for every agent a in that state. The agents who hear the announcement retain all arrows pointing to states where φ holds and delete all arrows pointing to states where φ does not hold, and that is independent of the truth of φ in the actual state; whereas the agents who do not hear the announcement think that nothing has happened, i.e., also independent of the truth of φ they think that the trivial action with precondition $\top$ happened. There is an economic way to define such an action model (in the sense of producing a resulting model with a minimal duplication of states into bisimilar states). Definition 5 spells out the inductive clause for attention-based announcement of an obviously inductively defined translation.

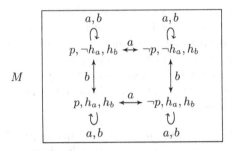

$\Downarrow$ announcement of p

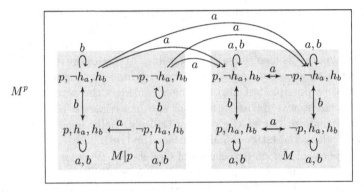

Fig. 1. Example of an attention-based announcement

Definition 5 (Action model for attention-based announcements). *Given a formula φ, the action model for the attention-based announcement of φ is the multi-pointed action model $\mathcal{A}_\varphi = (\mathbb{A}, R, Pre, P)$ where:*

- $\mathbb{A} = \{(i, J) \mid i \in \{0, 1\} \text{ and } J \subseteq AGT\} \cup \{w_\top\};$
- *R maps each agent $a \in AGT$ to*

$$R_a = \{((i, J), (1, K)) \mid i \in \{0, 1\} \text{ and } a \in J\} \cup \\ \{((i, J), w_\top) \mid a \notin J\} \cup \{(w_\top, w_\top)\};$$

- *$Pre : \mathbb{A} \to \mathcal{L}$ is defined as follows:*
  - *$Pre((i, J)) = \overline{\varphi} \wedge \bigwedge_{a \in AGT} \overline{h_a}$ where $\overline{\varphi}$ is either φ if $i = 1$ or $\neg\varphi$ if $i = 0$ and $\overline{h_a}$ is either h_a if $a \in J$ or $\neg h_a$ if $a \notin J$ for all $a \in AGT;$*
  - *$Pre(w_\top) = \top;$*
- *$P = \{(i, J) \mid i \in \{0, 1\} \text{ and } J \subseteq AGT\}$ is the set of points.*

Informally, the action model for the attention-based announcement of φ consists of $2^{n+1}+1$ actions and has 2^{n+1} initial points (alias actual actions). Each of these points is identified by the complete and disjoint set of preconditions $\overline{\varphi} \wedge \bigwedge_{a \in AGT} \overline{h_a}$, where $\overline{\varphi}$ is either φ or $\neg\varphi$ and where $\overline{h_a}$ is either h_a or $\neg h_a$. Moreover, there is a 'nothing happens' alternative with precondition $\top$, that is not an initial point. An attentive agent believes that any action point with precondition entailing φ may be the actual action. An inattentive agent believes that the action with precondition $\top$ is the actual action.

The action model for attention-based announcements is depicted in Figure 2 for the example of two agents a and b and the announcement φ. For example, if φ is false, h_a is true, and h_b is false, agent a hears the announcement φ and believes it to be true, therefore she believes the real action to be the one where φ is true — regardless of the values of h_a and h_b in states wherein it can be executed (we take the case of general accessibility, i.e., logic K_n). So that makes for *four* arrows. On the other hand, agent b does not hear the announcement and believes that nothing at all happens: a single arrow to the alternative with precondition $\top$ (and with reflexive arrows for a and b: b believes, incorrectly, that all agents believe that nothing happened).

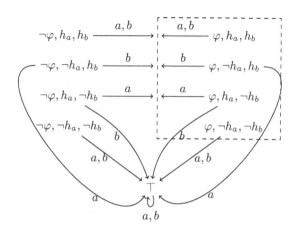

Fig. 2. The action model $\mathcal{A}$ corresponding to an attention-based announcements φ to two agents. An arrow pointing to a box points to all actions in the box.

This action model construction maps nicely with the semantics of $[\varphi]\psi$ that given a model M produces a model M^φ twice that size, consisting of a 'trivial' copy M plus a 'heard' copy $M|\varphi$. As all the 2^{n+1} different preconditions in the action model are exclusive, the product of that entire part of the action model produces a model of the same size as M, but with merely some removed arrows. A recursive translation defines an embedding from ABAL into action model logic.

Proposition 2. *Let M be an epistemic attention model. Let $\mathcal{A}_\varphi$ be the action model (according to Def. 5) corresponding to φ. Then $M^\varphi \;\underline{\leftrightarrow}\; M \otimes \mathcal{A}_\varphi$.*

4   Axiomatization and Complexity

Table 1 shows the axiomatization. It follows the pattern of believed announcements (arrow eliminating), not that of truthful announcements (state eliminating). The crucial axiom is

$$[\varphi]\mathbf{B}_a\psi \leftrightarrow ((h_a \to \mathbf{B}_a(\varphi \to [\varphi]\psi)) \land (\neg h_a \to \mathbf{B}_a\psi))$$

It says that the belief consequences of an attention-based announcement are either, if the agent pays attention, what the agent believes to be the consequences of the announcement in case it was true, or else, if the agent does not pay attention, what the belief consequences were before the announcement (i.e., an agent not hearing the announcement does not change her beliefs). Note that our axiom resembles Gerbrandy's axiom for private announcements. The axioms * formalize that agents have introspective beliefs and are not uncertain about what they hear. (Attention introspection is therefore just like awareness introspection in a the logic of awareness [8].) The axiomatization ABAL consists of all the derivation rules and axioms of Table 1. The axiomatization ABAL^{intro} consists of ABAL plus the *-ed axioms and rules. Soundness follows straightforwardly from the action model modelling of attention-based announcements that we will give in Section 3.

Table 1. The axiomatizations for ABAL and ABAL^{intro}

all propositional tautologies	$\mathbf{B}_a\varphi \to \mathbf{B}_a\mathbf{B}_a\varphi$ *
$\mathbf{B}_a(\varphi \to \psi) \to (\mathbf{B}_a\varphi \to \mathbf{B}_a\psi)$	$\neg\mathbf{B}_a\varphi \to \mathbf{B}_a\neg\mathbf{B}_a\varphi$ *
$[\varphi]\mathbf{B}_a\psi \leftrightarrow ((h_a \to \mathbf{B}_a(\varphi \to [\varphi]\psi)) \land (\neg h_a \to \mathbf{B}_a\psi))$	$h_a \to \mathbf{B}_a h_a$ *
$[\varphi](\psi \land \chi) \leftrightarrow ([\varphi]\psi \land [\varphi]\chi)$	$\neg h_a \to \mathbf{B}_a\neg h_a$ *
$[\varphi]\neg\psi \leftrightarrow \neg[\varphi]\psi$	From φ infer $\mathbf{B}_a\varphi$
$[\varphi]q \leftrightarrow q$	From φ infer $[\psi]\varphi$

Proposition 3. *The axiomatization of ABAL is sound and complete for the class of* K_n *models. The axiomatization of* ABAL^{intro} *is sound and complete for the class of* $\mathsf{K45}_n^h$ *models.*

Proof. The standard reduction argument applies: all axioms for the consequences of announcements push the announcement operator deeper into the formula on the right hand side, until one finally arrives at an announcement before a propositional variable, $[\varphi]p$, which is equivalent to p. Therefore the logic is equally expressive to the base modal logic — which is complete.

We note that by Proposition 1 the belief modality of ABAL^{intro} does not collapse into a knowledge modality; see [5] for an investigation of this issue.

In the remainder of the section we focus on the ABAL satisfiability problem on the class of all K_n frames. The satisfiability problem of a formula in the language of action model logic [6] plus the *union* operator over actions is NEXPTIME-complete [2]. ABAL is the fragment of action model logic for an action model of size exponential in the number of agents (see Section 3). So the satisfiability problem of ABAL is decidable.

It is difficult to turn the tableau method for action model logic into a PSPACE procedure because each node may contain an exponential amount of information in the length of the input formula. Surprisingly, we can adapt the tableau method of action model logic so that the amount of information in a node is polynomial in the size of the input.

Proposition 4. *Satisfiability of $\mathcal{L}$ formulas in the class of K_n models is PSPACE-complete.*

We leave open the complexity of satisfiability of formulas of the ABAL on the class of $\mathsf{K45}_n^h$ models. We conjecture that it equals that of the underlying epistemic logic, viz. PSPACE complete.

5   Joint Attention

The attention introspection axiom $h_a \to \mathbf{B}_a h_a$ of ABAL^{intro} only guarantees attention introspection for individuals, not for groups: it may happen that h_A is true while some $a \in A$ does not believe that h_A. We now investigate a condition under which attention introspection obtains in terms of common belief: *joint attention* or *joint attentional state*. That concept was widely discussed in the philosophical and in the cognitive science literature [15,13,7], and we show that it can be captured in our logic ABAL^{intro}.

We assume a common belief operator $\mathbf{C}_A$ for a subgroup A of the set of all agents, so that $\mathbf{C}_A \varphi$ stands for 'the agents in group A commonly believe φ', and which is interpreted in the usual way by the transitive closure of the union of all accessibility relations R_a for the agents in A.

Let $A \subseteq AGT$. The idea is that the agents in A have a joint attention (or are in a joint attentional state) if and only if they are looking at the source of information *together*, that is to say, every agent in A is looking at the source of information, every agent in A believes that every agent in A is looking at the source of information, and so on. More concisely, the agents in A are in a joint attentional state if and only if each of them is looking at the source of information and focusing his attention on it and they have common belief that each of them is looking at the source of information and focusing his attention on it. Formally:

$$\mathsf{JointAtt}_A \equiv_{def} h_A \wedge \mathbf{C}_A h_A.$$

Note that joint attention is closed under attention-based announcements: if $M, s \models \mathsf{JointAtt}_A$ then $M^\varphi, s \models \mathsf{JointAtt}_A$ for every φ such that $M, s \models \varphi$. Note moreover that when joint attention of all agents is satisfied then attention-based announcements are the same as public announcements.

As pointed out by [13,7], joint attention explains the genesis of common beliefs in the context of social interaction. Such genesis is often considered as related to public events in the sense that a common belief is either a consequence of an event whose occurrence is so evident (viz. public) that agents cannot but recognize it as when, during a soccer match, players mutually believe that they are playing soccer, or the product of a communication process as when the referee publicly announces that one player is expelled. From there on each player believes that each other player believes and so on that one of them has been expelled. Intuitively, an event is considered public as long as its occurrence is epistemically accessible by everybody such that it becomes common belief between them. But what are the intuitive conditions that make an event public? What are the reasons to believe that an occurring event is commonly believed? In a normal situation (what is announced is true, there is no noise in the communication channel, etc.) looking at the source of information and having a common belief that everyone is looking at the source of information (i.e., being in a joint attentional state) provide a sufficient condition for the formation of a common belief. This is captured by a validity of ABAL (to avoid Moorean phenomena we restrict ourselves to learning propositional variables):

$$\models \text{JointAtt}_A \rightarrow [p]\mathbf{C}_A p \tag{1}$$

We can actually characterize the formation of common belief of an atomic fact p as follows:

$$\models [p]\mathbf{C}_A p \leftrightarrow \mathbf{C}_A \bigwedge_{a \in A} (h_a \vee \mathbf{B}_a p) \tag{2}$$

Note that the equivalence is not valid if we replace p by h_a.

6   Attention Change

A good way to have your addressees pay attention is to clap your hands before making an announcement. Even if they were not paying attention, they now do. In other words, if $\neg h_a$ was true before, h_a is true now. And this is the case for all agents. This is a public way to make everybody pay attention to you. Even more, you have achieved their *joint* attention.

A less public way to make someone listen to you is to tap on her shoulder before you speak. This only makes that person attentive and not the other agents. Suppose that agents a and b are both not paying attention. If I tap on a's shoulder and then say something, only a and not b will hear it. If, on the other hand, I first tap a's shoulder and then b's then both a and b will hear the announcement, but they consider it possible that the other was not paying attention and does not hear it. The order does not make a difference. In contrast, clapping your hands is a way to ensure joint attention.

Drawing inspiration from [17,16], we model such fine-grained attention change by an assignment. Given a set of agents $A \subseteq AGT$, we distinguish the assignment $+A$ (merely shorthand for a simultaneous assignment $a_1 := \top, \ldots, a_n := \top$) that makes all agents $a \in A$ pay attention and hear subsequent announcements, from an assignment $-A$ that makes all h_a false. To the inductive definition of the language $\mathcal{L}$ (Def. 1) we add clauses for the modal operators $[+A]$ and $[-A]$, for $A \subseteq AGT$. We write $[+a_1, \ldots, a_n]$ instead of $[+\{a_1, \ldots, a_n\}]$. The semantics of attention-based assignment is then:

$$M, s \models [+A]\psi \text{ iff } M^{+A}, (s, 0) \models \psi$$
$$M, s \models [-A]\psi \text{ iff } M^{-A}, (s, 0) \models \psi$$

where $M^{+A} = (S', R', V')$ is defined as follows (the definition of M^{-A} is similar).

- $S' = S \times \{0, 1\}$;
- if $a \in A$ and $s, t \in S$ then $((s, i), (t, j)) \in R'_a$ iff $(s, t) \in R_a$ and
  1. (1) $i=0$ and $j=0$; or
  2. (2) $i=1$ and $j=1$;
- if $a \notin A$ and $s, t \in S$ then $((s, i), (t, j)) \in R'_a$ iff $(s, t) \in R_a$ and
  1. (1) $i=0$, $j=0$, and $(M, s) \models h_a$; or
  2. (2) $i=0$, $j=1$, and $(M, s) \not\models h_a$; or
  3. (3) $i=1$ and $j=1$;
- $(s, 0) \in V'(p)$ iff $s \in V(p)$, and $(s, 1) \in V'(p)$ iff $s \in V(p)$;
- if $a \in A$ then
  1. $(s, 0) \in V'(h_a)$ iff $s \in V(h_a)$, and
  2. $(s, 1) \in V'(h_a)$ iff $s \in V(h_a)$;
- if $a \notin A$ then
  1. $(s, 0) \in V'(h_a)$, and
  2. $(s, 1) \in V'(h_a)$ iff $s \in V(h_a)$.

In the case of the singleton attention assignment $+a$, agent a will now pay attention in the 0-copy of the initial model and may or may not be paying attention in the 1-copy (that copies the prior information state). If another agent b was already paying attention he will now know that a is now paying attention (the first item of the clause for agents not paying attention, above, wherein arrows point to other 0-worlds); else his knowledge of a's attention span is as before (the second item of the clause for agents not paying attention). The only factual information change takes place in the 0-copy, and only for h_a (this is the part $(s, 0) \in V'(h_a)$ in the last item, i.e., for all states s in the 0-copy h_a is now true).

Attention assignment preserves attention introspection. The order in successive change of attention does not matter, but it does not achieve joint attention:

- $\models [+a][+b]\varphi \leftrightarrow [+b][+a]\varphi$;
- $\not\models [+a, b]\varphi \leftrightarrow [+b][+a]\varphi$.

Just as attention-based announcements, an attention assignment correspond to an action model that is a function of (1) who is paying attention and who not before the assignment, and (2) the assignment. The logic to which attention assignment has been added can therefore also easily be axiomatized. (Details of this action model are omitted.)

The validities of the language extended by assignments can be axiomatized by means of reduction axioms for $[+A]$ and $[-A]$. Separate axioms are needed for all inductive cases $[+A]\varphi$ (again, just as for attention-based announcement, as a function of a rather complex action model). For the case of epistemic operators these are:

$$[+A]\mathbf{B}_a\varphi \leftrightarrow \begin{cases} \mathbf{B}_a[+A]\varphi & \text{if } a \in A \\ (h_a \rightarrow \mathbf{B}_a[+A]\varphi) \wedge (\neg h_a \rightarrow \mathbf{B}_a\varphi) & \text{if } a \notin A \end{cases}$$

$$[-A]\mathbf{B}_a\varphi \leftrightarrow \begin{cases} \mathbf{B}_a[-A]\varphi & \text{if } a \in A \\ (h_a \rightarrow \mathbf{B}_a[-A]\varphi) \wedge (\neg h_a \rightarrow \mathbf{B}_a\varphi) & \text{if } a \notin A \end{cases}$$

Finally, let $[[\varphi]]_A$ be the modal operator of private announcement to group A according to [9]. Then $[[\varphi]]_A \psi$ is equivalent to $[+A][-\overline{A}][\varphi]\psi$, where $\overline{A}$ is $AGT\backslash A$. It follows that the public announcement of φ can be captured in our logic by $[+AGT][\varphi]$.

7 Comparison and Further Research

Our proposal is related to several other logics in the DEL literature: arrow update logic [12], wherein a simple dynamic operator can have a large action model equivalent; reasoning about perception [18]; reasoning about perceptual beliefs [11]; reasoning about visually oriented agents [3]; action languages [1].

We plan to study the extension of attention-based announcement logic with common belief that we have sketched in Section 5. On a similar setting we intend to model trust-based announcement logic.

Acknowledgements. We thank the LORI reviewers for their comments and high praise. Hans van Ditmarsch is also affiliated to IMSc, Chennai, as research associate. We acknowledge support from European Research Council grant EPS 313360. Andreas Herzig and Emiliano Lorini acknowledge support of the the ANR project EmoTES.

References

1. Aucher, G.: BMS revisited. In: Proc. of 12th TARK, pp. 24–33 (2009)
2. Aucher, G., Schwarzentruber, F.: On the complexity of dynamic epistemic logic. In: Proc. of 13th TARK (2013)
3. Balbiani, P., Gasquet, O., Schwarzentruber, F.: Agents that look at one another. Logic Journal of IGPL (2012)
4. Balbiani, P., van Ditmarsch, H., Herzig, A., De Lima, T.: Tableaux for public announcement logic. Journal of Logic and Computation 20(1), 55–76 (2010)
5. Balbiani, P., van Ditmarsch, H., Herzig, A., de Lima, T.: Some truths are best left unsaid. In: Advances in Modal Logic (AiML). College Publications (2012)
6. Baltag, A., Moss, L.S., Solecki, S.: The logic of public announcements, common knowledge, and private suspicions. In: Proc. of 7th TARK, pp. 43–56 (1998)
7. Clark, H.H., Marshall, C.R.: Definite reference and mutual knowledge. In: Elements of Discourse Understanding. Cambridge University Press (1981)
8. Fagin, R., Halpern, J.Y.: Belief, awareness, and limited reasoning. Artificial Intelligence 34(1), 39–76 (1987)
9. Gerbrandy, J.D.: Bisimulations on Planet Kripke. PhD thesis, University of Amsterdam, ILLC Dissertation Series DS-1999-01 (1999)
10. Halpern, J.Y., Moses, Y.: A guide to completeness and complexity for modal logics of knowledge and belief. Artificial Intelligence 54, 319–379 (1992)
11. Herzig, A., Lorini, E.: A modal logic of perceptual belief. In: Lihoreau, F., Rebuschi, M. (eds.) Epistemology, Context and Formalism. Synthese Library. Springer (2013)
12. Kooi, B., Renne, B.: Arrow update logic. The Review of Symbolic Logic 4(04), 536–559 (2011)

13. Lorini, E., Tummolini, L., Herzig, A.: Establishing mutual beliefs by joint attention: towards a formal model of public events. In: Proc. of CogSci, pp. 1325–1330. Lawrence Erlbaum Associates (2005)
14. Plaza, J.A.: Logics of public communications. In: Proc. of the 4th ISMIS, pp. 201–216. Oak Ridge National Laboratory (1989)
15. Tommasello, M.: Joint attention as social cognition. In: Moore, C., Dunham, P. (eds.) Joint Attention: its Origins and Role in Development, pp. 103–130. Lawrence Erlbaum Associates (1995)
16. van Benthem, J., van Eijck, J., Kooi, B.: Logics of communication and change. Information and Computation 204(11), 1620–1662 (2006)
17. van Ditmarsch, H., van der Hoek, W., Kooi, B.: Dynamic epistemic logic with assignment. In: Proc. of 4th AAMAS, pp. 141–148. ACM (2005)
18. van Eijck, J.: Perception and change in update logic. In: van Eijck, J., Verbrugge, R. (eds.) Games, Actions and Social Software 2010. LNCS, vol. 7010, pp. 119–140. Springer, Heidelberg (2012)

Annex: Tableaux and Complexity

Let $\mathfrak{Lab}$ be a countable set of labels designed to represent worlds of the epistemic model (M, w). Our tableau method manipulates terms that we call *tableau terms*. They are of the following kind:

- $(\sigma\ \Sigma\ \varphi)$ where $\sigma \in \mathfrak{Lab}$ is a symbol (that represents a world in the initial model) and Σ is a sequence of formulas (where $[]$ denotes the empty list). This term means that φ is true after the announcements of the formulas of the sequence Σ (in that order) in the world denoted by σ;
- $(\sigma\ \Sigma\ \checkmark)$ means that the sequence Σ is executable in σ;
- $(\sigma\ \Sigma\ \otimes)$ means that the sequence Σ is not executable in the world denoted by σ;
- $(\sigma R_a \sigma_1)$ means that the world denoted by σ is linked by R_a to the world denoted by σ_1;
- $\perp$ denotes an inconsistency.

A *tableau rule* is represented by a *numerator* $\mathcal{N}$ above a line and a finite list of *denominators* $\mathcal{D}_1, \ldots, \mathcal{D}_k$ below this line, separated by vertical bars:

$$\frac{\mathcal{N}}{\mathcal{D}_1\ |\ \ldots\ |\ \mathcal{D}_k}$$

The numerator and the denominators are finite sets of tableau terms.

A *tableau tree* is a finite tree with a set of tableau terms at each node. A rule with numerator $\mathcal{N}$ is *applicable* to a node carrying a set Γ, if Γ contains an instance of $\mathcal{N}$. If no rule is applicable, Γ is said to be *saturated*. We call a node σ an *end node*, if the set of formulas Γ it carries is saturated or if $\perp \in \Gamma$. The tableau tree is extended as follows:

1. Choose a leaf node n carrying Γ where n is not an end node, and choose a rule ρ applicable to n.

2. (a) If ρ has only one denominator, add the appropriate instantiation to Γ.
   (b) If ρ has k denominators with $k>1$, create k successor nodes for n, where each successor i carries the union of Γ with an appropriate instantiation of denominator $\mathcal{D}_i$.

A branch in a tableau tree is a path from the root to an end node. A branch is *closed* if its end node contains $\bot$, otherwise it is *open*. A tableau tree is *closed* if all its branches are closed, otherwise it is *open*. The *tableau tree for a formula* $\varphi \in \mathcal{L}$ is the tableau tree obtained from the root $\{(\sigma_0 \,[\,]\, \varphi)\}$ when all leaves are end nodes.

The tableau rules are depicted in Figure 3. They contain the classical Boolean rules $(\wedge)$, $(\neg\neg)$ and a non-deterministic rule $(\neg\wedge)$ handling disjunctions. The rule $(\bot)$ makes the current execution fail. The rules $(\leftarrow_p)$ and $(\leftarrow_{\neg p})$ correspond to the fact that valuations are not changed by announcements. The rule *(hear)* decides non-deterministically for all atomic propositions h_a whether they are true or false. Note that this is a non-analytic rule: the formulas in its denominator are not necessarily subformulas of the input formula. Depending on the value of h_a, there are two versions of the rule for $\mathbf{B}_a$ and for $\neg\mathbf{B}_a$. The rules $(\checkmark)$, $(\otimes)$, $(clash_{\checkmark,\otimes})$ and $(\epsilon_{\otimes})$ deal with executability of the sequence Σ.

Proposition 5 (Soundness and Completeness of the Tableau Method).
A $\mathcal{L}$ formula φ is satisfiable iff there exists an open branch for φ.

Proof. $\boxed{\Leftarrow}$ If the formula φ is satisfiable, there exists a pointed model (M, w) such that $M, w \models \varphi$. We use the model M and the updated models from M and the announcement in φ as an oracle to guide the execution of the tableau method yielding to an open branch.

$\boxed{\Rightarrow}$ Given an open branch, we construct a model M where worlds are the nodes σ, relations are inferred from terms of the form $(\sigma \, R_a \, \sigma_1)$ and valuations are inferred from terms of the form $(\sigma \, \epsilon \, p)$ and $(\sigma \, \epsilon \, \neg p)$. We prove by induction over Σ, ψ that $(\sigma \, \Sigma \, \psi)$ is in the branch iff $M^\Sigma, \sigma \models \psi$, where M^Σ is the model obtained by updating M by the sequence Σ.

Proposition 6. *Satisfiability of $\mathcal{L}$ formulas in the class of K_n models is PSPACE-complete.*

Proof. As explained in [10], a tableau method leads to a PSPACE procedure if we can apply the rules by only keeping in memory the content of a branch. In our case the argument is essentially the same (see also [4]): we only keep in memory the information concerning the current node and its path to the root node, in order to be able to backtrack. We implicitly restrict the applicability of the *(hear)* rule to those h_a such that a occurs in the input formula. It is PSPACE-hard because we may reduce polynomially the satisfiability problem for K_n (the multi-agent version of the minimal modal logic K).

$$\frac{(\sigma\ \Sigma\ \varphi\wedge\psi)}{\begin{array}{c}(\sigma\ \Sigma\ \varphi)\\(\sigma\ \Sigma\ \psi)\end{array}}\ (\wedge)$$

$$\frac{(\sigma\ \Sigma\ \neg\neg\varphi)}{(\sigma\ \Sigma\ \varphi)}\ (\neg\neg)$$

$$\frac{(\sigma\ \Sigma\ \neg(\varphi\wedge\psi))}{(\sigma\ \Sigma\ \neg\varphi)\mid(\sigma\ \Sigma\ \neg\psi)}\ (\neg\wedge)$$

$$\frac{(\sigma\ \Sigma\ p)(\sigma\ \Sigma\ \neg p)}{\bot}\ (\bot)$$

$$\frac{(\sigma\ \Sigma\ [\varphi]\psi)}{(\sigma\ \Sigma\ ::\ \varphi\ \psi)}\ ([\varphi])$$

$$\frac{(\sigma\ \Sigma\ \neg[\varphi]\psi)}{(\sigma\ \Sigma\ :\ \varphi\ \neg\psi)}\ (\neg[\varphi])$$

$$\frac{(\sigma\ \Sigma\ p)}{(\sigma\ []\ p)}\ (\leftarrow_p)$$

$$\frac{(\sigma\ \Sigma\ \neg p)}{(\sigma\ []\ \neg p)}\ (\leftarrow_{\neg p})$$

$$\frac{(\sigma\ \Sigma\ ::\ \varphi\ \checkmark)}{\begin{array}{c}(\sigma\ \Sigma\ \varphi)\\(\sigma\ \Sigma\ \checkmark)\end{array}}\ (\checkmark)$$

$$\frac{(\sigma\ \Sigma\ ::\ \varphi\ \ \otimes)}{\begin{array}{c}(\sigma\ \Sigma\ \checkmark)\\(\sigma\ \Sigma\ \neg\varphi)\end{array}\Big|(\sigma\ \Sigma\ \otimes)}\ (\otimes)$$

$$\frac{(\sigma\ \Sigma\ \otimes)(\sigma\ \Sigma\ \checkmark)}{\bot}\ (clash_{\checkmark,\otimes})$$

$$\frac{(\sigma\ []\ \otimes)}{\bot}\ ([]_\otimes)$$

$$\frac{\begin{array}{c}(\sigma\ \Sigma\ \mathbf{B}_a\varphi)(\sigma\ []\ h_a)\\(\sigma\ R_a\ \sigma_1)\end{array}}{\begin{array}{c}(\sigma_1\ \Sigma\ \checkmark)\\(\sigma_1\ \Sigma\ \varphi)\end{array}\Big|(\sigma_1\ \Sigma\ \otimes)}\ (\mathbf{B}_a)$$

$$\frac{\begin{array}{c}(\sigma\ \Sigma\ \mathbf{B}_a\varphi)(\sigma\ []\ \neg h_a)\\(\sigma\ R_a\ \sigma_1)\end{array}}{(\sigma_1\ []\ \varphi)}\ (\mathbf{B}_a)$$

$$\frac{(\sigma\ []\ h_a)(\sigma\ \Sigma\ \neg\mathbf{B}_a\varphi)}{\begin{array}{c}(\sigma\ R_a\ \sigma_{\text{new}})\\(\sigma_{\text{new}}\ \Sigma\ \checkmark)\\(\sigma_{\text{new}}\ \Sigma\ \neg\varphi)\end{array}}\ (\neg\mathbf{B}_a)$$

$$\frac{(\sigma\ []\ \neg h_a)(\sigma\ \Sigma\ \neg\mathbf{B}_a\varphi)}{\begin{array}{c}(\sigma\ R_a\ \sigma_{\text{new}})\\(\sigma_{\text{new}}\ []\ \neg\varphi)\end{array}}\ (\neg\mathbf{B}_a)$$

$$\frac{}{(\sigma\ []\ h_a)\big|(\sigma\ []\ \neg h_a)}\ (hear)$$

Fig. 3. Tableau rules

An Offer You Cannot Refuse:
Obtaining Efficiency and Fairness in Preplay
Negotiation Games with Conditional Offers

Valentin Goranko[1] and Paolo Turrini[2]

[1] Technical University of Denmark and University of Johannesburg
vfgo@imm.dtu.dk
[2] Imperial College London
p.turrini@imperial.ac.uk

Abstract. We study a recently introduced extension of normal form games with a phase before the actual play of the game, where each player can make binding offers for payments of utility to the other players after the play of the game, contingent on the recipient playing the strategy indicated in the offer. Such offers transform the payoff matrix of the original game and allow for some degree of cooperation between rational players while preserving the non-cooperative nature of the game. We focus on 2-player negotiations games arising in the preplay phase when offers for payments are made conditional on a suggested matching offer of the same kind being made in return by the receiver. We study and analyze such bargaining games, obtain results describing their possible solutions and discuss the degrees of efficiency and fairness that can be achieved in such negotiation process depending on whether time is valuable or not.

1 Introduction

It is well-known that many solution concepts in non-cooperative games can induce outcomes that are far from being Pareto optimal. Some studies have considered various forms of preplay interaction between the players, aiming at improving the resulting payoffs. Such interaction range from cheap talk to signing contracts. Cheap talk affects neither the payoffs nor the non-cooperative rational behavior of the players, whereas by signing contracts the players pre-determine the outcome of the resulting normal form game, essentially playing as a coalition.

The problem in the focus of the present study is: *what can rational players achieve by means of interactive negotiations prior to playing a non-cooperative game?* In [3] we consider a version of preplay interaction between players, whereby they try to negotiate better outcomes in the forthcoming game by means of exchanging offers for additional (side) payments of utility conditional on the recipient playing the strategy indicated in the offer. More precisely, before the actual game is played any player, say A (Ann), can make a binding offer to any other player, say B (Bob), to pay him, after the end of the game, an explicitly declared amount of utility δ if B plays a strategy s specified in the offer by A. Such an offer

D. Grossi, O. Roy, and H. Huang (Eds.): LORI 2013, LNCS 8196, pp. 110–123, 2013.

effects a simple transformation of the payoff matrix of the game, by transferring the declared amount from the payoff of A to the payoff of B in every outcome corresponding to B playing δ. Players can exchange multiple such offers in attempt to transform the game into one where their expected payoffs, assuming rational behaviour of the other players according to a commonly adopted solution concept, would be better than those expected from the original game. Furthermore, players can make such offers *conditional on receiving desired matching offers*, which can be in turn accepted or rejected. Thus, a whole preplay negotiation phase emerges before the original normal form game is actually played, and it can be regarded as another game in which players bargain towards a mutually optimal transformation of the former.

This paper studies 2-player negotiations games arising in the preplay phase when conditional offers for payments are made on a suggested matching offer of the same kind made in return by the receiver. We study and analyze such bargaining games, obtain results describing their possible solutions and discuss the degrees of efficiency, in the sense of Pareto optimality of the resulting distribution, and fairness, in the sense of equitability of the resulting distribution, that can be achieved in such negotiation process in the cases where time is valuable or not. We focus on the ideas and intuitions behind preplay negotiation games with conditional offers, while, for space reasons, some core technical results are stated with brief proof sketches. Full proofs of those results are available in [4].

The paper is organized as follows. In Section 2 we introduce the preliminary game-theoretical notions. In Section 3 we discuss preplay offers in two-player normal form games and define the framework of preplay negotiation games. The analysis and main results are in Section 4. In Section 5 we discuss some related work and end with concluding remarks and further agenda in Section 6.

2   Preliminaries

Let $\mathcal{G} = (\{A, B\}, \{\Sigma_A, \Sigma_B\}, u)$ be a two player normal form game (henceforth NFG), where $\{A, B\}$ is a set of players, $\{\Sigma_A, \Sigma_B\}$ a set of finitely many strategies for each player and $u : \{A, B\} \times \Sigma_A \times \Sigma_B \to \mathbb{R}$ is a **payoff function** assigning to each player a utility for each strategy profile. The game is played by each player i choosing a strategy from Σ_i. The resulting strategy profile σ is the **outcome** of the play and $u_i(\sigma) = u(i, \sigma)$ is the associated payoff for i. An outcome of a play of the game $\mathcal{G}$ is called **maximal** if it is a Pareto optimal outcome with the highest sum of the payoffs of all players. Let $\mathbf{G_N}$ be the set of all normal form games for a set of players N. By **solution concept for $\mathbf{G_N}$** we mean a map $\mathfrak{S}$ that associates with each $\mathcal{G} \in \mathbf{G_N}$ a non-empty set $\mathfrak{S}(\mathcal{G})$ of outcomes of $\mathcal{G}$, called the $\mathfrak{S}$-**solution of the game**. For a player i, we denote $\mathfrak{S}_i$ the restriction of the mapping $\mathfrak{S}$ to i returning only the strategies of player i consistent with $\mathfrak{S}$, i.e., $\mathfrak{S}_i(\mathcal{G}) = \{\sigma_i \in \Sigma_i \mid \sigma \in \mathfrak{S}(\mathcal{G})\}$. We also use $-i$ for any $i \in \{A, B\}$ to denote i's opponent. In this work we do not commit to a specific solution concept for the normal form games but we assume that the one adopted by the players satisfies the necessary condition that *every outcome in any solution prescribed* *by that solution concept must survive iterated elimination of strictly dominated strategies.* We call such solution concepts **acceptable**.

Games for which the solution concept $\mathfrak{S}$ yileds a set of payoff equivalent outcomes will be called **uniformly $\mathfrak{S}$-solvable**. Games for which $\mathfrak{S}$ yileds only maximal outcomes will be called **optimally $\mathfrak{S}$-solvable**. Games that are both **optimally $\mathfrak{S}$-solvable** and **uniformly $\mathfrak{S}$-solvable** will be called **perfectly $\mathfrak{S}$-solvable**. $\mathfrak{S}$-solvable games for which the solution concept $\mathfrak{S}$ yields a single outcome will be called **$\mathfrak{S}$-solved**. For instance, every game with a strongly dominating strategy profile is $\mathfrak{S}$-solved for any acceptable solution concept $\mathfrak{S}$. Ideally, preplay negotiation games should transform the starting NFG into a perfectly $\mathfrak{S}$-solved, or at least perfectly $\mathfrak{S}$-solvable, one.

It is necessary for the preplay negotiation phase for each player to have an **expected value** of any NFG that can be played. For sake of definiteness we adopt here a conservative, risk-averse approach and will define for every acceptable solution concept $\mathfrak{S}$, game $\mathcal{G}$ and a player i, the expected value of $\mathcal{G}$ for i relative to the solution concept $\mathfrak{S}$ to be:

$$\mathsf{v}_i^{\mathfrak{S}}(\mathcal{G}) = \max_{\sigma_i \in \mathfrak{S}_i(\mathcal{G})} \min_{\sigma_{-i} \in \mathfrak{S}_{-i}(\mathcal{G})} u_i(\sigma)$$

We note that our further analysis does not depend essentially on this particular assumption; any other realistic notion of expected value of a NFG would yield similar results.

3   Two-Player Normal Form Games with Preplay Offers

3.1   Preplay Offers

Following [3] we use the notation $A \xrightarrow{\delta/\sigma_B} B$ to denote an offer made by player A to pay an amount δ to player B after the play of the game if player B plays strategy σ_B. Any preplay offer by A to B is assumed *binding for A*, upon B playing the specified strategy. However, such offer *does not* create any obligation for B, who is still at liberty to choose his strategy when the game is actually played. In particular, after her offer A does not know in advance whether B will play the desired by A strategy σ_B, and thus make use of the offer or not. The key observation applying here is that *after any preplay offer the game remains a non-cooperative normal form game, only the payoff matrix changes according to the offer.* We now illustrate preplay offers in a well-known scenario.

Motivating example: Prisoners' Dilemma. Consider a version of the Prisoner's Dilemma (PD) game in Figure 1, left. The only Nash Equilibrium (NE) of the game is (D, D), yielding a payoff of $(1, 1)$. Now, suppose player *Row* makes the offer *Row* $\xrightarrow{2/C}$ *Column* to the player *Column*. That offer transforms the game by transferring 2 utils from the payoff of *Row* to the payoff of *Column* in every entry of the column where *Column* plays C, as in Figure 1, middle.

	C	D
C	3,3	0,4
D	4,0	1,1

$Row \xrightarrow{2/C} Column$

	C	D
C	1,5	0,4
D	2,2	1,1

$Column \xrightarrow{2/C} Row$

	C	D
C	3,3	2,2
D	2,2	1,1

Fig. 1. From left to right: A Prisoner's Dilemma game; the game after the first offer by player Row; the game after the second offer by player Column

In this game player *Row* still has the incentive to play D, which strictly dominates C for him, but the dominant strategy for *Column* now is C, and thus the only NE is (D, C) with payoff $(2, 2)$ – strictly dominating the original payoff $(1, 1)$. Of course, *Column* can now realize that if player *Row* is to cooperate, an extra incentive is needed. That incentive *can be created* by an offer $Column \xrightarrow{2/C} Row$, that is, if *Column*, too, makes an offer to *Row* to pay him 2 utils after the game, if player *Row* cooperates. Then the game transforms as in Figure 1, right. In this game, the only Nash equilibrium is (C, C) with payoff $(3, 3)$, which is also Pareto optimal. Note that this is the same payoff for (C, C) as in the original PD game, but now both players have created incentives for each other to cooperate, thus escaping[1] from the original bad Nash equilibrium (D, D).

3.2 Conditional Offers

Consider now an instance of the Battle of the Sexes (Figure 2, left), with the column player called *Him* and the row player *Her*. It has two Nash equilibria: one preferred by *Her*: (*Ballet*, *Ballet*), and the other – by *Him*: (*Soccer*, *Soccer*).

Her \ Him	Ballet	Soccer
Ballet	5,3	1,1
Soccer	0,0	3,5

Her \ Him	Ballet	Soccer
Ballet	5,3	1,1
Soccer	2,−2	5,3

Her \ Him	Ballet	Soccer
Ballet	5,3	0,2
Soccer	2,−2	4,4

Fig. 2. From left to right: A Battle of the Sexes game; the game transformed by the offer $Him \xrightarrow{2/Soc} Her$ favouring *Her*; further transformed by an offer $Her \xrightarrow{1/Soc} Him$

An offer $Him \xrightarrow{2/Soccer} Her$[2] would transforms the game to one in Figure 2, middle. By doing so, *Him* makes the equilibrium (*Soccer*,*Soccer*) equally beneficial for *Her* as (*Ballet*,*Ballet*) and also sends the clear message that he intends to play *Soccer*, thus essentially breaking the coordination problem and deciding the game. However, this offer comes at a cost for *Him* and puts him

---

[1] Clearly, preplay offers can only work in case when at least part of the received payoff can actually be transferred from a player to another. They obviously cannot apply to scenarios such as the original PD, where one prisoner cannot offer to the other to stay in prison for him, even if they could communicate before the play.

[2] which can be made, for example, in the form of invitation to a dinner in a luxury restaurant after the soccer match, if *Her* pitches up there.

in a relatively disadvantaged position: with respect to the original game, he is worse off in one of the two Nash equilibria and he is not better off in the other.

The loss and disadvantage incurred by *Him* in the example above could be partly neutralized by an offer *Her* $\xrightarrow{1/Soccer}$ *Him*, which transforms the game to the one in Figure 2, right.

But, of course, *Her* has no incentive to make such an offer to *Him* in the middle game on Figure 2. So, the only realistic way for *Him* to force such matching offer by *Her* is to make his offer *Him* $\xrightarrow{2/Soccer}$ *Her conditional on Her making to Him the matching offer Her* $\xrightarrow{1/Soccer}$ *Him*[3]. This conditional offer is denoted hereafter as *Him* $\xrightarrow{2/Soccer \mid 1/Soccer}$ *Her*. The effect is that players reach an equitable redistribution of the payoffs in the expected (maxmin) outcome.

In practice, a conditional offer $A \xrightarrow{\alpha/\sigma_B \mid \beta/\rho_A} B$ enables the offering player to *suggest* a transformation of the game $\mathcal{G}$ into a game $\mathcal{G}(A \xrightarrow{\alpha/\sigma_B \mid \beta/\rho_A} B)$ that is updated according to the offer.

In other words, a suggested transformation updates the original game into a new game where only the payoff vectors change, according to the conditional offer that is made. At each profile, each player collects the positive reward received in the part of the conditional offer consistent with the profile, subtracting in the same fashion the payments given.

There are two possible responses to a conditional offer $A \xrightarrow{\alpha/\sigma_B \mid \beta/\rho_A} B$: it can be *accepted* or *rejected* by the receiving player. If rejected, the offer is immediately cancelled and does not commit any of the players to any payment, and therefore it does not induce any transformation of the game matrix. If accepted, the *actual transformation* induced by the offer is the suggested transformation defined above.

For space reasons we do not consider the possibility of withdrawing previously made offers, which is treated in [4].

3.3   Preplay Negotiation Games

Similarly to [6], our setting for normal form games with preplay offers begins with a given 'starting' normal form game $\mathcal{G}$ and consists of two phases:

- A *preplay negotiation phase*, where players negotiate on how to transform the game $\mathcal{G}$ by making offers, accepting or rejecting conditional offers they receive.
- An *actual play* phase where, after having agreed on some transformation X in the previous phase, the players play the game $\mathcal{G}$ updated with X.

We will call the resulting games *preplay negotiation games*.

In order to define a PNG we need to introduce some preliminaries: moves, histories and plays. Depending on some of the optional assumptions, the players

---

[3] For instance, *Her* could offer to bring a coolbox with cold beer and chips to the stadium if *Him* comes there.

can have several possible moves in a PNG. Let us consider the case where conditional offers are allowed. Then the moves available to the player whose turn is to play depend on whether or not he/she has received since his/her previous move any conditional offers. If so, we say that the player has **pending conditional offers**. The possible moves of the player in turn are as follows.

1. A player who has no pending conditional offers can:
   (a) *Make an offer* (conditional or not).
   (b) *Pass.*
2. A player who has pending conditional offers, can for each of them:
   (a) *Accept the pending offer* , and then make an offer of his/her own or pass.
   (b) *Reject the pending offer*, and then make an offer of his/her own or pass.

The PNG is over when all players have passed at their last move, or a player has opted out.

We now define the notion of a **history** in a PNG as a sequence of moves by the players who take their turns according to an externally set protocol (a detailed discussion of the possible external protocols is provided in [3]). Every finite history in such a game is associated with the current NFG: the result of the transformation of the starting game by all offers that are so far made and accepted. The current NFG of the empty history is the input NFG of the PNG. A **play** of a PNG is any finite history at the end of which the preplay negotiations game is over, or any infinite history.

In order to eventually define realistic solution concepts for preplay negotiations games we need to endow every history in such games with a value for every player. Intuitively, **the value of a history** is the value for the player of the current NFG associated with that history, in the case of non-valuable time, and the same value accordingly discounted in the case of valuable time.

Disagreements. The PNG may terminate if all players pass at some stage, in which case we say that the players have reached agreement, or may go on forever, in which case the players have failed to reach agreement; we call such situation a *(passive) disagreement* and we denote any such infinite history with D. We will not discuss disagreements and their consequences here, but will make the explicit assumption that *any agreement is better for every player than disagreement* in terms of the payoffs, e.g. by assigning payoffs of $-\infty$ in the entire game for each player if the PNG evolves as a disagreement. In [3] we also outline a more flexible and possibly more realistic alternative, whereby players can explicitly express tentative acceptance of the current NFG – the one on which they are currently negotiating by making offers or can terminate the negotiations by explicitly opting out, which would revert to the current game to the currently accepted by everyone NFG.

A **preplay negotiation game** (PNG) starts with an input NFG $\mathcal{G}$ and either ends with a transformed game $\mathcal{G}'$ or goes on forever, which we discuss further. The **outcome of a play of the PNG** is the resulting transformed game $\mathcal{G}'$ in the former case and 'Disagreement' (briefly D) in the latter case. By a **solution of a PNG** we mean *the set of all transformed normal form games that can be* *obtained as outcomes of plays induced by subgame perfect equilibrium strategy profiles in the PNG.* Finally, we say that a strategy in a PNG is **strongly efficient** if the vector of payoffs of the outcome it attains is a redistribution of the vector of payoffs of a maximal outcome.

4   Preplay Negotiations Games with Conditional Offers

First, let us state a useful general result, also valid in the case of many players PNG. An extensive form game is said to have the **One Deviation Property** (ODP) [8, Lemma 98.2] if, in order to check that a strategy profile is a Nash equilibrium in (some subgame of) that game, it suffices to consider the possible profitable deviations of each player not amongst *all* of its strategies (in that subgame), but only amongst the ones differing from the considered profile in the first subsequent move (in that subgame).

Lemma 1. *Every PNG has the One Deviation Property.*

Proof. It is easy to check that a strategy profile of a PNG is a subgame perfect equilibrium if and only if it is a subgame perfect equilibrium of the same PNG without disagreement histories as, notice, strategies leading to disagreement are cannot be used as credible threats. But, a PNG without disagreement histories is an extensive game of perfect information and finite horizon. By [8, Lemma 98.2] the PNG has the one deviation property.

4.1   The Case of Non-valuable Time

The value for a player of a history in a PNG is the value for the player of the current NFG associated with that history. When time is not valuable players assign the same value to the NFG associated with the current moment and the same game associated with any other moment in the future, which means that players can afford delaying offers at no extra cost.

   To analyze equilibrium strategies of PNG when time is not valuable we consider so called **stationary acceptance strategies** where players have a minimal acceptance threshold d and a minimal passing threshold $d' \geq d$ (both of which may vary among the players).

Proposition 1. *Every subgame perfect equilibrium strategy profile of a two-player PNG with non-valuable time consisting of stationary acceptance strategies is strongly efficient.*

Proof. Suppose not. Let d^- be a vector of expected values that is not the redistribution of a maximal outcome of the starting game associated to some subgame perfect equilibrium strategy profile. We know that such strategy profile yields a history h that ends with: 1) the proposal of d^-; 2) the acceptance of that proposal; 3) a pass; 4) a pass. Consider now some redistribution d^* of a maximal outcome where both players get more than in d^- and the history h where the last four steps are substituted by the following ones: 1) the proposal of d^*; 2) the acceptance of that proposal; 3) pass; 4) pass. By stationarity of strategies and the ODP, the player moving at step 1) is better off deviating from d^- and instead proposing d^*: a contradiction.

The condition of stationarity of acceptance strategies is needed if we want to avoid equilibrium strategies that lead to inefficiency. The example below provides a detailed instance of such cases.

Example 1 (Attaining inefficiency). Consider the following starting NFG.

$$
\begin{array}{c|c|c|}
 & L & R \\
\hline
U & 2,2 & 4,3 \\
\hline
D & 3,3 & 2,2 \\
\hline
\end{array}
$$

As there are no dominant strategy equilibria, there are acceptable solution concepts assigning 2 to each player.

We now construct a strategy profile of the PNG starting from that game, such that: (i) it is a SPE strategy profile and (ii) it attains an inefficient outcome.

1. At the root node player A *proposes* outcome (D, L) with payoff distribution $(3, 3)$ — i.e., makes a conditional offer where D, L is dominant strategy equilibrium and yields the payoff vector $(3, 3)$.
2. After such proposal player B accepts. However, if A had made a different offer (so, off the equilibrium path) B would reject and keep proposing outcome (U, R) with distribution of 5 for him and 2 for A and accepting (and passing on) maximal outcomes guaranteeing him at least 5. A, on the other hand, would not have better option than proposing the same distribution (5 for B and 2 for her) and accepting only maximal outcomes guaranteeing her at least 2. Notice that once they enter this subgame neither A nor B can profitably deviate from such distribution.
3. If, however, B did not accept the $(3, 3)$ deal then A would keep proposing outcome (U, R) with a redistribution of $(5, 2)$ (5 for her, 2 for him) and accepting at least that much. B on the other hand would also stick to the same distribution, accepting at least 2. Again, no player can profitably deviate from this stationary strategy profile starting from B's rejection.
4. After player B has accepted the deal $(3, 3)$, then A passes. If A did not pass, player B would go back to his $(2, 5)$ redistribution threat. Likewise with the next round. That eventually leads to the inefficient outcome $(3, 3)$.

It is easy to check that the strategy profile described above is a subgame perfect equilibrium. No player can at any point deviate profitably by proposing the outcome (U, L) with dominating payoff distribution, e.g., $(3.5, 3.5)$.

In general, in PNG with non-valuable time every redistribution of a maximal outcome can be attained as a solution.

Proposition 2. *Let $\mathcal{E}$ be PNG with non-valuable time starting from a NFG $\mathcal{G}$ and let $d = (x_A, x_B)$ be any redistribution of a maximal outcome of the starting NFG. The following strategy profile $\sigma = (\sigma_A, \sigma_B)$ is a subgame perfect equilbrium:*

For each player $i \in \{A, B\}$:
- *if i is the first player to move, he proposes a transformation of $\mathcal{G}$ where the vector of expected values in the transformed game is d;*
- *when i can make an offer and the previously made offer has not been accepted, he proposes a transformation of the current NFG where the vector of expected values in the transformed game is d;*
- *when i can make an offer and the previously made offer has been accepted, he passes;*
- *when i has a pending offer of a suggested transformation where the vector of expected values in the transformed game is d', he accepts it if and only if $x'_i \geq x_i$, and rejects it otherwise;*
- *when i can pass and the other player has just passed, he passes;*
- *when i can pass and the other player has not just passed, he proposes d;*
- *when i has just accepted a proposal he passes;*

Proof. We have to show that there is no subgame where a player i can profitably deviate from this strategy at its root. By Lemma 1 we can restrict ourselves to considering only first move deviations to the above described strategy.

Suppose the player has a pending offer that induces a transformation of the current NFG where the vector of expected values is d^*. If she accepts it then the outcome will be d^*, due to the definition of the strategy profile; if she rejects it, it will be the starting offer d. And she will accept if and only if she will get more from d^* than from d. So the acceptance component is optimal. For the remaining cases, if player i deviates from the prescribed strategy, due to the construction of the strategy and Lemma 1, the vector of payoffs associated to the outcome of $\mathcal{E}$ will be d anyway.

As a consequence of the previous proposition we obtain:

Corollary 1. *The game associated to the outcome of a subgame perfect equilibrium strategy profile consisting of stationary acceptance strategies in a two-player PNG with non-valuable time is optimally solvable.*

In summary, our analysis of two-player PNG with non-valuable time shows that efficiency can be attained when conditional offers are allowed and stationary acceptance strategies are followed. Indeed, any redistribution of the vector of payoffs of a maximal outcome can be made the unique solution of the final NFG by such SPE strategies. However, non-stationary acceptance strategies may lead to inefficient equilibria, as illustrated in the comment to Proposition 1.

To sum up, while SPE strategies in a two-player PNG can attain efficiency, some important issues are still remaining:

– SPE strategies with non-stationary acceptance need not be strongly efficient.

- players can keep making unfeasible moves as a part of a SPE strategy, i.e., there are forms of equilibria where some players strictly decrease their expected payoff with respect to the original game;
- even strongly efficient strategies do not always yield perfectly solved games, as there is no notion of *most fair* redistribution of the payoff vectors in the solution of the original game.

Thus, when time is of no value, even the option of making conditional offers is not sufficient to guarantee that fair and efficient outcomes are ever reached.

4.2 The Case of Valuable Time

We will see here that when time is of value all the problems mentioned above can be at least partially solved. To impose value on time we introduce, for each player i, a *payoff discounting factor* $\delta_i \in (0,1)$ applied at every round of the PNG *associated to offers that are made* to his payoffs. These factors measure the players' impatience, i.e., how much they value time, and reduce the payoffs accordingly as time goes by. The general intuition in this case, which we will justify further, is that for the sake of time efficiency, in a SPE strategy profile:

1. If any player is ever going to make an offer, she would never make any earlier offer that gives her, if accepted, a lesser value of the resulting game.
2. If any player is ever going to accept a given offer (or any other offer, at least as good for her) she should do it the first time when she receives such offer.

To facilitate the compatison we bargaining games we however restrict players' possible strategies, imposing some additional constraints:

- every game associated with a history of a PNG does not have outcomes *in the solution* (but, possibly elsewhere) that assign negative utility to players. Notice, that we do allow payoff vectors consisting of negative reals to be present in the game matrix, only we do not allow such vectors to be associated to outcomes in the solution. This constraint that we impose has several practical consequences:
  - players' expected payoffs *decrease* in time, i.e.,the discounting factor δ has always a negative effect on the expected payoff.
  - players can make offers that redistribute the payoff vectors associated with outcomes in the solution, leaving some nonnegative amount to each player and some strictly positive amount to some.
- the expected payoff of each player at any disagreement history can be assumed 0.

We will use the following notational conventions:

- (x,t) denotes the payoff vector x at time t, where each component x_i is discounted by δ_i^t; $(x,t)_i$ denotes the payoff of player i in vector x at time t.

- $\mathcal{G}_X$ will denote the set of all possible redistributions of payoffs of outcomes in a NFG $\mathcal{G}$ that assign nonnegative payoffs to all players. This set is compact, but generally not connected, as in the bargaining games of [4]. However, it is a finite union of compact and connected sets, and that will suffice to generalize the results from [4] that we need.

The following properties of every 2-person PNG with valuable time starting from a given NFG $\mathcal{G}$ are the four fundamental assumptions of Osborne and Rubinstein's bargaining model [8, p.122].

1. For each $x, y \in \mathcal{G}_X$ such that $x \neq y$, if $(x, 0)_i = (y, 0)_i$ then $(x, 0)_{-i} \neq (y, 0)_{-i}$. This holds because the set $\mathcal{G}_X$ is made by payoff vectors and subtracting some payoff to a player means adding it to the other.
2. $(b^i, 1)_{-i} = (b^i, 0)_{-i} = (D)_{-i}$, where b^i is the highest payoff that i obtains in $\mathcal{G}_X$ and $(D)_{-i}$ the payoff for $-i$ in any disagreement history. As b^i is the best agreement for player i it is also the worst one for player $-i$.
3. If x is Pareto optimal amongst the payoff vectors in $\mathcal{G}_X$ then, by definition of $\mathcal{G}_X$, there is no y with $(x, 0)_i \geq (y, 0)_i$ for each $i \in N$. Moreover, x is a redistribution of a maximal outcome in $\mathcal{G}$.
4. There is a unique pair (x^*, y^*) with $x^*, y^* \in \mathcal{G}_X$ such that $(x^*, 1)_A = (y^*, 0)_A$ and $(y^*, 1)_B = (x^*, 0)_B$ and both x^*, y^* are Pareto optimal amongst the payoff vectors in $\mathcal{G}_X$.

The first three statements above are quite straightforward. To see the last one, let $x^* = (x_A^*, x_B^*)$ and $y^* = (y_A^*, y_B^*)$ and let the sum of the payoffs in any maximal outcome in $\mathcal{G}$ be d. Then $(x_A^*, x_B^*, y_A^*, y_B^*)$ is the unique solution of the following, clearly consistent and determined system of equations:
$y_A = \delta_A x_A$, $x_B = \delta_B y_B$, $x_A + x_B = d$, $y_A + y_B = d$.
The solution (see also [8]) is:

$$x_A = d\frac{1 - \delta_B}{1 - \delta_A \delta_B}; \; y_A = \delta_A d\frac{1 - \delta_B}{1 - \delta_A \delta_B}$$

$$x_B = \delta_B d\frac{1 - \delta_A}{1 - \delta_A \delta_B}; \; y_B = d\frac{1 - \delta_A}{1 - \delta_A \delta_B}.$$

Relation with Bargaining Games. In the remaining part of the section we will explicitly view preplay negotiation as a bargaining process on how to play the starting normal form game. Using our observations and assumptions, we can adapt the results from [8] to show that when time is valuable not only all equilibria consisting of stationary acceptance strategies attain efficiency but they also do it by redistributing the payoff vector in relation to players' impatience. Stationary acceptance strategies will be needed to focus only on the maximal connected subspace of the set $\mathcal{G}_X$. To say it with a slogan, while in [8] efficiency and fairness can be obtained in scenarios that resemble the division of a cake, in our setting we prove similar results for a set of cakes, of possibly different size. We extend the efficiency and fairness results obtained in [8] for bargaining games of the type of 'division of a cake' to similar results for somewhat more general bargaining games of the type where players have to choose a cake from a set of cakes, of possibly different sizes and divide it. Our claim, in a nutshell, is that, when players employ stationary acceptance strategies, they immediately choose the largest cake and then bargain on how to divide it.

First, recall that in our framework time passes as new proposals are made. So, from a technical point if the PNG start with a game that is already perfectly solved, the player moving first will not be punished by passing immediately.

Then, without restriction of the generality of our analysis, we can assume a *unique* discounting factor for both players. Indeed, the discount factor of e.g., player A can be made equal to that of B while preserving the relative preferences of A on the set of outcomes by suitably re-scaling the payoffs of A in the input NFG, and therefore the expected value for A of that game; for technical details see [8, p.119] following an idea of Fishburn and Rubinstein quoted there.

Now we are ready to state the main result for this case:

Theorem 1. *Let (x^*, y^*) be the unique pair of payoff vectors defined above. Then, in a PNG with valuable time starting from a NFG $\mathcal{G}$ with a unique discounting factor δ for both players, the strategy of player A in every subgame perfect equilibrium consisting of stationary acceptance strategies satisfies the following (to obtain the strategy for B simply swap x^* and y^*):*

- *if A is the first player to move she 'proposes' outcome x^*, i.e., makes a conditional offer that, if accepted, would update the game into one with is dominant strategy equilibrium yielding the Pareto maximal outcome x^* as payoff vector;*
- *when A has a pending offer y', she accepts it if and only if the payoff she gets in y' is at least as much as in y^*;*
- *when A can pass, she passes if and only if the expected value associated to the proposed game y'_A is at least as much as y^*_A; otherwise she proposes x^*.*

Proof idea To prove the claim we use a variant of the argument given in [8] for bargaining games, summarized as follows.[4] We first show [Step 1] that the best SPE payoff for player A in any subgame $\mathcal{G}'_A$ starting with her proposal and where $\mathcal{G}'$ is the currently accepted game — let us denote it by $M_A(\mathcal{G}'_A)$ — yields the same utility as the worst one — $m_A(\mathcal{G}'_A)$ — which, in turn, is the payoff of A at x^*. The argument for B is symmetric. Then we show [Step 2] that in every SPE the initial proposal is x^*, which is immediately accepted by the other player, followed by each player passing. Finally, we show [Step 3] that the acceptance and the passing conditions are shared by every SPE strategy profile.

To summarize, when time is valuable and players' value of time (impatience) is measured by a vector of discount factors δ the SPEs following stationary acceptance strategies are essentially unique, efficient and redistribute a maximal

---

[4] A full proof can be found in [4].

payoff vector in a *fair* way, depending on players' impatience, viz. in each SPE play, players agree as soon as possible and divide (almost) evenly any of the maximal outcomes in the game. Thus, introducing value of time solves both problems of efficiency and fairness at once.

If a PNG starts with a game that is uniformly solvable but not optimally solvable (i.e. there is space for improvement), the player moving first can improve on the expected outcome of the initial game only on the condition to ensure to her opponent at least his expected value in the initial game.

5   Related Work

The present study has a rich pre-history, related to earlier work on bargaining and various pre-play negotiation procedures, notable examples of which are [2, 5, 7, 9, 10]; see [3] for a broad discussion. Here we only mention the most relevant recent work. To our knowledge, Jackson and Wilkie [6] are the first to have studied arbitrary transfers from a player to a player in a normal form game. Their framework bears essentiial similarities with ours, as it studies a two-stage transformation on a normal form game where players announce transfers of payments between each other on the initial normal form game and then play the updated game. Yet, there are substantial differences with our framework, the most important ones being that in [6] players make positive side payments to other players *conditional on the entire outcome of the game*, by announcing their offers simultaneously, and therefore time and its value do not play a role in the negotiations phase. Ellingsen and Paltseva [1] generalize [6] in several ways. In their framework each player specifies a (possibly negative) transfer to the other players for each (possibly mixed) strategy profile σ and, at the same time, specifies a signing decision for each contract of the other players. The authors show that their more general contracting game always has efficient equilibria. In particular, they show that all efficient outcomes guaranteeing to each player at least as much as the worst Nash-equilibrium payoff in the original game can be attained in some equilibrium. The message conveyed by this stream of works is that efficiency can be reached if the structure of players' offers is complex enough. Instead, we focus on the effects that additional factors in the preplay negotiation game, such as valuable time, withdrawals, and opting out have on attaining outcomes with desirable properties, such as efficiency and fairness.

6   Conclusion

We have analyzed the role and effect of conditional offers in preplay negotiation games under various assumptions concerning players' rationality, their value of time, possibility of revoking previous commitments and opting out from the negotiations. We have shown that when time is of no value efficiency can be attained, provided some coherence in players' behaviour is assumed (Prop. 1). Yet, there are cases in which inefficiency can occur in equilibrium outcomes (comment to Prop. 1). When time is not valuable, the outcomes reached in the PNG can be extremely unfair, even if efficient (Prop. 2). However, when time is valuable, under some natural assumptions players reach efficient and relatively fair outcomes (Theorem 1). The latter result draws an explicit connection between our preplay negotiation games and bargaining games studied in [8].

Some related issues remain still open, in particular the possibility of ruling out the existence of non-stationary acceptance strategies in all SPE profiles of PNGs with valuable time. As for the potential future developments, the framework can be extended in various ways, as also discussed in [3]. In particular, we conjecture that when players can only make *unconditional* offers, not contingent on matching offers by the recipients, the analysis of the preplay negotiations phase changes radically and in a way becomes even more challenging. Its complete analysis is the subject of an ongoing work. Lastly, the analysis of preplay negotiation phase in games with three and more players is substantially more complicated and is one of the main future directions in this research.

Acknowledgements. Valentin Goranko completed his work on this paper during his visit to the Centre International de Mathématiques et Informatique de Toulouse.

Paolo Turrini acknowledges the support of the IEF Marie Curie fellowship "Norms in Action: Designing and Comparing Regulatory Mechanisms for Multi-Agent Systems" (FP7-PEOPLE-2012-IEF, 327424 "NINA").

References

1. Ellingsen, T., Paltseva, E.: Non-cooperative contracting (2011) (submitted) http://www2.hhs.se/personal/ellingsen/pdf/Non-cooperativeContracting5.pdf
2. Farrell, J.: Communication, coordination and nash equilibrium. Economics Letters 27, 209–214 (1998)
3. Goranko, V., Turrini, P.: Non-cooperative games with preplay negotiations (2012), http://arxiv.org/abs/1208.1718 (submitted)
4. Goranko, V., Turrini, P.: Two-player preplay negotiation games with conditional offers. Working paper (2013), http://arxiv.org/abs/1304.2161
5. Guttman, J.M.: Understanding collective action: Matching behavior. American Economic Review 68(2), 251–255 (1978)
6. Jackson, M.O., Wilkie, S.: Endogenous games and mechanisms: Side payments among players. Review of Economic Studies 72(2), 543–566 (2005)
7. Kalai, E.: Preplay negotiations and the prisoner's dilemma. Mathematical Social Sciences 1, 375–379 (1981)
8. Osborne, M., Rubinstein, A.: A course in game theory. MIT Press (1994)
9. Rosenthal, R.W.: Induced outcomes in cooperative normal-form games. Review of Economic Studies, 975. Discussion Paper No. 178. Center for. Math. Studies, Northwestern University
10. Varian, H.R.: Varian. A solution to the problem of externalities when agents are well-informed. American Economic Review 84(5), 1278–1293 (1994)

Sequent Calculi for Multi-modal Logic with Interaction

Norbert Gratzl[*]

Munich Center for Mathematical Philosophy

Abstract. This paper studies Gentzen-style sequent calculi for multi-modal logics with interaction between the modalities. We prove cut elimination and some of its usual corollaries for two such logics: Standard Deontic Logic with the Ought-implies-Can principle, and a non-normal deontic logic where obligation, permissions and abilities interact in a complex way. The key insight of these results is to make rules sensitive to the shape of the formulas on either sides of the sequents. This way one can devise rules in a much more modular fashion. This feature of Hilbert-style systems is notoriously lost when one moves to sequent calculi. By partly restoring modularity the method proposed here can potentially provide a unified approach to the proof theory of multi-modal systems.

1 Introduction and Motivations

Inter-modal interaction is a key feature of many well-known multi-modal logics. In propositional dynamic logic [7] modalities for complex programmes are built compositionally from simpler ones. Modern logics for knowledge and beliefs (e.g. [3]) often contain principles such as "knowledge implies belief" ($KA \to BA$) and, when augmented with temporal modalities, principles such as "perfect recall" and "no miracle" [19].[1] Standard Deontic Logic (SDL) [10], one of our case studies for this paper, is usually taken to contain the Ought-implies-Can-principle:

$$OA \to \neg\square\neg A \qquad\qquad \text{(Ought-Can)}$$

Dealing with such interaction is relatively easy within Hilbert-style calculi. This is arguably due to the strong modular character of such axiom systems. This situation changes for Gentzen-style (sequent) calculi. Here extending a system with new (interaction) rules can break proof-theoretic properties of the original system.[2]

---

[*] This research is supported by the Alexander von Humboldt Foundation.

[1] The perfect recall principle states that if an agent can distinguish between two histories now she will be able to do so forever after. The no miracle principle states that if an agent cannot distinguish between two histories, then she cannot distinguish the result of executing the same action in these histories.

[2] This typically was true for standard Gentzen-style formulations of S5. The proof-theoretically "nice" properties of a hypersequent formulation of S5 is one of the main reasons for "going hypersequent" in this paper. An extension of 5 (stated below) to first order logic can be found in [6].

D. Grossi, O. Roy, and H. Huang (Eds.): LORI 2013, LNCS 8196, pp. 124–134, 2013.

In this paper we make a first step towards the development of a modular approach to sequent calculi for multi-modal systems. It is true that cut-free sequent calculi are known for the individual modalities constituting the systems just mentioned [8,13,21]. Known solutions include labelled sequents [11] or, as we will use here, *hypersequents* [15]. Likewise, some of these combined systems do enjoy a cut-free sequent calculus, for instance PDL [8], multi-agent epistemic logic [13] and temporal logic [4]. But these solutions do not generalize in a straightforward way to arbitrary multi-modal systems.

This paper takes a new route. The driving idea is to treat the different modalities modularly by formulating rules that are sensitive to the syntactic shape— intuitively, the main modal operator—of the formulas on which they operate.

We do so through two case studies. The first one is Standard Deontic Logic (SDL). This logic contains deontic and alethic modalities. They interact in a simple way, namely through the Ought-implies-Can principle. We propose such a syntax-sensitive rule—coined (DI)—that captures this interaction. This sensitivity is key to the admissibility of (Cut) and its corollaries. The alethic modal logic is S5. Arguably, S5 is typically taken to model *logical* necessity, but other interpretations are possible. The study of the first system lays the foundation for our second case study.

We then move to a more complex multi-modal logic with interaction. Like SDL, this is a deontic logic with alethic modalities. In this logic, however, obligation and permission are not dual notions, they are both non-normal modalities, and they interact in a sophisticated way with alethic notions. We choose to develop the proof theory of this system for two reasons. First, it provides a more complex testing ground for our modular approach, and thus shows its mathematical robustness. Second, this logic is well-suited to model obligations and permissions bearing on agents in games and interactive situations [16,17].

2   Deontic Language with an Alethic Modality

All through this paper we work in a propositional modal logic with "boxes" for deontic and alethic operators.

$$A := p \mid \neg A \mid A \vee A \mid \Box A \mid P A \mid O A$$

Formulas of the form OA and PA should be read, respectively, as "A is obligatory" and "A is permitted". Obligation and permission are duals in SDL, but not in our second case study, thus their introduction as primitives. $\Box A$ is an alethic modality, to be read "all actions available to the agent are A-actions." This modality has a dual, $\Diamond A$, to be read "it is possible for the agent to choose an A-action."

3   First Case Study: SDL with Ought-implies-Can

For reasons of easy generalization we start right away with hypersequents [1,2]. We use $\mid$ as a metalinguistic expression; Γ, Δ, Φ, and Ψ denote multisets of formulas; $\Sigma, \Sigma', \Sigma'' \ldots$ denote hypersequents.

Definition 1. *A hypersequent is a syntactic object of the form:*

$$\Gamma \Rightarrow \Delta \mid \Gamma_1 \Rightarrow \Delta_2 \mid \cdots \mid \Gamma_n \Rightarrow \Delta_n$$

We take the logic of the alethic modality to be S5, and take its hypersequent formulation directly from [14,12]; i.e. we take the proof-theoretic methods for 5 developed there, add and extend them further in order to cope with (some) multi-modal logics.

System 5

(Ax) $\Sigma \mid A, \Gamma \Rightarrow \Delta, A$ if A is atomic

$$(\neg L) \ \frac{\Sigma \mid \Gamma \Rightarrow \Delta, A}{\Sigma \mid \neg A, \Gamma \Rightarrow \Delta} \qquad\qquad (\neg R) \ \frac{\Sigma \mid A, \Gamma \Rightarrow \Delta}{\Sigma \mid \Gamma \Rightarrow \Delta, \neg A}$$

$$(\vee L) \ \frac{\Sigma \mid A, \Gamma \Rightarrow \Delta \quad \Sigma \mid B, \Gamma \Rightarrow \Delta}{\Sigma \mid A \vee B, \Gamma \Rightarrow \Delta} \qquad (\vee R) \ \frac{\Sigma \mid \Gamma \Rightarrow \Delta, A, B}{\Sigma \mid \Gamma \Rightarrow \Delta, A \vee B}$$

$$(\Box L1) \ \frac{\Sigma \mid A, \Box A, \Gamma \Rightarrow \Delta}{\Sigma \mid \Box A, \Gamma \Rightarrow \Delta} \qquad (\Box L2) \ \frac{\Sigma \mid \Box A, \Gamma \Rightarrow \Delta \mid A, \Gamma' \Rightarrow \Delta'}{\Sigma \mid \Box A, \Gamma \Rightarrow \Delta \mid \Gamma' \Rightarrow \Delta'}$$

$$(\Box R) \ \frac{\Sigma \mid \Gamma \Rightarrow \Delta \mid \ \Rightarrow A}{\Sigma \mid \Gamma \Rightarrow \Delta, \Box A}$$

We now extend 5 with additional rules. The basic rule for the deontic operator(s) is the Gentzen-style analogue of the K-axiom.[3]

$$\frac{\Sigma \mid \Gamma \Rightarrow A}{\Sigma \mid O\Gamma \Rightarrow OA} \ (K)$$

This rule indeed allows for a derivation of (a hypersequent analogue of) K:

$$\frac{B \Rightarrow B \quad \dfrac{A, B \Rightarrow A}{A, \neg A, B \Rightarrow}}{\dfrac{\neg A \vee B, A \Rightarrow B}{O(\neg A \vee B), OA \Rightarrow OB}} \ (K)$$

Now for the interaction principles. SDL contains in fact two of them. The first one is that obligation implies permission ($OA \rightarrow PA$) or equivalently, that obligations are consistent ($OA \rightarrow \neg O \neg A$). This is derivable using the following rule:

$$\frac{\Sigma \mid \Gamma \Rightarrow}{\Sigma \mid O\Gamma \Rightarrow} \ (D)$$

---

[3] Deontic logic is rarely studied in sequent calculi; a rare exception is [5].

The derivation goes as follows:

$$\cfrac{\cfrac{\cfrac{A \Rightarrow A}{A, \neg A \Rightarrow}}{OA, O\neg A \Rightarrow} \text{(D)}}{OA \Rightarrow \neg O\neg A}$$

Our main target is (Ought-Can). The next rule (DI) makes that principle derivable, and allows also for the proof of the admissibility of (Cut) in the next section.

$$\frac{\Sigma \mid \Gamma, \Delta^\square \Rightarrow}{\Sigma \mid O\Gamma, \Delta^\square \Rightarrow} \text{(DI)}$$

In this rule $\Delta^\square$ is a multi-set of "boxed" formulas. Each $A \in \Delta^\square$ has to be of the form $\square B$ with B a formula of our language. This syntactic restriction can be seen at work in the following derivation of (Ought-Can):

$$\cfrac{\cfrac{\cfrac{\cfrac{A, \square\neg A \Rightarrow A}{\neg A, A, \square\neg A \Rightarrow}}{A, \square\neg A \Rightarrow}}{OA, \square\neg A \Rightarrow} \text{(DI)}}{OA \Rightarrow \neg\square\neg A}$$

We call the system consisting of 5 and the rules just mentioned: 5DI. $\Delta^\square$ plays two roles. It restricts application of the rule. $\Delta^\square$ must contain only boxed formulas. This blocks derivations of (hyper)sequents as

$$OA \Rightarrow A$$

The other role is to enable the interaction between alethic and deontic modalities in a formally desirable way (i.e. admitting the admissibility of cut).

3.1 Proof Theory of 5DI

In this section we state the admissibility of some structural rules (merge, external and internal weakening, contraction) and cut in 5DI. First some preparatory lemmas.

Definition 2 (Height of a derivation, weight of a formula).
The height *of a derivation is the greatest number of successive applications of rules in it. The axiom (Ax) has height 0. A rule is* height-preserving admissible *whenever for each derivation of height n of a particular hypersequent Σ in which that rule is applied there is a derivation of that sequent that is of the same height and the rule (in question) is not applied.*

The weight *of a formula is defined as: (i) $w(A) = 0$ for A is an atom. (ii) $w(A \vee B) = w(A) + w(B) + 1$. (iii) $w(\neg A) = w(A) + 1$. (iv) $w(\square A) = w(OA) = w(A) + 1$.*

Lemma 1. *Every hypersequent of the form $\Sigma \mid A \Rightarrow A$ is derivable in 5DI.*

The proof of this is standard. For the alethic modal cases see [12, p.177] and [14]. By use of (K) the deontic case is dealt with.

Lemma 2. *(Merge), i.e.,*

$$(Merge) \frac{\Sigma \mid \Gamma \Rightarrow \Delta \mid \Phi \Rightarrow \Psi}{\Sigma \mid \Gamma, \Phi \Rightarrow \Delta, \Psi}$$

is height-preserving admissible in 5DI.

Lemma 3 (External and internal weakening).
Both (EW) and (IW), i.e.

$$(EW) \frac{\Sigma}{\Sigma \mid \Sigma'} \quad (IW) \frac{\Sigma \mid \Gamma \Rightarrow \Delta}{\Sigma \mid \Gamma, \Phi \Rightarrow \Delta, \Psi}$$

are height-preserving admissible in 5DI.

(EW) is proved straightforwardly by induction on the height of the derivation; the proof of (IW) also uses (Merge) and (EW).

Lemma 4. *The rules of contraction, i.e.*

$$(CL) \frac{\Sigma \mid A, A, \Gamma \Rightarrow, \Delta}{\Sigma \mid A, \Gamma \Rightarrow \Delta} \quad (CR) \frac{\Sigma \mid \Gamma \Rightarrow, \Delta, A, A}{\Sigma \mid \Gamma \Rightarrow, \Delta, A}$$

are admissible.

Definition 3. *The cut-height of an instance of the rule of (Cut) in a derivation is the sum of heights of derivation of the two premisses of (Cut).*

Theorem 1.

(Cut), i.e.

$$(Cut) \frac{\Sigma \mid \Gamma \Rightarrow \Delta, A \qquad \Sigma' \mid A, \Phi \Rightarrow \Psi}{\Sigma \mid \Sigma' \mid \Gamma, \Phi \Rightarrow \Delta, \Psi}$$

is admissible in 5DI.

The proof of the above theorem is by induction on the complexity of the cut formula plus a side-induction on the sum of the heights of the premisses. There are basically three main cases to consider. *Case 1*: The left premiss of Cut is an initial hypersequent. *Case 2*: The left premiss of Cut is inferred by a rule of 5DI but A is not principal in it. Cases 1 and 2 are dealt with as in [12, p.181f] for the alethic modal cases; the deontic logic cases are settled with ease as well. *Case 3*: The left premiss of Cut is inferred by a rule of 5DI and A is principal in it. In addition to [12, p.182] there are cases where the cut-formula is of the form OA.

There are two cases.

Case 3.1: OA is introduced by (K).

$$\dfrac{\dfrac{\varSigma \mid \varGamma \Rightarrow A}{\varSigma \mid O\varGamma \Rightarrow OA} \text{ (K)} \qquad \dfrac{\varSigma' \mid A, \varPhi \Rightarrow B}{\varSigma' \mid OA, O\varPhi \Rightarrow OB} \text{ (K)}}{\varSigma \mid \varSigma' \mid O\varGamma, O\varPhi \Rightarrow OB} \text{ (Cut)}$$

This transforms to:

$$\dfrac{\dfrac{\varSigma \mid \varGamma \Rightarrow A \qquad \varSigma' \mid A, \varPhi \Rightarrow B}{\varSigma \mid \varSigma' \mid \varGamma, \varPhi \Rightarrow B} \text{ (Cut)}}{\varSigma \mid \varSigma' \mid O\varGamma, O\varPhi \Rightarrow OB} \text{ (K)}$$

Case 3.2: OA is introduced by (K) in the left premiss and by (DI) in the right premiss of the (Cut).

$$\dfrac{\dfrac{\varSigma \mid \varGamma \Rightarrow A}{\varSigma \mid O\varGamma \Rightarrow OA} \text{ (K)} \qquad \dfrac{\varSigma' \mid A, \varPhi, \varDelta^{\square} \Rightarrow}{\varSigma' \mid OA, O\varPhi, \varDelta^{\square} \Rightarrow} \text{ (DI)}}{\varSigma \mid \varSigma' \mid O\varGamma, O\varPhi, \varDelta^{\square} \Rightarrow} \text{ (Cut)}$$

This transforms to:

$$\dfrac{\dfrac{\varSigma \mid \varGamma \Rightarrow A \qquad \varSigma' \mid A, \varPhi, \varDelta^{\square} \Rightarrow}{\varSigma \mid \varSigma' \mid \varGamma, \varPhi, \varDelta^{\square} \Rightarrow} \text{ (Cut)}}{\varSigma \mid \varSigma' \mid O\varGamma, O\varPhi, \varDelta^{\square} \Rightarrow} \text{ (DI)}$$

Cut-elimination theorems have a number of important corollaries, some of which we list for completeness. The *proofs* of these corollaries are immediate. We use a standard definition of sub-formulas.

Corollary 1 (Subformula property). *All formulas in a derivation of $\varSigma$ in 5DI are subformulas of $\varSigma$.*

Definition 4 (Consistency). *5DI is consistent iff the empty hypersequent is not derivable.*

Corollary 2 (Consistency). *5DI is consistent.*

The next theorem establishes the deductive equivalence of a SDL$^{\square}$ and 5DI. To this end we repeat the axioms and rules of inference of SDL$^{\square}$ as follows, with M either $\square$ or O:

(Ax0) All Tautologies.
(Ax1) $M(A \to B) \to (MA \to MB)$
(Ax2) $\Box A \to A$
(Ax3) $\Box A \to \Box\Box A$
(Ax4) $\neg\Box\neg A \to \Box\neg\Box\neg A$
(Ax5) $OA \to \neg O\neg A$
(Ax6) $OA \to \neg\Box\neg A$

Rules of inference

$$(\text{MP}) \ \frac{A \to B, A}{B} \quad (\text{Nec}) \ \frac{A}{\Box A}$$

The key for the proof of the deductive equivalence is the following translation of hypersequents into formulas of SDL [14]:

Definition 5 (Translation).

(i) $(\Gamma \Rightarrow \Delta)^\tau := \bigwedge \Gamma \to \bigvee \Delta$
(ii) $\Sigma_1 \mid \Sigma_2 \mid \cdots \mid \Sigma_n := \Box\Sigma_1^\tau \vee \Box\Sigma_2^\tau \vee \cdots \vee \Box\Sigma_n^\tau$

Theorem 2.

(i) *If* $SDL^\Box \vdash A$, *then* $5DI \vdash\Rightarrow A$.
(ii) *If* $5DI \vdash \Sigma$, *then* $SDL^\Box \vdash (\Sigma)^\tau$.

The proof is again standard. The crucial part is (ii). $5DI \vdash\Rightarrow \neg OA \vee \neg\Box\neg A$, given the translation τ this means really that $SDL^\Box \vdash \Box(\neg OA \vee \neg\Box\neg A)$ and of course it does so by (Nec); by (T), i.e. (Ax2), $SDL^\Box \vdash \neg OA \vee \neg\Box\neg A$.

4   Second Case Study: Obligation as Weakest Permission

The deontic logic of "obligation as weakest permission" (henceforth 5WP) has been proposed in [16,17] to analyse rational recommendations to players in game-theoretic situations. This logic departs in many ways from SDL. Its two deontic modalities are non-normal. They do not validate K nor the rule of necessitation. They are also two "boxes", while in SDL P is a "diamond". And they are not dual to each other. Finally, and more importantly, the main interaction principle between the deontic and the alethic modalities rests on the following idea: A is obligatory only if it is the logically weakest permitted action type that the agent can perform. Formally:

$$OA \to (PB \to \Box(B \to A)) \qquad\qquad \text{(Weakest Perm)}$$

5WP is thus a non-normal multi-modal system where the modalities interact in a subtle way. It is thus not only a conceptually interesting logical system— see again [16,17] for a detailed discussion of its motivations—but also a good benchmark case for the robustness of our modular approach to interaction.

As base rule for O and P we have the following, which essentially has K above but with a stronger antecedent [9], and with D either O or P:

$$\frac{\Sigma|A \Rightarrow B \qquad \Sigma|B \Rightarrow A}{\Sigma|DA \Rightarrow DB} \ (\mathsf{Ext(D)})$$

As in SDL, $\Box$ is a normal, S5 modality in 5WP. The rule for the interaction between O and P needs to be modified, since these are no longer interdefinable. The following rule clearly ensures the provability of $OA \Rightarrow PA$:

$$\frac{\Sigma|A \Rightarrow B}{\Sigma|OA \Rightarrow PB} \ (\mathsf{OP})$$

As above, our key rule (DI) takes care of the interaction rule between O and $\Box$, with the same derivation as of (Ought-Can).

Finally, (Weakest Perm) is derivable using the following, with 'D' either for 'O' or 'P'.

$$\frac{\Sigma|A \Rightarrow C \qquad \Sigma|B \Rightarrow E}{\Sigma|OA, DB \Rightarrow |C \Rightarrow E} \ (\mathsf{WP})$$

The derivation goes as follows:[4]

$$\frac{\dfrac{\dfrac{A \Rightarrow A \qquad B \Rightarrow B}{OA, PB \Rightarrow |B \Rightarrow A} \ (\mathsf{WP})}{\dfrac{OA, PB \Rightarrow | \Rightarrow B \to A}{OA, PB \Rightarrow \Box(B \to A)} \ ((\Box\mathsf{R})}}{\dfrac{OA \Rightarrow PB \to \Box(B \to A)}{\Rightarrow OA \to (PB \to \Box(B \to A))}}$$

The rules of the system are designed to allow for both the admissibility of *Cut* and furthermore 5WP and its Hilbert-style formulation are deductively equivalent (the corresponding theorem is stated below). The last fact establishes soundness and completeness of 5WP.

4.1   Proof Theory of 5WP

The lemmas under 2.3 do also hold in this system.

Theorem 3.

(Cut), i.e.

$$(Cut) \ \frac{\Sigma \mid \Gamma \Rightarrow \Delta, A|\Sigma^* \qquad \Sigma' \mid A, \Phi \Rightarrow \Psi|\Sigma^{**}}{\Sigma \mid \Sigma' \mid \Gamma, \Phi \Rightarrow \Delta, \Psi|\Sigma^*|\Sigma^{**}}$$

is admissible in 5WP.

The proof employs the same strategy as above. We highlight here some critical cases.

---

[4] We tacitly assume a derivable rule for the right introduction of the arrow.

– The cut-formula is PB introduced by (WP) in the left premiss and by (OP) in the right premiss of the (Cut).

$$\frac{\dfrac{\Sigma|A \Rightarrow C \qquad \Sigma|B \Rightarrow D}{\Sigma|OA, PB \Rightarrow |D \Rightarrow C} \text{ (WP)} \qquad \dfrac{\Sigma'|E \Rightarrow B}{\Sigma'|OE \Rightarrow PB}}{\Sigma|\Sigma'|OA, OE \Rightarrow |D \Rightarrow C}$$

This transforms to:

$$\frac{\dfrac{\Sigma|B \Rightarrow D \qquad \Sigma'|E \Rightarrow B}{\Sigma|\Sigma'|E \Rightarrow D} \qquad \Sigma|A \Rightarrow C}{\Sigma|\Sigma|OA, OE \Rightarrow |D \Rightarrow C}$$

– The cut-formula is PB, introduced by (WP) and (Ext(P)).

$$\frac{\dfrac{\Sigma|A \Rightarrow C \qquad \Sigma|B \Rightarrow D}{\Sigma|OA, PB \Rightarrow |D \Rightarrow C} \text{ (WP)} \qquad \dfrac{\Sigma'|E \Rightarrow B \qquad \Sigma'|B \Rightarrow E}{\Sigma'|PE \Rightarrow PB} \text{ (Ext(P))}}{\Sigma|\Sigma'|OA, PE \Rightarrow |D \Rightarrow C}$$

This transforms to:

$$\frac{\dfrac{\Sigma|B \Rightarrow D \qquad \Sigma'|E \Rightarrow B}{\Sigma|\Sigma'|E \Rightarrow D} \qquad \Sigma|A \Rightarrow C}{\Sigma|\Sigma'|OA, PE \Rightarrow |D \Rightarrow C}$$

– The cut-formula is OA, introduced by (WP) and (Ext(O)) is structurally the same as the previous one.

Given the definitions of sub-formula and consistency, the following two corollaries are provable, as well as having deductive equivalence with the axiomatic system presented in [16].

Corollary 3 (Subformula property). *All formulas in a derivation of Σ in 5WP are subformulas of Σ.*

Corollary 4 (Consistency).
5WP is consistent.

Let's call the following system: HD5. It is the Hilbert-style counterpart to 5WP.

– All propositional tautologies.
– S5 for $\Box$.

– For P and O and:
  (OiC) $OA \rightarrow \neg\Box\neg A$
  (OiP) $OA \rightarrow PA$
  (Weakest Perm) $OA \rightarrow (PB \rightarrow \Box(B \rightarrow A))$
– (MP), (Nec) for $\Box$, and the following for O and P:
  - (RP) From $A \leftrightarrow B$ infer $PA \leftrightarrow PB$.
  - (RO) From $A \leftrightarrow B$ infer $OA \leftrightarrow OB$.

The equivalence of both systems is established in the last theorem of this paper and it depends on the definition 5 above.

Theorem 4 (Deductive equivalence of HD5 and 5WP).

(i) If $HD5 \vdash A$, then $5WP \vdash\Rightarrow A$.
(ii) If $5WP \vdash \Sigma$, then $HD5 \vdash (\Sigma)^\tau$.

The proof is done by induction on the length of a derivation.

5   Conclusion

In this paper we studied two cases of multi-modal systems with interaction, for which we proved the admissibility of (Cut) and its corollaries. The driving idea was to split the hypersequents into their different modal parts, as exemplified by the rule (DI), which was crucial for these two systems. In the second system, in particular, it plays a key role in establishing the relation between normal and non-normal modalities.

The next obvious step is to see to what extent this splitting/modular methodology can be extended to analyse other known multi-modal systems with interaction. We would like to know for instance whether this methodology can help give a sequent calculi for logics that combine obligations and preferences [20], preferences and action [18], or different epistemic modalities [3]. Given how crucial the modularity of Hilbert-style systems is in completely axiomatizing complex interaction between modalities, we think that the results reported here promise many other successful applications.

References

1. Avron, A.: Hypersequents, logical consequence and intermediate logics for concurrency. Annals of Mathematics and Artifical Intelligence 4, 225–248 (1991)
2. Avron, A.: The methods of hypersequents in the proof theory of propositional and non-classical logics. In: Hodges, W., Hyland, M. (eds.) Logic: from Foundations to Applications. Proc. Logic Colloquium, pp. 1–32. Oxford University Press (1993)
3. Baltag, A., Smets, S.: Conditioanl doxastic models: A qualitative approach to dynamic belief revision. In: Electornic Notes in Theoretical Computer Science, vol. 165, pp. 5–21. Springer (2006)
4. Brünnler, K., Lange, M.: Cut-free sequent systems for temporal logic. Journal of Logic and Algebraic Programming 76(2), 216–225 (2008)

5. Governatori, G., Rotolo, A.: Labelled modal sequents. In: TABLEAUX 2000. Position Papers and Tutorials, pp. 3–21 (2000)
6. Gratzl, N.: A note on hypersequent first order s5 (submitted)
7. Harel, D., Kozen, D., Tiuryn, J.: Dynamic Logic (Foundations of Computing). MIT Press (2000)
8. Hill, B., Poggiolesi, F.: A contraction-free and cut-free sequent calculus for propositional dynamic logic. Studia Logica 94(1), 47–72 (2010)
9. Lavendhomme, R., Lucas, T.: Sequent calculi and decision procedures for weak modal systems. Studia Logica, 121–145 (2000)
10. McNamara, P.: Deontic logic. In: E. N. Zalta, editor, Standford Encyclopedia of Philosophy. Spring edn. (2006),
   http://plato.stanford.edu/archives/spr2006/entries/logic-deontic/
11. Negri, S.: Proof analysis in modal logic. Journal of Philosophical Logic 34(5), 507–544 (2005)
12. Poggiolesi, F.: Gentzen Calculi for Modal Propositional Logic. Springer (2011)
13. Poggiolesi, F.: From single agent to multi-agent via hypersequents. Logica Universalis, 1–20 (2012)
14. Poggiolesi, F.: A cut-free simple sequent calculus for modal logic s5. Review of Symbolic Logic 1(1), 3–15 (2008)
15. Restall, G.: Proofnets for s5: sequents and circuits for modal logic. In: Logic Colloquium, vol. 28, pp. 151–172 (2005)
16. Roy, O., Anglberger, A.J.J., Gratzl, N.: The logic of obligation as weakest permission. In: Ågotnes, T., Broersen, J., Elgesem, D. (eds.) DEON 2012. LNCS, vol. 7393, pp. 139–150. Springer, Heidelberg (2012)
17. Roy, O., Anglberger, A., Gratzl, N.: The logic of best action from a deontic perspective. In: Baltag, A., Smets, S. (eds.) Johan F.A.K. van Benthem on Logical and Informational Dynamics (2013)
18. Sun, X.: Conditional ought, a game theoretical perspective. In: van Ditmarsch, H., Lang, J., Ju, S. (eds.) LORI 2011. LNCS, vol. 6953, pp. 356–369. Springer, Heidelberg (2011)
19. Van Benthem, J., Gerbrandy, J., Hoshi, T., Pacuit, E.: Merging frameworks for interaction. Journal of Philosophical Logic 38(5), 491–526 (2009)
20. van Benthem, J., Grossi, D., Liu, F.: Deontics = betterness + priority. In: Governatori, G., Sartor, G. (eds.) DEON 2010. LNCS (LNAI), vol. 6181, pp. 50–65. Springer, Heidelberg (2010)
21. Wansing, H.: Displaying modal logic. Kluwer Academic Publishers (1998)

Dynamic Epistemic Logic Displayed

Giuseppe Greco[1], Alexander Kurz[2], and Alessandra Palmigiano[3]

[1] ILLC, Amsterdam, NL
[2] University of Leicester, UK
[3] TBM, Delft, NL

Abstract. We introduce a display calculus for the logic of Epistemic Actions and Knowledge (EAK) of Baltag-Moss-Solecki. This calculus is *cut-free* and *complete* w.r.t. the standard Hilbert-style presentation of EAK, of which it is a *conservative extension*, given that—as is common to display calculi—it is defined on an expanded language in which all logical operations have adjoints. The additional dynamic operators do not have an interpretation in the standard Kripke semantics of EAK, but do have a natural interpretation in the final coalgebra. This proof-theoretic motivation revives the interest in the global semantics for dynamic epistemic logics pursued among others by Baltag [4], Cîrstea and Sadrzadeh [8].

1 Introduction

Dynamic logics form a large family of nonclassical logics, and perhaps the one enjoying the widest range of applications. Indeed, they are designed to formalize *change* caused by *actions* of diverse nature: updates on the memory state of a computer, displacements of moving robots in an environment, measurements in models of quantum physics, belief revisions, knowledge updates, etc. In each of these areas, formulas express properties of the model encoding the present state of affairs, as well as the pre- and post-conditions of a given action. Actions are semantically represented as *transformations* of one model into another, encoding the state of affairs after the action has taken place. Languages for dynamic logics are expansions of classical propositional logic with *dynamic operators*, parametrized with actions; dynamic operators are modalities interpreted in terms of the transformation of models corresponding to their action-parameters.

However, when dynamic logics feature both dynamic and 'static' modalities, as in the case of the Dynamic Epistemic Logics, they typically lose many desirable properties, such as the closure under uniform substitution. This and other difficulties make their algebraic and proof-theoretic treatment not straightforward, and indeed, the existing proposals appeal to technical solutions which do not meet some of the requirements commonly sought for in proof-theoretic semantics [21, 22]. In [2], a tableaux calculus is introduced, which is labelled, and restricted to the logic of Public Announcements (PAL); in [15] and [16], sequent calculi are presented, covering truthful and arbitrary public announcements respectively, which are again labelled. In [5] and [9], sequent calculi are defined, which are nested; these calculi are sound and complete w.r.t. a certain algebraic semantics which is more general than the standard Kripke semantics for the logic of Baltag-Moss-Solecki; they manipulate sequents whose succedents are unary,

D. Grossi, O. Roy, and H. Huang (Eds.): LORI 2013, LNCS 8196, pp. 135–148, 2013.

and in which three types of objects feature on a par (formulas, agents and actions); finally, two different entailment relations occur, for actions and propositions, respectively, which need to be brought together by means of rules of hybrid type.

In the present paper, we bring into focus that (at least one aspect of) the difficulties hinted at above is the following. Whereas the interpretation of the *adjoints* of static modal operators is equally available in standard models and in the final coalgebra, this is no longer the case for dynamic modalities. In particular, Section 2 will emphasize that dynamic modalities do not in general come in adjoint pairs w.r.t. the standard Kripke semantics. In other words, display postulates (cf. Section 2) are not sound for dynamic modalities w.r.t. to the standard semantics. However, the soundness of these display postulates will be shown w.r.t. the final coalgebra semantics.

After reviewing dynamic epistemic logic·(EAK) in Section 3, we define the Belnap's style display calculus D.EAK in Section 4. In Section 5, we outline the proofs that D.EAK is sound w.r.t. the final coalgebra semantics, complete w.r.t. the well known Hilbert-style presentation of EAK, and that the cut rule is eliminable. In Section 6 we briefly discuss why D.EAK is a conservative extension of EAK, and we outline some ongoing research directions.

2   Coalgebraic Semantics of Dynamic Logics

Modal formulas A are interpreted in Kripke models $\mathbb{M} = (W, R, V)$ as subsets of their domains W, and we write $[\![A]\!]_M \subseteq W$ for their interpretation. Equivalently, we can describe the interpretation of A in each Kripke model via the final coalgebra[1] $\mathbb{Z}$ first by defining $[\![A]\!]_\mathbb{Z}$ to be the set of elements of $\mathbb{Z}$ satisfying A, and then by recovering $[\![A]\!]_M \subseteq W$ as

$$[\![A]\!]_\mathbb{M} = f^{-1}([\![A]\!]_\mathbb{Z}), \tag{1}$$

where f is the unique homomorphism $\mathbb{M} \to \mathbb{Z}$. This construction works essentially because, in the category of models/Kripke structures/coalgebras, homomorphisms (i.e. functional bisimulations) preserve the satisfaction/validity of modal formulas. Bisimulation invariance is also enjoyed by formulas of such dynamic logics as EAK (cf. Section 3). Hence, for these dynamic logics, both Kripke semantics and the final coalgebra semantics are equivalently available. However, so far the community has not warmed up to adopting the final coalgebra semantics for dynamic logic, Baltag's [4], and Cîrstea and Sadrzadeh's [8] being among the few proposals exploring this setting. This is unlike the case of standard modal logic, in which the coalgebraic option has taken off, to the point that it has given rise to a field in its own right. In the present section, we offer new reasons to consider the final coalgebra semantics for dynamic logic; indeed, we bring to the fore one aspect in which the final coalgebra semantics for dynamic logics is more advantageous than the standard semantics.

The interpretation of dynamic modalities is given in terms of the *actions* parametrizing them. Actions can be semantically represented as transformations of Kripke models,

---

[1] Here we rely on the theorem of [1] that the final coalgebra $\mathbb{Z}$ exists. Moreover, even if the carrier of $\mathbb{Z}$ is a proper class, it is still the case that subsets of $\mathbb{Z}$ correspond precisely to 'modal predicates', that is, predicates that are invariant under bisimilarity, see [14].

i.e., as relations between states of different Kripke models. From the viewpoint of the final coalgebra, any action symbol α can then be interpreted as a binary relation α_Z on the final coalgebra Z. In this way, the following well known fact becomes immediately applicable to the final coalgebra model:

Proposition 1. *Every relation $R \subseteq X \times Y$ gives rise to the modal operators*

$$\langle R \rangle, [R] : PY \rightarrow PX \text{ and } \langle R^\circ \rangle, [R^\circ] : PX \rightarrow PY$$

defined as follows: for every $V \subseteq X$ and every $U \subseteq Y$,

$$\langle R \rangle U = \{x \in X \mid \exists y \,.\, xRy \;\&\; y \in U\} \qquad [R]U = \{x \in X \mid \forall y \,.\, xRy \;\Rightarrow\; y \in U\}$$
$$\langle R^\circ \rangle V = \{y \in Y \mid \exists x \,.\, xRy \;\&\; x \in V\} \qquad [R^\circ]V = \{y \in Y \mid \forall x \,.\, xRy \;\Rightarrow\; x \in V\}.$$

These operators come in adjoint pairs:

$$\langle R \rangle U \subseteq V \text{ iff } U \subseteq [R^\circ]V \tag{2}$$
$$\langle R^\circ \rangle V \subseteq U \text{ iff } V \subseteq [R]U. \tag{3}$$

Let $\langle \alpha_Z \rangle, [\alpha_Z], \langle \alpha_Z{}^\circ \rangle, [\alpha_Z{}^\circ]$ be the semantic modal operators given by Proposition 1 in the special case where $X = Y$ is the carrier Z of $\mathbb{Z}$; they respectively provide a natural interpretation in the final coalgebra Z for the four connectives $\langle \alpha \rangle, [\alpha], \widehat{\alpha}, \overline{\alpha}$, parametric in the action symbol α. As a direct consequence of the adjunctions (2), (3), the following display postulates, which are so crucial for the present work, are sound under this interpretation (cf. Section 5.1 for more details on this interpretation).

$$\frac{\{\alpha\}X \vdash Y}{X \vdash \overline{\alpha}\, Y} \;\; {\scriptstyle \frac{\{\alpha\}}{\overline{\alpha}}} \qquad \frac{X \vdash \{\alpha\}Y}{\widehat{\alpha}\, X \vdash Y} \;\; {\scriptstyle \frac{\widehat{\alpha}}{\{\alpha\}}}$$

On the other hand, standard Kripke models are not in general closed under (the interpretations of) α and α°. As a direct consequence of this fact, we can show that e.g. the display postulate $\left({\scriptstyle \frac{\{\alpha\}}{\overline{\alpha}}} \right)$ is not sound in some Kripke models M for any interpretation of formulas of the form $\overline{\alpha}\, B$ in M.

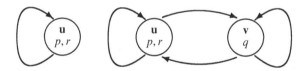

Fig. 1. The models M^α and M

Indeed, consider the model M represented on the right-hand side of the picture above; let the action α be the public announcement (cf. [3]) of the atomic proposition r, and let $A := \Box p$ and $B := q$; hence M^α is the submodel on the left-hand side of the picture. Let $i : M^\alpha \hookrightarrow M$ be the submodel injection map. Clearly, $[\![\Box p]\!]_M = \varnothing$, which implies that the inclusion $[\![A]\!]_M \subseteq [\![\overline{\alpha}\, B]\!]_M$ trivially holds for any interpretation of $\overline{\alpha}\, B$ in M; however, $i[\![\![\Box p]\!]_{M^\alpha}] = \{u\}$, hence $[\![\langle \alpha \rangle \Box p]\!]_M = [\![\alpha]\!]_M \cap i[\![\![\Box p]\!]_{M^\alpha}] = V(r) \cap \{u\} = \{u\} \not\subseteq \{v\} = [\![q]\!]_M$, which falsifies the inclusion $[\![\langle \alpha \rangle A]\!]_M \subseteq [\![B]\!]_M$. This proves our claim.

3   The Logic of Epistemic Actions and Knowledge

In the present section, the relevant preliminaries on the syntax and semantics of the logic of epistemic actions and knowledge (EAK) [3] will be given, which are different but equivalent to the original version [3], and follow the presentation in [13, 17].

Let AtProp be a countable set of proposition letters. The set $\mathcal{L}$ of formulas A of (the single-agent[2] version of) the logic of epistemic actions and knowledge (EAK) and the set $\mathsf{Act}(\mathcal{L})$ of the *action structures* α over $\mathcal{L}$ are built simultaneously as follows:

$$A := p \in \mathsf{AtProp} \mid \neg A \mid A \vee A \mid \Diamond A \mid \langle \alpha \rangle A \quad (\alpha \in \mathsf{Act}(\mathcal{L})).$$

An *action structure over* $\mathcal{L}$ is a tuple $\alpha = (K, k, \alpha, Pre_\alpha)$, such that K is a finite nonempty set, $k \in K$, $\alpha \subseteq K \times K$ and $Pre_\alpha : K \to \mathcal{L}$. Notice that α denotes *both* the action structure *and* the accessibility relation of the action structure. Unless explicitly specified otherwise, occurrences of this symbol are to be interpreted contextually: for instance, in $j\alpha k$, the symbol α denotes the relation; in M^α, the symbol α denotes the action structure. Of course, in the multi-agent setting, each action structure comes equipped with *a collection* of accessibility relations indexed in the set of agents, and then the abuse of notation disappears.

Sometimes we will write $Pre(\alpha)$ for $Pre_\alpha(k)$. Let $\alpha_i = (K, i, \alpha, Pre_\alpha)$ for every action structure $\alpha = (K, k, \alpha, Pre_\alpha)$ and every $i \in K$. Intuitively, the actions α_i for $k\alpha i$ encode the uncertainty of the (unique) agent about the action that is actually taking place. The standard stipulations hold for the defined connectives $\top$, $\bot$, $\wedge$, $\to$ and $\leftrightarrow$.

Models for EAK are relational structures $M = (W, R, V)$ such that W is a nonempty set, $R \subseteq W \times W$, and $V : \mathsf{AtProp} \to \mathcal{P}(W)$ is a map. The evaluation of the static fragment of the language is standard. For every Kripke frame $\mathcal{F} = (W, R)$ and every $\alpha \subseteq K \times K$, let the Kripke frame $\coprod_\alpha \mathcal{F} := (\coprod_K W, R \times \alpha)$ be defined[3] as follows: $\coprod_K W$ is the $|K|$-fold coproduct of W (which is set-isomorphic to $W \times K$), and $R \times \alpha$ is the binary relation on $\coprod_K W$ defined as

$$(w, i)(R \times \alpha)(u, j) \quad \text{iff} \quad wRu \text{ and } i\alpha j.$$

For every model $M = (W, R, V)$ and every action structure $\alpha = (K, k, \alpha, Pre_\alpha)$, let

$$\coprod_\alpha M := (\coprod_K W, R \times \alpha, \coprod_K V)$$

be such that its underlying frame is defined as detailed above, and $(\coprod_K V)(p) := \coprod_K V(p)$ for every $p \in \mathsf{AtProp}$. Finally, the *update* of M with the action structure α is the submodel $M^\alpha := (W^\alpha, R^\alpha, V^\alpha)$ of $\coprod_\alpha M$ the domain of which is the subset

$$W^\alpha := \{(w, j) \in \coprod_K W \mid M, w \Vdash Pre_\alpha(j)\}.$$

---

[2] The multi-agent generalization of this simpler version is straightforward, and consists in taking the indexed version of the modal operators, axioms, and interpreting relations (both in the models and in the action structures) over a set of agents.

[3] We will of course apply this definition to relations α which are part of the specification of some action structure; in these cases, the symbol α in $\coprod_\alpha \mathcal{F}$ will be understood as the action structure. This is why the abuse of notation turns out to be useful.

Given this preliminary definition, formulas of the form $\langle\alpha\rangle A$ are evaluated as follows:

$$M, w \Vdash \langle\alpha\rangle A \quad \text{iff} \quad M, w \Vdash Pre_\alpha(k) \text{ and } M^\alpha, (w, k) \Vdash A.$$

Proposition 2 ([3, Theorem 3.5]). *EAK is axiomatized completely by the axioms and rules for the minimal normal modal logic K plus the following axioms:*

1. $\langle\alpha\rangle p \leftrightarrow (Pre(\alpha) \wedge p)$;
2. $\langle\alpha\rangle\neg A \leftrightarrow (Pre(\alpha) \wedge \neg\langle\alpha\rangle A)$;
3. $\langle\alpha\rangle(A \vee B) \leftrightarrow (\langle\alpha\rangle A \vee \langle\alpha\rangle B)$;
4. $\langle\alpha\rangle\Diamond A \leftrightarrow (Pre(\alpha) \wedge \bigvee\{\Diamond\langle\alpha_i\rangle A \mid k\alpha i\})$.

An immediate and well known consequence of the theorem above is that every $\mathcal{L}$-formula is EAK-equivalent to some formula in the static fragment of $\mathcal{L}$. This implies in particular that $\mathcal{L}$-formulas are invariant under standard bisimulation, and this fact extends of course to the multi-agent version.

The representation of actions as action structures is just one possible approach. Here we prefer to keep a black-box perspective on actions, and to identify agents a with the indistinguishability relation they induce on actions; so, in the remainder of the paper, the role of the action-structures α_i for $k\alpha i$ will be played by actions β such that $\alpha a \beta$.

4 EAK Displayed

In the present section, the display calculus D.EAK for the logic EAK (cf. section 3) is introduced piecewise: in the next subsection, display calculi will be presented which are multi-modal versions of display-style sequent calculi proposed in the literature for the (bi-)intuitionistic versions of basic and tense normal modal logic [11,21]. This presentation is modular w.r.t. intuitionistic logic: namely, for the sake of a more straightforward extension to the intuitionistic counterparts of PAL and EAK [13,17], it takes the connectives in the language of IEAK as first-class citizens; the classical base is captured by adding the two *Grishin rules* (see below) to the system. In section 4.2, the rules for the dynamic connectives are introduced. The calculus D.EAK consists of all the rules in the two subsections.

The language $\mathcal{L}(\text{m-IK})$ of the multi-modal version of Fischer Servi's intuitionistic modal logic IK features one pair of modal connectives for each element a in a set A of agents, and consists of *formulas* built from a set of atomic propositions $\{p, q, r, \ldots\}$ and one constant $\bot$, according to the following BNF grammar:

$$A := p \mid \bot \mid A\wedge A \mid A\vee A \mid A \rightarrow A \mid \Diamond_a A \mid \Box_a A.$$

The language $\mathcal{L}(\text{tm-IK})$ of the "tense-like" version of m-IK is obtained by expanding $\mathcal{L}(\text{m-IK})$ with one pair of (adjoint) modalities $\blacklozenge_a$ and $\blacksquare_a$, for each a in A.

The language $\mathcal{L}(\text{btm-IK})$ of the bi-intuitionistic version of tm-IK is obtained by expanding the language of tm-IK with $\top$ and one extra propositional connective $\succ$, referred to as *subtraction* or *disimplication*,[4] which behaves as the dual intuitionistic implication. The reader is referred to [18] for an axiomatic presentation of bi-intuitionistic logic and to [11,12] for its relative display calculi.

---

[4] Formulas $A \succ B$ are classically equivalent to $\neg A \wedge B$.

4.1 The Static Fragment

Display calculi typically involve sequents $X \vdash Y$, where X and Y are *structures*, built from formulas A (in the present case, $A \in \mathcal{L}(\text{m-IK})$ (resp. $\mathcal{L}(\text{tm-IK})$, $\mathcal{L}(\text{btm-IK})$)) and the structural constant I by means of *structural connectives* (or *proxies*), according to the following BNF grammar[5]:

$$X := I \mid A \mid X > X \mid X\,;X \mid \bullet_a X \mid \circ_a X.$$

Each structural connective is associated with a pair of logical connectives, as follows:

Proxies	$>$	$;$	I		$\circ_a$	$\bullet_a$	Structural symbols
Connectives	$\succ$ $\rightarrow$	$\wedge$ $\vee$	$\top$	$\bot$	$\Diamond_a$ $\Box_a$	$\blacklozenge_a$ $\blacksquare_a$	Operational symbols

moreover, structural connectives form adjoint pairs by definition (which will be witnessed in the ensuing display postulates), as follows:

$$;\ \dashv > \qquad > \dashv\ ; \qquad \circ_a \dashv \bullet_a \qquad \bullet_a \dashv \circ_a$$

The display calculi D.m-IK, D.tm-IK and D.btm-IK are defined by means of rules which are classified as *structural* and as *operational* rules. The structural rules below only concern structural connectives, and are common to the three of them (where the structures $X^{-\alpha}$ and $Y^{-\alpha}$ are dynamic-proxy-free):[6]

$$\frac{}{p \vdash p}\,Id \qquad \frac{X \vdash Y}{I \vdash X > Y}\,I_L \qquad \frac{X \vdash Z}{Y^{-\alpha} \vdash X > Z}\,W_L \qquad \frac{X\,;X \vdash Y}{X \vdash Y}\,C_L \qquad \frac{I \vdash X}{I \vdash \circ_a X}\,I_{\circ_a}$$

$$\frac{X \vdash A \quad A \vdash Y}{X \vdash Y}\,Cut \qquad \frac{Y \vdash X}{X > Y \vdash I}\,I_R \qquad \frac{Z \vdash Y}{Y > Z \vdash X^{-\alpha}}\,W_R \qquad \frac{Y \vdash X\,;X}{Y \vdash X}\,C_R \qquad \frac{I \vdash X}{I \vdash \bullet_a X}\,I_{\bullet_a}$$

$$\frac{Y\,;X \vdash Z}{X\,;Y \vdash Z}\,E_L \qquad \frac{X\,;(Y\,;Z) \vdash W}{(X\,;Y)\,;Z \vdash W}\,A_L \qquad \frac{Y \vdash \circ_a X > \circ_a Z}{Y \vdash \circ_a(X > Z)}\,{}_{\circ_a}> \qquad \frac{X\,;Y \vdash Z}{Y \vdash X > Z}\,; > \qquad \frac{\circ_a X \vdash Y}{X \vdash \bullet_a Y}\,{}^{\circ_a}{}_{\bullet_a}$$

$$\frac{Z \vdash X\,;Y}{Z \vdash Y\,;X}\,E_R \qquad \frac{W \vdash (Z\,;Y)\,;X}{W \vdash Z\,;(Y\,;X)}\,A_R \qquad \frac{Y \vdash \bullet_a X > \bullet_a Z}{Y \vdash \bullet_a(X > Z)}\,{}_{\bullet_a}> \qquad \frac{Z \vdash Y\,;X}{Y > Z \vdash X}\,> ; \qquad \frac{X \vdash \circ_a Y}{\bullet_a X \vdash Y}\,{}_{\bullet_a}{}^{\circ_a}$$

The operational rules govern the introduction of the logical connectives: here below are the ones which are common to the three calculi:

$$\frac{}{\bot \vdash I}\,\bot_L \qquad \frac{I \vdash X}{\top \vdash X}\,\top_L \qquad \frac{A\,;B \vdash Z}{A \wedge B \vdash Z}\,\wedge_L \qquad \frac{B \vdash Y \quad A \vdash X}{B \vee A \vdash Y\,;X}\,\vee_L \qquad \frac{X \vdash A \quad B \vdash Y}{A \rightarrow B \vdash X > Y}\,\rightarrow_L$$

$$\frac{X \vdash I}{X \vdash \bot}\,\bot_R \qquad \frac{}{I \vdash \top}\,\top_R \qquad \frac{X \vdash A \quad Y \vdash B}{X\,;Y \vdash A \wedge B}\,\wedge_R \qquad \frac{Z \vdash B\,;A}{Z \vdash B \vee A}\,\vee_R \qquad \frac{Z \vdash A > B}{Z \vdash A \rightarrow B}\,\rightarrow_R$$

---

[5] Notice that, in the context of the full calculus, the variables X, Y, Z, W appearing in the rules in the present subsection are to be interpreted as structures of the full language of D.EAK, unless explicitly indicated otherwise with symbols such as $X^{-\alpha}$.

[6] The weakening rules W_L and W_R are equivalent to the standard ones via the Display Postulates $\binom{;}{>}$ and $\binom{>}{;}$; in these rules, the principal structure appears 'in display'; besides making an easier life in the proof of the cut elimination, we believe that this feature of W_L and W_R is more in line with the general design principles of display calculi. Notice also that the presence of the rules E_L and E_R makes it possible for us to dispense with the structural connective $<$ and its relative rules, such as $A\,;B \vdash C/A \vdash C < B$.

Here below, from the left to right, are the operational rules completing D.m-IK (1st and 2nd column), D.tm-IK (4th and 5th column), and D.btm-IK (3rd column):

$$\dfrac{\circ_a A \vdash X}{\Diamond_a A \vdash X} \Diamond_{aL} \qquad \dfrac{A \vdash X}{\Box_a A \vdash \circ_a X} \Box_{aL} \qquad \dfrac{A > B \vdash Z}{A \succ B \vdash Z} \succ_L \qquad \dfrac{\bullet_a A \vdash X}{\blacklozenge_a A \vdash X} \blacklozenge_{aL} \qquad \dfrac{A \vdash X}{\blacksquare_a A \vdash \bullet_a X} \blacksquare_{aL}$$

$$\dfrac{X \vdash A}{\circ_a X \vdash \Diamond_a A} \Diamond_{aR} \qquad \dfrac{X \vdash \circ_a A}{X \vdash \Box_a A} \Box_{aR} \qquad \dfrac{A \vdash X \qquad Y \vdash B}{X > Y \vdash A \succ B} \succ_R \qquad \dfrac{X \vdash A}{\bullet_a X \vdash \blacklozenge_a A} \blacklozenge_{aR} \qquad \dfrac{X \vdash \bullet_a A}{X \vdash \blacksquare_a A} \blacksquare_{aR}$$

Finally, the classical versions of each of these calculi can be obtained from the above ones e.g. by adding the following *Grishin's* structural rules [12, 21]:

$$\dfrac{X > (Y\,;Z) \vdash W}{(X > Y)\,;Z \vdash W} Gri_L \qquad \dfrac{W \vdash X > (Y\,;Z)}{W \vdash (X > Y)\,;Z} Gri_R$$

4.2   The Dynamic Fragment

The calculi introduced in the present subsection involve sequents $X \vdash Y$, where X and Y are *structures*, built from formulas $A \in \mathcal{L}(\text{m-IEAK})$ (resp. $\mathcal{L}(\text{tm-IEAK})$, $\mathcal{L}(\text{btm-IEAK})$) and the structural constant I according to the following BNF grammar:

$$X := \text{I} \mid A \mid X\,;X \mid X > X \mid \bullet_a X \mid \circ_a X \mid \{\alpha\}X \mid \widetilde{\underline{\alpha}}\,X.$$

Hence, the structural language above expands the one of the previous subsection with structural connectives $\{\alpha\}$ and $\widetilde{\underline{\alpha}}$ for each action α; these are by definition adjoint to each other as follows: $\{\alpha\} \dashv \widetilde{\underline{\alpha}}$ and $\widetilde{\underline{\alpha}} \dashv \{\alpha\}$. The proxy $\{\alpha\}$ is associated with the logical connectives $[\alpha]$ and $\langle\alpha\rangle$, and thus it occurs in the operational rules concerning them. Likewise, new logical connectives $\widehat{\underline{\alpha}}$ and $\underline{\alpha}$ are introduced which stand in an analogous relation with $\widetilde{\underline{\alpha}}$, and which are adjoint to $[\alpha]$ and $\langle\alpha\rangle$ as follows: $\langle\alpha\rangle \dashv \underline{\alpha}$ and $\widehat{\underline{\alpha}} \dashv [\alpha]$. As discussed in section 2, these new connectives have a natural interpretation in the final coalgebra, but not in the standard semantics.

$\{\alpha\}$	$\widehat{\underline{\alpha}}$	
$[\alpha]$	$\langle\alpha\rangle$	$\widehat{\underline{\alpha}}$	$\underline{\alpha}$

The display calculi D.m-IEAK, D.tm-IEAK and D.btm-IEAK are defined by adding the rules of the present subsection to D.m-IK, D.tm-IK and D.btm-IK, respectively; the display calculus D.EAK is obtained by adding the Grishin rules to D.m-IK. The rules in the present subsection come in four groups: pure and contextual structural and operational rules. Here follow the pure structural rules; the dynamic *display postulates* appear in the 5th column below:

$$\dfrac{}{\{\alpha\} p \vdash p} atom_L \qquad \dfrac{}{p \vdash \{\alpha\} p} atom_R \qquad \dfrac{X \vdash Y}{\{\alpha\} X \vdash \{\alpha\} Y} balance$$

$$\dfrac{\{\alpha\}Y > \{\alpha\}Z \vdash X}{\{\alpha\}(Y > Z) \vdash X} {}_{\{\alpha\}} > \quad \dfrac{\{\alpha\}X\,;\{\alpha\}Y \vdash Z}{\{\alpha\}(X\,;Y) \vdash Z} {}_{\{\alpha\}}; \quad \dfrac{\widehat{\underline{\alpha}}\,Y > \widehat{\underline{\alpha}}\,X \vdash Z}{\widehat{\underline{\alpha}}\,(Y > X) \vdash Z} {}_{\widehat{\underline{\alpha}}} > \quad \dfrac{\widehat{\underline{\alpha}}\,X\,;\widehat{\underline{\alpha}}\,Y \vdash Z}{\widehat{\underline{\alpha}}\,(X\,;Y) \vdash Z} {}_{\widehat{\underline{\alpha}}}; \quad \dfrac{\{\alpha\}X \vdash Y}{X \vdash \widetilde{\underline{\alpha}}\,Y} {}_{\{\alpha\}}^{\widetilde{\underline{\alpha}}}$$

$$\dfrac{Y \vdash \{\alpha\}X > \{\alpha\}Z}{Y \vdash \{\alpha\}(X > Z)} {}_{> \{\alpha\}} \quad \dfrac{Z \vdash \{\alpha\}Y\,;\{\alpha\}X}{Z \vdash \{\alpha\}(Y\,;X)} {}_{;\{\alpha\}} \quad \dfrac{Y \vdash \widehat{\underline{\alpha}}\,X > \widehat{\underline{\alpha}}\,Z}{Y \vdash \widehat{\underline{\alpha}}\,(X > Z)} {}_{> \widehat{\underline{\alpha}}} \quad \dfrac{Z \vdash \widehat{\underline{\alpha}}\,Y\,;\widehat{\underline{\alpha}}\,X}{Z \vdash \widehat{\underline{\alpha}}\,(Y\,;X)} {}_{;\widehat{\underline{\alpha}}} \quad \dfrac{Y \vdash \{\alpha\}X}{\widetilde{\underline{\alpha}}\,Y \vdash X} {}_{\{\alpha\}}^{\widetilde{\underline{\alpha}}}$$

Here below are the pure operational rules:

$$\frac{\{\alpha\}A \vdash X}{\langle\alpha\rangle A \vdash X} \; \langle\alpha\rangle_L \qquad \frac{X \vdash A}{\{\alpha\}X \vdash \langle\alpha\rangle A} \; \langle\alpha\rangle_R \qquad \frac{A \vdash X}{[\alpha]A \vdash \{\alpha\}X} \; [\alpha]_L \qquad \frac{X \vdash \{\alpha\}A}{X \vdash [\alpha]A} \; [\alpha]_R$$

The *contextual* rules encode inferences which can be performed only in the presence of a given assumption (in the case at hand, the preconditions of the action parametrizing a dynamic proxy). Here below the contextual structural rules:

reduce   swap-in   swap-out

$$\frac{Pre(\alpha)\,;\{\alpha\}A \vdash X}{\{\alpha\}A \vdash X} \; r_L \qquad \frac{Pre(\alpha)\,;\{\alpha\}\circ_a X \vdash Y}{Pre(\alpha)\,;\circ_a\{\beta\}_{\alpha a\beta} X \vdash Y} \; s\text{-}in_L \qquad \frac{\big(Pre(\alpha)\,;\circ_a\{\beta\} X \vdash Y \mid \alpha a\beta\big)}{Pre(\alpha)\,;\{\alpha\}\circ_a X \vdash \;\text{\bf ;}\big(Y \mid \alpha a\beta\big)} \; s\text{-}out_L$$

$$\frac{X \vdash Pre(\alpha) > \{\alpha\}A}{X \vdash \{\alpha\}A} \; r_R \qquad \frac{Y \vdash Pre(\alpha) > \{\alpha\}\circ_a X}{Y \vdash Pre(\alpha) > \circ_a\{\beta\}_{\alpha a\beta} X} \; s\text{-}in_R \qquad \frac{\big(Y \vdash Pre(\alpha) > \circ_a\{\beta\} X \mid \alpha a\beta\big)}{\text{\bf ;}\big(Y \mid \alpha a\beta\big) \vdash Pre(\alpha) > \{\alpha\}\circ_a X} \; s\text{-}out_R$$

The *swap-in* rules are unary and should be read as follows: if the premise holds, then the conclusion holds relative to any action β such that $\alpha a\beta$. The *swap-out* rules do not have a fixed arity; they have as many premises[7] as there are actions β such that $\alpha a\beta$; in the conclusion, the symbol $\text{\bf ;}\big(Y \mid \alpha a\beta\big)$ refers to a string $(\cdots(Y\,;Y)\,;\cdots\,;Y)$ with n occurrences of Y, where n is the number of actions β such that $\alpha a\beta$. Finally, the contextual operational rules:

reverse

$$\frac{Pre(\alpha)\,;\{\alpha\}A \vdash X}{Pre(\alpha)\,;[\alpha]A \vdash X} \; rev_L \qquad \frac{X \vdash Pre(\alpha) > \{\alpha\}A}{X \vdash Pre(\alpha) > \langle\alpha\rangle A} \; rev_R$$

5   Soundness, Completeness and Cut Elimination

5.1   Soundness in the Final Coalgebra

In the present section, we outline the soundness of D.EAK w.r.t. the final coalgebra semantics. Structures will be translated into formulas, and formulas will be interpreted as subsets of the final coalgebra, as discussed in section 2. In order to translate structures as formulas, proxies need to be translated as logical connectives; to this effect, any given occurrence of a proxy is translated as one or the other of its associated logical connectives, according to which side of the sequent the given occurrence can be displayed on as main connective [6, 21], as reported in Table 1.

Sequents $A \vdash B$ will be interpreted as inclusions $[\![A]\!]_Z \subseteq [\![B]\!]_Z$; rules $(A_i \vdash B_i \mid i \in I)/C \vdash D$ will be interpreted as implications of the form "if $[\![A_i]\!]_Z \subseteq [\![B_i]\!]_Z$ for every $i \in I$, then $[\![C]\!]_Z \subseteq [\![D]\!]_Z$". As for rules not involving $\widetilde{\alpha}$, we will rely on the following observation, which is based on the invariance of EAK-formulas under bisimulation (cf. Section 3):

---

[7] The *swap-out* rule could indeed be infinitary if action structures were allowed to be infinite, which in the present setting, as in [3], is not the case.

Table 1. Translation of proxies into logical connectives

Main connective	if displayed in antecedent	if displayed in succedent
I	⊤	⊥
$A \, ; B$	$A \wedge B$	$A \vee B$
$A > B$	$A \succ B$	$A \rightarrow B$
$\circ A$	$\Diamond A$	$\Box A$
$\bullet A$	$\blacklozenge A$	$\blacksquare A$
$\{\alpha\}A$	$\langle \alpha \rangle A$	$[\alpha]A$
$\widehat{\underline{\alpha}} \, A$	$\widehat{\underline{\alpha}} \, A$	$\overline{\underline{\alpha}} \, A$

Lemma 1. *The following are equivalent for all EAK-formulas A and B:*
(1) $[\![A]\!]_Z \subseteq [\![B]\!]_Z$*;*
(2) $[\![A]\!]_M \subseteq [\![B]\!]_M$ *for every model M.*

Proof. The direction from (2) to (1) is clear; conversely, fix a model M, and let $f : M \rightarrow Z$ be the unique arrow; then (1) immediately implies that $[\![A]\!]_M = f^{-1}([\![A]\!]_Z) \subseteq f^{-1}([\![B]\!]_Z) = [\![B]\!]_M$.

In the light of the lemma above, and using the translations provided in Table 1, the soundness of unary rules $A \vdash B / C \vdash D$ not involving $\widehat{\underline{\alpha}}$, such as *balance*, $\langle \alpha \rangle_R$ and $[\alpha]_L$, can be straightforwardly checked as implications of the form "if $[\![A]\!]_M \subseteq [\![B]\!]_M$ on every model M, then $[\![C]\!]_M \subseteq [\![D]\!]_M$ on every model M". As an example, let us check the soundness of *balance*: Let A, B be EAK-formulas such that $[\![A]\!]_M \subseteq [\![B]\!]_M$ on every model M. Let us fix a model M, and show that $[\![\langle \alpha \rangle A]\!]_M \subseteq [\![[\alpha]B]\!]_M$. As discussed in [13, Subsection 4.2], the following identities hold in any standard model:

$$[\![\langle \alpha \rangle A]\!]_M = [\![Pre(\alpha)]\!]_M \cap \iota_k^{-1}[i[[\![A]\!]_{M^\alpha}]], \tag{4}$$

$$[\![[\alpha]A]\!]_M = [\![Pre(\alpha)]\!]_M \Rightarrow \iota_k^{-1}[i[[\![A]\!]_{M^\alpha}]], \tag{5}$$

where the map $i : M^\alpha \rightarrow \coprod_\alpha M$ is the submodel embedding, and $\iota_k : M \rightarrow \coprod_\alpha M$ is the embedding of M into its k-colored copy. Letting $g(-) := \iota_k^{-1}[i[-]]$, we need to show

$$[\![Pre(\alpha)]\!]_M \cap g([\![A]\!]_{M^\alpha}) \subseteq [\![Pre(\alpha)]\!]_M \Rightarrow g([\![B]\!]_{M^\alpha}).$$

This is a direct consequence of the Heyting-valid implication "if $b \leq c$ then $a \wedge b \leq a \rightarrow c$", the monotonicity of g, and the assumption that $[\![A]\!]_M \subseteq [\![B]\!]_M$ holds on every model, hence on M^α.

Actually, for all rules $(A_i \vdash B_i \mid i \in I) / C \vdash D$ not involving $\widehat{\underline{\alpha}}$ except *balance*, $\langle \alpha \rangle_R$ and $[\alpha]_L$, stronger soundness statements can be proven of the form "for every model M, if $[\![A_i]\!]_M \subseteq [\![B_i]\!]_M$ for every $i \in I$, then $[\![C]\!]_M \subseteq [\![D]\!]_M$" (this amounts to the soundness w.r.t. the standard semantics). This is the case for all display postulates not involving $\widehat{\underline{\alpha}}$, the soundness of which boils down to the well known adjunction conditions holding in every model M. As to the remaining rules not involving $\widehat{\underline{\alpha}}$, thanks to the following general principle of *indirect (in)equality*, the stronger soundness condition above boils down to the verification of inclusions which interpret validities of IEAK [13], and hence, a fortiori, of EAK. Same arguments hold for the Grishin rules, except that their soundness boils down to classical but not intuitionistic validities.

Lemma 2. *(Principle of indirect inequality)* Tfae for any preorder P and all $a, b \in P$:
(1) $a \leq b$;
(2) $x \leq a$ implies $x \leq b$ for every $x \in P$;
(3) $b \leq y$ implies $a \leq y$ for every $y \in P$.

As an example, let us verify $s\text{-}out_L$: fix a model M, fix EAK-formulas A and B, and assume that for every action β, if $\alpha a \beta$ then $[\![Pre(\alpha)]\!]_M \cap [\![\Diamond_a \langle \beta \rangle A]\!]_M \subseteq [\![B]\!]_M$, i.e., that $[\![Pre(\alpha)]\!]_M \cap \bigcup \{[\![\Diamond_a \langle \beta \rangle A]\!]_M \mid \alpha a \beta\} \subseteq [\![B]\!]_M$; we need to show that $[\![Pre(\alpha)]\!]_M \cap [\![\langle \alpha \rangle \Diamond_a A]\!]_M \subseteq [\![B]\!]_M$. By the principle of indirect inequality, it is enough to show that

$$[\![\langle \alpha \rangle \Diamond_a A]\!]_M \subseteq [\![Pre(\alpha)]\!]_M \cap \bigcup \{[\![\Diamond_a \langle \beta \rangle A]\!]_M \mid \alpha a \beta\},$$

which is true (cf. Proposition 2). Finally, the soundness of the rules which do involve $\widehat{\alpha}$ remains to be shown. The soundness of the display postulates immediately follows from Proposition 1. As an example, let us verify the soundness of $\left(\frac{\widehat{\alpha}}{>} \right)$: translating the structures into formulas, and applying the principle of indirect inequality, it boils down to verifying that $[\![\widehat{\alpha} (A \succ B)]\!]_Z \subseteq [\![\overline{\alpha} A]\!]_Z \succ [\![\widehat{\alpha} B]\!]_Z$ for all EAK-formulas A and B. Since, in Z, $\widehat{\alpha}$ and $\overline{\alpha}$ are respectively interpreted as $\langle \alpha^\circ \rangle$ and $[\alpha^\circ]$, this inclusion can be rewritten as

$$\langle \alpha^\circ \rangle ([\![A]\!]_Z \succ [\![B]\!]_Z) \subseteq [\alpha^\circ][\![A]\!]_Z \succ \langle \alpha^\circ \rangle [\![B]\!]_Z,$$

where $A \succ B$ can be interpreted classically, i.e. as $\neg A \wedge B$. The straightforward verification that this is an instance of a principle valid in every frame is left to the reader.

5.2 Completeness

For the completeness of D.EAK, it is enough to show that all the axioms of EAK are derivable in D.EAK. Due to space restrictions, here we only report on the derivations of $\langle \alpha \rangle \Diamond_a A \leftrightarrow Pre(\alpha) \wedge \bigvee \{\Diamond_a \langle \beta \rangle A \mid \alpha a \beta\}$. For ease of notation, we assume that the actions β such that $\alpha a \beta$ form the set $\{\beta_i \mid 1 \leq i \leq n\}$.

$$
\begin{array}{ccc}
\dfrac{\dfrac{\dfrac{A \vdash A}{\{\beta_1\}A \vdash \langle \beta_1 \rangle A}}{\dfrac{\Diamond_a \{\beta_1\}A \vdash \Diamond_a \langle \beta_1 \rangle A}{Pre(\alpha) \,;\, \Diamond_a \{\beta_1\}A \vdash \Diamond_a \langle \beta_1 \rangle A}} & \cdots & \dfrac{\dfrac{\dfrac{A \vdash A}{\{\beta_n\}A \vdash \langle \beta_n \rangle A}}{\dfrac{\Diamond_a \{\beta_n\}A \vdash \Diamond_a \langle \beta_n \rangle A}{Pre(\alpha) \,;\, \Diamond_a \{\beta_n\}A \vdash \Diamond_a \langle \beta_n \rangle A}}
\end{array}
$$

(schematic derivation)

$$Pre(\alpha) \,;\, \{a\}\Diamond_a A \vdash \,\stackrel{\bullet}{\mathbf{;}} \left(\Diamond_a \langle \beta_i \rangle A \right)$$

$$Pre(\alpha) \,;\, \{a\}\Diamond_a A \vdash \bigvee \left(\Diamond_a \langle \beta_i \rangle A \right)$$

$$\{a\}\Diamond_a A \vdash \bigvee \left(\Diamond_a \langle \beta_i \rangle A \right)$$

$$\dfrac{Pre(\alpha) \vdash Pre(\alpha) \qquad Pre(\alpha) \,;\, \{a\}\Diamond_a A \vdash Pre(\alpha) \wedge \bigvee \left(\Diamond_a \langle \beta_i \rangle A \right)}{\{a\}\Diamond_a A \vdash Pre(\alpha) \wedge \bigvee \left(\Diamond_a \langle \beta_i \rangle A \right)}$$

$$\Diamond_a A \vdash \overline{\alpha} Pre(\alpha) \wedge \bigvee \left(\Diamond_a \langle \beta_i \rangle A \right)$$

$$\Diamond_a A \vdash \overline{\alpha} Pre(\alpha) \wedge \bigvee \left(\Diamond_a \langle \beta_i \rangle A \right)$$

$$\{a\}\Diamond_a A \vdash Pre(\alpha) \wedge \bigvee \left(\Diamond_a \langle \beta_i \rangle A \right)$$

$$\langle \alpha \rangle \Diamond_a A \vdash Pre(\alpha) \wedge \bigvee \left(\Diamond_a \langle \beta_i \rangle A \right)$$

$$
\begin{array}{c}
\cfrac{
\cfrac{
\cfrac{
\cfrac{
\cfrac{
\cfrac{
\cfrac{
\cfrac{
\cfrac{
\cfrac{A \vdash A}{\circ_a A \vdash \Diamond_a A}
}{\{\alpha\}\circ_a A \vdash \langle\alpha\rangle\Diamond_a A}
}{\cfrac{Pre(\alpha)\,;\{\alpha\}\circ_a A \vdash \langle\alpha\rangle\Diamond_a A}{Pre(\alpha)\,;\circ_a\{\beta_1\}A \vdash \langle\alpha\rangle\Diamond_a A}\ \text{s-in}}
}{\circ_a\{\beta_1\}A \vdash Pre(\alpha) > \langle\alpha\rangle\Diamond_a A}
}{\{\beta_1\}A \vdash \bullet_a(Pre(\alpha) > \langle\alpha\rangle\Diamond_a A)}
}{\langle\beta_1\rangle A \vdash \bullet_a(Pre(\alpha) > \langle\alpha\rangle\Diamond_a A)}
}{\circ_a\langle\beta_1\rangle A \vdash Pre(\alpha) > \langle\alpha\rangle\Diamond_a A}
}{\Diamond_a\langle\beta_1\rangle A \vdash Pre(\alpha) > \langle\alpha\rangle\Diamond_a A}
& \cdots & \cdots
}{}
\end{array}
$$

Right derivation:

$$
\cfrac{
\cfrac{
\cfrac{
\cfrac{
\cfrac{
\cfrac{
\cfrac{
\cfrac{\cfrac{A \vdash A}{\circ_a A \vdash \Diamond_a A}}{\{\alpha\}\circ_a A \vdash \langle\alpha\rangle\Diamond_a A}
}{\cfrac{Pre(\alpha)\,;\{\alpha\}\circ_a A \vdash \langle\alpha\rangle\Diamond_a A}{Pre(\alpha)\,;\circ_a\{\beta_n\}A \vdash \langle\alpha\rangle\Diamond_a A}\ \text{s-in}}
}{\circ_a\{\beta_n\}A \vdash Pre(\alpha) > \langle\alpha\rangle\Diamond_a A}
}{\{\beta_n\}A \vdash \bullet_a(Pre(\alpha) > \langle\alpha\rangle\Diamond_a A)}
}{\langle\beta_n\rangle A \vdash \bullet_a(Pre(\alpha) > \langle\alpha\rangle\Diamond_a A)}
}{\circ_a\langle\beta_n\rangle A \vdash Pre(\alpha) > \langle\alpha\rangle\Diamond_a A}
}{\Diamond_a\langle\beta_n\rangle A \vdash Pre(\alpha) > \langle\alpha\rangle\Diamond_a A}
$$

$$
\cfrac{
\cfrac{
\cfrac{
\bigvee\!\left(\Diamond_a\langle\beta_i\rangle A\right) \vdash \begin{array}{c}\bullet\\[-3pt]\bullet\end{array}_a\!\left(Pre(\alpha) > \langle\alpha\rangle\Diamond_a A\right)
}{\bigvee\!\left(\Diamond_a\langle\beta_i\rangle A\right) \vdash Pre(\alpha) > \langle\alpha\rangle\Diamond_a A}
}{Pre(\alpha)\,;\bigvee\!\left(\Diamond_a\langle\beta_i\rangle A\right) \vdash \langle\alpha\rangle\Diamond_a A}
}{Pre(\alpha)\wedge\bigvee\!\left(\Diamond_a\langle\beta_i\rangle A\right) \vdash \langle\alpha\rangle\Diamond_a A}
$$

5.3 Cut-Elimination

In the present subsection, we outline the proof of the cut eliminability of D.EAK following the original strategy devised by Gentzen (cf. [20]). Without loss of generality, we consider a derivation π of the sequent $X \vdash Y$ in D.EAK which contains a unique application of Cut as the last rule (let us refer to this application as Cut*), and we show that a derivation of the same sequent exists in which Cut is not applied. We proceed by induction on the set of tuples (ρ, δ), ordered lexicographically, where ρ is the complexity of the cut formula in Cut* (the *rank* of Cut*), and δ is the sum of the maximal lengths of branches in the subdeductions of the premises of Cut* (the *degree* of Cut*). In the base case, Cut* can be directly eliminated by exhibiting a cut-free proof π' with the same conclusion. This is more in general the case when the cut formula is atomic.

The inductive step consists in transforming π into a derivation π' with the same conclusion and one or more applications of Cut with lower rank or with same rank but lower degree. The typical situation in the original Gentzen proof is that, when the cut formula is not atomic and is not *principal*[8] in at least one of the premises, the transformation involves one or more Cut-applications of same rank and lower degree than Cut*, whereas when the cut formula is not atomic and is principal in both premises, the transformation involves one or more Cut-applications of lower rank than Cut*, as illustrated, e.g., in the following transformation:

$$
\cfrac{
\cfrac{\vdots\ \pi_1}{\cfrac{X \vdash A}{\{\alpha\}X \vdash \langle\alpha\rangle A}}
\quad
\cfrac{\vdots\ \pi_2}{\cfrac{\{\alpha\}A \vdash Y}{\langle\alpha\rangle A \vdash Y}}
}{\{\alpha\}X \vdash Y}
\quad\rightsquigarrow\quad
\cfrac{
\cfrac{\vdots\ \pi_1}{X \vdash A}
\quad
\cfrac{\cfrac{\vdots\ \pi_2}{\{\alpha\}A \vdash Y}}{A \vdash \overline{\underline{\alpha}}\,Y}
}{\cfrac{X \vdash \overline{\underline{\alpha}}\,Y}{\{\alpha\}X \vdash Y}}
$$

This regularity breaks down when the Cut-formula is principal in both premises and has been introduced by means of an application of either contextual rules *reverse*. In

---

[8] An occurrence of a formula in a node of a derivation is *principal* if that occurrence has been introduced by means of the last rule applied in the subdeduction ending in that node.

this case, such a simple transformation as the one above is not available, and we need to consider all the possible ways in which the *proxy* $\{\alpha\}$ has been introduced in the subdeduction of each premise of Cut*. The proxy $\{\alpha\}$ might have been introduced by $\langle\alpha\rangle_L$, *atom*$_R$, *balance*, and $\left(\frac{\overline{\alpha}}{\{\alpha\}}\right)$ (applied bottom-up) on the left premise, and by $[\alpha]_R$, *atom*$_L$, *balance*, and $\left(\frac{\langle\alpha\rangle}{\overline{\alpha}}\right)$ (applied bottom-up) on the right premise. This creates 16 sub-cases (each of which can be subdivided into simpler and more complicated instances), of which we illustrate just two (in their least complicated incarnations), as examples: the following one produces a Cut application of lower rank:

$$
\begin{array}{c}
\vdots\,\pi_1 \qquad\qquad \vdots\,\pi_2 \\
\dfrac{B \vdash A \qquad A \vdash C}{B \vdash C} \\
\{\alpha\}B \vdash \langle\alpha\rangle C \\
\vdots\,\pi_2^* \\
\dfrac{Pre(\alpha)\,;\{\alpha\}B \vdash Y}{Pre(\alpha)\,;[\alpha]B \vdash Y} \\
[\alpha]B \vdash Pre(\alpha) > Y \\
\vdots\,\pi_1^* \\
X \vdash Pre(\alpha) > Y
\end{array}
\quad \rightsquigarrow \quad
\begin{array}{c}
\vdots\,\pi_1 \qquad\qquad \vdots\,\pi_2 \\
\dfrac{B \vdash A}{[\alpha]B \vdash \{\alpha\}A} \qquad \dfrac{A \vdash C}{\{\alpha\}A \vdash \langle\alpha\rangle C} \\
\vdots\,\pi_2^* \\
\dfrac{Pre(\alpha)\,;\{\alpha\}A \vdash Y}{Pre(\alpha)\,;[\alpha]A \vdash Y} \\
[\alpha]A \vdash Pre(\alpha) > Y \\
\dfrac{X \vdash \{\alpha\}A}{X \vdash [\alpha]A} \\
X \vdash Pre(\alpha) > Y
\end{array}
$$

the next one produces a Cut application of same rank and lower degree:

$$
\begin{array}{c}
\vdots\,\pi_1 \qquad\qquad \vdots\,\pi_2 \\
\dfrac{A \vdash A}{A \vdash \{\alpha\}A} \qquad \dfrac{A \vdash C}{\{\alpha\}A \vdash \langle\alpha\rangle C} \\
\vdots\,\pi_1^* \qquad \vdots\,\pi_2^* \\
\dfrac{X \vdash \{\alpha\}A}{X \vdash [\alpha]A} \quad \dfrac{Pre(\alpha)\,;\{\alpha\}A \vdash Y}{Pre(\alpha)\,;[\alpha]A \vdash Y} \\
[\alpha]A \vdash Pre(\alpha) > Y \\
X \vdash Pre(\alpha) > Y
\end{array}
\quad \rightsquigarrow \quad
\begin{array}{c}
\vdots\,\pi_1 \\
\dfrac{A \vdash A}{A \vdash \{\alpha\}A} \\
\vdots\,\pi_1^* \qquad \vdots\,\pi_2 \\
\dfrac{X \vdash \{\alpha\}A}{X \vdash [\alpha]A} \quad \dfrac{A \vdash C}{[\alpha]A \vdash \{\alpha\}C} \\
X \vdash \{\alpha\}C \\
X \vdash Pre(\alpha) > \{\alpha\}C \\
X \vdash Pre(\alpha) > \langle\alpha\rangle C \\
Pre(\alpha)\,;X \vdash \langle\alpha\rangle C \\
\vdots\,\pi_2^* \\
\dfrac{Pre(\alpha)\,;(Pre(\alpha)\,;X) \vdash Y}{(Pre(\alpha)\,;Pre(\alpha))\,;X \vdash Y} \\
\dfrac{Pre(\alpha)\,;Pre(\alpha) \vdash Y < X}{Pre(\alpha) \vdash Y < X} \\
\dfrac{Pre(\alpha)\,;X \vdash Y}{X \vdash Pre(\alpha) > Y}
\end{array}
$$

6   Conclusions, Conservativity, and Further Directions

Besides the cut-elimination, the results in the present paper can be summarized by the following chain of inclusions between consequence relations, where K is the class of standard Kripke models:

$$\vDash_K = \vdash_{EAK} \subseteq \vdash_{D.EAK} \subseteq \vDash_Z .$$

D.EAK Conservatively Extends EAK. Of course, the language of the latter two consequence relations is an expansion of the language of the former two. To be able to claim that D.EAK adequately captures EAK, we need to show that $\vdash_{D.EAK}$ is a conservative extension of $\vdash_{EAK}$. To see this, let A, B be EAK-formulae such that $A \vdash_{D.EAK} B$. By the soundness of D.EAK w.r.t. the final coalgebra semantics, this implies that $[\![A]\!]_Z \subseteq [\![B]\!]_Z$, which, by Lemma 1, implies that $[\![A]\!]_M \subseteq [\![B]\!]_M$ for every Kripke model M, which, by the completeness of EAK w.r.t. the standard Kripke semantics, implies that $A \vdash_{EAK} B$.

Proof-Theoretic Semantics for EAK. The rules of EAK enjoy the following requirements, which are well known in the literature of proof-theoretic semantics [21, 22]: the fundamental structural rules of D.EAK are 'eliminable': i.e., *Id* can be restricted to atomic formulas, and *Cut* can be removed without affecting the set of theorems. The operational rules enjoy the properties of *separation*: each of them introduces exactly one connective, and of *symmetry*: for each connective, its left-introduction rules and its right-introduction rules form nonempty and disjoint sets. All of them but the *reverse* rules also enjoy *explicitness*, which can be reformulated as follows: the side structures occur unrestricted. However, the offending side substructure is limited to the formula $Pre(\alpha)$, which can always be derived, e.g. via weakening. Hence, we conjecture that this offense is essentially harmless. An entirely satisfactory motivation that D.EAK provides proof-theoretic semantics for the connectives of EAK is work in progress.

Intuitionistic Coalgebraic Semantics. We wish to develop the intuitionistic version of these results. This requires to work in the setting of the final coalgebra for the Vietoris functor on discrete Esakia spaces (S4-frames and p-morphisms).

Cut-Elimination *á la* Belnap. Our proof of cut elimination, which is very lengthy and could only be sketched in the present paper, follows the methodology of Gentzen's original proof. A shorter and more insightful route to the same result consists in either applying Belnap's meta-theorem for cut elimination [6] for display calculi, or some suitable extension of it. In the latter case, this strengthening would be essentially analogous to extension of Belnap's meta-theorem to linear logic [7, 19], and is the focus of current investigation.

References

1. Aczel, P., Mendler, N.: A Final Coalgebra Theorem. In: Pitt, D.H., Rydeheard, D.E., Dybjer, P., Pitts, A.M., Poigné, A. (eds.) Category Theory and Computer Science. LNCS, vol. 389, pp. 357–365. Springer, Heidelberg (1989)
2. Balbiani, P., van Ditmarsch, H.P., Herzig, A., de Lima, T.: Tableaux for public announcement logic. Journal of Logic and Computation 20(1), 55–76 (2010)
3. Baltag, A., Moss, L., Solecki, S.: The Logic of Public Announcements, Common Knowledge, and Private Suspicions. CWI technical report SEN-R9922 (1999)
4. Baltag, A.: A Coalgebraic Semantics for Epistemic Programs. Electronic Notes in Theoretical Computer Science 82(1) (2003)
5. Baltag, A., Coecke, B., Sadrzadeh, M.: Epistemic actions as resources. Journal of Logic and Computation 17, 555–585 (2007)
6. Belnap, N.: Display logic. J. Philos. Logic 11, 375–417 (1982)

7. Belnap, N.: Linear Logic Displayed. Notre Dame J. Formal Logic 31(1), 14–25 (1989)
8. Cîrstea, C., Sadrzadeh, M.: Coalgebraic Epistemic Update Without Change of Model. In: Mossakowski, T., Montanari, U., Haveraaen, M. (eds.) CALCO 2007. LNCS, vol. 4624, pp. 158–172. Springer, Heidelberg (2007)
9. Dyckhoff, R., Sadrzadeh, M., Truffaut, J.: Algebra, Proof Theory and Applications for an Intuitionistic Logic of Propositions, Actions and Adjoint Modal Operators. ACM Transactions on Computational Logic 666 (2013)
10. Fischer-Servi, G.: Axiomatizations for Some Intuitionistic Modal Logics. Rend. Sem. Mat Polit. di Torino 42, 179–194 (1984)
11. Goré, R., Postniece, L., Tiu, A.: Cut-elimination and Proof Search for Bi-Intuitionistic Tense Logic. In: Proc. Advances in Modal Logic, pp. 156–177 (2010)
12. Goré, R.: Dual Intuitionistic Logic Revisited. In: Dyckhoff, R. (ed.) TABLEAUX 2000. LNCS (LNAI), vol. 1847, pp. 252–267. Springer, Heidelberg (2000)
13. Kurz, A., Palmigiano, A.: Epistemic Updates on Algebras. Logical Methods in Computer Science (forthcoming, 2013)
14. Kurz, A., Rosický, J.: Mathematical Structures in Computer Science, vol. 15, pp. 149–166. Cambridge University Press (2005)
15. Maffezzoli, P., Negri, S.: A Gentzen-style analysis of Public Announcement Logic. In: Arrazola, X., Ponte, M. (eds.) Proceedings of the International Workshop on Logic and Philosophy of Knowledge, Communication and Action, pp. 293–313. University of the Basque Country Press (2010)
16. Maffezzoli, P., Negri, S.: A proof theoretical perspective on Public Announcement Logic. Logic and Philosophy of Science 9, 49–59 (2011)
17. Ma, M., Palmigiano, A., Sadrzadeh, M.: Algebraic Semantics and Model Completeness for Intuitionistic Public Announcement Logic. Annals of Pure and Applied Logic (to appear, 2013)
18. Rauszer, C.: A Formalization of The Propositional Calculus of H-B Logic. Studia Logica 33, 23–34 (1974)
19. Restall, G.: An Introduction to Substructural Logics. Routledge, London (2000)
20. Troelstra, A.S., Schwichtenberg, H.: Basic Proof Theory. Cambridge University Press (2000)
21. Wansing, H.: Displaying Modal Logic. Kluwer Academic Publisher, Dordrecht (1998)
22. Wansing, H.: The Idea of a Proof-Theoretic Semantics and the Meaning of the Logical Operations. Studia Logica 64(1), 3–20 (2000)

Cellular Games, Nash Equilibria, and Fibonacci Numbers

Kristine Harjes and Pavel Naumov

Department of Mathematics and Computer Science
McDaniel College, Westminster, Maryland 21157, USA
{keh013,pnaumov}@mcdaniel.edu

Abstract. The paper introduces a notion of cellular game that is intended to represent rationally behaving cells of a cellular automaton. The focus is made on studying properties of functional dependence between strategies of different cells in a Nash equilibrium of such games. The main result is a sound and complete axiomatization of these properties. The construction in the proof of completeness is based on the Fibonacci numbers.

1 Introduction

Cellular Games. A (one-dimensional) cellular automaton is a two-way-infinite row of cells that transition from one state to another under certain rules. The rules are assumed to be identical for all cells. Usually, rules are chosen in such a way that the next state of each cell is determined by the current states of the cell itself and its two neighboring cells.

In this paper we consider an object similar to cellular automaton that we call *cellular game*. Each cell of the row is now viewed as a player, whose pay-off function only depends on the strategy of the cell itself and the strategies of its two neighbors. The cellular game is *homogeneous* in the sense that all players have the same pay-off function. Such games can model linearly-spaced homogeneous economic agents who only interact with their neighbors.

As an example, consider a cellular game G in which each player has only two strategies: 0 and 1. Let $\{s_i\}_{i \in \mathbb{Z}}$ be any strategy profile of this game. Let the pay-off of player i be defined as follows: the pay-off is positive if $s_{i-1} + s_i + s_{i+1} \equiv 0$ (mod 2) and is zero otherwise. A Nash equilibrium of this game will be any strategy profile for which condition $s_{i-1} + s_i + s_{i+1} \equiv 0$ (mod 2) is satisfied for each $i \in \mathbb{Z}$. Hence, Nash equilibria in this game have one of the following two forms:

$$\ldots 000000000000000000 \ldots$$

$$\ldots 110110110110110110 \ldots$$

D. Grossi, O. Roy, and H. Huang (Eds.): LORI 2013, LNCS 8196, pp. 149–161, 2013.

Functional Dependence. The main focus of this work is on properties of Nash equilibria of cellular games. Specifically, we study if knowing the strategy of one of the players in a Nash equilibria one can predict the strategy of the other player. If knowing the strategy of a player $a \in \mathbb{Z}$ one can predict the strategy of player $b \in \mathbb{Z}$, then we say that the strategy of b is functionally dependent on the strategy of player a and denote it by $a \rhd b$. For example, in any Nash equilibrium of the game G above, strategies of two players that are three-cells apart are always the same. Thus, $G \vDash 1 \rhd 4$, $G \vDash 2 \rhd 5$, etc. In more general terms, $G \vDash a \rhd a + 3$ for each $a \in \mathbb{Z}$. Note that the property $a \rhd a + 3$ is true for cellular game G, but is not true for many other cellular games. For example, it is not true for the cellular game where each player has a constant pay-off. In this paper we are interested in the universal properties of functional dependence that are true for all cellular games. The trivial examples of such properties are Reflexivity and Transitivity:

$$a \rhd a, \tag{1}$$

$$a \rhd b \to (b \rhd c \to a \rhd c). \tag{2}$$

Since all players have the same pay-off function, we also have Homogeneity:

$$a \rhd b \to (a + c) \rhd (b + c). \tag{3}$$

Just like it is common to assume that cellular automata have a finite number of states, we assume that in cellular games each player has only finitely many strategies. As we will show in Lemma 4, this, perhaps unexpectedly, implies Symmetry:

$$a \rhd b \to b \rhd a. \tag{4}$$

In this paper we will answer the question whether there are any other universal properties of functional dependence in cellular games in addition to properties (1), (2), (3), and (4). To state a hypothetical example of such property, let us get back to the previously discussed game G. As we have seen, $G \vDash a \rhd a + 3$ for each $a \in \mathbb{Z}$. At the same time, by simply analyzing the listed above Nash equilibria of this game one can observe that $G \nvDash a \rhd a + 1$ and $G \nvDash a \rhd a + 2$. In other words,

$$a \rhd (a + 3) \to (a \rhd (a + 1) \lor a \rhd (a + 2))$$

is *not* a universal property of the functional dependence, because game G provides a counterexample. Note that this counterexample appears to heavily rely on the fact that the pay-off function of each player takes into account the strategies of exactly three players: the player herself and her two neighbors. Thus, one might expect that there will be no game G' for which the following property is false:

$$a \rhd (a + 100) \to (a \rhd (a + 1) \lor a \rhd (a + 2) \lor \cdots \lor a \rhd (a + 99)). \tag{5}$$

This would make formula (5) a valid universal principle of functional dependence in cellular games.

Surprisingly, however, such game G' does exist. Indeed, if F_{100} is the 100th Fibonacci number, then one can consider a "Fibonacci" cellular game in which the set of strategies of each player is $\mathbb{Z}_{F_{100}}$. Let the pay-off of player i be a fixed positive number if $s_{i-1} + s_i = s_{i+1}$ in $\mathbb{Z}_{F_{100}}$ and be zero otherwise. As we will see later, for this game the assumption of formula (5) is true and each disjunct in the conclusion is false.

Furthermore, the main result of this paper is that there are no other universal properties of the cellular games except for Reflexivity (1), Transitivity (2), Homogeneity (3), and Symmetry (4). In other words, the set of these axioms is complete with respect to the cellular game semantics.

Related Work. Assumptions on linear structure of the game could be relaxed to an arbitrary "dependency graph" setting where a pay-off function of each player is determined only by its own strategy and the strategies of its neighbors. Various properties of such games have been studied before [1,2,3,4]. In particular, in our work [5], we axiomatized properties of functional dependency between *sets* of players universal to all games that share the same dependence graph. The key difference of this work is the "homogeneity" assumption that all players have the same pay-off function. This assumption leads to Symmetry and Homogeneity axioms that were not present in any form in [5]. What is possibly even more interesting, this assumption leads to the use of Fibonacci numbers in the proof of completeness, which was not needed in [5].

One might argue that this work is not really about cellular games, but rather about general information flow properties in cellular-automaton-like structures. We would agree, except that the information flow setting that we study seems to be the most natural when described in terms of Nash equilibria. Properties of functional dependence relation in another network flow setting has been studied by More and Naumov [6]. Information flow properties of linear communication chains expressible in modal epistemic language were axiomatized by Kane and Naumov [7].

2   Syntax and Semantics

Definition 1. *Let Φ be the minimal set of formulas that satisfies the following conditions:*

1. $\perp \in \Phi$,
2. $a \triangleright b \in \Phi$ for each $a, b \in \mathbb{Z}$,
3. if $\varphi \in \Phi$ and $\psi \in \Phi$, then $\varphi \rightarrow \psi \in \Phi$.

Definition 2. *Cellular game is a pair (S, u), where*

1. S is a finite set of "strategies",
2. u is a "pay-off" function from S^3 to $\mathbb{R}$.

In the above definition domain of the function u is S^3 because the pay-off of each player is determined by her own strategy and the strategies of her two neighbors. By a strategy profile of a cellular game (S, u) we mean any set $\{s_i\}_{i \in \mathbb{Z}}$ such that $s_i \in S$ for each $i \in \mathbb{Z}$.

Definition 3. *A Nash Equilibrium $\{e_i\}_{i \in \mathbb{Z}}$ of a game (S, u) is any strategy profile such that*

$$u(e_{i-1}, s, e_{i+1}) \leq u(e_{i-1}, e_i, e_{i+1}),$$

for each $i \in \mathbb{Z}$ and each $s \in S$.

Lemma 1. *For each $k \in \mathbb{Z}$, if $\{e_i\}_{i \in \mathbb{Z}}$ is a Nash equilibrium of a cellular game, then $\{e_{i+k}\}_{i \in \mathbb{Z}}$ is a Nash equilibrium of the same cellular game.*   □

The set of all Nash equilibria of a cellular game $G = (S, u)$ will be denoted by $NE(G)$. The next definition is one of the key definitions of this paper. In part 2 we formally specify the semantics of the predicate $\rhd$.

Definition 4. *For any formula $\varphi \in \Phi$ and any cellular game G, relation $G \vDash \varphi$ is defined recursively as follows:*

1. $G \nvDash \bot$,
2. $G \vDash a \rhd b$ iff for every $\{e'_i\}_{i \in \mathbb{Z}} \in NE(G)$ and $\{e''_i\}_{i \in \mathbb{Z}} \in NE(G)$, if $e'_a = e''_a$, then $e'_b = e''_b$,
3. $G \vDash \psi \to \chi$ iff $G \nvDash \psi$ or $G \vDash \chi$.

3   Axioms

Our logical system, in addition to propositional tautologies in the language Φ and the Modus Ponens inference rule, contains the following axioms:

1. Reflexivity: $a \rhd a$,
2. Transitivity: $a \rhd b \to (b \rhd c \to a \rhd c)$,
3. Homogeneity: $a \rhd b \to (a + c) \rhd (b + c)$,
4. Symmetry: $a \rhd b \to b \rhd a$.

We write $X \vdash \varphi$ if formula ϕ is provable in our system extended by the set of additional axioms X. We write $\vdash \varphi$ instead of $\varnothing \vdash \varphi$.

4   Example

Soundness of our logical system will be shown in the next section. Here we give an example of a non-trivial property provable in our logical system. We will later use this result in the proof of the completeness theorem.

Lemma 2. *For any $a, b \in \mathbb{Z}$,*

$$\vdash 0 \rhd a \to 0 \rhd a \cdot b.$$

Proof. If $b = 0$, then $0 \rhd a \cdot b$ is an instance of Reflexivity axiom.

Suppose next that $b > 0$. From the assumption $0 \rhd a$ by Homogeneity axiom,

$$0 \rhd a \qquad a \rhd 2a \qquad 2a \rhd 3a \qquad 3a \rhd 4a \qquad \ldots \qquad (b-1)a \rhd ab$$

Thus, by multiple applications of Transitivity axiom, $0 \rhd a \cdot b$.

Finally, assume that $b < 0$. As we have just shown above, assumption $0 \rhd a$ implies $0 \rhd a \cdot |b|$. Hence, by Homogeneity axiom, $-a \cdot |b| \rhd 0$. In other words, $a \cdot b \rhd 0$. Therefore, $0 \rhd a \cdot b$ by Symmetry axiom. $\qquad \square$

5 Soundness

Theorem 1 (soundness). *If $\vdash \phi$, then $G \vDash \phi$ for each cellular game G.*

We prove soundness of each of the axioms as a separate lemma using Definition 4.

Lemma 3 (reflexivity). $G \vDash a \rhd a$.

Proof. For every $\{e'_i\}_{i \in \mathbb{Z}}, \{e''_i\}_{i \in \mathbb{Z}} \in NE(G)$, if $e'_a = e''_a$, then $e'_a = e''_a$. $\qquad \square$

Lemma 4 (symmetry). *If $G \vDash a \rhd b$, then $G \vDash b \rhd a$.*

Proof. Let $V(a) = \{e_a \mid \{e_i\}_{i \in \mathbb{Z}} \in NE(G)\}$ for each $a \in \mathbb{Z}$. Assume now that $G \vDash a \rhd b$. Thus, there is a function $f : V(a) \to V(b)$ such that $e_b = f(e_a)$ for each $\{e_i\}_{i \in \mathbb{Z}} \in NE(G)$. We will show that function f is a surjection from $V(a)$ onto $V(b)$. Let $y \in V(b)$. Hence, $y = e_b$ for some $\{e_i\}_{i \in \mathbb{Z}} \in NE(G)$ due to the definition of $V(b)$. Then, $f(e_a) = e_b = y$. Therefore, f is a surjection of $V(a)$ onto $V(b)$.

By Lemma 1, $V(a) = V(b)$. Thus, f is a surjection of a finite set into a finite set of the same size. It is well-known in set theory that any such function is a bijection. Hence, $e_a = f^{-1}(e_b)$ for each $e \in NE(G)$. Therefore, $G \vDash b \rhd a$. $\qquad \square$

Lemma 5 (transitivity). *If $G \vDash a \rhd b$ and $G \vDash b \rhd c$, then $G \vDash a \rhd c$.*

Proof. Consider any $\{e'_i\}_{i \in \mathbb{Z}}, \{e''_i\}_{i \in \mathbb{Z}} \in NE(G)$. Suppose that $e'_a = e''_a$. We will show that $e'_c = e''_c$. Indeed, by the first assumption of the lemma, $e'_b = e''_b$. Therefore, by the second assumption of the lemma, $e'_c = e''_c$. $\qquad \square$

Lemma 6 (homogeneity). *If $G \vDash a \rhd b$, then $G \vDash (a + c) \rhd (b + c)$.*

Proof. Consider any $\{e'_i\}_{i \in \mathbb{Z}}, \{e''_i\}_{i \in \mathbb{Z}} \in NE(G)$. Suppose that $e'_{a+c} = e''_{a+c}$. We will need to show that $e'_{b+c} = e''_{b+c}$. Indeed, by Lemma 1, $\{e'_{i+c}\}_{i \in \mathbb{Z}}, \{e''_{i+c}\}_{i \in \mathbb{Z}} \in NE(G)$. Hence, by the assumption of the lemma, $e'_{b+c} = e''_{b+c}$. $\qquad \square$

6 Completeness

In Theorem 2, given in the end of this section, we prove the completeness theorem for our logical system. We start, however, with several technical definitions and lemmas.

6.1   Rank

For any nonempty set of integers A, by $gcd(A)$ we mean the greatest common divisor of all integers in the set A. If set A contains only zeros, then $gcd(A)$ is assumed to be equal to the smallest infinite ordinal ω.

The following lemma is well-known result in elementary number theory which is commonly referred to as Bézout identity or Bézout lemma. It is usually proven through an analysis of Euclidian algorithm [8, p. 7].

Lemma 7. *For any integers a and b there are integers u and v such that* $gcd(a,b) = ua + vb$. $\square$

In this paper, in addition to Bézout identity, we also refer to a lesser-known more general result whose proof is reproduced below.

Lemma 8. *For any non-empty set of integers $\{a_1, a_2, \ldots, a_n\}$, there are integers $c_1, \ldots, c_n$ such that $gcd(\{a_1, a_2, \ldots, a_n\}) = c_1 a_1 + \cdots + c_n a_n$.*

Proof. Induction on n. If $n = 1$, then let c_1 be 1. Thus, $gcd(\{a_1\}) = a_1 = c_1 a_1$. If $n > 1$, then by Lemma 7, there must exist u and v such that

$$gcd(\{a_1, \ldots, a_n\}) = gcd(\{gcd(\{a_1, \ldots, a_{n-1}\}, a_n\}) = u \cdot gcd(\{a_1, \ldots, a_{n-1}\}) + v a_n.$$

At the same time, by the Induction Hypothesis, there are integers $c_1, \ldots, c_{n-1}$ such that $gcd(\{a_1, a_2, \ldots, a_{n-1}\}) = c_1 a_1 + \cdots + c_{n-1} a_{n-1}$. Therefore,

$$gcd(\{a_1, \ldots, a_n\}) = u \cdot gcd(\{a_1, \ldots, a_{n-1}\}) + v a_n = u c_1 a_1 + \cdots + u c_{n-1} a_{n-1} + v a_n.$$

$\square$

Lemma 9. *For any set of integer numbers $A = \{a_1, a_2, \ldots\}$, if $gcd(A)$ is finite, then there are an integer $k > 0$ and integers $c_1, \ldots, c_k$ such that*

$$c_1 a_1 + \cdots + c_k a_k = gcd(A).$$

Proof. Consider monotonic sequence

$$gcd\{a_1\} \geq gcd\{a_1, a_2\} \geq gcd\{a_1, a_2, a_3\} \geq \cdots$$

Due to the well-ordering principle, this sequence must have a smallest element, which, thus, is equal to $gcd(A)$. In other words, there is k such that $gcd(A) = gcd(\{a_1, a_2, a_3, \ldots, a_k\})$. Finally, by Lemma 8, there are integers $c_1, \ldots, c_k$ such that

$$c_1 a_1 + \cdots + c_k a_k = gcd\{a_1, a_2, a_3, \ldots, a_k\} = gcd(A).$$

$\square$

Definition 5. *For any set of statements X in language Φ, let*

$$rank(X) = gcd\{ d \mid X \vdash 0 \rhd d \}.$$

Since we have only defined $gcd(A)$ for nonempty set A, for the above definition to be valid we need to show that set $\{\, d \mid X \vdash 0 \rhd d \,\}$ is not empty, which is true since $X \vdash 0 \rhd 0$ by Reflexivity axiom.

Lemma 10. *If $rank(X)$ is finite, then $X \vdash 0 \rhd rank(X)$.*

Proof. By Lemma 9, there are $k \geq 1$ and $c_1, \ldots, c_k, d_1, \ldots, d_k \in \mathbb{Z}$ such that

$$c_1 d_1 + \cdots + c_k d_k = gcd\{d \mid X \vdash 0 \rhd d\} \tag{6}$$

and

$$X \vdash 0 \rhd d_1, \qquad X \vdash 0 \rhd d_2, \qquad \ldots \qquad X \vdash 0 \rhd d_k.$$

By Lemma 2,

$$X \vdash 0 \rhd c_1 d_1, \qquad X \vdash 0 \rhd c_2 d_2, \qquad \ldots \qquad X \vdash 0 \rhd c_k d_k.$$

By the Homogeneity axiom,

$$X \vdash 0 \rhd c_1 d_1,$$
$$X \vdash c_1 d_1 \rhd c_1 d_1 + c_2 d_2,$$
$$\ldots$$
$$X \vdash c_1 d_1 + \cdots + c_{k-1} d_{k-1} \rhd c_1 d_1 + \cdots + c_k d_k.$$

By the Transitivity axiom, applied $k - 1$ times,

$$X \vdash 0 \rhd c_1 d_1 + \cdots + c_k d_k.$$

Therefore, $X \vdash 0 \rhd rank(X)$ due to equation (6). $\qquad\qquad\square$

Lemma 11. *If $rank(X)$ is finite, then $X \vdash a \rhd b$ if and only if $rank(X) \mid (b-a)$.*

Proof. ($\Rightarrow$). Suppose that $X \vdash a \rhd b$. Thus, $X \vdash 0 \rhd (b - a)$ by the Homogeneity axiom. Therefore, $rank(X) \mid (b - a)$ by Definition 5. ($\Leftarrow$). Suppose that $rank(X) \mid (b - a)$. Thus, $X \vdash 0 \rhd (b - a)$ due to Lemma 2 and Lemma 10. Therefore, $X \vdash a \rhd b$ by the Homogeneity axiom. $\qquad\qquad\square$

Lemma 12. *If $rank(X) = \omega$, then $X \vdash a \rhd b$ if and only if $a = b$.*

Proof. ($\Rightarrow$). If $X \vdash a \rhd b$, then, by the Homogeneity axiom, $X \vdash 0 \rhd (b - a)$. Hence, $b - a = 0$ due to the assumption $rank(X) = \omega$. Therefore, $a = b$. ($\Leftarrow$). $X \vdash a \rhd a$ by Reflexivity axiom. $\qquad\qquad\square$

6.2 Game G_d for $2 < d < \omega$

Definition 6. *For any integer $d > 2$, let G_d be the game $(\mathbb{Z}_{F_d}, u)$, where*

$$u(x, y, z) = \begin{cases} 1 & \text{if } x + y = z \text{ in } \mathbb{Z}_{F_d}, \\ 0 & \text{otherwise.} \end{cases}$$

Lemma 13. $e_{n-1} + e_n = e_{n+1}$ *for each* $\{e_i\}_{i \in \mathbb{Z}} \in NE(G_d)$ *and each* $n \in \mathbb{Z}$. □

By $F_0, F_1, F_2, F_3, F_4, F_5, \ldots$ we mean Fibonacci numbers $0, 1, 1, 2, 3, 5, \ldots$ and by $[F_0], [F_1], [F_2], \ldots$ their congruence classes in $\mathbb{Z}_{F_d}$.

Lemma 14. *For each* $n, k \in \mathbb{Z}$ *and each* $\{e_i\}_{i \in \mathbb{Z}} \in NE(G_d)$, *if* $k > 0$, *then*

$$e_{n+k} = [F_{k-1}]e_n + [F_k]e_{n+1}.$$

Proof. Induction on k. If $k = 1$, then

$$e_{n+1} = [0] \cdot e_n + [1] \cdot e_{n+1} = [F_0] \cdot e_n + [F_1] \cdot e_{n+1}.$$

If $k = 2$, then, by Lemma 13,

$$e_{n+2} = e_n + e_{n+1} = [1] \cdot e_n + [1] \cdot e_{n+1} = [F_1] \cdot e_n + [F_2] \cdot e_{n+1}.$$

If $k > 2$, then by Lemma 13, the Induction Hypothesis, and the recurrence relation for Fibonacci numbers,

$$
\begin{aligned}
e_{n+k} &= e_{n+k-2} + e_{n+k-1} \\
&= ([F_{k-3}] \cdot e_n + [F_{k-2}] \cdot e_{n+1}) + ([F_{k-2}] \cdot e_n + [F_{k-1}] \cdot e_{n+1}) \\
&= ([F_{k-3}] + [F_{k-2}]) \cdot e_n + ([F_{k-2}] + [F_{k-1}]) \cdot e_{n+1} \\
&= [F_{k-1}] \cdot e_n + [F_k] \cdot e_{n+1}.
\end{aligned}
$$

□

Lemma 15. *For each* $n, q, d \in \mathbb{Z}$, *if* $q \geq 0$, *then*

$$e_{n+qd} = [F_{d-1}^q] \cdot e_n,$$

where F_{d-1}^q *is* F_{d-1} *raised to power* q.

Proof. Induction on q. If $q = 0$, then

$$e_{n+qd} = e_n = [1] \cdot e_n = [F_{d-1}^0] \cdot e_n.$$

Note that in the above we rely on the fact that $F_{d-1}^0 = 1$, which is true because $d > 2$ and, thus, $F_{d-1} > 0$.

Assume now that $q > 0$. Thus, by Lemma 14, the Induction Hypothesis, and due to the fact that $F_d = [0]$ in $\mathbb{Z}_{F_d}$,

$$
\begin{aligned}
e_{n+qd} &= e_{n+(q-1)d+d} \\
&= [F_{d-1}] \cdot e_{n+(q-1)d} + [F_d] \cdot e_{n+(q-1)d+1} \\
&= [F_{d-1}] \cdot e_{n+(q-1)d} + [0] \cdot e_{n+(q-1)d+1} \\
&= [F_{d-1}] \cdot e_{n+(q-1)d} = [F_{d-1}] \cdot [F_{d-1}^{q-1}]e_n = [F_{d-1}^q] \cdot e_n.
\end{aligned}
$$

□

Lemma 16. *If $d \mid (b - a)$, then $G_d \vDash a \triangleright b$.*

Proof. Due to Lemma 4, we can assume that $a < b$. Suppose that $b = a + qd$ and that $e'_a = e''_a$ for some $\{e'_i\}_{i \in \mathbb{Z}}, \{e''_i\}_{i \in \mathbb{Z}} \in NE(G_d)$. We will show that $e'_b = e''_b$. Indeed, by Lemma 15,

$$e'_b = e'_{a+qd} = [F^q_{d-1}] \cdot e'_a.$$

Similarly, $e''_b = [F^q_{d-1}] \cdot e''_a$. Thus,

$$e'_b = [F^q_{d-1}] \cdot e'_a = [F^q_{d-1}] \cdot e''_a = e''_b.$$

$\square$

By **0** we mean the constant function from $\mathbb{Z}$ to $\mathbb{Z}_{F_d}$ equal to $[0]$ on all integer numbers.

Lemma 17. $\mathbf{0} \in NE(G_d)$.

Proof. See Definition 6. $\square$

Note that the Fibonacci sequence

$$F_0 = 0, F_1 = 1, F_2 = 1, F_3 = 2, \ldots$$

can be expanded to negative subscripts in a way that preserves the recurrence relation $F_n = F_{n-1} + F_{n-2}$ for the Fibonacci numbers:

$$F_{-1} = 1, F_{-2} = 0, F_{-3} = 1, F_{-4} = -1, F_{-5} = 2, \ldots$$

By **F** we mean the function from $\mathbb{Z}$ to $\mathbb{Z}_{F_d}$ such that $\mathbf{F}(z) = [F_z]$.

Lemma 18. $\mathbf{F} \in NE(G_d)$.

Proof. See Definition 6. $\square$

Lemma 19. *Element $[F_{d-1}]$ is invertible in $\mathbb{Z}_{F_d}$ for each $d > 2$.*

Proof. Due to the identity $F_n = F_{n-1} + F_{n-2}$, the Euclidian algorithm, when applied to any two consecutive Fibonacci numbers, generates the complete sequence of Fibonacci numbers in reverse order:

$$F_n, F_{n-1}, F_{n-2}, F_{n-3}, \ldots, 13, 8, 5, 3, 2, 1, 1, 0.$$

Hence, $gcd(F_n, F_{n-1}) = 1$ for each $n \geq 1$. In particular, $gcd(F_d, F_{d-1}) = 1$. Thus, due to Lemma 7, there are integers a and b such that

$$a \cdot F_d + b \cdot F_{d-1} = 1.$$

Therefore, $b \cdot F_{d-1} \equiv 1 \pmod{F_d}$. In other words, $[b] \cdot [F_{d-1}] = [1]$ in $\mathbb{Z}_{F_d}$. $\square$

Lemma 20. *If $d > 2$ and $G_d \vDash a \triangleright b$, then $d \mid (b - a)$.*

Proof. Assume that $G_d \models a \triangleright b$. Due to Lemma 4, we can also assume that $a < b$. Suppose that $b - a = qd + r$ for $0 \leq r < d$. We will show that $r = 0$.

Consider strategy profiles $\{e'_i\}_{i \in \mathbb{Z}}$ and $\{e''_i\}_{i \in \mathbb{Z}}$ such that

$$e'_i = \mathbf{0}(i - a)$$

and

$$e''_i = \mathbf{F}(i - a)$$

for each $i \in \mathbb{Z}$. By Lemma 1, $\{e'_i\}_{i \in \mathbb{Z}}, \{e''_i\}_{i \in \mathbb{Z}} \in NE(G)$. Note that

$$e'_a = \mathbf{0}(a - a) = [0] = \mathbf{F}(a - a) = e''_a \pmod{F_d}.$$

Thus, $e'_b = e''_b$ due to the assumption $G_d \models a \triangleright b$. Hence, $\mathbf{0}(b - a) = \mathbf{F}(b - a)$. In other words, $\mathbf{F}(b - a) = [0]$. Hence, $\mathbf{F}(r + qd) = [0]$. At the same time, by Lemma 15,

$$\mathbf{F}(r + qd) = [F_{d-1}^q]\mathbf{F}(r) = [F_{d-1}^q \cdot F_r] = [F_{d-1}]^q \cdot [F_r].$$

Then, $[F_{d-1}]^q \cdot [F_r] = [0]$. By Lemma 19, element $[F_{d-1}]$ is invertible in $\mathbb{Z}_{F_d}$. Thus,

$$[F_{d-1}]^{-q} \cdot [F_{d-1}]^q \cdot [F_r] = [F_{d-1}]^{-q} \cdot [0].$$

Hence, $[F_r] = [0]$. In other words, $F_r \equiv 0 \pmod{F_d}$. Recall that $0 \leq r < d$ by the choice of r. Thus, $0 \leq F_r < F_d$. Taking into account $F_r \equiv 0 \pmod{F_d}$, we can conclude that $F_r = 0$. Recall now that $r \geq 0$. Therefore, $r = 0$. □

6.3   Special Cases: Game G_ω, G_1, and G_2

Definition 7. *Let G_ω be the game $(\{0, 1\}, u)$, where $u(x, y, z) = 0$ for all $x, y, z \in \mathbb{Z}$.*

Lemma 21. $G_\omega \models a \triangleright b$ *if and only if $a = b$.*

Proof. Any strategy profile of game G_ω is a Nash equilibrium. □

Definition 8. *let G_1 be the game $(\{0, 1\}, u)$, where*

$$u(x, y, z) = \begin{cases} 1 & \text{if } y = z \\ 0 & \text{otherwise.} \end{cases}$$

In other words, in game G_1 each player is paid to be equal to her right neighbor.

Lemma 22. $G_1 \models a \triangleright b$ *for any $a, b \in \mathbb{Z}$.*

Proof. Game G_1 has only two Nash equilibria: $\ldots 000 \ldots$ and $\ldots 111 \ldots$. □

Definition 9. *let G_2 be the game $(\{0, 1, panic\}, u)$, where $u(x, y, z)$ is defined as follows: if $x = z \neq panic$, then y is not paid no matter what its value is; otherwise y is paid a positive amount if it is equal to panic.*

Lemma 23. *Game G_2 has five Nash equilibria:*

1. *all players are in the state of panic,*
2. *all players are in the state 0,*
3. *all players are in the state 1,*
4. *even-indexed players are in state 0, odd-indexed players are in state 1,*
5. *even-indexed players are in state 1, odd-indexed players are in state 0.*

$\square$

Lemma 24. $G_2 \vDash a \rhd b$ *if and only if* $2 \mid (a - b)$.

Proof. Follows from Lemma 23. $\square$

7   Completeness: Final Steps

Lemma 25. *For any set of formulas X in the language Φ, there is a game G such that, $X \vdash a \rhd b$ if and only if $G \vDash a \rhd b$, for each $a, b \in \mathbb{Z}$.*

Proof. Case I: $rank(X) = 1$. Consider game G_1. ($\Rightarrow$): Note that $G_1 \vDash a \rhd b$ by Lemma 22. ($\Leftarrow$): By the assumption of this case, $rank(X) = 1$. Thus, $X \vdash 0 \rhd 1$ by Lemma 10. Hence, $X \vdash 0 \rhd (b - a)$, by Lemma 2. Therefore, $X \vdash a \rhd b$, by the Homogeneity axiom.

Case II: $rank(X) = 2$. Consider game G_2. ($\Rightarrow$) : Let $G_2 \nvDash a \rhd b$. Thus, by Lemma 24, $b - a = 2q + 1$ for some $q \in \mathbb{Z}$. We need to show that $X \nvdash a \rhd b$. Suppose the opposite, $X \vdash a \rhd b$. Hence, $X \vdash 0 \rhd (b - a)$ by the Homogeneity axiom. Thus, $X \vdash 0 \rhd 2q + 1$. This is a contradiction with the assumption $rank(X) = 2$, because $2q + 1$ is not divisible by 2. ($\Leftarrow$): Assume that $G_2 \vDash a \rhd b$. Thus, $b - a = 2q$, by Lemma 24. Recall the assumption $rank(X) = 2$. Thus, by Lemma 10, $X \vdash 0 \rhd 2$. Hence, $X \vdash 0 \rhd 2q$ by Lemma 2. In other words, $X \vdash 0 \rhd (b - a)$. Thus, by the Homogeneity axiom, $X \vdash a \rhd b$.

Case III: $2 < d = rank(X) < \omega$. Consider game G_d. ($\Rightarrow$) : Let $X \vdash a \rhd b$. Hence, by the Homogeneity axiom, $X \vdash 0 \rhd (b - a)$. Thus, $d \mid (b - a)$ due to Definition 5. Thus, $G_d \vDash a \rhd b$, by Lemma 16. ($\Leftarrow$) : Assume that $G_d \vDash a \rhd b$. Hence, $d \mid (b - a)$ by Lemma 20. Thus, $b - a = qd$ for some $q \in \mathbb{Z}$. Recall that $d = rank(X) < \omega$. Thus, $X \vdash 0 \rhd d$ by Lemma 10. Hence, $X \vdash 0 \rhd qd$, by Lemma 2. In other words, $X \vdash 0 \rhd (b - a)$. Therefore, $X \vdash a \rhd b$, by the Homogeneity axiom.

Case IV: $rank(X) = \omega$. Consider game G_ω. ($\Rightarrow$) : Let $X \vdash a \rhd b$. Thus, by the Homogeneity axiom, $X \vdash 0 \rhd (b - a)$. Hence, $b - a = 0$, due to the fact that $rank(X) = \omega$. Then $G_\omega \vDash a \rhd b$ by Lemma 21. ($\Leftarrow$) : Assume that $G_\omega \vDash a \rhd b$. Thus, $a = b$, by Lemma 21. Therefore, $X \vdash a \rhd b$ by the Reflexivity axiom. $\square$

Lemma 26. *For any maximal consistent set of formulas $X \subset \Phi$, there is a game G such that $X \vdash \psi$ if and only if $G \vDash \psi$ for each $\psi \in \Phi$.*

Proof. Consider any maximal consistent set $X \subset \Phi$. Let G be the game that exists by Lemma 25. We will prove that $X \vdash \psi$ if and only if $G \vDash \psi$ by induction on the structural complexity of formula ψ. The case when ψ is an atomic proposition follows from Lemma 25. If ψ is constant $\bot$, then $G \nvDash \psi$ by Definition 4 and $X \nvdash \psi$ due to consistency of the set X. The case when ψ is an implication follows from the assumption of maximality and consistency of X in the standard way.   □

Theorem 2 (completeness). *If $G \vDash \phi$ for each cellular game G, then $\vdash \phi$.*

Proof. Suppose that $\nvdash \phi$. Let X be a maximal consistent subset of Φ containing formula $\neg\phi$. By Lemma 26, there is a cellular game G such that $G \vDash \neg\phi$.   □

8   Conclusion

In this paper we gave a complete axiomatization of functional dependence in linear cellular games with finite number of strategies. The two natural extensions of this work are games with infinite number of strategies and cellular games on a plane.

If players are allowed to have infinite number of strategies, then the Symmetry axiom is no longer sound. Indeed, consider a game in which strategies are infinite boolean sequences. Let $a_1, a_2, a_3, \ldots$ be the strategies of a player u and $b_1, b_2, b_3, \ldots$ be the strategies of her right neighbor $u + 1$. The player u gets a fixed positive pay-off if and only if

$$a_i = b_{i-1} \quad \text{for all } i \geq 2. \tag{7}$$

Note that equation (7) puts no restriction on the value of a_1. It is easy to see that the Nash equilibria of this game are all strategy profiles in which equation (7) is satisfied for all adjacent players u and $u+1$. Thus, for this game $u \triangleright u+1$ is true. At the same time, equation (7) puts no restrictions on a_1 and, thus, formula $u + 1 \triangleright u$ is false. The complete axiomatization of functional dependence in linear cellular games with infinite number of strategies remains an open problem.

By a cellular game on the plane we mean a game on a square grid where the pay-off of each player is determined by her own strategy and the strategies of her eight neighbors. Our attempts, made together with Jeffrey Kane, to generalize results of this paper to such games were unsuccessful due to the fact that we were not able to find the right "two-dimensional" version of of Fibonacci numbers. We even do not know, for example, if there is a game (see Figure 1) in which formula $a \triangleright c$ is true, but formulas $b_k \triangleright c$ are false for each $k \in \{1, 2, \ldots, 10\}$.

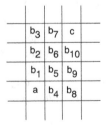

Fig. 1. Plane Game

References

1. Kearns, M.J., Littman, M.L., Singh, S.P.: Graphical models for game theory. In: Breese, J.S., Koller, D. (eds.) UAI, pp. 253–260. Morgan Kaufmann (2001)
2. Littman, M.L., Kearns, M.J., Singh, S.P.: An efficient, exact algorithm for solving tree-structured graphical games. In: Dietterich, T.G., Becker, S., Ghahramani, Z. (eds.) NIPS, pp. 817–823. MIT Press (2001)
3. Elkind, E., Goldberg, L.A., Goldberg, P.W.: Computing good Nash equilibria in graphical games. In: MacKie-Mason, J.K., Parkes, D.C., Resnick, P. (eds.) ACM Conference on Electronic Commerce, pp. 162–171. ACM (2007)
4. Elkind, E., Goldberg, L.A., Goldberg, P.W.: Nash equilibria in graphical games on trees revisited. In: Electronic Colloquium on Computational Complexity (ECCC) (005) (2006)
5. Harjes, K., Naumov, P.: Functional dependence in strategic games. In: 1st International Workshop on Strategic Reasoning, Rome, Italy. Electronic Proceedings in Theoretical Computer Science, vol. 112, pp. 9–15 (March 2013)
6. More, S.M., Naumov, P.: The functional dependence relation on hypergraphs of secrets. In: Leite, J., Torroni, P., Ågotnes, T., Boella, G., van der Torre, L. (eds.) CLIMA XII 2011. LNCS, vol. 6814, pp. 29–40. Springer, Heidelberg (2011)
7. Kane, J., Naumov, P.: Epistemic logic for communication chains. In: 14th Conference on Theoretical Aspects of Rationality and Knowledge (TARK 2013), Chennai, India, pp. 131–137 (January 2013)
8. Jones, G., Jones, J.: Elementary Number Theory. Springer Undergraduate Mathematics Series. Springer (1998)

Reasoning about Actions Meets Strategic Logics

Andreas Herzig[1], Emiliano Lorini[1], and Dirk Walther[2]

[1] University of Toulouse, IRIT-CNRS Toulouse, France
[2] TU Dresden, Theoretical Computer Science
Center for Advancing Electronics Dresden Dresden, Germany

Abstract. We introduce ATLEA, a novel extension of Alternating-time Temporal Logic with explicit actions in the object language. ATLEA allows to reason about abilities of agents under commitments to play certain actions. Pre- and postconditions as well as availability and unavailability of actions can be expressed. We show that the multiagent extension of Reiter's solution to the frame problem can be encoded into ATLEA. We also consider an epistemic extension of ATLEA. We demonstrate that the resulting logic is sufficiently expressive to reason about uniform choices of actions. Complexity results for the satisfiability problem of ATLEA and its epistemic extension are given in the paper.

1 Introduction

Several formalisms for reasoning about actions were suggested in AI, including situation calculus [19], event calculus [20], fluent calculus [21,22], and so-called action languages such as A and C [9,14]. These formalisms provide languages to describe actions in terms of pre- and postconditions. We are interested in reasoning about actions within the framework of Alternating-time Temporal Logic (ATL) [2], a logic for reasoning about strategic abilities. In ATL there are no names for actions and there is no obvious way to describe the behaviour of actions. We therefore extend ATL to ATLEA: ATL with Explicit Actions in the object language. We demonstrate that the resulting logic allows to reason about multiagent actions. In particular, we show that ATLEA allows us to specify the pre- and post-conditions of actions and to check whether in a given situation an agent or coalition of agents has the capability to ensure a given outcome.

The paper is organised as follows. Section 2 introduces ATLEA, and Section 3 illustrates how pre- and postconditions of actions can be specified. We then consider an epistemic extension of ATLEA and demonstrate that it is sufficiently expressive to reason about the conditions under which an agent has a uniform choice to ensure a given state of affairs (Section 4).

2 ATL with Explicit Actions

An *action commitment* is a pair (a, ω) consisting of an agent a and an action name ω, also written $a \mapsto \omega$: a is committed to perform ω at the current state. An *action*

D. Grossi, O. Roy, and H. Huang (Eds.): LORI 2013, LNCS 8196, pp. 162–175, 2013.

commitment function is a finite set ρ of action commitments such that ρ is a partial function in its first argument: for every two (a, ω) and (a, ω') in ρ we have $\omega = \omega'$. We write $\rho(a) = \omega$ if $(a, \omega) \in \rho$; otherwise we say that $\rho(a)$ is undefined. The partial function ρ describes the commitments of the agents a in $\mathrm{dom}(\rho)$ to play action $\rho(a)$ at the current state.

Action commitment functions parameterise ATL path quantifiers. A formula of the form $\langle\!\langle A \rangle\!\rangle_\rho \psi$ is read: "while the agents in $\mathrm{dom}(\rho)$ perform the actions as specified in ρ, the agents in A have a strategy to ensure the temporal property ψ, no matter what the agents in $\Sigma \setminus A$ do." Just as in ATL, there is an existential quantification over the strategies of the agents in coalition A and a universal quantification over the strategies of the agents outside of A. The selection of strategies occurs simultaneously, without interdependencies between the agents. The novel part in ATLEA is that we only quantify over strategies respecting ρ. Note that in the path quantifier $\langle\!\langle A \rangle\!\rangle_\rho$, the function ρ may commit both members of the coalition A (the proponents) and its opponents outside A. A special case is when $\rho = \emptyset$: then $\langle\!\langle A \rangle\!\rangle_\rho$ is nothing but the ATL operator $\langle\!\langle A \rangle\!\rangle$. For example, the formula $\langle\!\langle a, c \rangle\!\rangle_{\{a \mapsto \omega_a, b \mapsto \omega_b\}} \psi$ holds at a state w if, and only if, there is a strategy for coalition $\{a, c\}$ where a performs ω_a at w, such that for all strategies for $\Sigma \setminus \{a, c\}$ where b performs ω_b at w, all paths resulting from the chosen strategies satisfy the temporal property ψ.

We fix a set Π of *atomic propositions*, a set Σ of *agents*, and a set Ω of *action names*. We assume that these three sets are countably infinite and disjoint.[1] The language of ATLEA is defined over the signature $\langle \Pi, \Sigma, \Omega \rangle$.

Definition 1 (ATLEA **syntax**). *The following grammar defines state formulas φ and path formulas ψ:*

$$\varphi ::= p \mid \neg\varphi \mid \varphi \vee \varphi \mid \langle\!\langle A \rangle\!\rangle_\rho \psi$$
$$\psi ::= \neg\psi \mid \bigcirc\varphi \mid \varphi \, \mathcal{U} \, \varphi$$

where p ranges over Π, A ranges over finite subsets of Σ and ρ ranges over action commitment functions with action names from Ω. The language of ATLEA *consists of state formulas.*

We sometimes omit set parentheses as in $\langle\!\langle a \rangle\!\rangle_{a \mapsto \omega} \bigcirc\varphi$. For state formulas, the Boolean operators $\wedge$, $\rightarrow$, $\leftrightarrow$, and the logical constants $\top$ and $\bot$ are defined as usual by means of $\neg$ and $\vee$. The commonly used temporal operators 'sometime' and 'forever' are defined as the path formulas $\Diamond\varphi = (\top \, \mathcal{U} \, \varphi)$ and $\Box\varphi = \neg(\top \, \mathcal{U} \, \neg\varphi)$, respectively.

Formulas are evaluated on concurrent game structures that additionally interpret action names as moves of players.

Definition 2 (CGSN). *Let $S = \{1, \ldots, n\} \subset \Sigma$, $n \geq 1$, be a finite set of agents, $P \subset \Pi$ a finite set of atomic propositions, and $O \subset \Omega$ be a finite set of action names. A Concurrent Game Structure with action Names (CGSN) $\mathcal{C}$ for the signature $\langle S, P, O \rangle$ is a tuple $\mathcal{C} = \langle W, V, M, Mov, E, \| \cdot \| \rangle$, where:*

- *W is a finite, non-empty set of worlds (alias states);*
- *$V : W \longrightarrow 2^P$ is a valuation function;*
- *M is a finite, non-empty set of moves;*

---

[1] Infinite signatures are relevant for the analysis of the complexity of the satisfiability problem.

- $Mov : W \times S \longrightarrow 2^M \setminus \emptyset$ maps a world w and an agent a to the non-empty set $Mov(w, a)$ of moves available to a at w;
- $E : W \times M^S \longrightarrow W$ is a transition function mapping a world w and a move profile $\boldsymbol{m} = \langle m_1, \ldots, m_n \rangle$ (one move for each agent) to the world $E(w, \boldsymbol{m})$;
- $\| \cdot \| : O \longrightarrow M$ is a denotation function mapping action names in O to moves in M.

CGSNs are finite objects. We obtain infinitely many classes of CGSNs, one per signature. In a CGSN, an action name is interpreted as a move (which may interpret several action names). $Mov(w, a)$ determines which of the moves from M are available to a at state w. We say that *action ω_a is available to agent a at w* if $\|\omega_a\| = m_a$ and $m_a \in Mov(w, a)$.

A *strategy for an agent a* is a function f_a that maps every world w to a move $f_a(w) \in Mov(w, a)$ available to a at w.[2] A *strategy for a coalition $A \subseteq S$* is a function F_A that maps every agent a from A to a strategy $F_A(a)$ for a. Given an action commitment function ρ, a strategy F_A for A is called *compatible with ρ at w* if for all $a \in A \cap \mathrm{dom}(\rho)$,

$$F_A(a)(w) = \|\rho(a)\|.$$

Clearly, when $A \cap \mathrm{dom}(\rho) = \emptyset$ then any strategy F_A for coalition A is compatible with ρ. We denote with $\mathsf{strat}(A, \rho, w)$ the set of all strategies for A that are compatible with ρ at w. When the interpretation of agent a's commitment is not among the moves available at w, i.e., when $\|\rho(a)\| \notin Mov(w, a)$, then no strategy for a is compatible with ρ at w. This holds more generally for coalitions containing a: if $\|\rho(a)\| \notin Mov(w, a)$ for some $a \in A$ then $\mathsf{strat}(A, \rho, w) = \emptyset$.

A move profile is used to determine a successor of a state using the transition function E. We define the set of available move profiles at state w as follows:
$$\mathsf{prof}(w) = \{\langle m_1, \ldots, m_n \rangle \mid m_i \in Mov(w, i)\}.$$
The set of *possible successors of w* is the set of states $E(w, \boldsymbol{m})$ where $\boldsymbol{m}$ ranges over $\mathsf{prof}(w)$. An infinite sequence $\lambda = x_0 x_1 x_2 \cdots$ of worlds from W is called a *computation* if x_{i+1} is a successor of x_i for all positions $i \geq 0$. $\lambda[i]$ denotes the i-th component x_i in λ, and with $\lambda[0, i]$ the initial sequence $x_0 \cdots x_i$ of λ.

The set $\mathsf{out}(w, F_A)$ of *outcomes* of a strategy F_A for A starting at a world w is the set of all computations $\lambda = x_0 x_1 x_2 \cdots$ such that $x_0 = w$ and, for every $i \geq 0$, there is a move profile $\boldsymbol{m} = \langle m_1, \ldots, m_n \rangle \in \mathsf{prof}(x_i)$ such that:

- $m_a = F_A(a)(x_i)$, for all $a \in A$; and
- $x_{i+1} = E(x_i, \boldsymbol{m})$.

A strategy F_S for all agents in the signature specifies exactly one play: $\mathsf{out}(w, F_S)$ is a singleton. A CGSN $\mathcal{C}$ for $\langle S, P, O \rangle$ allows to interpret an **ATLEA** formula φ if S contains all agents, P all atomic propositions, and O all action names occurring in φ. The satisfaction relation is defined as follows:[3]

---

[2] The logic is defined for memoryless strategies. The extension to perfect recall strategies is straightforward.

[3] We skip the cases for atomic propositions, Boolean and temporal operators; they are defined as in **ATL** [2].

$$\mathcal{C}, w \models \langle\!\langle A \rangle\!\rangle_\rho \psi \text{ iff there exists } F_A \in \mathsf{strat}(A, \rho, w) \text{ such that}$$
$$\text{for all } F_{S\setminus A} \in \mathsf{strat}(S\setminus A, \rho, w) \text{ it holds that}$$
$$\mathcal{C}, \lambda \models \psi, \text{ where } \{\lambda\} = \mathsf{out}(w, F_A \cup F_{S\setminus A}).$$

Validity and satisfiability are defined as expected: φ is valid if $\mathcal{C}, w \models \varphi$ for every state w of every CGSN $\mathcal{C}$ whose signature contains that of φ; φ is satisfiable if $\neg\varphi$ is not valid.

With **ATLEA** we can express the (un-)availability of actions. A formula of the form $\langle\!\langle a \rangle\!\rangle_{a\mapsto w_a} \bigcirc\top$ states that action w_a is available to agent a at the current state. More generally, we have that $\mathcal{C}, w \models \langle\!\langle A \rangle\!\rangle_\rho \bigcirc\top$ iff $\|\rho(a)\| \in Mov(w, a)$ for all $a \in A \cap \mathrm{dom}(\rho)$. The other way round, to express the unavailability of w_a to a, we have that $\mathcal{C}, w \models \neg\langle\!\langle A \rangle\!\rangle_\rho \bigcirc\top$ iff there is an $a \in \mathrm{dom}(\rho) \cap A$ such that $\|\rho(a)\| \notin Mov(w, a)$.

ATL is the fragment of **ATLEA** where every action commitment function is empty.[4] Without commitments **ATLEA** formulas can be interpreted in CGSNs with empty denotation functions, which are essentially concurrent game structures as used in **ATL**. A crucial difference to **ATL**, however, is the fact that **ATLEA** can detect the difference between memoryless and perfect recall strategies. Consider a CGSN for one agent with two states x and y such that $p \in V(x)$ but $p \notin V(y)$, $Mov(x, a) = \{1, 2\}$ and $Mov(y, a) = \{1\}$, $x = E(x, 1)$, $y = E(x, 2) = E(y, 1)$, and $\|\omega\| = 1$. The formula $\langle\!\langle a \rangle\!\rangle_{a\mapsto w}\bigcirc p \wedge \langle\!\langle a \rangle\!\rangle_{a\mapsto w}(\top\,\mathcal{U}\,\neg p)$ is false at a state x under memoryless strategies, but it holds for strategies that allow a recall of at least one predecessor.

The proposition below illustrates that the status of some **ATLEA** counterparts of **ATL** axioms [12] depends on the interplay of the two arguments in the **ATLEA** operator $\langle\!\langle A \rangle\!\rangle_\rho$.

Proposition 1. *The following formulas are* **ATLEA** *valid.*

1. $\langle\!\langle A \rangle\!\rangle_\rho\bigcirc\top$ *for* $\mathrm{dom}(\rho) \cap A$ *empty*
2. $\neg\langle\!\langle A \rangle\!\rangle_\rho\bigcirc\bot$ *for* $\mathrm{dom}(\rho) \setminus A$ *empty*
3. $(\langle\!\langle A \rangle\!\rangle_\rho\bigcirc\varphi \wedge \langle\!\langle B \rangle\!\rangle_\rho\bigcirc\psi) \to \langle\!\langle A \cup B \rangle\!\rangle_\rho\bigcirc(\varphi \wedge \psi)$ *for* $A \cap B \subseteq \mathrm{dom}(\rho)$
4. $\langle\!\langle A \rangle\!\rangle_\rho\bigcirc\varphi \to \langle\!\langle A \rangle\!\rangle_{\rho'}\bigcirc\varphi$ *for* $\rho' = \rho \cup \{a \mapsto w\}, a \notin A$
5. $\langle\!\langle A \rangle\!\rangle_{\rho'}\bigcirc\varphi \to \langle\!\langle A \rangle\!\rangle_\rho\bigcirc\varphi$ *for* $\rho' = \rho \cup \{a \mapsto w\}, a \in A$
6. $\langle\!\langle A \cup \{a\} \rangle\!\rangle_\rho\bigcirc\varphi \to \langle\!\langle A \rangle\!\rangle_\rho\bigcirc\varphi$ *for* $a \in \mathrm{dom}(\rho)$
7. $\langle\!\langle A \rangle\!\rangle_\rho(\varphi\,\mathcal{U}\,\psi) \leftrightarrow ((\psi \wedge \langle\!\langle A \rangle\!\rangle_\rho\bigcirc\top) \vee (\varphi \wedge \langle\!\langle A \rangle\!\rangle_\rho\bigcirc\langle\!\langle A \rangle\!\rangle_\emptyset(\varphi\,\mathcal{U}\,\psi)))$

Item 1 generalises the **ATL** axiom ($\top$) (obtained when ρ is empty). Another particular case is when A is empty: then both $\langle\!\langle\emptyset\rangle\!\rangle_\rho\bigcirc\top$ is valid. Item 2 generalises the **ATL** axiom ($\bot$) (obtained when ρ is empty). To see that Item 1 is invalid when $\mathrm{dom}(\rho)$ and A are not disjoint, suppose $a \in \mathrm{dom}(\rho) \cap A$: then we can always find a CGSN $\mathcal{C}$ and a state w such that $\|\rho(a)\| \notin Mov(w, a)$, and then $\mathcal{C}, w \not\models \langle\!\langle A \rangle\!\rangle_\rho\bigcirc\top$. Item 3 generalises **ATL**'s superadditivity axiom (S), relaxing the constraint of disjointness of A and B. Intuitively it says that when the actions of the agents that are in both, A and B, are fixed by ρ then these agents cannot have different strategies to enforce φ and ψ. For that reason, the powers of the two coalitions can be combined. Consider the case where

---

[4] We note in passing: **ATL** does not allow for negated path formulas, while **ATLEA** does (cf. Def. 1).

$A \subseteq \text{dom}(\rho)$. It then follows from the monotony of $\langle\!\langle A \rangle\!\rangle_\rho$ and our superadditivity axiom (Item 3) that $\langle\!\langle A \rangle\!\rangle_\rho$ is a normal modal box operator. Items 4 and 5 are respectively about increasing commitment of opponents and releasing commitment of proponents. Item 6 is about dismissing committed proponents. Item 7 is a fixpoint axiom of ATL. It allows to rewrite formulas in a way such that ρ is empty in all modal operators of the form $\langle\!\langle A \rangle\!\rangle_\rho(\varphi\,\mathcal{U}\,\psi)$.[5] Moreover, the generalisations of the ATL inference rules of Modus Ponens and Necessitation all preserve validity. However, we leave a complete axiomatisation of ATLEA for future work.

Remark 1. As an extension of ATLEA, we may consider PDL program operators such as sequential composition, iteration and test over action descriptions. For the one-agent case, this is related to CTL with Path Relativisation [15]. It would also be interesting to study complements of actions, as well as the loop construct, which allows to formulate action commitments of the form $a \mapsto \omega^\infty$ stating that a plays the action denoted with ω at all states. We can also view $a \mapsto \omega^\infty$ as a commitment of a to play ω in all situations. In other words, ω^∞ is a strategy. This means that we can specify entire (memoryless) strategies within such an extension of ATLEA. This motivates a study of the relationships between the extension of ATLEA and other logics with representations of strategies in the object language such as ATLES [28] and Strategy Logic [6], which we leave for furture work. The equivalence in Item 7 of Proposition 1 becomes invalid if we generalise commitments from atomic actions to sequences of actions. The extension of ATLEA by the program operators of PDL is subject of ongoing work.

Theorem 1. *The satisfiability problem for* ATLEA *is ExpTime-complete.*

The ExpTime lower bound carries over from the fragment ATL [29]. The matching upper complexity bound can be shown by adapting the decision procedure for ATL [29], which is a type elimination constructions inspired from [8].

3   Reasoning about Actions

We now put ATLEA to work and demonstrate its usefulness in reasoning about multiagent actions. We start by encoding in ATLEA Reiter's action descriptions in terms of complete conditions for the executability and the effects of actions. We build on the mapping of Reiter's solution to the frame problem into dynamic epistemic logics with assignments as done in [7]. We take the multiagent context into account by integrating ideas stemming from logics of propositional control. There, the set of propositional variables is partitioned among the agents, and an agent controlling a variable is the only one able to change its truth value [26].

---

[5] We note that ATL's S-maximality axiom $\neg\langle\!\langle S \rangle\!\rangle_\emptyset \bigcirc \neg\psi \leftrightarrow \langle\!\langle \emptyset \rangle\!\rangle_\emptyset \bigcirc \psi$ (which relates the empty coalition with the set of all agents) does not make sense in our setting: as formulas φ are evaluated in CGSNs whose signature *contains* that of φ, there is no way of 'grasping' the set of all agents S of a given model. Our ATLEA (and also the underlying version of ATL) is more general than ATL as defined in [12]. The latter is actually a *family* of logics: each member of the family is defined for a finite set of agents, yielding uncountably many axiomatisations.

3.1 Action Descriptions

Let $\langle S, P, O \rangle$ be a finite signature. Let Φ be the set of propositional formulas over P. An *action description for* $\langle S, P, O \rangle$ is a tuple $\mathcal{T} = \langle \mathsf{agt}, \mathsf{poss}, \mathsf{eff} \rangle$ where

- $\mathsf{agt} : O \longrightarrow S$ associates to each action name ω an agent agt_ω;
- $\mathsf{poss} : O \longrightarrow \Phi$ associates to each action name ω a propositional formula poss_ω such that for every agent a, $\bigvee_{\omega | \mathsf{agt}_\omega = a} \mathsf{poss}_\omega$ is valid in propositional logic;
- $\mathsf{eff} : O \longrightarrow P \longrightarrow \Phi$ is a mapping associating to each action name ω a partial function $\mathsf{eff}_\omega : P \longrightarrow \Phi$, such that if poss_{ω_1} and poss_{ω_2} are satisfiable in propositional logic and $\mathsf{agt}_{\omega_1} \neq \mathsf{agt}_{\omega_2}$ then the domains of eff_{ω_1} and eff_{ω_2} are disjoint.

The function agt associates actions to agents who can perform them. No two agents have the same action. The propositional formula poss_ω characterises the conditions under which ω is executable by agt_ω; the constraint says that at every state, each agent has at least one action that is executable. The intuition of the function eff is that when eff_ω is defined for p then one of the things ω does is to assign to p the truth value of $\mathsf{eff}_\omega(p)$: if φ is true before ω then p is true after ω, and if φ is false before ω then p is false after ω. When eff_ω is undefined for p then ω leaves the truth value of p unchanged. The disjointness constraint guarantees that there is no state where two different agents have executable actions changing the truth value of p. This is more liberal a condition than exclusive control[6] that is common in logics of propositional control [26,10]. We call ours *local exclusive control*.

Example 1. Consider a light that is controlled by two switches. The position of these switches is described by two propositional variables; moreover, there are variables describing whether agent a is close to switch k or not: $P = \{up_1, up_2\} \cup \{close_{a,k} \mid a, k \in \{1,2\}\}$. The light is on if the switches are either both up ($up_1 \wedge up_2$) or both down ($\neg up_1 \wedge \neg up_2$); in other words, the light is on if and only if $up_1 \leftrightarrow up_2$. There are two agents: $S = \{1, 2\}$. Each agent a can toggle each switch k ($toggle_{a,k}$) or do nothing: $O = \{toggle_{a,k} \mid a, k \in \{1,2\}\} \cup \{nop_a \mid a \in \{1,2\}\}$. Therefore the action description $\mathcal{T} = \langle \mathsf{agt}, \mathsf{poss}, \mathsf{eff} \rangle$ is as follows.

- $\mathsf{agt}_{toggle_{a,k}} = \mathsf{agt}_{nop_a} = a$, for all agents a and switches k;
- $\mathsf{poss}_{toggle_{a,k}} = close_{a,k} \wedge \neg close_{a',k}$, for agents $a \neq a'$ (in order to flip a switch the agent has to be the only one close to it);
- $\mathsf{poss}_{nop_a} = \top$;
- eff_{nop_a} is undefined for all $p \in P$ (the action nop_a does not change any variable);
- $\mathsf{eff}_{toggle_{a,k}}$ is defined for up_k, and $\mathsf{eff}_{toggle_{a,k}}(up_k) = \neg up_k$.

Observe that the function eff obeys our constraints on action descriptions: for the conjunction $\mathsf{poss}_{toggle_{1,1}} \wedge \mathsf{poss}_{toggle_{2,2}}$ to be propositionally satisfiable, the domains of eff, $\mathrm{dom}(\mathsf{eff}_{toggle_{1,1}}) = \{up_1\}$ and $\mathrm{dom}(\mathsf{eff}_{toggle_{2,2}}) = \{up_2\}$, have to be disjoint, which is indeed the case.

---

[6] According to [10], control is exclusive when $\mathsf{agt}_{\omega_1} \neq \mathsf{agt}_{\omega_2}$ implies that the domains $\mathrm{dom}(\mathsf{eff}_{\omega_1})$ and $\mathrm{dom}(\mathsf{eff}_{\omega_2})$ are disjoint, whatever poss_{ω_1} and poss_{ω_2} are. (We have adapted the notation.)

Action descriptions are an economic description of a domain and 'count as a solution to the frame problem' [19]: the descriptions only talk about what changes and do not contain frame axioms. A given $\mathcal{T} = \langle \text{agt}, \text{poss}, \text{eff} \rangle$ determines what Reiter calls a successor state axiom for each $p \in P$; in the situation calculus this takes the following form:

$$p(do(x, s)) \leftrightarrow \Big(\bigvee_{\omega \mid p \in \text{dom}(\text{eff}_\omega)} (x = \omega \wedge \text{eff}_\omega(p)) \Big) \vee \Big(p(s) \wedge \neg \bigvee_{\omega \mid p \in \text{dom}(\text{eff}_\omega)} x = \omega \Big)$$

where x is an action variable and s is a situation variable, both universally quantified. It says that action x makes p true iff either x is an action whose precondition for making p true holds, or p was true before and x is not an action changing p.

3.2   CGSNs for $\mathcal{T}$

We now associate concurrent game structures with action names to a given action description.

Let $\langle S, P, O \rangle$ be a signature. Let $\mathcal{T} = \langle \text{agt}, \text{poss}, \text{eff} \rangle$ be an action description and let $\mathcal{C} = \langle W, V, M, Mov, E, \| \cdot \| \rangle$ be CGSN. $\mathcal{C}$ is a *CGSN for $\mathcal{T}$* iff:

- $M = O$;
- $Mov(w, a) = \{\omega \in O \mid \text{agt}_\omega = a \ \& \ V(w) \models \text{poss}_\omega\}$;
- $V(E(w, \boldsymbol{m})) =$

$$\{p \mid \exists i \in S, \text{eff}_{m_i} \text{ defined for } p \ \& \ V(w) \models \text{eff}_{m_i}(p)\} \cup$$
$$\{p \mid p \in V(w) \ \& \ \forall i \in S, \text{eff}_{m_i} \text{ undefined for } p\};$$

- $\|\omega\| = \omega$.

In the clause for Mov, the condition $V(w) \models \text{poss}_\omega$ has to be understood as truth of poss_ω in the propositional interpretation $V(w)$. Note that the clause for E corresponds to Reiter's successor state axiom.

A state formula φ of the language of **ATLEA** is *valid in the class of CGSNs for $\mathcal{T}$* iff $\mathcal{C}, w \models \varphi$ for every state w of every CGSNs $\mathcal{C}$ for $\mathcal{T}$ whose signature contains that of φ. Moreover, φ is *satisfiable in a CGSN for $\mathcal{T}$* iff $\neg\varphi$ is not satisfiable.

We can now formulate two important problems in reasoning about actions. Suppose given a signature $\langle S, P, O \rangle$, an action description $\mathcal{T}$, a formula describing the initial state φ_i and a formula describing the goal state φ_g. The *prediction problem* for a sequence of multiagent actions $\rho_1, \ldots, \rho_n$ is to decide whether it the case that

$$\varphi_i \rightarrow \langle\!\langle \text{dom}(\rho_1) \rangle\!\rangle_{\rho_1} \bigcirc \cdots \langle\!\langle \text{dom}(\rho_n) \rangle\!\rangle_{\rho_n} \bigcirc \varphi_g$$

is valid in the class of CGSNs for $\mathcal{T}$; the *planning problem* for a set of agents A is to decide whether it the case that

$$\varphi_i \rightarrow \langle\!\langle A \rangle\!\rangle_\emptyset \Diamond \varphi_g$$

is valid in the class of CGSNs for $\mathcal{T}$.

Example 2. Let us take up Example 1. Whether

$$(close_{1,1} \wedge \neg close_{2,1} \wedge close_{2,2} \wedge \neg close_{1,2} \wedge up_1 \wedge \neg up_2) \rightarrow$$
$$\langle\!\langle 1,2 \rangle\!\rangle_{1 \mapsto nop_1, 2 \mapsto toggle_{2,2}} \bigcirc(up_1 \leftrightarrow up_2)$$

is valid in the CGSNs for $\mathcal{T}$ is a prediction problem. Whether

$$(close_{1,1} \wedge \neg close_{2,1} \wedge close_{2,2} \wedge \neg close_{1,2}) \rightarrow$$
$$\langle\!\langle 1,2 \rangle\!\rangle_\emptyset \bigcirc(up_1 \leftrightarrow up_2)$$

is valid in the CGSNs for $\mathcal{T}$ is a planning problem. Both implications are valid in the class of CGSNs for $\mathcal{T}$.

3.3   Reduction to ATLEA Satisfiability

We now show that for finite signatures, satisfiability in a CGSN for $\mathcal{T}$ can be reduced to ATLEA satisfiability.

Proposition 2. *Let $\langle S, P, O \rangle$ be a finite signature. Let $\mathcal{T}$ be an action description in $\langle S, P, O \rangle$ and let φ be a formula in $\langle S, P, O \rangle$. φ is satisfiable in a CGSN for $\mathcal{T}$ iff $\varphi \wedge \langle\!\langle \emptyset \rangle\!\rangle \Box(\bigwedge \Gamma)$ is ATLEA satisfiable, where Γ collects the following sets of formulas, for every $a \in S$, $p \in P$, and $\omega \in O$:*

1. $\mathsf{poss}_\omega \leftrightarrow \langle\!\langle \mathsf{agt}_\omega \rangle\!\rangle_{\mathsf{agt}_\omega \mapsto \omega} \bigcirc \top$
2. $\mathsf{eff}_\omega(p) \rightarrow \langle\!\langle \emptyset \rangle\!\rangle_{\mathsf{agt}_\omega \mapsto \omega} \bigcirc p$, *for $p \in \mathsf{dom}(\mathsf{eff}_\omega)$*
3. $\neg\mathsf{eff}_\omega(p) \rightarrow \langle\!\langle \emptyset \rangle\!\rangle_{\mathsf{agt}_\omega \mapsto \omega} \bigcirc\neg p$, *for $p \in \mathsf{dom}(\mathsf{eff}_\omega)$*
4. $\left(\bigwedge_{\omega|p \in \mathsf{dom}(\mathsf{eff}_\omega)} \neg\mathsf{poss}_\omega\right) \rightarrow (p \rightarrow \langle\!\langle \emptyset \rangle\!\rangle_\emptyset \bigcirc p)$
5. $\left(\bigwedge_{\omega|p \in \mathsf{dom}(\mathsf{eff}_\omega)} \neg\mathsf{poss}_\omega\right) \rightarrow (\neg p \rightarrow \langle\!\langle \emptyset \rangle\!\rangle_\emptyset \bigcirc\neg p)$
6. $\mathsf{poss}_\omega \rightarrow (p \rightarrow \langle\!\langle \emptyset \rangle\!\rangle_{\mathsf{agt}_\omega \mapsto \omega'} \bigcirc p)$

   for $p \in \mathsf{dom}(\mathsf{eff}_\omega)$ and $p \notin \mathsf{dom}(\mathsf{eff}_{\omega'})$;
7. $\mathsf{poss}_\omega \rightarrow (\neg p \rightarrow \langle\!\langle \emptyset \rangle\!\rangle_{\mathsf{agt}_\omega \mapsto \omega'} \bigcirc\neg p)$

   for $p \in \mathsf{dom}(\mathsf{eff}_\omega)$ and $p \notin \mathsf{dom}(\mathsf{eff}_{\omega'})$.

Formula 1 translates the information specified in $\mathcal{T}$ about the executability of ω. Formulas 2 and 3 translate the information about the effects of ω. The last four clauses are about the frame axioms and basically express that those variables p for which eff_ω is undefined are left unchanged by the execution of ω. Formulas 4 and 5 say that when none of the actions changing p is executable then the truth value of p remains unchanged. Consider formulas 6 and 7: suppose p is one of the effects of ω ($p \in \mathsf{dom}(\mathsf{eff}_\omega)$) and suppose at the present state ω is executable (poss_ω is true); then due to the local exclusive control constraint on the eff function of $\mathcal{T}$, at that state p can only be changed by agt_ω. Therefore, when agt_ω performs a different action ω' not affecting p then the truth value of p remains unchanged, whatever the other agents do.

Observe that the cardinality of Γ is polynomial in the number of symbols in the signature (more precisely: cubic). As the length of every formula in Γ is bound by the cardinality of Γ (because of items 4 and 5), the length of the formula $\bigwedge \Gamma$ is polynomial in the number of symbols in the signature, too. We can therefore polynomially embed the reasoning problems of prediction and planning into ATLEA.

4   Epistemic Extension

We now sketch an epistemic extension of **ATLEA** along the lines of [25]. We call our logic Alternating-time Temporal Epistemic Logic with Explicit Actions, **ATELEA**.

4.1   ATELEA

We add knowledge modalities K_a to the language, one per agent a in Σ, and as well as common knowledge modalities C_A, one per finite subset A of Σ. We read the formula $K_a\varphi$ as "a knows that φ is true" and the formula $C_A\varphi$ as "the agents in A have common knowledge that φ is true".

Concurrent Epistemic Game Structures with action Names CEGSNs) are of the form

$$\mathcal{C}^+ = \langle W, V, M, Mov, E, \|\cdot\|, \{R_a\}_{a\in S}\rangle$$

where $\langle W, V, M, Mov, E, \|\cdot\|\rangle$ is a CGSN (cf. Def. 2) and where every $R_a \subseteq W \times W$ is an equivalence relation.

Given a CEGSN $\mathcal{C}^+ = \langle W, V, M, Mov, E, \|\cdot\|, \{R_a\}_{a\in S}\rangle$, the satisfaction relation $\models$ is defined as follows:

$$\mathcal{C}^+, w \models K_a\varphi \text{ iff } \mathcal{C}^+, v \models \varphi \text{ for all } v \in W \text{ with } wR_av$$

$$\mathcal{C}^+, w \models C_A\varphi \text{ iff } \mathcal{C}^+, v \models \varphi \text{ for all } v \in W \text{ with } wR_A^+v$$

where $R_A = \bigcup_{a\in A}R_a$ and where R_A^+ is the transitive closure of R_A. For the **ATLEA** operators the definition is as before.

We can extend the decision procedure for **ATLEA** to allow for the epistemic operators. This is done similarly to **ATEL** compared to **ATL** [27]. We obtain the following result.

Theorem 2. *The satisfiability problem for* **ATELEA** *is ExpTime-complete.*

Let us take over the concrete semantics for **ATLEA** given in Section 3 and consider the class of CEGSNs structures induced by an action specification. Let $\mathcal{T} = \langle \text{agt}, \text{poss}, \text{eff}\rangle$ be an action specification and $\mathcal{C}^+ = \langle \mathcal{C}, \{R_a\}_{a\in S}\rangle$ a CEGSN for a finite signature $\langle S, P, O\rangle$. We say that $\mathcal{C}^+$ is a *CEGSN for $\mathcal{T}$* if $\mathcal{C}$ is a CGSN for $\mathcal{T}$ as defined in Section 3.2.

As the following proposition highlights, satisfiability in a CEGSN for an action specification $\mathcal{T}$ can be reduced to **ATELEA** satisfiability: satisfiability with respect to the general class of CEGSNs.

Let $dg(\varphi)$ be the maximal number of nestings of ATLEA operators $\langle\!\langle A\rangle\!\rangle_\rho$ and ATELEA epistemic operators K_a or C_A within φ. Let $(\langle\!\langle\emptyset\rangle\!\rangle\Box C_A)^n\psi$, for $n \geq 0$, be the formula where $\langle\!\langle\emptyset\rangle\!\rangle\Box C_A$ is iterated n times. (So $(\langle\!\langle\emptyset\rangle\!\rangle\Box C_A)^0\psi$ is ψ.)

Proposition 3. *Let $\mathcal{T}$ be an action specification in the finite signature $\langle S, P, O\rangle$ and let $dg(\varphi) = n$. Let φ be a formula of the language of* **ATELEA** *in $\langle S, P, O\rangle$. φ is satisfiable in a CEGSN for $\mathcal{T}$ iff $\varphi \wedge (\langle\!\langle\emptyset\rangle\!\rangle\Box C_S)^n(\bigwedge \Gamma) \wedge (C_S\langle\!\langle\emptyset\rangle\!\rangle\Box)^n(\bigwedge \Gamma)$ is* **ATELEA** *satisfiable, where Γ is the finite set of formulas defined in Proposition 2.*

The proof can be done in a way similar to that of Prop. 2.

4.2   Reasoning about Uniform Choices in ATELEA

An interesting aspect of our logic is that it allows us to express the concept of *uniform choice*. Specifically, we say that agent a has a *uniform choice* from the finite set of actions O to ensure that φ will be true in the next state when there exists an action in O such that a knows that by choosing this action she will ensure φ in the next state, no matter what the other agents will do. This can be expressed in ATELEA as follows:

$$\mathsf{UC}_a(O,\varphi) \stackrel{\text{def}}{=} \bigvee_{w \in O} \mathsf{K}_a \langle\!\langle \{a\} \rangle\!\rangle_{\{a \mapsto w\}} \bigcirc \varphi$$

$\mathsf{UC}_i(O,\varphi)$ has to be read "agent a has a uniform choice from the finite set of actions O to ensure φ in the next state". This concept of uniform choice is closely related to the concept of power. In fact, a given agent a's power of achieving a certain result φ involves not only a's capability of achieving a but also a's knowledge about this capability. For example, for a thief to have the power of opening a safe, he must know the safe's combination. (See [16] for a detailed analysis of the distinction between capability and power.)

Furthermore, in ATELEA we can draw non-trivial inferences showing that, given certain initial conditions, an agent has (or has not) a uniform choice to ensure φ in the next state. Consider the following continuation of Example 1.

Example 3 (cont.). Remember that the light is on if the switches are either both up $(up_1 \wedge up_2)$ or both down $(\neg up_1 \wedge \neg up_2)$. Let us therefore abbreviate the equivalence $up_1 \leftrightarrow up_2$ by *lightOn*. Suppose that in the initial situation agent 1 knows that the light is off. Moreover, suppose that agent 1 knows that he is close to switch 1. Finally, let us assume that agent 1 knows that agent 2 cannot perform the action of toggling switch 1 or switch 2 because he is far away from both switches. In other words, agent 1 knows that agent 2 cannot interfere with his actions. Then we can prove that agent 1 has a uniform choice to ensure that the light is on in the next state. Indeed, it is easy to show the following formula is valid in the class of CEGSN determined by the action description $\mathcal{T}$ of Example 1:

$$(\mathsf{K}_1 \neg lightOn \wedge \mathsf{K}_1 close_{1,1} \wedge \mathsf{K}_1 (\neg close_{2,1} \wedge \neg close_{2,2})) \rightarrow$$
$$\mathsf{UC}_1(\{toggle_{1,1}, toggle_{1,2}, nop_1\}, lightOn)$$

Thanks to the common knowledge operator we can generalize the previous notion of uniform choice to coalitions of agents. It is reasonable to assume that the agents in a coalition A have the power to ensure a given outcome φ only if they can *coordinate* their actions in such a way that φ will be true in the next state. In order to achieve this level of coordination, the agents in A must have common knowledge that by performing a given joint action they will together make φ true, that is, the agents in A must have a *uniform collective choice* to ensure φ. Uniform collective choice can be formally expressed as follows. Let $A = \{1, \ldots, k\}$. Then:

$$\mathsf{UC}_A(O,\varphi) \stackrel{\text{def}}{=} \bigvee_{\omega_1, \ldots, \omega_k \in O} \mathsf{C}_A \langle\!\langle A \rangle\!\rangle_{\{1 \mapsto \omega_1, \ldots, k \mapsto \omega_k\}} \bigcirc \varphi$$

$\mathsf{UC}_A(O,\varphi)$ has to be read "coalition A has a uniform collective choice from the set of actions O to ensure φ in the next state".

Example 4 (cont.). Let us continue our running example and suppose that agents 1 and 2 have common knowledge that: (1) the light is off, and (2) agent 1 is close to switch 1 and far from switch 2 while agent 2 is close to switch 2 and far from switch 1. Then we can prove that the coalition $\{1,2\}$ has a uniform collective choice to ensure that the light is on in the next state. Indeed, it is easy to show that the following formula is valid in the class of CEGSN determined by the action description $\mathcal{T}$ of Example 1:

$$(\mathsf{C}_{\{1,2\}}\neg lightOn \wedge \mathsf{C}_{\{1,2\}}(close_{1,1} \wedge close_{2,2}) \wedge$$
$$\mathsf{C}_{\{1,2\}}(\neg close_{1,2} \wedge \neg close_{2,1})) \to \mathsf{UC}_{\{1,2\}}(O, lightOn)$$

with $O = \{toggle_{a,k} \mid a, k \in \{1,2\}\} \cup \{nop_a \mid a \in \{1,2\}\}$. Furthermore, we can also prove that if e.g. the agents do not have common knowledge whether the light is on then there is no uniform collective choice ensuring that the light is on. That is,

$$(\neg \mathsf{C}_{\{1,2\}} lightOn \wedge \neg \mathsf{C}_{\{1,2\}} \neg lightOn) \to$$
$$\neg \mathsf{UC}_{\{1,2\}}(O, lightOn)$$

5   Related Work

Several authors have noted that while strategic logics provide an interesting abstract formalism to reason about actions and strategies, it would nevertheless be useful to have actions or strategies as first-class objects. This was tried for Coalition Logic (for example by [4,13]) and for some very expressive logics that turned out to be undecidable (for example [17,5,23]). We here only overview extensions of ATL.

Alternating-time temporal logic with Actions (ATL-A) together with its epistemic extension was introduced in [1] to obtain a strategic logic for describing actions as well as their interaction with knowledge, and to solve problems with previous approaches. ATL-A corresponds to a version of ATLEA with commitment functions ρ defined over non-deterministic composition of action names and in which any such ρ can only occur in formulas of the form $\langle\!\langle A \rangle\!\rangle_\rho \bigcirc \varphi$. While we appreciate ATL-A as an interesting contribution to incorporate actions in strategic logics, we argue that the better design lies with ATLEA. The syntax of ATL-A is unwieldy as each alternative action for every agent has to be mentioned in the formula. This makes it impossible to express a's commitment $a \mapsto w_a$ to use action w_a in ATL-A with a general (infinite) action signature; and even if we restrict the logic to a finite action signature the resulting ATL-A formula will be huge. Abbreviations were suggested (already in [1]) for ATL-A to be more friendly to modellers. ATL-A defines the temporal operators 'forever' and 'until' with action specifications in terms of 'next-time' and the respective fixpoint equation from ATL (cf. Item 7 in Proposition 1). While coupling one-step actions with 'next-time' formulas is conceptually clear, using fixpoint equations to define other temporal operators involves an exponential blowup in formula size, which may be an issue with reasoning complexity. Extending ATL-A to plans of actions appears to require major changes of its semantics, whereas extending ATLEA this way requires defining what it means for a coalitional strategy to be compatible with a complex action description (cf. the set strat(.) in Section 2). In [1], model checking for ATL-A was studied, while the satisfiability problem, which is relevant for synthesis and mechanism design, is not considered.

Commitment ATL, CATL, [24] is an extension of ATL with ternary operators of the form $C_i(\sigma_i, \varphi)$ with the intended reading "if it were the case that agent i committed to the strategy σ_i, then φ". The interpretation of this operator is in terms of model updates: $C_i(\sigma_i, \varphi)$ is true at world w of a given model M if and only if φ holds at w of model M' that results from eliminating from M all moves that are not consistent with agent i's strategy σ_i.[7] The complexity of the satisfiability problem for CATL has not been studied, whereas the complexity result for ATLEA and its epistemic extension is one of our main contributions here. There is also an important conceptual difference: the former considers commitments to play *strategies* while the latter considers commitments to play *actions*. From this point of view, CATL is much closer to ATL with Explicit Strategies (ATLES) by [28], where ATL-path quantifiers are parameterised with commitment functions for strategies [28], than to our ATLEA.

As for the differences between our ATLEA and Walther *et al.*'s ATLES, it is worth noting that with ATLEA we can formalise the (un-)availability of actions at states (cf. the side conditions of items 1 and 2 in Prop. 1), whereas with ATLES one can reference and reason with existing strategies but not reason about their availability. Another difference is the local nature of commitments in ATLEA, i.e., commitments to atomic actions are released after one time step (cf. Item 7 in Prop. 1).

The integration of game-theoretic concepts into the situation calculus was a subject of recent research. Belle and Lakemeyer [3] study games in extensive form (in its imperfect information version), where only one agent can act per state. Consequently no interactions have to be accounted for. They don't have path quantifiers, which allows them to define regression. De Giacomo, Lespérance and Pearce [11] have studied a multiagent version of the situation calculus in order to reason about extensive games where at most one agent can act at a given state. That agent is identified by a predicate $Control(a)$ indicating that a controls the current state. Concurrency is simulated by interleaving. They have a (first-order) language with ATL path quantifiers. For a given signature, the quantifier $\langle\!\langle A \rangle\!\rangle \bigcirc \varphi$ is basically regressed to

$$\left(\bigvee_{a \in A} Control(a) \wedge \bigvee_{\omega \in O} \langle \omega \rangle \varphi \right) \vee \left(\bigvee_{a \notin A} Control(a) \wedge \bigwedge_{\omega \in O} [\omega] \varphi \right)$$

where $\langle \omega \rangle$ and $[\omega]$ are the dynamic operators of PDL. This relies on finiteness of the set of agent and action symbols. While all these approaches do not really allow for 'true' concurrency, Reiter [18] had proposed to extend his solution to the frame problem to concurrent actions. Different from us, he allows for several actions to be performed simultaneously by the same agent and does not assume exclusive control of propositional variables. This comes with the problem of interacting preconditions: there are states where two actions ω_1 and ω_2 with inconsistent postconditions are performed concurrently. This is avoided by our condition of (local) exclusive control.

---

[7] CATL models are called Action-based Alternating Transition Systems (AATSs) and are closely related to CGSNs.

6   Conclusion

We have introduced a variant of Alternating-time Temporal Logic that has explicit actions. The interesting aspect of our logic is that it combines ATL's strategic reasoning with reasoning about actions in terms of pre- and postconditions as traditionally done in AI.

In future research, we will investigate the extension by regular expressions over actions. This will allow to talk not only about uniform choices, but also about uniform strategies. Moreover, we intend to provide sound and complete axiomatizations both for ATLEA and for its epistemic extension ATELEA.

Acknowledgements. We thank the LORI reviewers for their comments. Andreas Herzig and Emiliano Lorini acknowledge the support of the LabEx project CIMI, and Dirk Walther the support of the German Research Foundation (DFG) within the Cluster of Excellence 'Center for Advancing Electronics Dresden'.

References

1. Ågotnes, T.: Action and knowledge in alternating-time temporal logic. Synthese 149(2), 375–407 (2006)
2. Alur, R., Henzinger, R., Kupferman, O.: Alternating-time temporal logic. Journal of the ACM 49(5), 672–713 (2002)
3. Belle, V., Lakemeyer, G.: Reasoning about imperfect information games in the epistemic situation calculus. In: Proceedings of AAAI 2010. AAAI Press (2010)
4. Borgo, S.: Coalitions in action logic. In: Proceedings of IJCAI 2007, pp. 1822–1827 (2007)
5. Brihaye, T., Da Costa, A., Laroussinie, F., Markey, N.: ATL with strategy contexts and bounded memory. In: Artemov, S., Nerode, A. (eds.) LFCS 2009. LNCS, vol. 5407, pp. 92–106. Springer, Heidelberg (2008)
6. Chatterjee, K., Henzinger, T.A., Piterman, N.: Strategy logic. Information and Computation 208(6), 677–693 (2010)
7. van Ditmarsch, H., Herzig, A., de Lima, T.: From Situation Calculus to Dynamic Logic. Journal of Logic and Computation 21(2), 179–204 (2011)
8. Emerson, E.A.: Temporal and modal logic. In: Handbook of Theoretical Computer Science. Formal Models and Semantics, vol. B, pp. 995–1072. MIT Press (1990)
9. Gelfond, M., Lifschitz, V.: Representing action and change by logic programs. The Journal of Logic Programming 17, 301–321 (1993)
10. Gerbrandy, J.: Logics of propositional control. In: Proceedings of AAMAS 2006, pp. 193–200. ACM (2006)
11. De Giacomo, G., Lespérance, Y., Pearce, A.R.: Situation calculus based programs for representing and reasoning about game structures. In: Proceedings of KR 2010. AAAI Press (2010)
12. Goranko, V., van Drimmelen, G.: Complete axiomatization and decidability of alternating-time temporal logic. Theoretical Computer Science 353(1-3), 93–117 (2006)
13. Herzig, A., Lorini, E.: A dynamic logic of agency I: STIT, capabilities and powers. Journal of Logic, Language and Information 19(1), 89–121 (2010)
14. Kartha, N., Lifschitz, V.: A simple formalizations of actions using circumscription. In: Proceedings of IJCAI 1995, pp. 1970–1975 (1995)

15. Lange, M., Latte, M.: A CTL-based logic for program abstractions. In: Dawar, A., de Queiroz, R. (eds.) WoLLIC 2010. LNCS (LNAI), vol. 6188, pp. 19–33. Springer, Heidelberg (2010)
16. Lorini, E., Troquard, N., Herzig, A., Broersen, J.: Grounding power on actions and mental attitudes. Logic Journal of the IGPL 21(3), 311–331 (2013)
17. Mogavero, F., Murano, A., Vardi, M.Y.: Reasoning about strategies. In: Proceedings of FSTTCS 2010, pp. 133–144 (2010)
18. Reiter, R.: Natural actions, concurrency and continuous time in the situation calculus. In: Proceedings of KR 1996, pp. 2–13. Morgan Kaufmann (1996)
19. Reiter, R.: Knowledge in Action: Logical Foundations for Specifying and Implementing Dynamical Systems. MIT Press (2001)
20. Shanahan, M.: Solving the frame problem: a mathematical investigation of the common sense law of inertia. MIT Press (1997)
21. Thielscher, M.: Representing the knowledge of a robot. In: Proceedings of KR 2000, pp. 109–120. Morgan Kaufmann (2000)
22. Thielscher, M.: Reasoning Robots - The Art and Science of Programming Robotic Agents. Applied Logic, vol. (33). Springer, Dordrecht (2005)
23. Troquard, N., Walther, D.: On satisfiability in ATL with strategy contexts. In: del Cerro, L.F., Herzig, A., Mengin, J. (eds.) JELIA 2012. LNCS, vol. 7519, pp. 398–410. Springer, Heidelberg (2012)
24. van der Hoek, W., Jamroga, W., Wooldridge, M.: A logic for strategic reasoning. In: Proceedings of AAMAS 2005, pp. 157–164. ACM Press (2005)
25. van der Hoek, W., Wooldridge, M.: Cooperation, knowledge, and time: Alternating-time temporal epistemic logic and its applications. Studia Logica 75, 125–157 (2003)
26. van der Hoek, W., Wooldridge, M.: On the logic of cooperation and propositional control. Artificial Intelligence 164(1-2), 81–119 (2005)
27. Walther, D.: ATEL with common and distributed knowledge is exptime-complete. In: Proceedings of M4M-4 (2005)
28. Walther, D., van der Hoek, W., Wooldridge, M.: Alternating-time Temporal Logic with Explicit Strategies. In: Proceedings of TARK 2007, pp. 269–278. ACM (2007)
29. Walther, D., Lutz, C., Wolter, F., Wooldridge, M.: ATL satisfiability is indeed Exptime-complete. Journal of Logic and Computation 16(6), 765–787 (2006)

A Dynamic Deontic Logic Based on Histories

Fengkui Ju and Li Liang

Department of Philosophy, Beijing Normal University, Beijing, China
fengkui.ju@bnu.edu.cn,
liangli_logic@163.com

Abstract. We aim to present a deontic logic with updates as an extension of Boolean Modal Logic. The features of this logic include the following: (a) deontic relations are defined on sets of finite sequences of states, called histories, and consequently, formulas are evaluated at histories, not states; and (b) it has two dynamic operators, which tend to update the obligation states of agents in different ways. This logic reflects the distinction between the descriptive and prescriptive use of norm sentences.

1 Introduction

One fundamental issue of deontic logic is *Jorgensen's dilemma*, as noted by [4]. This dilemma was originally about imperatives. There are inferences involving imperatives in our lives. However, imperatives express orders and do not have truth values, so it is hard to say that there is a logic of imperatives. A dilemma arises. Traditionally, deontic logic does not consider imperatives. However, norm sentences such as *"you should stay"* or *"you may leave"* are similar to imperatives in many cases: they can also be used to change agents' behaviors, and therefore do not have truth values. Hence, this dilemma is also a serious problem in deontic logic. There are two puzzles attached to this dilemma: *Ross's Paradox* and the *Free Choice Permission Paradox*, both of which were identified by [7]. The first puzzle can be illustrated by the inference *"you should mail this letter; therefore, you should mail it or burn it"*. This inference is intuitively strange, but valid according to classical logic. The second puzzle is opposite to the first one; it notes that the inference *"you may drink coffee or tea; therefore, you may drink coffee"* is not valid in the classical logic but is intuitively plausible.

In order to solve Jorgensen's dilemma, as mentioned in [3], many philosophers have proposed a distinction between two different uses of norm sentences: descriptive and prescriptive uses. Norm sentences are descriptively used to state what the agent ought to do or what he is allowed to do, among other actions, etc. These sentences can be true or false in these cases. In the prescriptive way, norm sentences are used to generate norms and do not have truth values. Jorgensen's dilemma would disappear if the prescriptive use of norm sentences were not relevant to deontic logic. Deontic logic is "legalized" this way. We consider this distinction reasonable. However, we do not think that prescriptive norm sentences are irrelevant to deontic logic. We believe that for any moral agent,

D. Grossi, O. Roy, and H. Huang (Eds.): LORI 2013, LNCS 8196, pp. 176–189, 2013.

there is an obligation state regarding his obligations and freedoms. Descriptive norm sentences describe these states, while prescriptive norm sentences change them. In this paper, we present a dynamic deontic logic to realize this concept.

There is a "dynamic" direction in deontic logic, in which works are based on dynamic logics. A fundamental work is that of [6], which provided a deontic logic as an extension of Propositional Dynamic Logic. Influenced by Anderson-Kanger Deontic Logic, this work introduced a propositional constant, that intuitively means that the requirements of morality are violated. Deontic operators are defined by this constant, but they are applied to actions, not propositions. There is also a "dynamic" direction in semantics for imperatives and permissions starting from [8]. This work is based on update semantics. It proposed a notion *plans*, i.e., a set of to-do lists, which can be viewed as sets of actions. Imperatives and permissions update plans in different ways: the former tend to "strengthen" them, while the latter tend to "weaken" them.

This paper attempts to combine the spirits of these two research lines. As a propositional dynamic logic, Boolean Modal Logic contains these three action constructors: *complement, intersection* and *choice*. Our work is an extension of this logic in both language and semantics. The extended language contains a deontic operator, applied to actions, and two dynamic operators, corresponding to the descriptive utterance of obligations and the prescriptive utterance of permissions. The prescriptive utterance of obligations is derived from other utterances. A model is a labeled transition system plus a deontic relation, which is defined on the set of finite sequences of states, called *histories*, not on the set of states. The truth of a formula is defined against histories, not states. Histories represent what the agent has done. In this way, the idea of *what you have done affects what you can do* is reflected semantically. Descriptive norm sentences describe models, while prescriptive norm sentences update models by changing deontic relations. In this logic, Ross's Paradox is not valid, but the Free Choice Permission Paradox is. At the end, we axiomatize the logic not containing any dynamic operators.

2   Language and Semantics

2.1   Language

Let Π_0 be a countable set of atomic actions and Φ_0 a countable set of atomic propositions. Let a range over Π_0 and p over Φ_0. The sets Π of actions and Φ of propositions are defined as follows:

$$\alpha ::= a \,|\, 1 \,|\, \overline{\alpha} \,|\, (\alpha \cap \alpha) \,|\, (\alpha \cup \alpha)$$
$$\phi ::= p \,|\, \top \,|\, O\alpha \,|\, \neg\phi \,|\, (\phi \wedge \phi) \,|\, \langle\alpha\rangle\phi \,|\, [\downarrow \alpha]\phi \,|\, [\uparrow \alpha]\phi$$

The empty action 0 is defined as $\overline{1}$. Other routine propositional connectives, the falsity $\bot$, and the dual $[\alpha]\phi$ of $\langle\alpha\rangle\phi$ are defined in the usual way. To perform $\overline{\alpha}$ is to do something that is not α. To perform $\alpha \cap \beta$ is to perform α and β at the same time. To perform $\alpha \cup \beta$ is to perform α or β. This language does not have compositions of actions, and all actions are just *one unit deep*. $O\alpha$ means that the agent ought to do α. As the dual of $O\alpha$, $P\alpha$ is defined as $\neg O\bar{\alpha}$, which means that the agent may do α. For any α, $O\alpha$ is called a *pure deontic formula*. $\downarrow\alpha$ denotes the action of descriptive utterance of "*you should do α*", and $[\downarrow\alpha]\phi$ means after this utterance, ϕ is true. $\uparrow\alpha$ denotes the prescriptive utterance of "*you may do α*", and $[\uparrow\alpha]\phi$ indicates that ϕ is true after the utterance.

2.2   Models

Let W be a set of states. Let Δ_W denote the set of finite non-empty sequences of states in W. Each element of Δ_W is called a *history* of W. Capitals like H, J and K denote histories. For any $H \in \Delta_W$, let $\overset{\circ}{H}$ denote the last state of H. A *model* is a tuple $\mathfrak{M} = (W, \{R_\alpha \mid \alpha \in \Pi\}, \mathcal{D}, V)$ where

(1) W is a non-empty set of states;
(2) $R_\alpha \subseteq W \times W$;
(3) $\mathcal{D} \subseteq \Delta_W \times \Delta_W$ and for any $(H, J) \in \mathcal{D}$, $J = (H, w)$ for some $w \in W$;
(4) V is a function from Φ_0 to 2^W.

$\mathcal{D}$ is called the *deontic relation*. There is no loop for $\mathcal{D}$, i.e., any history H can not reach itself in finite steps. This intuitively means that agents' histories are always going to be their histories. A model $\mathfrak{M}$ is standard if it meets such constraints:

(1) $R_1 = W \times W$;
(2) $R_{\bar{\alpha}} = W \times W - R_\alpha$;
(3) $R_{\alpha\cap\beta} = R_\alpha \cap R_\beta$;
(4) $R_{\alpha\cup\beta} = R_\alpha \cup R_\beta$;
(5) $\mathcal{D}$ is serial.

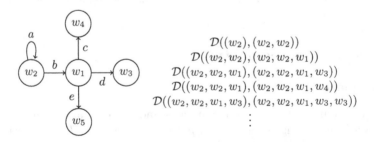

$$\mathcal{D}((w_2), (w_2, w_2))$$
$$\mathcal{D}((w_2, w_2), (w_2, w_2, w_1))$$
$$\mathcal{D}((w_2, w_2, w_1), (w_2, w_2, w_1, w_3))$$
$$\mathcal{D}((w_2, w_2, w_1), (w_2, w_2, w_1, w_4))$$
$$\mathcal{D}((w_2, w_2, w_1, w_3), (w_2, w_2, w_1, w_3, w_3))$$
$$\vdots$$

Fig. 1. A Standard Model

Figure 1 depicts what a standard model looks like, where valuations are omitted. A labeled transition system is on the left, and the deontic relation is on the right. Histories are sequences of states. Since actions are transitions of states, histories represent what the agent has done. Suppose he is standing in w_2 with a *blank* history (w_2) behind him, which means that he has done nothing. According to the deontic relation, he now must perform a. After a is done, he is still in w_2, however, his history is now (w_2, w_2), and he must perform b, which will take him to w_1. What he is allowed to do is dependent on what he has done. There are three actions possible for the agent to perform in w_1: c, d and e. However, given the history (w_2, w_2, w_1), as a moral agent, he is not allowed to do e, and he must perform c or d, although he can freely choose which one.

We require the deontic relation to be serial. We make this requirement for the following reasons: we believe that for any action, no matter what the world is and what the agent has done, he is allowed to perform it or the opposite of it, and we do not think a coherent legal system could tolerate the existence of situations in which the agent is forbidden to do *anything*. In some cases, performing an action in a state might not change this state. For example, consider an agent pushing a revolving door. It is not reasonable to think that what the agent has to do never changes before and after performing this sort of actions, because otherwise if the agent has to push this door, he might have to push it forever. This is one reason we introduce histories as parameters in defining deontic relations. A second reason will be explained later.

2.3   Updates of Models

Let $\mathfrak{M} = (W, \{R_\alpha \mid \alpha \in \Pi\}, \mathcal{D}, V)$ be a model, H a history, and α an action.

Definition 1 (Two Updates of Deontic Relations).

(1) $\mathcal{D}_\alpha^H = \mathcal{D} - \{(H, (H, w)) \mid \neg R_\alpha(\mathring{H}, w)\}$;
(2) $\mathcal{D}_H^\alpha = \mathcal{D} \cup \{(H, (H, w)) \mid R_\alpha(\mathring{H}, w)\}$.

The only difference among $\mathcal{D}$, $\mathcal{D}_\alpha^H$ and $\mathcal{D}_H^\alpha$ lies in that H might "see" less in $\mathcal{D}_\alpha^H$ and "see" more in $\mathcal{D}_H^\alpha$ than in $\mathcal{D}$. For any $\mathcal{D}$, let $g_H(\mathcal{D}) = \{w \in W \mid \mathcal{D}(H, (H, w))\}$, which is called the *goodness set* of H in $\mathcal{D}$. Let $R_\alpha^{\mathring{H}} = \{w \mid R_\alpha(\mathring{H}, w)\}$. It can be verified $g_H(\mathcal{D}_\alpha^H) = g_H(\mathcal{D}) \cap R_\alpha^{\mathring{H}}$ and $g_H(\mathcal{D}_H^\alpha) = g_H(\mathcal{D}) \cup R_\alpha^{\mathring{H}}$. Essentially, the two updates are two different ways of changing the goodness sets of H in $\mathcal{D}$. If $\mathcal{D}$ is serial, then $\mathcal{D}_H^\alpha$ is serial, but $\mathcal{D}_\alpha^H$ might not be. However, given that $\mathcal{D}$ is serial, if there is a w such that $\mathcal{D}(H, (H, w))$ and $R_\alpha(\mathring{H}, w)$, then $\mathcal{D}_\alpha^H$ is serial. The following proposition includes some results about manipulating updates, which will be used later:

Proposition 1.

(1) $(\mathcal{D}_\alpha^H)_\beta^J = (\mathcal{D}_\beta^J)_\alpha^H$;
(2) $(\mathcal{D}_H^\alpha)_J^\beta = (\mathcal{D}_J^\beta)_H^\alpha$;
(3) $(\mathcal{D}_\alpha^H)_J^\beta = (\mathcal{D}_J^\beta)_\alpha^H$, where $J \neq H$.

Based on the updates of deontic relations, we define updates of models:

Definition 2 (Two Updates of Models).

(1) $\mathfrak{M}_\alpha^H = (W, \{R_\alpha \mid \alpha \in \Pi\}, \mathcal{D}_\alpha^H, V)$;
(2) $\mathfrak{M}_H^\alpha = (W, \{R_\alpha \mid \alpha \in \Pi\}, \mathcal{D}_H^\alpha, V)$.

The two updates only change the deontic relations of models. We see if $\mathfrak{M}$ is standard, then $\mathfrak{M}_H^\alpha$ is standard, but $\mathfrak{M}_\alpha^H$ might not be. $\mathfrak{M}_\alpha^H$ can be viewed as the result of updating $\mathfrak{M}$ with the descriptive utterance of "*you should do* α" at the history H, and $\mathcal{D}_H^\alpha$ as the result of updating $\mathcal{D}$ with the prescriptive utterance of "*you may do* α" at H. The first update tends to "stop" some transitions, while the second tends to "free" some links. We take the model illustrated in **Figure 1** as an example. Uttering "*you should do* c" in the descriptive way at (w_2, w_2, w_1) would cut the deontic link between (w_2, w_2, w_1) and (w_2, w_2, w_1, w_3). This means the agent is not allowed to transition to w_3 and must perform c. Prescriptively uttering "*you may do* e" at (w_2, w_2, w_1) would generate a link between (w_2, w_2, w_1) and (w_2, w_2, w_1, w_5), which means he can do e now.

2.4   Semantics

Let $\mathfrak{M} = (W, \{R_\alpha \mid \alpha \in \Pi\}, \mathcal{D}, V)$ be a model, and H a history. Here we do not require $\mathfrak{M}$ to be standard. Truth of formulas at H is defined as follows:

(1) $\mathfrak{M}, H \Vdash p \Leftrightarrow \mathring{H} \in V(p)$;
(2) $\mathfrak{M}, H \Vdash \top$ always holds;
(3) $\mathfrak{M}, H \Vdash O\alpha \Leftrightarrow$ for any $w \in W$, if $\mathcal{D}(H, (H, w))$, then $R_\alpha(\mathring{H}, w)$;
(4) $\mathfrak{M}, H \Vdash \neg\phi \Leftrightarrow$ not $\mathfrak{M}, H \Vdash \phi$;
(5) $\mathfrak{M}, H \Vdash (\phi \wedge \psi) \Leftrightarrow \mathfrak{M}, H \Vdash \phi$ and $\mathfrak{M}, H \Vdash \psi$;
(6) $\mathfrak{M}, H \Vdash \langle\alpha\rangle\phi \Leftrightarrow$ there is a $w \in W$ such that $R_\alpha(\mathring{H}, w)$ and $\mathfrak{M}, (H, w) \Vdash \phi$;
(7) $\mathfrak{M}, H \Vdash [\downarrow \alpha]\phi \Leftrightarrow \mathfrak{M}, H \Vdash P\alpha$ implies $\mathfrak{M}_\alpha^H, H \Vdash \phi$;
(8) $\mathfrak{M}, H \Vdash [\uparrow \alpha]\phi \Leftrightarrow \mathfrak{M}_H^\alpha, H \Vdash \phi$.

It can be verified that

(9) $\mathfrak{M}, H \Vdash P\alpha \Leftrightarrow$ there is a $w \in W$ such that $\mathcal{D}(H, (H, w))$ and $R_\alpha(\mathring{H}, w)$;
(10) $\mathfrak{M}, H \Vdash [\alpha]\phi \Leftrightarrow$ for any $w \in W$, if $R_\alpha(\mathring{H}, w)$, then $\mathfrak{M}, (H, w) \Vdash \phi$.

The formula $\langle\alpha\rangle\phi$ being true at H means that there is a way to perform α such that after α is done, ϕ is true at the new history. It can be verified that $\mathfrak{M}, H \Vdash O\alpha$ if $g_H(\mathcal{D}) \subseteq R_\alpha^{\mathring{H}}$. This intuitively means that α is obligatory for the agent if whatever he does without violating morality, α would be performed. We can also verify that $\mathfrak{M}, H \Vdash P\alpha$ if $g_H(\mathcal{D}) \cap R_\alpha^{\mathring{H}} \neq \emptyset$. This means that he is allowed to perform α if there is a way to perform α without violating morality. Similar ideas can be found in [3].

We consider only standard models reasonable. As discussed, given that $\mathfrak{M}$ is standard, $\mathfrak{M}_\alpha^H$ might not be standard, unless there is a $w \in W$ such that

$\mathcal{D}(H, (H, w))$ and $R_\alpha(\overset{\circ}{H}, w)$, that is, $\mathfrak{M}, H \Vdash P\alpha$. This is why we define the truth condition of $[\downarrow \alpha]\phi$ as conditional. Those updates resulting in non-standard models are unsuccessful ones. The truth of formulas is defined at histories in general models, not just in standard models, so the definition is well-defined. This semantics would collapse to classical relational semantics if the deontic part were ignored; hence, it is a genuine extension of Boolean Modal Logic.

A formula ϕ is *valid* if for any *standard* model $\mathfrak{M}$ and history H, $\mathfrak{M}, H \Vdash \phi$.

3 Valid Formulas

Proposition 2. *The following formulas are valid:*

(1) $O\alpha \to P\alpha$;
(2) $P\alpha \to \langle\alpha\rangle\top$;
(3) $[\downarrow \alpha]O\alpha$;
(4) $\langle\alpha\rangle\top \to [\uparrow \alpha]P\alpha$.

It is easy to verify this proposition. From the first two items in it, we obtain that *Kant's Law*, expressed as $O\alpha \to \langle\alpha\rangle\top$, is valid. The third indicates that the agent ought to do α after the descriptive utterance of *"you should do α"*. The last item expresses that he is allowed to do α after the prescriptive utterance of *"you may do α"*, given that α is possible to perform.

[3] proposed a principle to explain why Ross's Paradox seems invalid: in our intuitions, if a norm sentence N_1 entails N_2, then the normative effects of N_1 entail the normative effects of N_2. The prescriptive utterance of *"you should mail the letter or burn it"* gives the agent the permission to burn the mail, but the utterance of *"you should mail the letter"* does not; therefore, the normative effects of the former do not entail the normative effects of the latter. Then Ross's Paradox is not valid. We consider this principle plausible. Even further, we believe that its converse is also reasonable. In fact, the bi-implication version of this principle underlies update semantics in defining validity. According to the stronger version, the Free Choice Permission Paradox is valid, as the prescriptive utterance of *"you may drink coffee"* just gives the agent the freedom to drink coffee, whereas the utterance of *"you may drink coffee or tea"* gives him the freedom to drink tea, in addition to the freedom to drink coffee.

Our language contains dynamic operators and models contain normative factors; thus, the normative effects of utterances can be expressed in this setting. We believe that prescriptive norm sentences generate not only obligations but also permissions. In [5], we have argued that in the aspect of normative effects, prescriptively uttering *"you should do α"* is equivalent to prescriptively uttering *"you may do α"* and then descriptively uttering *"you should do α"*. Define $[\uparrow\downarrow \alpha]$ as $[\uparrow \alpha][\downarrow \alpha]$, which represents the action of prescriptively uttering *"you should do α"*. Ross's Paradox fails here. Let c denote the action *mailing the letter*, and e the action *burning the letter*. We look at the model illustrated in **Figure 1**. It can be verified that $[\uparrow\downarrow c][\uparrow\downarrow (c \cup e)]Pe$ is true at the history (w_2, w_2, w_1), but $[\uparrow\downarrow c]Pe$ is false at it. Therefore, the normative effects of *"you should mail the* *letter"* do not entail the normative effects of *"you should mail it or burn it"*. One may wonder why $[\downarrow \alpha][\uparrow \alpha]$ is not used to denote prescriptive utterances of *"you should do α"*. Here is the reason. The update sequence $[\uparrow \alpha][\downarrow \alpha]$ might be different from $[\downarrow \alpha][\uparrow \alpha]$, and the only difference is this: given that α is possible to perform, $[\uparrow \alpha][\downarrow \alpha]$ would always be successful, but $[\downarrow \alpha][\uparrow \alpha]$ might not, as $[\downarrow \alpha]$ might make a standard model not serial. We believe that in real life, given α is possible to perform, prescriptive utterance of *"you should do α"* is always meaningful. The Free Choice Permission Paradox, $[\uparrow (\alpha \cup \beta)][\uparrow \alpha]\phi \leftrightarrow [\uparrow (\alpha \cup \beta)]\phi$, is valid in this semantics, which is easy to check.

The following lemma says that the two updates do not change a model much:

Lemma 1. *J is a proper super-sequence of H.*

(1) $\mathfrak{M}_\alpha^H, J \Vdash \phi$ if and only if $\mathfrak{M}, J \Vdash \phi$;
(2) $\mathfrak{M}_H^\alpha, J \Vdash \phi$ if and only if $\mathfrak{M}, J \Vdash \phi$.

Proposition 1 is used in proving this lemma. By this lemma, the following two propositions hold:

Proposition 3. *The following formulas are valid:*

(1) $[\downarrow \alpha]p \leftrightarrow (P\alpha \to p)$;
(2) $[\downarrow \alpha]\top \leftrightarrow (P\alpha \to \top)$;
(3) $[\downarrow \alpha]O\beta \leftrightarrow (P\alpha \to O(\overline{\alpha} \cup \beta))$;
(4) $[\downarrow \alpha]\neg\phi \leftrightarrow (P\alpha \to \neg[\downarrow \alpha]\phi)$;
(5) $[\downarrow \alpha](\phi \wedge \psi) \leftrightarrow ([\downarrow \alpha]\phi \wedge [\downarrow \alpha]\psi)$;
(6) $[\downarrow \alpha]\langle\beta\rangle\phi \leftrightarrow (P\alpha \to \langle\beta\rangle\phi)$;
(7) $[\downarrow \alpha][\downarrow \beta]\phi \leftrightarrow [\downarrow (\alpha \cap \beta)]\phi$.

Proposition 4. *The following formulas are valid:*

(1) $[\uparrow \alpha]p \leftrightarrow p$;
(2) $[\uparrow \alpha]\top \leftrightarrow \top$;
(3) $[\uparrow \alpha]O\beta \leftrightarrow (O\beta \wedge [\alpha \cap \overline{\beta}]\bot)$;
(4) $[\uparrow \alpha]\neg\phi \leftrightarrow \neg[\uparrow \alpha]\phi$;
(5) $[\uparrow \alpha](\phi \wedge \psi) \leftrightarrow ([\uparrow \alpha]\phi \wedge [\uparrow \alpha]\psi)$;
(6) $[\uparrow \alpha]\langle\beta\rangle\phi \leftrightarrow \langle\beta\rangle\phi$;
(7) $[\uparrow \alpha][\uparrow \beta]\phi \leftrightarrow [\uparrow (\alpha \cup \beta)]\phi$.

From these propositions, we obtain that the formulas containing only one dynamic operator can be equivalently reduced to the formulas not containing any. By introducing histories, we can obtain valid formulas $[\downarrow \alpha]\langle\beta\rangle\phi \leftrightarrow \langle\beta\rangle\phi$ and $[\uparrow \alpha]\langle\beta\rangle\phi \leftrightarrow \langle\beta\rangle\phi$, and consequently obtain the reduction of dynamic operators. This is the above mentioned second motivation for using the notion of histories.

4   Axiomatization

There is a complete axiomatization of this logic. For brevity, we only axiomatize the logic restricted to Φ', the sub-language of Φ not containing any dynamic operators, and leave the full axiomatization for another occasion.

4.1 Axiomatization

Let Φ_{PC} be the language generated from $\Phi_0 \cup \{\top\}$ under $\neg, \wedge$ and $\vee$, where Φ_0 is the set of atomic propositions. Let f be a *natural* bijective function from the set Π of actions to Φ_{PC}. We say α and β are *equivalent* if $f(\alpha) \leftrightarrow f(\beta)$ is a tautology. For instance, $\overline{a \cap b}$ is equivalent to $\overline{a} \cup \overline{b}$. The axiomatization of the logic consists of six classes of axioms:

A. Basic axioms of normal modal logics:
  (1) *all propositional tautologies*;
  (2) $[\alpha](\phi \to \psi) \to ([\alpha]\phi \to [\alpha]\psi)$.
B. The axiom for choice: $\langle \alpha \cup \beta \rangle \phi \leftrightarrow \langle \alpha \rangle \phi \vee \langle \beta \rangle \phi$.
C. Axioms for the universal modality:
  (1) $\phi \to \langle 1 \rangle \phi$;
  (2) $\phi \to [1]\langle 1 \rangle \phi$;
  (3) $\langle 1 \rangle \langle 1 \rangle \phi \to \langle 1 \rangle \phi$;
  (4) $\langle \alpha \rangle \phi \to \langle 1 \rangle \phi$.
D. The axiom for the empty modality: $[0]\bot$.
E. Axioms for equivalence of actions: $\langle \alpha \rangle \phi \leftrightarrow \langle \alpha' \rangle \phi$, if α and α' are equivalent.
F. Axioms for the deontic operator O:
  (1) $(O\alpha \wedge O\beta) \leftrightarrow O(\alpha \cap \beta)$;
  (2) $\neg O\alpha \to P\overline{\alpha}$;
  (3) $(O\alpha \wedge P\beta) \to P(\alpha \cap \beta)$;
  (4) $O\alpha \to \langle \alpha \rangle \top$;
  (5) $P\alpha \to \langle \alpha \rangle \top$.

and two inference rules:

(1) *Modus Ponens*: given ϕ and $\phi \to \psi$, prove ψ;
(2) *Generalization*: given ϕ, prove $[\alpha]\phi$.

The soundness of the logic is easy to show. We use the Henkin method to show the completeness. For the part of Boolean Modal Logic, a technique from [2], called the *copy method*, is borrowed.

   To show the completeness, it suffices to show for any consistent formulas ϕ, there is a standard model $\mathfrak{M}$ and a history H such that $\mathfrak{M}, H \Vdash \phi$. Let G be a consistent formula. Let Σ be the smallest set of formulas such that $G \in \Sigma$ and Σ is closed under subformulas.

4.2 CINF

Let $a_1, \ldots, a_n$ be all the atomic actions of Σ and $A = \{a_1, \ldots, a_n, \overline{a_1}, \ldots, \overline{a_n}\}$. Each element of A is called a literal action. Define Θ_A as such a set: $\{X \subseteq A \mid$ *for any $i \leq n$, exactly one of a_i and $\overline{a_i}$ is in X*$\}$. Θ_A has 2^n members. For any $X \in \Theta_A$, $\gamma = \bigcap X$ is called a path relative to Σ, which is an intersection of some literal actions. There are 2^n paths, if we do not consider orders of literal actions. Enumerate these paths as $\gamma_1, \ldots, \gamma_{2^n}$. In any standard model,

$R_{\gamma_1}, \ldots, R_{\gamma_{2^n}}$ are pairwise disjoint blocks and the union of them is $W \times W$. In other words, $\{R_{\gamma_1}, \ldots, R_{\gamma_{2^n}}\}$ is a partition of $W \times W$. By some refections we can get that for any α built from $a_1, \ldots, a_n$, if α is not equivalent to 0, R_α is the union of some of these blocks. These blocks are like atomic parts of $W \times W$. Here is an example. Suppose a, b and c are all the atomic actions under considerations. There are 8 paths: $a \cap b \cap c, \ldots, \bar{a} \cap \bar{b} \cap \bar{c}$, and $W \times W$ is divided into 8 parts: $R_{a \cap b \cap c}, \ldots R_{\bar{a} \cap \bar{b} \cap \bar{c}}$. Each non-empty action whose atomic actions occur in a, b and c is the union of some of these parts. For example, $R_{a \cap \overline{(b \cap c)}} = R_{a \cap b \cap \bar{c}} \cup R_{a \cap \bar{b} \cap c} \cup R_{a \cap \bar{b} \cap \bar{c}}$.

The classes **D** and **E** of axioms guarantee this result:

Lemma 2. *For any α occurring in Σ, if α is not equivalent to 0, there are paths $\gamma_{n_1}, \ldots, \gamma_{n_m}$ such that α is equivalent to $\gamma_{n_1} \cup \ldots \cup \gamma_{n_m}$.*

For any α not equivalent to 0, we call $\gamma_{n_1} \cup \ldots \cup \gamma_{n_m}$ the *choice-intersection normal form* (CINF) of α relative to Σ. Actions equivalent to 0 such as $a \cap \bar{a}$ do not have corresponding CINFs.

Lemma 3.

(1) Let $\gamma_{h_1} \cup \ldots \cup \gamma_{h_i}$ and $\gamma_{j_1} \cup \ldots \cup \gamma_{j_k}$ be the CINFs of β and $\bar{\beta}$. Then $\{\gamma_{j_1}, \ldots, \gamma_{j_k}\} = \{\gamma_1, \ldots, \gamma_{2^n}\} - \{\gamma_{h_1}, \ldots, \gamma_{h_i}\}$;

(2) Let $\gamma_{h_1} \cup \ldots \cup \gamma_{h_i}$, $\gamma_{j_1} \cup \ldots \cup \gamma_{j_k}$ and $\gamma_{l_1} \cup \ldots \cup \gamma_{l_m}$ be the CINFs of β, π and $\beta \cap \pi$. Then $\{\gamma_{l_1}, \ldots, \gamma_{l_m}\} = \{\gamma_{h_1}, \ldots, \gamma_{h_i}\} \cap \{\gamma_{j_1}, \ldots, \gamma_{j_k}\}$;

(3) Let $\gamma_{h_1} \cup \ldots \cup \gamma_{h_i}$, $\gamma_{j_1} \cup \ldots \cup \gamma_{j_k}$ and $\gamma_{l_1} \cup \ldots \cup \gamma_{l_m}$ be the CINFs of β, π and $\beta \cup \pi$. Then $\{\gamma_{l_1}, \ldots, \gamma_{l_m}\} = \{\gamma_{h_1}, \ldots, \gamma_{h_i}\} \cup \{\gamma_{j_1}, \ldots, \gamma_{j_k}\}$;

(4) If α is equivalent to 1, the CINF of α is $\gamma_1 \cup \ldots \cup \gamma_{2^n}$.

The axiom **D** is used in proving the second item.

4.3   An Incomplete Model and Its Generated Submodel

Let $\mathfrak{M}^C = (W^C, \{R_\alpha^C \mid \alpha \in \Pi\}, V^C)$ be the structure where

(1) W^C is the set of maximal consistent sets;
(2) $R_\alpha uv$ if and only if for any ϕ, $\phi \in v$ implies $\langle \alpha \rangle \phi \in u$;
(3) For any p, $V(p) = \{u \in W^C \mid p \in u\}$.

This structure is not a model, as the deontic relation is missing. Actually, if we ignore the *deontic part* of the language, it is the *canonical model*.

Lemma 4.

(1) If $\langle \alpha \rangle \phi \in u$, there is a $v \in W^C$ such that $\phi \in v$ and $R_\alpha^C uv$;
(2) $R_{\alpha \cup \beta}^C = R_\alpha^C \cup R_\beta^C$;
(3) For any α equivalent to 0, $R_\alpha^C = \emptyset$.

Let w be a maximal consistent set containing G. Let $\mathfrak{M} = (W, \{R_\alpha \mid \alpha \in \Pi\}, V)$ be the substructure of $\mathfrak{M}^C$ generated from w under the relation R_1^C. The class **C** of axioms guarantee that R_1 is the universal relation on W. Here a similar lemma with **Lemma 4**:

Lemma 5.

(1) If $\langle\alpha\rangle\phi \in u$, there is a $v \in W$ such that $\phi \in v$ and $R_\alpha uv$;
(2) $R_{\alpha\cup\beta} = R_\alpha \cup R_\beta$;
(3) For any α equivalent to 1, $R_\alpha = W \times W$;
(4) For any α equivalent to 0, $R_\alpha = \emptyset$.

4.4   The Filtration

Define a relation $\approx_\Sigma$ on W as this: $u \approx_\Sigma v$ if and only if $u \cap \Sigma = v \cap \Sigma$. This is an equivalence relation. Let $\mathfrak{M}^f = (W^f, \{R_\alpha^f \mid \alpha \in \Pi\}, V^f)$ be such a structure:

(1) W^f is the partition of W under $\approx_\Sigma$;
(2) $R_\alpha^f|u||v|$ if and only if there are $x \in |u|$ and $y \in |v|$ such that $R_\alpha xy$;
(3) for any p, $V^f(p) = \{|u| \mid p \in u\}$.

For any $x, y \in |u|$ and $\phi \in \Sigma$, $\phi \in x$ if and only if $\phi \in y$. We use $\phi \triangleright |u|$ to express that $\phi \in x$ for any $x \in |u|$. Here is a similar lemma with **Lemma 5**:

Lemma 6.

(1) If $\langle\alpha\rangle\phi \triangleright |u|$, there is a $v \in W$ such that $\phi \triangleright |v|$ and $R_\alpha^f|u||v|$;
(2) $R_{\alpha\cup\beta}^f = R_\alpha^f \cup R_\beta^f$;
(3) For any α equivalent to 1, $R_\alpha^f = W^f \times W^f$;
(4) For any α equivalent to 0, $R_\alpha^f = \emptyset$.

With the help of **Lemma 3** and **6**, it is not hard to show the following lemma:

Lemma 7.

(1) For any α of Σ not equivalent to 0, $R_\alpha^f = R_{\gamma_{n_1}\cup...\cup\gamma_{n_m}}^f$, where $\gamma_{n_1}\cup...\cup\gamma_{n_m}$ is the CINF of α;
(2) $R_{\gamma_1}^f \cup ... \cup R_{\gamma_{2^n}}^f = W^f \times W^f$;
(3) For any $\langle\alpha\rangle\top$ and $u \in W$, if $\langle\alpha\rangle\top \triangleright |u|$, there is a $v \in W$ such that $R_\alpha^f|u||v|$.

We present some observations on the situation confronting us. Our purpose is to show the consistent formula G is satisfiable in a standard model. Two things are important: a standard model and satisfiability. The structure $\mathfrak{M}^f = (W^f, \{R_\alpha^f \mid \alpha \in \Pi\}, V^f)$ might not be standard even if we ignore the deontic part, as $R_{\overline{\alpha}}^f = W^f \times W^f - R_\alpha^f$ and $R_{\alpha\cap\beta}^f = R_\alpha^f \cap R_\beta^f$ might not be satisfied. We hope to transform it to a standard model. Those actions not built from $a_1, \ldots, a_n$ are irrelevant, and we can freely manipulate the interpretations of their parts not involving $a_1, \ldots, a_n$; therefore, these actions do not present a problem. However, we are using the Henkin method, so at least to some extent, we should "respect" the interpretations of the actions built from $a_1, \ldots, a_n$, if we want to obtain satisfiability. Via some reflections with the help of **Lemma 3** and **6**, we can see that if $R^f_{\gamma_1}, \ldots, R^f_{\gamma_{2^n}}$ are real atomic blocks, $\mathfrak{M}^f$ is standard. That $R^f_{\gamma_1}, \ldots, R^f_{\gamma_{2^n}}$ are real atomic blocks means that $R^f_{\gamma_1}, \ldots, R^f_{\gamma_{2^n}}$ are pairwise disjoint and $R^f_{\gamma_1} \cup \ldots \cup R^f_{\gamma_{2^n}} = W^f \times W^f$. The second condition holds by **Lemma 7**. The situation is now clear: to achieve our goal, we only need to achieve two things: making $R^f_{\gamma_1}, \ldots, R^f_{\gamma_{2^n}}$ pairwise disjoint and respecting the interpretations of the actions built from $a_1, \ldots, a_n$. The copy method given in [2] can perform both at the same time, although it made a few mistakes, which will be explained later in a footnote.

4.5   A Standard Model

Let $\mathfrak{M}_1 = (W_1, \{R^1_\alpha \mid \alpha \in \Pi\}, V_1), \ldots, \mathfrak{M}_{2^n} = (W_{2^n}, \{R^{2^n}_\alpha \mid \alpha \in \Pi\}, V_{2^n})$ be 2^n pairwise disjoint structures which are isomorphic to $\mathfrak{M}^f$. We now build up a standard model from these structures.

For any $i \leq 2^n$, let f_i be an isomorphism from $\mathfrak{M}_i$ to $\mathfrak{M}^f$. Let $f = f_1 \cup \ldots \cup f_{2^n}$ and $U = W_1 \cup \ldots \cup W_{2^n}$. Let $g : U \to \{1, \ldots, 2^n\}$ be this function: for any $s \in U$, $g(s)$ is the index of the set from which s comes, i.e., for any $s \in U$, $s \in W_{g(s)}$. For any $s \in U$, let $\phi \ltimes s$ denote $\phi \triangleright f(s)$. For any $i \leq 2^n$, we define a relation B_{γ_i} on U:

Definition 3 (Atomic Relations). *Let $s, t \in U$. Let $\gamma_{k_1}, \ldots, \gamma_{k_m}$ be the sequence such that (i) it consists of all paths γ such that $R^f_\gamma f(s) f(t)$ and (ii) $k_1 < \ldots < k_m$. $B_{\gamma_i} st$ if and only if there is a $j \leq m$ such that $i = k_j$ and $j = (g(t) \bmod m) + 1$.*

The sequence $\gamma_{k_1}, \ldots, \gamma_{k_m}$ is never empty, which is guaranteed by **Lemma 7**. If a path γ_i is not occurring in $\gamma_{k_1}, \ldots, \gamma_{k_m}$, there is no $j \leq m$ such that $i = k_j$, and so not $B_{\gamma_i} st$. Suppose γ_i is occurring in $\gamma_{k_1}, \ldots, \gamma_{k_m}$. Then there is one and only one $j \leq m$ such that $i = k_j$, which means that γ_i is the j-th element in $\gamma_{k_1}, \ldots, \gamma_{k_m}$. In this case, if $j = (g(t) \bmod m) + 1$, then $B_{\gamma_i} st$, or else not.

Our purpose is to produce atomic relations. To do this, we must get this: for any $s, t \in U$, (s, t) belongs to one and only one path. This definition gives a way to assign (s, t) to the "right" path[1]. Here is a concrete example. Let a, b, c be all the atomic actions of Σ. Paths are $\gamma_1 = a \cap b \cap c, \ldots, \gamma_8 = \overline{a} \cap \overline{b} \cap \overline{c}$. Let $s, t \in U$ and $g(t) = 4$. Then $t \in W_4$. Suppose $R^f_{\gamma_2} f(s) f(t), R^f_{\gamma_5} f(s) f(t), R^f_{\gamma_7} f(s) f(t)$, and no other paths can do this. The sequence satisfying the two conditions in **Definition 3** is $\gamma_2, \gamma_5, \gamma_7$. Then $k_1 = 2, k_2 = 5$ and $k_3 = 7$. As $j = 2$ satisfies that $5 = k_j$ and $j = (4 \bmod 3) + 1$, we get $B_{\gamma_5} st$. For any $i \leq 8$, if $i \neq 5$, there

---

[1] [2] made a mistake at this point: the definition of atomic relations given by it can not guarantee that for any $s, t \in U$, (s, t) belongs to only one path. There are two other mistakes in this work: (i) By **Lemma 2**, those actions equivalent to 0 do not correspond to any CINF, but this paper did not notice this; (ii) **Lemma 3** is necessary to the proof of completeness, but it is not mentioned in this paper at all.

is no j satisfying that $i = k_j$ and $j = (4 \bmod 3) + 1$, and so not $B_{\gamma_i} st$. By the following lemma, $B_{\gamma_1}, \ldots, B_{\gamma_{2^n}}$ are real atomic relations:

Lemma 8.

(1) $B_{\gamma_1}, \ldots, B_{\gamma_{2^n}}$ are pairwise disjoint;
(2) $B_{\gamma_1} \cup \ldots \cup B_{\gamma_{2^n}} = U \times U$.

Proof. (1) Assume there are $i, j \leq 2^n$ such that $i \neq j$ and $B_{\gamma_i} \cap B_{\gamma_j} \neq \emptyset$. Then there are $s, t \in U$ such that $B_{\gamma_i} st$ and $B_{\gamma_j} st$. Let $\gamma_{k_1}, \ldots, \gamma_{k_m}$ be the sequence such that it consists of all paths γ such that $R_\gamma^f f(s) f(t)$ and $k_1 < \ldots < k_m$. Let $x, y \leq m$ be such that $i = k_x$ and $j = k_y$. As $k_x \neq k_y$, $x \neq y$. By the definitions of B_{γ_i} and B_{γ_j}, we get that $x = (g(t) \bmod m) + 1$ and $y = (g(t) \bmod m) + 1$. This is impossible. Then $B_{\gamma_1}, \ldots, B_{\gamma_{2^n}}$ are pairwise disjoint.

(2) Trivially, we get $B_{\gamma_1} \cup \ldots \cup B_{\gamma_{2^n}} \subseteq U \times U$. Let $s, t \in U$. Let $\gamma_{k_1}, \ldots, \gamma_{k_m}$ be the sequence such that it consists of all paths γ such that $R_\gamma^f f(s) f(t)$ and $k_1 < \ldots < k_m$. Let $x \leq m$ be such that $x = (g(t) \bmod m) + 1$. By the definition of $B_{\gamma_{k_x}}$, we have $B_{\gamma_{k_x}} st$.

Definition 4 (A Model). $\mathfrak{N} = (U, \{S_\alpha \mid \alpha \in \Pi\}, \mathcal{E}, Z)$ *is the model where*

(1) U is defined as above;
(2) For any atomic action a occurring in Σ, $S_a = B_{\gamma_{n_1}} \cup \ldots \cup B_{\gamma_{n_m}}$, where $\gamma_{n_1} \cup \ldots \cup \gamma_{n_m}$ is the CINF of a^2; For any atomic action b not occurring in Σ, $S_b = U \times U$; Interpretations of compound actions are defined from interpretations of atomic actions by corresponding operations;
(3) $\mathcal{E}(H, J)$ if and only if there is a $s \in U$ such that $J = (H, s)$ and for any $O\alpha \in \Sigma$, if $O\alpha \ltimes \mathring{H}$, then $S_\alpha(\mathring{H}, s)$;
(4) For any p, $Z(p) = \bigcup_{i \leq 2^n} V_i(p)$.

$S_{\overline{\alpha}} = U \times U - S_\alpha$, $S_{\alpha \cap \beta} = S_\alpha \cap S_\beta$ and $S_{\alpha \cup \beta} = S_\alpha \cup S_\beta$. As $S_1 = B_{\gamma_1} \cup \ldots \cup B_{\gamma_{2^n}}$, $S_1 = U \times U$ by **Lemma 8**. If $\mathcal{E}$ is serial, this is a standard model.

By the following lemma, in this model, the interpretations of the actions not equivalent to 0 are unions of some atomic relations.

Lemma 9. *For any α of Σ not equivalent to 0, $S_\alpha = B_{\gamma_{n_1}} \cup \ldots \cup B_{\gamma_{n_m}}$, where $\gamma_{n_1} \cup \ldots \cup \gamma_{n_m}$ is the CINF of α.*

In proving this lemma, we have to use **Lemma 3**. Now we show a crucial result:

Lemma 10. α *occurs in Σ.*

(1) For any $s, t \in U$, if $S_\alpha st$, $R_\alpha^f f(s) f(t)$;
(2) For any $s \in U$ and $y \in W^f$, if $R_\alpha^f f(s) y$, there is a $t \in U$ such that $f(t) = y$ and $S_\alpha st$.

---

[2] Here a might be the universal action 1.

Proof. (1) Assume $S_\alpha st$. By **Definition 4**, S_α is the result of operating on atomic actions. Thus, if α is equivalent to 0, S_α is empty. Then α is not equivalent to 0. Let $\gamma_{n_1} \cup \ldots \cup \gamma_{n_m}$ be the CINF of α. By **Lemma 9**, $S_\alpha = B_{\gamma_{n_1}} \cup \ldots \cup B_{\gamma_{n_m}}$. There is an $i \leq m$ such that $B_{\gamma_{n_i}} st$. By the definition of $B_{\gamma_{n_i}}$, $R^f_{\gamma_{n_i}} f(s) f(t)$. By **Lemma 7** and **6**, $R^f_\alpha = R^f_{\gamma_{n_1}} \cup \ldots \cup R^f_{\gamma_{n_m}}$. Then $R^f_\alpha f(s)(t)$.

(2) Assume $R^f_\alpha f(s) y$. By **Lemma 6**, α is not equivalent to 0. Let $\gamma_{n_1} \cup \ldots \cup \gamma_{n_m}$ be the CINF of α. Since $R^f_\alpha = R^f_{\gamma_{n_1}} \cup \ldots \cup R^f_{\gamma_{n_m}}$, there is an $i \leq m$ such that $R^f_{\gamma_{n_i}} f(s) y$. Let $\gamma_{k_1}, \ldots, \gamma_{k_h}$ be the sequence such that it consists of all the paths γ such that $R^f_\gamma f(s) y$ and $k_1 < \ldots < k_h$. Then γ_{n_i} occurs in $\gamma_{k_1}, \ldots, \gamma_{k_h}$. Let $j \leq h$ be such that $k_j = n_i$. Suppose $j = h$. Let $t \in U$ be such that $g(t) = 1$ and $f(t) = y$. It can be verified that $B_{\gamma_{k_h}} st$, that is, $B_{\gamma_{n_i}} st$. Since $S_\alpha = B_{\gamma_{n_1}} \cup \ldots \cup B_{\gamma_{n_m}}$, $S_\alpha st$. Suppose $j < h$. Let $t \in U$ be such that $g(t) = j + 1$ and $f(t) = y$. It can also be verified that $B_{\gamma_{k_j}} st$, that is, $B_{\gamma_{n_i}} st$. Since $S_\alpha = B_{\gamma_{n_1}} \cup \ldots \cup B_{\gamma_{n_m}}$, $S_\alpha st$.

Lemma 11 (Existence Lemmas for $\langle \alpha \rangle \phi$ and $\langle \alpha \rangle \top$).

(1) For any $\langle \alpha \rangle \phi \in \Sigma$, if $\langle \alpha \rangle \phi \ltimes s$, there is a t such that $\phi \ltimes t$ and $S_\alpha st$;
(2) For any $\langle \alpha \rangle \top$, if $\langle \alpha \rangle \top \ltimes s$, there is a t such that $S_\alpha st$.

By **Lemma 6 and 7**, this result is not hard to prove. By **Axiom F1, F4** and **Lemma 11**, this is the case:

Lemma 12. $\mathcal{E}$ *is serial.*

We now know that $\mathfrak{N}$ is a standard model.

Lemma 13 (Existence Lemma for $P\alpha$). *If $P\alpha \ltimes \mathring{H}$, there is a $t \in U$ such that $\mathcal{E}(H, (H, t))$ and $S_\alpha(\mathring{H}, t)$.*

This lemma can be shown by **Axiom F1, F3, F5** and **Lemma 11**. Then by **Axiom F2, Lemma 10, 11** and **13**, we can show the truth lemma. Finally, we have the completeness.

Lemma 14 (Truth Lemma). *For any $\phi \in \Sigma$, $\phi \ltimes \mathring{H}$ if and only if $\mathfrak{N}, H \Vdash \phi$.*

Proposition 5 (Completeness). *The logic restricted to Φ' is complete with respect to the class of standard models.*

From **Definition 4** we see that actually the deontic relation in the standard model collapses to be on the set of states. This implies that if dynamic operators were ignored, a classical relational semantics would be enough for the logic, and it would not be necessary to introduce the notion of histories.

Here are some words on the axiomatization of the whole logic. It can be shown that the rule of *equivalence replacement* for the two dynamic operators holds in this semantics: if $\phi \leftrightarrow \phi'$ is valid, both $[\downarrow \alpha]\phi \leftrightarrow [\downarrow \alpha]\phi'$ and $[\uparrow \alpha]\phi \leftrightarrow [\uparrow \alpha]\phi'$ are valid. Based on **Propositions 3, 4, 5** and this rule, we can obtain a complete axiomatization in a similar way as [1] for Public Announcement Logic.

5   Future Work

Our semantics does not work for descriptive permissions. The descriptive utterances of sentences such as *"you may do α or β"* might only inform the agent that he is allowed to do something, but not specify it. After such utterances, the agent might still not know how to act. This sort of utterances raises uncertainties, but our semantics does not have any settings to handle them.

Next on our agenda are two questions. The language in this work contains three action operators: complement, intersection and choice. It is natural to add test and composition to obtain a more powerful language in which conditional and sequential obligations and permissions can be expressed. This is an issue we want to pursue in the future. According to our semantics, the last update always overrides the previous ones. For example, given that α is possible to perform, the prescriptive permission *"you may do α"* always gives the agent the freedom to do α, regardless of what obligations have been put on him. This is the case only if there is only one speaker or moral source. In real life, there are many moral sources whose authorities are ranked, and only orders and permissions from speakers with higher authorities can overwhelm those from speakers with lower authorities. Introducing prioritized speakers into this framework is another direction of future work for us.

Acknowledgment. This research is supported by the National Social Science Foundation of China (No. 12CZX053) and the Major Bidding Project of the National Social Science Foundation of China (No. 10&ZD073). We would like to thank Maria Aloni, Davide Grossi, Fenrong Liu, Johan van Benthem, Hans van Ditmarsch, Frank Veltman, Yanjing Wang, Yi Wang, and the anonymous referees for their helpful comments and suggestions.

References

1. van Ditmarsch, H., van Der Hoek, W., Kooi, B.: Dynamic Epistemic Logic. Springer, Heidelberg (2007)
2. Gargov, G., Passy, S.: A note on boolean modal logic. In: Petkov, P.P. (ed.) Mathematical Logic, pp. 311–321. Plenum Press (1990)
3. Hilpinen, R.: Deontic logic. In: Goble, L. (ed.) The Blackwell Guide to Philosophical Logic, pp. 159–182. Blackwell Publishing (2001)
4. Jorgensen, J.: Imperatives and logic. Erkenntnis 7(1), 288–296 (1937)
5. Ju, F., Liu, F.: Prioritized imperatives and normative conflicts. European Journal of Analytic Philosophy 7(2), 33–54 (2011)
6. Meyer, J.J.C.: A different approach to deontic logic: Deontic logic viewed as a variant of dynamic logic. Notre Dame Journal of Formal Logic 29(1), 109–136 (1988)
7. Ross, A.: Imperatives and logic. Philosophy of Science 11(1), 30–46 (1944)
8. Veltman, F.: Imperatives at the borderline of semantics and pragmatics. Manuscript (2009)

Sequent Systems for Nondeterministic Propositional Logics without Reflexivity

Louwe B. Kuijer

University of Groningen

Abstract. In order to deal with ambiguity in statements made in a natural language I introduce nondeterministic semantics for propositional logic with an arbitrary set C of connectives. The semantics are based on the idea that Γ entails Δ if and only if every possible deterministic disambiguation of Γ entails every possible deterministic disambiguation of Δ. I also introduce a cut-free sequent style proof system S_C that is sound and complete for the given semantics. Finally I show that while the semantics and proof system do not satisfy reflexivity they do allow certain kinds of substitution of equivalents.

1 Introduction

When attempting to apply logical methods to sentences in natural language we often run into problems related to ambiguity. This ambiguity could be caused by ambiguous predicates, but it could also be caused by ambiguous connectives. Take for example a sentence of the form "A or B". Such a statement is ambiguous; the "or" could be inclusively ($\vee$) or exclusively ($\oplus$).

One approach for dealing with such ambiguity is to require disambiguation before allowing the sentence to be phrased in a logic. So "A or B" would have to be represented by either the formula $A \vee B$ or by the formula $A \oplus B$. Unfortunately this approach sometimes cannot be used, as it is not always possible to determine which unambiguous sentence was meant by the speaker. Sometimes the speaker is unable or unwilling to clarify their utterance and it is even possible that the speaker is uncertain about what they meant.[1]

Another approach is to consider an ambiguous statement as nondeterministic, where all possible disambiguations of the statement could be the meaning of the statement. See for example [1] for an example of this approach applied to ambiguous predicates. This approach has the advantage of being applicable even if no good choice of disambiguation is available, at the cost of resulting in a weaker logic. This nondeterministic approach is the one I use in this paper.

It should be noted that this approach is also to some extent usable for connectives that are not merely ambiguous but not truth-functional. Consider a

---

[1] For an example of a situation where the speaker is uncertain about what he means consider the first paragraph of the introduction, where I state that an "or" can be interpreted inclusively ($\vee$) *or* exclusively ($\oplus$). I do not know whether this *or* should be considered inclusively or exclusively.

D. Grossi, O. Roy, and H. Huang (Eds.): LORI 2013, LNCS 8196, pp. 190–203, 2013.

connective $\rightsquigarrow$ representing a causal implication. The causal part of the implication is not truth-functional; the truth value of $p \rightsquigarrow q$ is not known if either p doesn't hold or if both p and q hold. However, the truth value of $p \rightsquigarrow q$ is known if p holds and q does not, in that case it is false. Treating $p \rightsquigarrow q$ as a formula that is false if p is true and q is false and nondeterministically true or false in other cases allows some reasoning about $\rightsquigarrow$ in the framework of nondeterministic propositional logic.

In this paper I define a type of nondeterministic propositional semantics that allows all occurrences of nondeterministic connectives to be interpreted independently. In order to do this I start with deterministic semantics and then define the nondeterministic semantics based on all possible (deterministic) disambiguations. I also give a sound and complete proof system for the nondeterministic logic. Because of their pleasing properties with respect to automated reasoning I use a cut-free sequent style proof system. The proof system also turns out to be quite elegant, despite the complicated semantics.

The structure of the paper is as follows. First, in Section 1.1 I compare my approach to a somewhat similar existing approach using Nmatrices (see [2–4]). Then in Section 2 I define the deterministic logic $\mathcal{L}_{\overline{C}}$ and in Section 3 I define the nondeterministic logic $\mathcal{L}_C$. In Section 4 I give a sequent style proof system S_C which I prove to be complete for $\mathcal{L}_C$ in Section 5. Finally, in Section 6 I consider a few properties of $\mathcal{L}_C$ and S_C.

1.1   Comparison to Nmatrices

Semantics for nondeterministic propositional logic have been introduced in [2–4] using so-called *Nmatrices*. The semantics I use here for $\mathcal{L}_C$ are in many ways very similar to those using Nmatrices, but with one important difference. When using Nmatrices different occurrences of a single nondeterministic connective are allowed to have different interpretations, *but only insofar as identical (sub)formulas have the same interpretation everywhere.*

For example, let $*$ represent the "or" connective that can mean either $\vee$ or $\oplus$. Then in the Nmatrices approach the formula $(p * q) \wedge (p * q)$ can mean two things; either $(p \vee q) \wedge (p \vee q)$ or $(p \oplus q) \wedge (p \oplus q)$. The mixed disambiguations $(p \vee q) \wedge (p \oplus q)$ and $(p \oplus q) \wedge (p \vee q)$ are not allowed because $(p * q)$ and $(p * q)$ are the same formula and therefore must have the same choice of disambiguation.

The importance this approach gives to identity of formulas has a few unusual consequences, such as substitution of equivalents being unsound, even if the equivalents are provably equivalent. When using Nmatrices the formula $(p * q) \rightarrow (p * q)$ is a tautology but the formula $((p \wedge p) * q) \rightarrow (p * q)$ is not.

In the semantics for $\mathcal{L}_C$ I therefore allow different occurrences of a single nondeterministic connective to have different interpretations without restriction. The nondeterministic formula $(p * q) \wedge (p * q)$ then allows four disambiguations; $(p \vee q) \wedge (p \vee q)$, $(p \oplus q) \wedge (p \oplus q)$, $(p \vee q) \wedge (p \oplus q)$ or $(p \oplus q) \wedge (p \vee q)$.

The approach taken in this paper does lead to different unusual consequences. In particular, the inference relation for $\mathcal{L}_C$ is not reflexive since for example $p * q \models p * q$ might mean $p \vee q \models p \oplus q$.

2 Deterministic Semantics

First we need a few notational preliminaries. Let a nonempty set $\mathcal{P}$ of propositional variables be given. Furthermore, let $\overline{C}$ be a set of truth-functional connectives. For each connective $\circ \in \overline{C}$ let $r_\circ \in \mathbb{N}$ be the arity of $\circ$ and $f_\circ : \{0,1\}^{r_\circ} \to \{0,1\}$ the truth function of $\circ$.

For binary connectives I use the standard infix notation. Nullary connectives are denoted $\top$ if the associated truth function is the constant function 1 and $\bot$ if the truth function is the constant function 0.

I use p, q as variables for propositional variables, lower case Greek letters for formulas and uppercase Greek letters for multisets of formulas. Now let us define the language $L_{\overline{C}}$ and the models for $\mathcal{L}_{\overline{C}}$.

Definition 1 (Language $L_{\overline{C}}$). *The language $L_{\overline{C}}$ using the connectives in $\overline{C}$ is the smallest set such that if $p \in \mathcal{P}$ then $p \in L_{\overline{C}}$ and if $\circ \in \overline{C}$ and $\varphi_1, \cdots, \varphi_{r_\circ} \in L_{\overline{C}}$ then $\circ(\varphi_1, \cdots, \varphi_{r_\circ}) \in L_{\overline{C}}$.*

Definition 2 (Models). *A model $\mathcal{M}$ is a valuation function $\mathcal{M} : \mathcal{P} \to \{0,1\}$ that assigns to each propositional variable the value 'true' (1) or 'false' (0).*

Now we can define the semantics for $\mathcal{L}_{\overline{C}}$.

Definition 3 (Satisfaction relation $\models$). *The satisfaction relation $\models$ between models and formulas is defined inductively by*

- $\mathcal{M} \models p$ *if and only if* $\mathcal{M}(p) = 1$,
- *for every* $\circ \in \overline{C}$ *we have* $\mathcal{M} \models \circ(\varphi_1, \cdots, \varphi_{r_\circ})$ *if and only if* $f_\circ(v_1, \cdots, v_{r_\circ}) = 1$ *where* $v_i = 1$ *if and only if* $\mathcal{M} \models \varphi_i$.

Furthermore, $\forall \gamma \in \Gamma : \mathcal{M} \models \gamma$ is denoted by $\mathcal{M} \models_\wedge \Gamma$ and $\exists \delta \in \Delta : \mathcal{M} \models \delta$ is denoted by $\mathcal{M} \models_\vee \Delta$.

From this satisfaction relation we define an entailment relation.

Definition 4 (Entailment relation $\models_{\overline{C}}$). *The entailment relation $\models_{\overline{C}}$ is a relation between multisets $\Gamma, \Delta \subseteq L_{\overline{C}}$ of formulas. We have $\Gamma \models_{\overline{C}} \Delta$ if and only if for every model $\mathcal{M}$ such that $\mathcal{M} \models_\wedge \Gamma$ it holds that $\mathcal{M} \models_\vee \Delta$.*

Lemma 1. *Let $\Gamma, \Delta \subseteq L_{\overline{C}}$ be any finite multisets of $\mathcal{L}_{\overline{C}}$ formulas such that $\Gamma \models_{\overline{C}} \Delta$, and let Φ be any set of propositional variables that contains all variables that occur in Γ or Δ. Then for any partition Φ_1, Φ_2 of Φ one of the following statements holds:*

1. *there is a $\gamma \in \Gamma$ such that $\Phi_1, \gamma \models_{\overline{C}} \Phi_2$,*
2. *there is a $\delta \in \Delta$ such that $\Phi_1 \models_{\overline{C}} \delta, \Phi_2$.*

Proof. All connectives in $\overline{C}$ are truth-functional, so the value of any formula in a model is fully determined by the values of the propositional variables that occur in the formula have in that model. Fix any partition Φ_1, Φ_2 of Φ and let $\mathfrak{M}$ be the set of models that satisfy all of Φ_1 and none of Φ_2. Then for every $\psi \in \Gamma \cup \Delta$ we either have $\mathcal{M} \models \psi$ for every model $\mathcal{M} \in \mathfrak{M}$ or $\mathcal{M} \not\models \psi$ for every $\mathcal{M} \in \mathfrak{M}$.

Suppose there is a $\gamma \in \Gamma$ such that the second possibility holds for that formula, so $\mathcal{M} \not\models \gamma$ for every $\mathcal{M} \in \mathfrak{M}$. Then every model that satisfies all of Φ_1 and none of Φ_2 does not satisfy γ, so every model that satisfies all of Φ_1 and γ must satisfy some of Φ_2. We therefore have $\Phi_1, \gamma \models_{\overline{C}} \Phi_2$.

Suppose then that for every $\gamma \in \Gamma$ the first possibility holds, so $\mathcal{M} \models \gamma$ for every $\mathcal{M} \in \mathfrak{M}$. We have $\Gamma \models_{\overline{C}} \Delta$ so for every $\mathcal{M} \in \mathfrak{M}$ there is a $\delta \in \Delta$ such that $\mathcal{M} \models \delta$. But then $\mathcal{M} \models \delta$ for every such model, as the value of δ is constant on $\mathfrak{M}$. This implies that every model that satisfies all of Φ_1 and none of Φ_2 also satisfies δ, so every model that satisfies all of Φ_1 must either satisfy δ or one of Φ_2. We therefore have $\Phi_1 \models_{\overline{C}} \delta, \Phi_2$.

Note that in particular Lemma 1 implies that if Φ contains all the propositional variables of φ then $\Phi_1, \varphi \models_{\overline{C}} \Phi_2$ or $\Phi_1 \models_{\overline{C}} \varphi, \Phi_2$ since $\varphi \models_{\overline{C}} \varphi$.

3 Nondeterministic Semantics

Let us start by defining nondeterministic connectives.

Definition 5 (Nondeterministic connective). *A nondeterministic connective $\circ$ is a connective with arity $r_\circ$ and partial truth function $f_\circ : \{0,1\}^{r_\circ} \to \{0,1,?\}$.*

Note that besides $\top$ and $\bot$ there is now a third possible nondeterministic nullary connective that has the constant function $?$ as partial truth function. I denote this connective by $?$ as well.

Definition 6 (Language L_C). *The language L_C using the connectives in C is the smallest set such that if $p \in \mathcal{P}$ then $p \in L_C$ and if $\circ \in C$ and $\varphi_1, \cdots, \varphi_{r_\circ} \in L_C$ then $\circ(\varphi_1, \cdots, \varphi_{r_\circ}) \in L_C$.*

The logic $\mathcal{L}_C$ is then based on the idea that a nondeterministic connective $\circ$ can be disambiguated as one of several deterministic connectives $\overline{\circ}$.

Definition 7 (Disambiguation of a connective). *Let $\circ$ be a nondeterministic connective with arity $r_\circ$ and associated nondeterministic truth function $f_\circ$. A truth-functional connective $\overline{\circ}$ is a disambiguation of $\circ$ if $r_{\overline{\circ}} = r_\circ$ and for each $v \in \{0,1\}^{r_\circ}$ such that $f_\circ(v) \in \{0,1\}$ it holds that $f_\circ(v) = f_{\overline{\circ}}(v)$.*

So a disambiguation $\overline{\circ}$ of a connective $\circ$ is a truth-functional connective with a truth function $f_{\overline{\circ}}$ where every $?$ from $f_\circ$ is replaced by either a 0 or a 1.

Example 1. Suppose $*$ is the nondeterministic connective with the following truth table.

φ_1	φ_1	$\varphi_1 * \varphi_2$
1	1	?
1	0	1
0	1	1
0	0	0

Then there are two possible disambiguations $\bar{*}$ and $\bar{*}'$ of $*$, namely

φ_1	φ_1	$\varphi_1\bar{*}\varphi_2$	φ_1	φ_1	$\varphi_1\bar{*}'\varphi_2$
1	1	1	1	1	0
1	0	1	1	0	1
0	1	1	0	1	1
0	0	0	0	0	0

So the disambiguations of $*$ are a disjunction ($\vee$) and an "exclusive or" ($\oplus$).

Definition 8 (Disambiguation of a formula). *The disambiguations of formulas are given inductively by the following.*

- *If $p \in \mathcal{P}$ then p is a disambiguation of p.*
- *If $\varphi = \circ(\varphi_1, \cdots, \varphi_{r_\circ})$, $\overline{\varphi_i}$ is a disambiguation of φ_i for each $1 \leq 1 \leq r_\circ$ and $\bar{\circ}$ is a disambiguation of $\circ$ then $\bar{\circ}(\overline{\varphi_1}, \cdots, \overline{\varphi_{r_\circ}})$ is a disambiguation of φ.*

If Γ is a multiset of formulas then $\overline{\Gamma}$ is a disambiguation of Γ if $\overline{\Gamma}$ can be obtained from Γ by replacing formulas with one of their disambiguations.

Definition 9 (Entailment relation $\models_C$). *Let $\overline{C}$ be the set of all connectives that are the disambiguation of a connective in C. The entailment relation $\models_C$ is a relation between multisets $\Gamma, \Delta \subseteq L_C$ of $\mathcal{L}_C$ formulas. We have $\Gamma \models_C \Delta$ if and only if $\overline{\Gamma} \models_{\overline{C}} \overline{\Delta}$ for every disambiguations $\overline{\Gamma}$ of Γ and $\overline{\Delta}$ of Δ.*

Note that this is a very strong standard for entailment. If we have $\Gamma \models_C \Delta$ and Γ holds under some disambiguation then Δ must hold under every disambiguation. Using the terminology of [1] the relation $\models_C$ not only preserves *truth on some disambiguation* (truth-osd) and *truth on every disambiguation* (truth-osd only), it also requires a true-osd antecedent to have a true-osd only consequent. This is a stronger condition than the ones discussed in [1], but it has a very clear game-theoretical or dialectical interpretation: $\Gamma \models_C \Delta$ if and only if you can conclude Δ from Γ without any possibility for an opponent to give a disambiguation that proves you wrong. Equivalently, $\Gamma \models_C \Delta$ if and only if one cannot rationally reject Δ while accepting Γ.

For the completeness theorems given later in the paper it is very useful to be able to work with finite multisets of formulas. I therefore give a compactness proof for $\models_C$ here. The compactness of deterministic propositional logic is quite well known, see for example [5] for two different versions of the proof. The proof for the compactness of nondeterministic propositional logic can be obtained by some small (but notationally complicated) modifications to the existing proofs. Here I give a topological proof using Tychonoff's theorem [6, 7], which states that every product of compact sets is compact.

Lemma 2 (Compactness of $\mathcal{L}_C$). *Let Γ, Δ be any multisets of $\mathcal{L}_C$ formulas. If $\Gamma \models_C \Delta$ then there are finite sub-multisets $\Gamma' \subseteq \Gamma$ and $\Delta' \subseteq \Delta$ such that $\Gamma' \models_C \Delta'$.*

Proof. In order to keep our notation simple I treat the multisets as sets in this proof. This is purely a matter of notation, it amounts to the same thing as adding a label to every occurrence of a formula in a multiset to make them distinct.

Let Γ, Δ be any sets of formulas such that $\Gamma \models_C \Delta$ and suppose towards a contradiction that there are no finite subsets $\Gamma' \subseteq \Gamma$ and $\Delta' \subseteq \Delta$ such that $\Gamma' \models_C \Delta'$. Let G and D be partitions of Γ and Δ respectively that contain only finite sets. By padding G or D with multiple copies of the empty set if necessary we can guarantee the existence of a bijection $f : G \to D$.

Consider the set $\{0, 1\}$ as a finite topological space (with the discrete topology). This topological space is compact. Then by Tychonoff's theorem the set $\{0, 1\}^{\mathcal{P}}$ (with the product topology) is compact. The elements of $\{0, 1\}^{\mathcal{P}}$ are exactly the models of $\mathcal{L}_{\overline{C}}$.

For any $\Phi \subseteq \Gamma$ and $\Psi \subseteq \Delta$ let $V(\Phi, \Psi)$ be the set of models $\mathcal{M} \in \{0, 1\}^{\mathcal{P}}$ such that for some disambiguations $\overline{\Phi}$ and $\overline{\Psi}$ of Φ and Ψ we have $\mathcal{M} \models_\wedge \overline{\Phi}$ and $\mathcal{M} \not\models_\vee \overline{\Psi}$. So $V(\Phi, \Psi)$ is the set of countermodels to $\Phi \models_C \Psi$. Take any $G' \subseteq G$ and let $D' = f(G')$. We will show that

$$\bigcap_{\Phi \in G'} V(\Phi, f(\Phi)) = V\left(\bigcup G', \bigcup D'\right). \tag{1}$$

The important step is that, since G and D are partitions of Γ and Δ, the elements of G' are mutually disjunct, as are those of D'. This implies that if for every $\Phi \in G'$ the set $\overline{\Phi}$ is a disambiguation of Φ then $\bigcup_{\Phi \in G'} \overline{\Phi}$ is a disambiguation of $\bigcup G'$, because the disambiguations of different elements of G' cannot disagree about how a formula should be disambiguated. We can also go the other way, every disambiguation $\overline{\bigcup G'}$ of G' induces unique disambiguations $\overline{\Phi}$ for every $\Phi \in G'$ such that $\overline{\bigcup G'} = \bigcup_{\Phi \in G'} \overline{\Phi}$.

The same holds for the disambiguations of $\bigcup D'$ and the disambiguations for every $f(\Phi) = \Psi \in D'$; if given $\overline{\Psi}$ for all $\Psi \in D'$ then $\bigcup_{\Psi \in D'} \overline{\Psi}$ is a disambiguation of $\bigcup D'$ and for each disambiguation $\overline{\bigcup D'}$ there are unique disambiguations $\overline{\Psi}$ for all $\Psi \in D'$ such that $\overline{\bigcup D'} = \bigcup_{\Psi \in D'} \overline{\Psi}$.

Take any $\mathcal{M} \in \bigcap_{\Phi' \in G'} V(\Phi', f(\Phi'))$. This $\mathcal{M}$ has the property that for every $\Phi \in G'$ and $\Psi = f(\Phi)$ there are disambiguations $\overline{\Phi}$ and $\overline{\Psi}$ such that $\mathcal{M} \models_\wedge \overline{\Phi}$ and $\mathcal{M} \not\models_\vee \overline{\Psi}$. The sets $\bigcup_{\Phi \in G'} \overline{\Phi}$ and $\bigcup_{\Phi \in G'} \overline{\Phi}$ are disambiguations of $\bigcup G'$ and $\bigcup D'$. Furthermore, $\mathcal{M}$ satisfies all of $\bigcup_{\Phi \in G'} \overline{\Phi}$ and none of $\bigcup_{\Phi \in G'} \overline{\Phi}$ so $\mathcal{M} \in V(\bigcup G', \bigcup D')$. We therefore have

$$\bigcap_{\Phi \in G'} V(\Phi, f(\Phi)) \subseteq V\left(\bigcup G', \bigcup D'\right). \tag{2}$$

Now take any $\mathcal{M} \in V(\bigcup G', \bigcup D')$. This $\mathcal{M}$ has the property that there are disambiguations $\overline{\bigcup G'}$ and $\overline{\bigcup D'}$ of $\bigcup G'$ and $\bigcup D'$ such that $\mathcal{M}$ satisfies all of $\overline{\bigcup G'}$ and none of $\overline{\bigcup D'}$. For each $\Phi \in G'$ and $\Psi = f(\Phi) \in D'$ let $\overline{\Phi}$ and $\overline{\Psi}$ be the disambiguations such that $\overline{\bigcup G'} = \bigcup_{\Phi \in G'} \overline{\Phi}$ and $\overline{\bigcup D'} = \bigcup_{\Psi \in D'} \overline{\Psi}$. Then for each $\Phi \in G'$ and $\Psi = f(\Phi) \in D'$ the model $\mathcal{M}$ satisfies all of $\overline{\Phi}$ and none of $\overline{\Psi}$ so $\mathcal{M} \in \bigcap_{\Phi \in G'} V(\Phi, f(\Phi))$. We therefore have

$$V\left(\bigcup G', \bigcup D'\right) \subseteq \bigcap_{\Phi \in G'} V(\Phi, f(\Phi)) \tag{3}$$

which together with (2) implies (1).

For any $\Gamma' \in G$ and $\Delta' \in D$ the set $V(\Gamma', \Delta')$ is closed because all subsets of $\{0,1\}^p$ are clopen since we started with the discrete topology. Furthermore if follows from (1) that for any $n \in \mathbb{N}$ and any $\Gamma'_1, \cdots, \Gamma'_n \in G$ we have

$$\bigcap_{1 \leq i \leq n} V(\Gamma'_i, f(\Gamma'_i)) = V\left(\bigcup_{1 \leq i \leq n} \Gamma'_i, \bigcup_{1 \leq i \leq n} f(\Gamma'_i)\right).$$

The set $V(\bigcup_{1 \leq i \leq n} \Gamma'_i, \bigcup_{1 \leq i \leq n} f(\Gamma'_i))$ is nonempty because $\bigcup_{1 \leq i \leq n} \Gamma'_i$ and $\bigcup_{1 \leq i \leq n} f(\Gamma'_i)$ are finite subsets of Γ and Δ so by assumption $\bigcup_{1 \leq i \leq n} \Gamma'_i \not\models_C \bigcup_{1 \leq i \leq n} f(\Gamma'_i)$. Then $\bigcap_{\Gamma' \in G} V(\Gamma', f(\Gamma'))$ is an intersection of closed sets with the finite intersection property so it is nonempty by the compactness of $\{0,1\}^p$. We have

$$\bigcap_{\Gamma' \in G} V(\Gamma', f(\Gamma')) = V\left(\bigcup_{\Gamma' \in G} \Gamma', \bigcup_{\Gamma' \in G} f(\Gamma')\right) = V(\Gamma, \Delta),$$

so $V(\Gamma, \Delta)$ is also nonempty. But this implies that $\Gamma \not\models_C \Delta$, which contradicts the choice of Γ and Δ. The assumption that there are no finite subsets $\Gamma' \subseteq \Gamma$ and $\Delta' \subseteq \Delta$ such that $\Gamma' \models_C \Delta'$ must therefore be false, which proves the lemma.

We also need a nondeterministic variant of Lemma 1.

Lemma 3. *Let* Γ, Δ *be any finite multisets of* $\mathcal{L}_C$ *formulas such that* $\Gamma \models_C \Delta$, *and let* Φ *be any set of propositional variables that contains all variables that occur in* Γ *or* Δ. *Then for any partition* Φ_1, Φ_2 *of* Φ *one of the following statements holds:*

1. *there is a* $\gamma \in \Gamma$ *such that* $\Phi_1, \gamma \models_C \Phi_2$,
2. *there is a* $\delta \in \Delta$ *such that* $\Phi_1 \models_C \delta, \Phi_2$.

Proof. Suppose towards a contradiction that $\Phi_1, \gamma \not\models_C \Phi_2$ for all $\gamma \in \Gamma$ and $\Phi_1 \not\models_C \delta, \Phi_2$ for all $\delta \in \Delta$. Then for each $\gamma_i \in \Gamma$ there is a disambiguation $\overline{\gamma}_i$ such that $\Phi_1, \overline{\gamma}_i \not\models_{\overline{C}} \Phi_2$ and for each $\delta_i \in \Delta$ there is a disambiguation $\overline{\delta}_i$ such that $\Phi_1 \not\models_{\overline{C}} \overline{\delta}_i, \Phi_2$.

Now let $\overline{\Gamma} = \{\overline{\gamma}_i \mid \gamma_i \in \Gamma\}$ and $\overline{\Delta} = \{\overline{\delta}_i \mid \delta_i \in \Delta\}$. These $\overline{\Gamma}$ and $\overline{\Delta}$ are disambiguations of Γ and Δ so from $\Gamma \models_C \Delta$ it follows that $\overline{\Gamma} \models_{\overline{C}} \overline{\Delta}$. But $\overline{\Gamma}$ and $\overline{\Delta}$ live in a deterministic logic, so from Lemma 1 it follows that there either is a $\overline{\gamma} \in \overline{\Gamma}$ such that $\Phi_1, \overline{\gamma}_i \models_{\overline{C}} \Phi_2$ or a $\overline{\delta} \in \overline{\Delta}$ such that $\Phi_1 \models_{\overline{C}} \overline{\delta}_i, \Phi_2$. This contradicts the choice of $\overline{\Gamma}$ and $\overline{\Delta}$, so the initial assumption must have been wrong, which proves the lemma.

4   Proof System

The proof system consists of a few structural rules and some rules that are based on the abbreviated partial truth tables of the connectives.

Definition 10 (Rules of $S_{\varnothing}$). *The rules of $S_{\varnothing}$ are the rules Axiom (Ax), Left Contraction (CL), Right Contraction (CR), Left Weakening (WL) and Right Weakening (WR), given by*

$$\frac{}{p \vdash p}\ Ax \qquad \frac{\Gamma_1, \Gamma_2, \Gamma_2 \vdash \Delta}{\Gamma_1, \Gamma_2 \vdash \Delta}\ CL \qquad \frac{\Gamma \vdash \Delta_1, \Delta_2, \Delta_2}{\Gamma \vdash \Delta_1, \Delta_2}\ CR$$

$$\frac{\Gamma_1 \vdash \Delta}{\Gamma_1, \Gamma_2 \vdash \Delta}\ WL \qquad \frac{\Gamma \vdash \Delta_1}{\Gamma \vdash \Delta_1, \Delta_2}\ WR$$

The formula p in Ax is called principal, as are all elements of Γ_2 in CL and WL and all elements of Δ_2 in CR and WR.

Note that the Axiom used here results in a very limited form of reflexivity. This is because $\models_C$ is not reflexive. The logical rules correspond to the abbreviated truth tables of the connectives and can be obtained by a multi-step procedure.

Definition 11 (Rules R_C). *The set R_C rules for the abbreviated truth tables are obtained using the following procedure.*

1. *Start with $R_C = \varnothing$.*
2. *For any $\circ \in C$, $v \in \{0,1\}^{r_\circ}$ and $1 \leq i \leq r_\circ$ let U_i be the sequent $\Gamma, \varphi_i \vdash \Delta$ if the i-th entry of v is 0 and let U_i be the sequent $\Gamma, \varphi_i \vdash \Delta$ if the i-th entry of v is 1. Now for every $\circ \in C$ and $v \in \{0,1\}^{r_\circ}$ add the rule*

$$\frac{U_1 \qquad \cdots \qquad U_{r_\circ}}{\Gamma, \circ(\varphi_1, \cdots, \varphi_{r_\circ}) \vdash \Delta}\ R_{\circ,v}$$

to R_C if $f_\circ(v) = 0$, add the rule

$$\frac{U_1 \qquad \cdots \qquad U_{r_\circ}}{\Gamma \vdash \circ(\varphi_1, \cdots, \varphi_{r_\circ}), \Delta}\ R_{\circ,v}$$

to R_C if $f_\circ(v) = 1$ and add no rule to R_C if $f_\circ(v) = ?$.
3. *If there are two rules*

$$\frac{U_1 \quad \cdots \quad U_{j-1} \quad \Gamma, \varphi_i \vdash \Delta \quad U_{j+1} \quad \cdots \quad U_k}{W}\ R_{\circ,v}$$

and

$$\frac{U_1 \quad \cdots \quad U_{j-1} \quad \Gamma \vdash \varphi_i, \Delta \quad U_{j+1} \quad \cdots \quad U_k}{W}\ R_{\circ,v'}$$

in R_C then add the rule

$$\frac{U_1 \quad \cdots \quad U_{j-1} \quad U_{j+1} \quad \cdots \quad U_k}{W}\ R_{\circ,v''}$$

where $v'' \in \{0, 1, ?\}^{r_\circ}$ is the vector with the same value as v and v' where they agree and ? where they do not agree. Repeat this step until no more rules can be added.

4. *If there are two rules*

$$\frac{U_1 \quad \cdots \quad U_k}{W} R_1 \qquad \frac{U_1' \quad \cdots \quad U_{k'}'}{W} R_2$$

in R_C with $\{U_1, \cdots, U_k\} \subset \{U_1', \cdots, U_{k'}'\}$ then remove the rule R_2. Repeat this step until no more rules can be removed.

The rules for $\top, \bot$ and ? are degenerate cases, for $\top$ and $\bot$ we have the rules

$$\frac{}{\Gamma \vdash \top, \Delta} \top \qquad \frac{}{\Gamma, \bot \vdash \Delta} \bot$$

and for ? we have no rules at all. The procedure given in Definition 11 terminates for any finite set of connectives and gives a unique set of rules for a given set of connectives. Let us consider a simple example of how this procedure works.

Example 2. Let $*$ be the binary connective with the following partial truth table.

φ_1	φ_2	$\varphi_1 * \varphi_2$
0	0	?
0	1	1
1	0	1
1	1	1

so $*$ behaves either like a disjunction or like guaranteed truth. Then in step 2 of the procedure the following three rules are added.

$$\frac{\Gamma \vdash \varphi_1, \Delta \qquad \Gamma, \varphi_2 \vdash \Delta}{\Gamma \vdash \varphi_1 * \varphi_2, \Delta} R_{*,(1,0)}$$

$$\frac{\Gamma, \varphi_1 \vdash \Delta \qquad \Gamma \vdash \varphi_2, \Delta}{\Gamma \vdash \varphi_1 * \varphi_2, \Delta} R_{*,(0,1)} \qquad \frac{\Gamma \vdash \varphi_1, \Delta \qquad \Gamma \vdash \varphi_2, \Delta}{\Gamma \vdash \varphi_1 * \varphi_2, \Delta} R_{*,(1,1)}$$

In step 3 we then combine the rule $R_{*,(1,1)}$ with both the rule $R_{*,(1,0)}$ and the rule $R_{*,(0,1)}$ to obtain the following two rules

$$\frac{\Gamma \vdash \varphi_2, \Delta}{\Gamma \vdash \varphi_1 * \varphi_2, \Delta} R_{*,(?,1)} \qquad \frac{\Gamma \vdash \varphi_1, \Delta}{\Gamma \vdash \varphi_1 * \varphi_2, \Delta} R_{*,(1,?)}$$

Finally, in step 4 we remove the rules $R_{*,(1,0)}$, $R_{*,(0,1)}$ and $R_{*,(1,1)}$ because their premises are supersets of those of $R_{*,(?,1)}$ and $R_{*,(1,?)}$. In the end the rules for $*$ are therefore only the rules $R_{*,(?,1)}$ and $R_{*,(1,?)}$. These rules represent all we know about $*$, namely that $\varphi_1 * \varphi_2$ is true if at least one of φ_1 and φ_2 is true.

Definition 12 (Rule Cut). *The rule Cut is given by*

$$\frac{\Gamma_1 \vdash \varphi, \Delta_1 \qquad \Gamma_2, \varphi \vdash \Delta_2}{\Gamma_1, \Gamma_2 \vdash \Delta_1, \Delta_2} Cut$$

Definition 13 (Proof system S_C). *The proof system S_C consists of the rules $S_\varnothing$ together with the rules R_C. The proof system $S_C + Cut$ consists of the rules of S_C together with the rule Cut.*

Definition 14 (Derivation). *A derivation in a proof system S is a finite labeled tree T such that:*

- *every node of T is labeled by either a sequent or an empty label,*
- *if a node s of T with label V has child nodes $t_1, \cdots, t_n$ with labels $U_1, \cdots U_n$ then*

$$\frac{U_1 \qquad \cdots \qquad U_n}{V}$$

is an instance of a rule of S.

The non-empty labels of nodes that do not have child nodes are called the premises of the derivation and the label of the root is called the conclusion of the derivation.

Definition 15 (Derivable). *A sequent U is derivable in a proof system S if there is a derivation in S that has no premises and U as conclusion.*

Definition 16 (Admissible). *A rule*

$$\frac{U_1 \qquad \cdots \qquad U_n}{V} R$$

is admissible in S if every sequent that is derivable in $S+R$ is derivable in S.

For most of the rules of $S_C + Cut$ it should be immediately clear that they are sound for $\models_C$. The only rules for which there could be some doubt about the soundness are the contraction and Cut rules. For these rules it can also quite easily be seen that they are sound. Consider for example left contraction. If we have $\Gamma_1, \Gamma_2, \Gamma_2 \models_C \Delta$ then for any disambiguations $\overline{\Gamma_1}$ of Γ_1, $\overline{\Gamma_2}$ of Γ_2, $\overline{\Gamma_2}'$ of Γ_2 and $\overline{\Delta}$ of Δ we have $\overline{\Gamma_1}, \overline{\Gamma_2}, \overline{\Gamma_2}' \models_{\overline{C}} \overline{\Delta}$. In particular this is the case if $\overline{\Gamma_2} = \overline{\Gamma_2}'$, so we have $\overline{\Gamma_1}, \overline{\Gamma_2}, \overline{\Gamma_2} \models_{\overline{C}} \overline{\Delta}$. CL is sound for deterministic propositional logic so $\overline{\Gamma_1}, \overline{\Gamma_2} \models_{\overline{C}} \overline{\Delta}$. This holds for any disambiguations $\overline{\Gamma_1}, \overline{\Gamma_2}$ and $\overline{\Delta}$ so $\Gamma_1, \Gamma_2 \models_C \Delta$. Soundness for CR and Cut is obtained in the same way.

5   Completeness

I prove the completeness of S_C by showing the completeness of $S_C + Cut$ and showing that Cut is admissible in S_C. I start with a very limited form of completeness and then use it to show full completeness.

Lemma 4. *Let φ be any $\mathcal{L}_C$ formula and let Φ be a set of propositional variables that includes all the variables that occur in φ and let Φ_1, Φ_2 be any partition of Φ. Then $\Phi_1, \varphi \models_C \Phi_2$ implies that $\Phi_1, \varphi \vdash \Phi_2$ is derivable in $S_C + Cut$ and $\Phi_1 \models_C \varphi, \Phi_2$ implies that $\Phi_1 \vdash \varphi, \Phi_2$ is derivable in $S_C + Cut$.*

Proof. I give the proof for the case where $\Phi_1, \varphi \models_C \Phi_2$. The other case is analogous by the duality of the left and right side. Suppose that φ, Φ, Φ_1 and Φ_2 are as in the lemma and $\Phi_1, \varphi \models_C \Phi_2$. To show is that $\Phi_1, \varphi \vdash \Phi_2$ is derivable.

The proof now proceeds by induction on the construction of φ. First suppose φ is atomic, so $\varphi = p$ for some $p \in \mathcal{P}$. Then $\varphi \in \Phi_2$ so $\Phi_1, \varphi \vdash \Phi_2$ is derivable by using Ax to obtain $p \vdash p$ and subsequently weakening.

Suppose therefore as induction hypothesis that φ is not atomic, and that the lemma holds for all subformulas of φ. Then $\varphi = \circ(\varphi_1, \cdots, \varphi_{r_\circ})$ for some $\circ \in C$ and $\mathcal{L}_C$ formulas $\varphi_1, \cdots, \varphi_{r_\circ}$. Let $N_1 = \{i \in \{1, \cdots, r_\circ\} \mid \Phi_1, \varphi_i \models \Phi_2\}$, $N_2 = \{i \in \{1, \cdots, r_\circ\} \mid \Phi_1 \models \varphi_i, \Phi_2\}$ and $N_3 = \{1, \cdots, r_\circ\} \setminus (N_1 \cup N_2)$.

For $i \in N_1$ let U_i be the sequent $\Phi_1, \varphi_i \vdash \Phi_2$ and for $i \in N_2$ let U_i be the sequent $\Phi_1 \vdash \varphi_i, \Phi_2$. Then by the induction hypothesis U_i is derivable for $i \in N_1 \cup N_2$. Now take any $i \in N_3$. Then the value of φ_i under the partition Φ_1, Φ_2 depends on the chosen disambiguation of φ_i. But for every disambiguation $\overline{\varphi}$ of φ we have $\Phi_1, \overline{\varphi} \models_{\overline{C}} \Phi_2$. This implies that, given the (fixed) values of φ_j with $j \in (N_1 \cup N_2)$ the value of φ is determinate and independent of the values of φ_i with $i \in N_3$.

Let $v = (v_1, \cdots, v_{r_\circ}) \in \{0, 1\}^{r_\circ}$ be any vector such that $v_i = 1$ if $i \in N_1$ and $v_i = 0$ if $i \in N_2$ and $v' = (v_1', \cdots, v_{r_\circ}') \in \{0, 1, ?\}^{r_\circ}$ the vector such that $v_i' = 1$ if $i \in N_1$, $v_i' = 0$ if $i \in N_2$ and $v_i' = ?$ if $i \in N_3$. The value of φ is independent of the values of φ_i with $i \in N_3$ so the rule

$$\frac{\{U_i \mid 1 \le i \le r_\circ\}}{\Phi_1, \varphi \vdash \Phi_2} \text{ R}_{\circ,v}$$

was added in step 2 of Definition 11. This is true regardless of the choice of v_i for $i \in N_3$, so in step 3 of Definition 11 a rule

$$\frac{\{U_i \mid i \in N_1 \cup N_2\}}{\Phi_1, \varphi \vdash \Phi_2} \text{ R}_{\circ,v'}$$

is generated. It is possible that $\text{R}_{\circ,v'}$ is removed in step 4, but then there is a rule $\text{R}_{\circ,v''}$ that takes a subset of $\{U_i \mid i \in N_1 \cup N_2\}$ as premises and has $\Phi_1, \varphi \vdash \Phi_2$ as conclusion. So whether or not $\text{R}_{\circ,v'}$ gets removed in step 4 it follows from the fact that U_i is derivable for $i \in N_1 \cup N_2$ that $\Phi_1, \varphi \vdash \Phi_2$ is derivable. This completes the induction step and thereby the proof.

Theorem 1 (Weak completeness of S_C+Cut). *For every finite multisets Γ, Δ of $\mathcal{L}_C$ formulas we have that $\Gamma \models_C \Delta$ implies that $\Gamma \vdash \Delta$ is derivable in S_C+Cut.*

Proof. Let $\Phi = \{p_1, \cdots, p_n\}$ be the set of propositional variables that occur in either Γ or Δ, and Φ_1, Φ_2 any partition of Φ. Then from Lemma 3 it follows that there either is a $\gamma \in \Gamma$ such that $\Phi_1, \gamma \models_C \Phi_2$ or a $\delta \in \Delta$ such that $\Phi_1 \models_C \delta, \Phi_2$.

From Lemma 4 it follows that in the first case the sequent $\Phi_1, \gamma \vdash \Phi_2$ is derivable and in the second case the sequent $\Phi_1 \vdash \delta, \Phi_2$ is derivable. In either case the sequent $\Gamma, \Phi_1 \vdash \Phi_2, \Delta$ can then be derived by weakening.

So for every partition Φ_1, Φ_2 of Φ the sequent $\Gamma, \Phi_1 \vdash \Phi_2, \Delta$ is derivable. For $0 \leq m \leq n$ let $\Phi^m = \{p_1, \cdots, p_{n-m}\}$. The proof now proceeds by induction on m. I just showed that if $m = 0$ then for every partition Φ_1^m, Φ_2^m of Φ^m the sequent $\Gamma, \Phi_1^m \vdash \Phi_2^m, \Delta$ is derivable.

Suppose then as induction hypothesis that $m > 0$ and that for partitions $\Phi_1^{m-1}, \Phi_2^{m-1}$ of Φ^{m-1} the sequent $\Gamma, \Phi_1^{m-1} \vdash \Phi_2^{m-1}, \Delta$ is derivable. Let Φ_1^m, Φ_2^m be any partition of Φ^m. Then both $\Phi_1^m \cup \{p_{n-m+1}\}, \Phi_2$ and $\Phi_1^m, \Phi_2 \cup \{p_{n-m+1}\}$ are partitions of Φ^{m-1} so $\Gamma, \Phi_1^m, p_{n-m+1} \vdash \Phi_2^m, \Delta$ and $\Gamma, \Phi_1^m \vdash p_{n-m+1}, \Phi_2^m, \Delta$ are both derivable. By using Cut (followed by CL and CR to get rid of extra copies of Γ, Δ, Φ_1^m and Φ_2^m) the sequent $\Gamma, \Phi_1^m \vdash \Phi_2^m, \Delta$ is then also derivable. This completes the induction step, so $\Gamma, \Phi_1^m \vdash \Phi_2^m, \Delta$ is derivable for any m and any partition Φ_1^m, Φ_2^m of Φ^m. Taking $m = n$ we then get $\Gamma \vdash \Delta$ being derivable, which is what was to be shown.

Left to show now is that Cut is admissible in S_C. The proof I give here is very similar to existing proofs for Cut-elimination as given in for example [8–10].

Theorem 2 (Cut elimination). *The rule Cut is admissible in S_C.*

Proof. The proof is by a case distinction on the rule R preceding the application of Cut. In all possible cases the application of Cut could be "moved up"; that is, it would have been possible to either apply Cut before R or to eliminate the Cut entirely. Since Cut cannot be applied before the first step of a proof this implies that at some point the Cut must be removed, so Cut is admissible.

Most of the cases are as in the existing Cut-elimination proofs. I omit those cases, for details see the proofs in for example [8–10]. The case that is different from the existing proofs is if both premises for the Cut rule are obtained using a $R_{o,v}$ rule where the Cut formula is principal. The last few steps of T are then

$$
\cfrac{
\cfrac{\Gamma_1, \pm\varphi_1 \vdash \mp\varphi_1, \Delta_1 \quad \cdots \quad \Gamma_1, \pm\varphi_{r_o} \vdash \mp\varphi_{r_o}, \Delta_1}{\Gamma_1 \vdash \varphi, \Delta_1} R_{o,v} \quad \cfrac{\Gamma_2, \pm\varphi_1 \vdash \mp\varphi_1, \Delta_2 \quad \cdots \quad \Gamma_2, \pm\varphi_{r_o} \vdash \mp\varphi_{r_o}, \Delta_2}{\Gamma_2, \varphi \vdash \Delta_2} R_{o,v'}
}{\Gamma_1, \Gamma_2 \vdash \Delta_1, \Delta_2} \text{Cut}
$$

where $\varphi = o(\varphi_1, \cdots, \varphi_{r_o})$. The application of $R_{o,v}$ adds a φ on the right side of the $\vdash$, the application of $R_{o,v'}$ adds a φ on the left side of the $\vdash$. The rules $R_{o,v}$ and $R_{o,v'}$ must therefore be different, so $v \neq v'$. This implies that there is at least one i such that φ_i occurs on one side of the $\vdash$ in a premise $\Gamma_1, \pm\varphi_i \vdash \mp\varphi_1, \Delta_1$ and on the other side in a premise $\Gamma_2, \mp\varphi_1 \vdash \pm\varphi_1, \Delta_2$. So

$$
\cfrac{\Gamma_1, \pm\varphi_i \vdash \mp\varphi_1, \Delta_1 \quad \Gamma_2, \mp\varphi_1 \vdash \pm\varphi_1, \Delta_2}{\Gamma_1, \Gamma_2 \vdash \Delta_1, \Delta_2} \text{Cut}
$$

is an alternative derivation of $\Gamma_1, \Gamma_2 \vdash \Delta_1, \Delta_2$. The Cut could therefore have been applied before the $R_{o,v}$ rules, which is what was to be shown.

Corollary 1 (Strong completeness of S_C). *For every multisets Γ, Δ of $\mathcal{L}_C$ formulas we have that $\Gamma \models_C \Delta$ implies that $\Gamma \vdash \Delta$ is derivable in S_C.*

Proof. By the compactness Lemma 2 there are finite multisets $\Gamma' \subseteq \Gamma$ and $\Delta' \subseteq \Delta$ such that $\Gamma' \models_C \Delta'$, so by Theorems 1 and 2 the sequent $\Gamma' \vdash \Delta'$ is derivable in S_C. Then $\Gamma \vdash \Delta$ is also derivable in S_C by weakening from $\Gamma' \vdash \Delta'$.

6 Properties and Applications of S_C

Let us consider a few of the properties of S_C. Reflexivity is not admissible in S_C, because it is not sound for $\models_C$. Likewise, if $\leftrightarrow$ is a classical bi-implication and $[\varphi/\psi]$ represents the substitution of ψ for φ the rule

$$\frac{\Gamma \vdash \Delta}{\varphi \leftrightarrow \psi, \Gamma[\varphi/\psi] \vdash \Delta[\varphi/\psi]}$$

representing a very strong kind of substitution of equivalents is not admissible. Two weaker kinds of substitution of equivalents are admissible though.

Lemma 5. *If $\leftrightarrow$ is the classical bi-implication and $\leftrightarrow \in C$ the rule Substitution of Deterministic Equivalents (EqDet) given by*

$$\frac{\varphi \vdash \varphi \quad \psi \vdash \psi \quad \Gamma \vdash \Delta}{\varphi \leftrightarrow \psi, \Gamma[\varphi/\psi] \vdash \Delta[\varphi/\psi]} \; EqDet$$

and the rule Substitution of Provably Equivalents (EqPr) given by

$$\frac{\varphi \vdash \psi \quad \psi \vdash \varphi \quad \Gamma \vdash \Delta}{\Gamma[\varphi/\psi] \vdash \Delta[\varphi/\psi]} \; EqPr$$

are admissible.

Proof. The easiest way to see that these rules are admissible is to use the soundness and completeness of S_C. A rule is admissible in S_C if the conclusion of the rule is derivable in S_C if all the premises are. Let us first consider the rule EqDet. Suppose that $\varphi \vdash \varphi$, $\psi \vdash \psi$ and $\Gamma \vdash \Delta$ are derivable. To show is that $\varphi \leftrightarrow \psi, \Gamma[\varphi/\psi] \vdash \Delta[\varphi/\psi]$ is derivable.

By the soundness of S_C we know that $\varphi \models_C \varphi$, $\psi \models_C \psi$ and $\Gamma \models_C \Delta$. Now let $\mathcal{M}$ be a model such that for some disambiguations $\overline{\varphi}$ of φ, $\overline{\psi}$ of ψ and $\overline{\Gamma[\varphi/\psi]}$ of $\Gamma[\varphi/\psi]$ we have $\mathcal{M} \models \overline{\varphi} \leftrightarrow \overline{\psi}$ and $\mathcal{M} \models_\wedge \overline{\Gamma[\varphi/\psi]}$. From $\varphi \models_C \varphi$ it follows that for every model all disambiguations of φ have the same value. Likewise, from $\psi \models_C \psi$ it follows that all disambiguations of ψ have the same value.

Since some disambiguations of φ and ψ have the same value of $\mathcal{M}$ this implies that every disambiguation of φ has the same value as every disambiguation of ψ in $\mathcal{M}$. The disambiguations live in a deterministic truth-functional logic so we can replace any occurrence of any disambiguation of ψ by any disambiguation of φ without changing the value on $\mathcal{M}$. So from $\mathcal{M} \models_\wedge \overline{\Gamma[\varphi/\psi]}$ it follows that $\mathcal{M} \models_\wedge \overline{\Gamma}$ for some disambiguation $\overline{\Gamma}$ of Γ.

Then by $\Gamma \vdash \Delta$ we know that $\mathcal{M} \models_\vee \overline{\Delta}$ for each disambiguation $\overline{\Delta}$ of Δ. We can replace any disambiguation of φ by any disambiguation of ψ without changing the value on $\mathcal{M}$ so $\mathcal{M} \models_\vee \overline{\Delta[\varphi/\psi]}$ for any disambiguation $\overline{\Delta[\varphi/\psi]}$.

We started with any model $\mathcal{M}$ satisfying $\overline{\varphi} \leftrightarrow \overline{\psi}$ and $\overline{\Gamma[\varphi/\psi]}$ for some disambiguations and found that $\mathcal{M}$ satisfies $\overline{\Delta[\varphi/\psi]}$ for all disambiguations, so $\varphi \leftrightarrow \psi, \Gamma[\varphi/\psi] \models_C \Delta[\varphi/\psi]$. By completeness this implies that $\varphi \leftrightarrow \psi, \Gamma[\varphi/\psi] \vdash \Delta[\varphi/\psi]$ is derivable which is what was to be shown.

Left to show is that ExPr is admissible. Suppose towards a contradiction that $\varphi \vdash \psi$, $\psi \vdash \varphi$ and $\Gamma \vdash \Delta$ are derivable in S_C but $\Gamma[\varphi/\psi] \vdash \Delta[\varphi/\psi]$ is not. Then by the soundness of S_C we have $\varphi \models_C \psi$, $\psi \models_C \varphi$ and $\Gamma \models_C \Delta$ while by the completeness of S_C we have $\Gamma[\varphi/\psi] \not\models_C \Delta[\varphi/\psi]$.

So for some disambiguations $\overline{\Gamma[\varphi/\psi]}$ of $\Gamma[\varphi/\psi]$ and $\overline{\Delta[\varphi/\psi]}$ of $\Delta[\varphi/\psi]$ we have $\overline{\Gamma[\varphi/\psi]} \not\models_{\overline{C}} \overline{\Delta[\varphi/\psi]}$. There are disambiguations $\overline{\Gamma}$ of Γ, $\overline{\Delta}$ of Δ, $\overline{\varphi}$ of φ and $\overline{\psi}$ of ψ such that $\overline{\Gamma[\varphi/\psi]} = \overline{\Gamma}[\overline{\varphi}/\overline{\psi}]$ and $\overline{\Delta[\varphi/\psi]} = \overline{\Delta}[\overline{\varphi}/\overline{\psi}]$.

From $\varphi \models_C \psi$ and $\psi \models_C \varphi$ it follows that any disambiguations of φ and ψ are equivalent, so in particular $\overline{\varphi}$ and $\overline{\psi}$ are equivalent. But from $\Gamma \models_C \Delta$ it follows that $\overline{\Gamma} \models_{\overline{C}} \overline{\Delta}$ and by substitution of equivalents in deterministic propositional logic this implies that $\overline{\Gamma}[\overline{\varphi}/\overline{\psi}] \models_{\overline{C}} \overline{\Delta}[\overline{\varphi}/\overline{\psi}]$, which contradicts $\overline{\Gamma[\varphi/\psi]} \not\models_{\overline{C}} \overline{\Delta[\varphi/\psi]}$. Our initial assumption that $\varphi \vdash \psi$, $\psi \vdash \varphi$ and $\Gamma \vdash \Delta$ are derivable and $\Gamma[\varphi/\psi] \vdash \Delta[\varphi/\psi]$ is not must therefore be false. So if $\varphi \vdash \psi$, $\psi \vdash \varphi$ and $\Gamma \vdash \Delta$ are derivable then so is $\Gamma[\varphi/\psi] \vdash \Delta[\varphi/\psi]$.

7   Conclusion

I introduced nondeterministic semantics for propositional logic that do not satisfy reflexivity. The main idea of the semantics is to use deterministic disambiguations of nondeterministic formulas and to say that $\Gamma \models_C \Delta$ if and only if $\overline{\Gamma} \models_{\overline{C}} \overline{\Delta}$ for all possible disambiguations $\overline{\Gamma}$ of Γ and $\overline{\Delta}$ of Δ. I also introduced a sequent-style proof system S_C that is sound and complete for $\models_C$ and showed that S_C allows some types of substitution of equivalents.

References

1. Lewis, D.: Logic for equivocators. Noûs 16(3), 431–441 (1982)
2. Avron, A., Lev, I.: Canonical propositional gentzen-type systems. In: Goré, R.P., Leitsch, A., Nipkow, T. (eds.) IJCAR 2001. LNCS (LNAI), vol. 2083, pp. 529–544. Springer, Heidelberg (2001)
3. Avron, A., Lev, I.: Non-deterministic matrices. In: Proceedings of the 34th International Symposium on Multiple-Valued Logic, pp. 282–287 (2004)
4. Avron, A., Lev, I.: Non-deterministic multiple-valued structures. Journal of Logic and Computation 15(3), 241–261 (2005)
5. Barwise, J.: An introduction to first-order logic. In: Barwise, J. (ed.) Handbook of Mathematical Logic. Elsevier Science Publishers (1977)
6. Tychonoff, A.: Ein Fixpunktsatz. Mathematische Annalen 111(1), 767–776 (1935)
7. Tychonoff, A.: Über die topologische Erweiterung von Räumen. Mathematische Annalen 102(1), 544–561 (1930)
8. Troelstra, A.S., Schwichtenberg, H.: Basic Proof Theory. Cambridge University Press (1996)
9. Negri, S., von Plato, J.: Structural Proof Theory. Cambridge University Press (2001)
10. von Plato, J.: A proof of gentzen's hauptsatz without multicut. Archive for Mathematical Logic 40(1), 9–18 (2001)

How to Update Neighborhood Models

Minghui Ma[1] and Katsuhiko Sano[2]

[1] Center for the Study of Logic and Intelligence, Southwest University, China
mmh.thu@gmail.com
[2] School of Information Science, Japan Advanced Institute of Science and Technology, Japan
v-sano@jaist.ac.jp

Abstract. This paper studies two ways of updating neighborhood models: update by taking the intersection of the neighborhoods with the announced proposition, and update by selecting all the neighborhoods that can entail the announced proposition. For each of these two ways, we establish reduction axioms and some basic model-theoretic results on public announcement logic and dynamic epistemic logic of product update. We also study various notions of group knowledge such as common and distributed knowledge over neighborhood models.

Public announcement logic (**PAL**) and dynamic epistemic logic of product update (**DEL**) have been studied for a long time[1]. **PAL** was first presented by Plaza in [10], and later **DEL** was explored by Baltag, Moss and Solecki in [2], and recently by van Benthem in [11] and other logicians. **DEL** provides us with a general theory for interpreting update of multi-agent models. For some basics of **DEL** (under Kripke semantics), the reader is referred to [11] and [15]. In this paper, we study neighborhood semantics for both **PAL** and **DEL**. There were several papers (cf. [12]) on updating operation on neighborhood models (or its generalization), while there is no study on **DEL** over neighborhood models, to the best of the authors' knowledge. What we would like to propose in this paper is the two ways of update the neighborhood $\tau(w)$ of the current state: taking 'intersection' and 'subset'.

Fig. 1. $\tau(w)$ **Fig. 2.** Intersection-update by φ **Fig. 3.** Subset-update by φ

Let us explain the intuitive idea behind two ways of updating. Let the shaded circles of Fig. 1 be neighborhoods around w and $[\![\varphi]\!]$ be the denotation of φ in the model. The readers may regard $\tau(w)$ as the set of propositions that an agent does know at the

---

[1] We would like to the anonymous reviewers for their invaluable comments. The work of the first author was supported by the project of China National Social Sciences Fund (Grant no. 12CZX054) and the work of the second author was supported by JSPS KAKENHI, Grant-in-Aid for Young Scientists (B) 24700146.

D. Grossi, O. Roy, and H. Huang (Eds.): LORI 2013, LNCS 8196, pp. 204–217, 2013.

state w. By taking intersection of all the neighborhoods, we obtain Fig. 2. This notion was studied, e.g., in [12, pp.69-70]. On the other hand, by restricting our attention to all the neighborhoods 'inside' the denotation $[\![\varphi]\!]$, we obtain Fig. 3. It is clear that the second notion is stronger than the first one. We can regard this subset-update as restricting our knowledge to the propositions that 'derive' φ. In other words, the notion of subset update by φ gives us the assumptions or evidences that explain φ. We should note that the subset update is an 'opposite' operation to the notion of evidence removal in [12]. We should also note that a similar idea to our subset update also appears in a recent study of public announcements over subset space logics [1, p.239].

This paper explores the two ways of updating neighborhood models for both **PAL** and **DEL** in terms of axiomatization and basic model theory. We also demonstrate when the two ways of updating neighborhood models coincide with each other.

1 Neighborhood Semantics for Epistemic Logic

Now we introduce the neighborhood semantics for epistemic logic (**EL**), the static part of dynamic epistemic logics. Henceforce, we fix a set G of agents and **Prop** of propositional variables. The language $\mathcal{L}_{EL}$ of **EL** is defined as the expansion of the propositional syntax with $\Box_i\varphi$ ($i \in G$), where $\Box_i\varphi$ is read as: the agent i knows that φ.

A *neighborhood frame* is a pair $\mathfrak{F} = (W, \{\tau_i\}_{i \in G})$ where W is a non-empty set and $\tau_i : W \rightarrow \wp(\wp(W))$ is a mapping. Moreover, we define some types of neighborhood frames as follows:

- $\mathfrak{F}$ is *monotone* if $X \in \tau_i(w)$ and $X \subseteq Y$ jointly imply $Y \in \tau_i(w)$, for all $w \in W$, $X, Y \in \wp(W)$ and $i \in G$.
- $\mathfrak{F}$ is *non-empty* if $W \in \tau_i(w)$ for all $w \in W$ and $i \in G$.
- $\mathfrak{F}$ is *closed under intersections* if $X \in \tau_i(w)$ and $Y \in \tau_i(w)$ jointly imply $X \cap Y \in \tau_i(w)$, for all $w \in W$, $X, Y \in \wp(W)$ and $i \in G$.
- $\mathfrak{F}$ is *regular* if $\mathfrak{F}$ is monotone and closed under intersections.
- $\mathfrak{F}$ is *normal* if $\mathfrak{F}$ is regular and non-empty.

Given any neighborhood frame $\mathfrak{F} = (W, \{\tau_i\}_{i \in G})$, for each τ_i, we define the unary operation $\Box_{\tau_i} : \wp(W) \rightarrow \wp(W)$ by: $\Box_{\tau_i}X = \{ w \in W \mid X \in \tau_i(w) \}$. We will use $\Box_i$ as a modal operator in the syntax of $\mathcal{L}_{EL}$ and $\Box_{\tau_i}$ as a unary operator in the semantic side, i.e., the semantic operation to interpret the knowledge operator $\Box_i$.

We say that $\mathfrak{F} = (W, \{\tau_i\}_{i \in G})$ is a *topological space*, if $\mathfrak{F}$ is normal and satisfies the following conditions: for all $w \in W$, $X \subseteq W$, and $i \in G$,

(T) $X \in \tau_i(w)$ implies $w \in X$;   (4) $X \in \tau_i(w)$ implies $\Box_{\tau_i}X \in \tau_i(w)$.

In what follows in this paper, we concentrate on *monotone* neighborhood frames. We define that a *neighborhood model* is a pair of neighborhood frame and a valuation $V : \mathsf{Prop} \rightarrow \wp(W)$. Given any neighborhood model $\mathfrak{M} = (W, \{\tau_i\}_{i \in G}, V)$, any $w \in W$ and any formula φ of $\mathcal{L}_{EL}$, we define the satisfaction relation $\mathfrak{M}, w \models \varphi$ as follows:

$$\mathfrak{M}, w \models p \quad \text{iff} \quad w \in V(p), \quad \mathfrak{M}, w \models \varphi \vee \psi \text{ iff } \mathfrak{M}, w \models \varphi \text{ or } \mathfrak{M}, w \models \psi,$$
$$\mathfrak{M}, w \models \neg\varphi \text{ iff } \mathfrak{M}, w \not\models \varphi, \quad \mathfrak{M}, w \models \Box_i\varphi \quad \text{iff} \quad [\![\varphi]\!]_\mathfrak{M} \in \tau_i(w),$$

where $[\![\varphi]\!]_{\mathfrak{M}} = \{v \in W \mid \mathfrak{M}, v \models \varphi\}$ is the truth set of φ in $\mathfrak{M}$. If $(W, \{\tau_i\}_{i\in G})$ is monotone, the truth clause for $\square_i\varphi$ can be rewritten as follows:

$$\mathfrak{M}, w \models \square_i\varphi \text{ iff } X \subseteq [\![\varphi]\!]_{\mathfrak{M}} \text{ for some } X \in \tau_i(w).$$

Definition 1. *Given any monotone $\mathfrak{F} = (W, \{\tau_i\}_{i\in G})$ and $\mathfrak{F}' = (W', \{\tau'_i\}_{i\in G})$, $f : W \to W'$ is a* bounded morphism, *if for all $w \in W$ and $i \in G$, the following hold:* (Forth) $\forall X \in \tau_i(w)(f[X] \in \tau'_i(f(w)))$; (Back) $\forall X' \in \tau'_i(f(w))(f^{-1}[X'] \in \tau_i(w))$.

Definition 2. *Given monotone neighborhood frames $\mathfrak{F} = (W, \{\tau_i\}_{i\in G})$ and $\mathfrak{F}' = (W', \{\tau'_i\}_{i\in G})$, we say that $\mathfrak{F}'$ is a* generated subframe *of $\mathfrak{F}$, if the inclusion map $i : W' \to W$ is a bounded morphism, i.e., for all $w' \in W$ and all $X \subseteq W$, $X \cap W' = i^{-1}[X] \in \tau'_i(w')$ iff $X \in \tau_i(w')$. We also say that $\mathfrak{F}'$ is a* relational subframe *of $\mathfrak{F}$, if $W' \subseteq W$ and $\tau'_i(w') = \{X \subseteq W' \mid X \in \tau_i(w')\}$ for all $w' \in W'$ and $i \in G$.*

The notion of relational subframe was first considered by Hansen [4] under a different name. We can explain her underlying idea on the notion as follows: if we regard τ_i as a *relation* between W and $\wp(W)$, then τ'_i in the above definition satisfies $\tau'_i = \tau_i \cap (W' \times \wp(W'))$, i.e., τ'_i is the restriction of τ_i to the domain W'. Moreover, a subframe and a relational subframe coincide for certain classes of neighborhood frames.

Proposition 1. *Let $\mathfrak{F} = (W, \{\tau_i\}_{i\in G})$ and $\mathfrak{F}' = (W', \{\tau'_i\}_{i\in G})$ be regular neighborhood frames. Assume that $W' = \square_{\tau_i} W'$ for all $i \in G$. Then $\mathfrak{F}'$ is a generated subframe of $\mathfrak{F}$ iff $\mathfrak{F}'$ is a relational subframe of $\mathfrak{F}$.*

Proof. The left-to-right direction is easy. For the other direction, assume that $W' = \square_{\tau_i} W'$ for all $i \in G$. It suffices to show that $\{X \cap W' \mid X \in \tau_i(w')\} = \{X \subseteq W' \mid X \in \tau_i(w')\}$, for all $w' \in W'$ and $i \in G$. The $\supseteq$-direction holds clearly for all monotone neighborhood frames. For the other direction, take any $X \in \tau'_i(w)$. By assumption, $W' \in \tau_i(w')$. Since $\mathfrak{F}$ is regular, $X \cap W' \in \tau'_i(w)$. $\qquad\square$

Definition 3. *Let $\mathfrak{M} = (W, \{\tau_i\}_{i\in G}, V)$ be a monotone neighborhood model, and $\varnothing \neq X \subseteq W$. Define the* intersection submodel *$\mathfrak{M}^{\cap X} = (X, \{\tau_i^{\cap X}\}_{i\in G}, V^X)$ and the* subset submodel *$\mathfrak{M}^{\subseteq X} = (X, \{\tau_i^{\subseteq X}\}_{i\in G}, V^X)$ induced from X by:*

$$\tau_i^{\cap X}(w) = \{P \cap X \mid P \in \tau_i(w)\} \text{ and } \tau_i^{\subseteq X}(w) = \{P \subseteq X \mid P \in \tau_i(w)\}$$

for every $w \in X$ and $i \in G$, and $V^X(p) = V(p) \cap X$ for any $p \in$ Prop.

Note that this definition allows the case where $\varnothing \in \tau_i^{\cap X}(w)$, the case where $\varnothing \in \tau_i^{\subseteq X}(w)$, or the case where $\tau_i^{\subseteq X}(w) = \varnothing$.

Proposition 2. *Let $\mathfrak{M}$ be a monotone neighborhood model. Then for any $X \subseteq W$, the intersection submodel $\mathfrak{M}^{\cap X}$ and the subset submodel $\mathfrak{M}^{\subseteq X}$ are monotone.*

Proof. First, we establish monotonicity of $\tau_i^{\cap X}(w)$. Assume $P \in \tau_i^{\cap X}(w)$. Then $P = Q \cap X$ for some $Q \in \tau_i(w)$. Fix any P' such that $P \subseteq P' \subseteq X$. Since τ_i is monotone, $Q \cup P' \in \tau_i(w)$. Then $(Q \cup P') \cap X = P'$. Hence $P' \in \tau_i^{\cap X}(w)$. Second, we move to $\tau_i^{\subseteq X}(w)$. Assume $P \in \tau_i^{\subseteq X}(w)$. Then $P \subseteq X$ and $P \in \tau_i(w)$. Fix any P' such that $P \subseteq P' \subseteq X$. By monotonicity of $\tau_i(w)$, $P' \in \tau_i(w)$. Then $P' \in \tau_i^{\subseteq X}(w)$. $\qquad\square$

Proposition 3. *Let $\mathfrak{M}$ be a monotone neighborhood model. The frame properties of closure under intersections, (T), and (4), are preserved under taking the intersection submodel $\mathfrak{M}^{\cap X}$ and the subset submodel $\mathfrak{M}^{\subseteq X}$. The frame property of non-emptiness is also preserved under taking the intersection submodel $\mathfrak{M}^{\cap X}$.*

In general, however, we cannot assure that the property of non-emptiness is preserved under taking *subset* submodels[2]. We leave the detailed investigation of preservation of non-emptiness as a further direction, and focus on the study of $\tau^{\cap X}$ and $\tau^{\subseteq X}$ over a class of monotone neighborhood frames in this paper.

2   Neighborhood Semantics for PAL

2.1   Subset and Intersection Semantics for $\mathcal{L}_{PAL}$

The language $\mathcal{L}_{PAL}$ is the extension of $\mathcal{L}_{EL}$ by adding public announcement formulas $[\varphi]\psi$. We define $\langle\varphi\rangle\psi$ as $\neg[\varphi]\neg\psi$. Recall that Kripke semantics for $\mathcal{L}_{PAL}$ interprets an announcement operator $[\varphi]$ in terms of a (Kripke) submodel induced by truth set of φ in the given model. Here we introduce two kinds of neighborhood semantics for public announcement operators. For this purpose, we employ the notions of intersection submodel and subset submodel (definition 3), which are based on the notions of generated subframe and relational subframe (definition 2), respectively. These notions allow us to define two kinds of neighborhood semantics for $[\varphi]$: intersection semantics and subset semantics. Let us first introduce our intersection semantics, and then, move to our subset semantics.

Definition 4 (Intersection Semantics). *Given a monotone neighborhood model $\mathfrak{M} = (W, \{\tau_i\}_{i \in G}, V)$, the notion of truth of an $\mathcal{L}_{PAL}$-formula φ (notation: $\mathfrak{M}, w \models_\cap \varphi$) is defined recursively as usual, except the following clause for $[\varphi]\psi$:*

$$\mathfrak{M}, w \models_\cap [\varphi]\psi \text{ iff } \mathfrak{M}, w \models_\cap \varphi \text{ implies } \mathfrak{M}^{\cap\varphi}, w \models_\cap \psi$$

where $\mathfrak{M}^{\cap\varphi}$ is the intersection submodel $\mathfrak{M}^{\cap[\![\varphi]\!]_\mathfrak{M}}$.

For the update of intersection submodel, we can also provide the reduction axioms which are used for eliminating public announcements in all $\mathcal{L}_{PAL}$-formulas, where the reduction axioms for propositional letters and connectives are the same as public announcement logic under Kripke semantics.

---

[2] Let us consider the single agent case. Let $\mathfrak{F} = (\mathbb{R}, \tau_\mathbb{R})$ be the real line with the ordinary Euclidean topology. Fix any point $x \in \mathbb{R}$. It is clear that $\mathfrak{F}$ satisfies the non-emptiness condition. Then, $\tau_\mathbb{R}^{\cap\mathbb{N}}(x) = \{X \subseteq \mathbb{N} \mid x \in X\}$ but $\tau_\mathbb{R}^{\subseteq\mathbb{N}}(x) = \varnothing$, since there is no open set O around x such that $O \subseteq \mathbb{N}$. One way to avoid this difficulty is that, when we are interested in normal neighborhood frames or topological spaces, we revise the neighborhood mapping of the subset submodel into: $\tau_i^{\subseteq X; \cup\{X\}}(w) = \{P \subseteq X \mid P \in \tau_i(w)\} \cup \{X\}$. The underlying idea of this definition is to add the trivial evidence (i.e., X) to the set of evidences around each of the states in the subset-updated model. The update by $\tau_i^{\subseteq X; \cup\{X\}}$ preserves non-emptiness as well as monotonicity, closure under intersections, (T) and (4).

Table 1. List of Axioms for $[\varphi]$ for Intersection and Subset Semantics

(RAtom) $[\varphi]p \leftrightarrow (\varphi \to p)$	(R$\Box_i$int) $[\varphi]\Box_i\psi \leftrightarrow (\varphi \to \Box_i[\varphi]\psi)$
(R$\neg$) $\quad [\varphi]\neg\psi \leftrightarrow (\varphi \to \neg[\varphi]\psi)$	(R[·]) $\quad [\varphi][\psi]\xi \leftrightarrow [\varphi \wedge [\varphi]\psi]\xi$
(R$\vee$) $\quad [\varphi](\psi \vee \xi) \leftrightarrow ([\varphi]\psi \vee [\varphi]\xi)$	(R$\Box_i$sub) $[\varphi]\Box_i\psi \leftrightarrow (\varphi \to \Box_i\langle\varphi\rangle\psi)$

Theorem 1. *For any complete epistemic logic Λ under monotone neighborhood semantics, the $\mathcal{L}_{\mathbf{PAL}}$-extension under intersection semantics is completely axiomatized by Λ plus* (RAtom), (R$\neg$), (R$\vee$), (R$\Box_i$int), *and* (R[·]) *of Table 1.*

Proof. We only check the validity of (R$\Box_i$int) and (R[·]). Then, we can employ the same proof-strategy as in the proof of Kripke completeness of **PAL** in [11] and [15], i.e., the completeness of **PAL**-extension of Λ is reduced to that of Λ. We can establish (R[·]) by the fact $(\mathfrak{M}^\varphi)^\psi = \mathfrak{M}^{\varphi\wedge[\varphi]\psi}$. So, we concentrate on (R$\Box_i$int). (1) From left to right, assume $\mathfrak{M}, w \models_\cap [\varphi]\Box_i\psi$ and $\mathfrak{M}, w \models_\cap \varphi$. It suffices to show $[[\varphi]\psi]_{\mathfrak{M}} \in \tau_i(w)$. By assumption, $\mathfrak{M}^{\cap\varphi}, w \models_\cap \Box_i\psi$. Then $[\![\psi]\!]_{\mathfrak{M}^{\cap\varphi}} \in \tau_i^{\cap\varphi}(w)$. Then $[\![\psi]\!]_{\mathfrak{M}^{\cap\varphi}} \supseteq X \cap [\![\varphi]\!]_{\mathfrak{M}}$ for some $X \in \tau_i(w)$. It follows that $X \subseteq \overline{[\![\varphi]\!]_{\mathfrak{M}}} \cup [\![\psi]\!]_{\mathfrak{M}^{\cap\varphi}} = [\![[\varphi]\psi]\!]_{\mathfrak{M}}$, where $\overline{[\![\varphi]\!]_{\mathfrak{M}}}$ is the complement of $[\![\varphi]\!]_{\mathfrak{M}}$ in $\mathfrak{M}$. This implies $[\![[\varphi]\psi]\!]_{\mathfrak{M}} \in \tau_i(w)$, as desired. (2) For the other direction, assume $\mathfrak{M}, w \models_\cap \varphi \to \Box_i[\varphi]\psi$ and $\mathfrak{M}, w \models_\cap \varphi$. Our goal is to show $\mathfrak{M}, w \models [\varphi]\Box_i\psi$. By assumption, it suffices to show that $[\![\psi]\!]_{\mathfrak{M}^{\cap\varphi}} \in \tau_i^{\cap\varphi}(w)$. By assumption, $\mathfrak{M}, w \models_\cap \Box_i[\varphi]\psi$. Hence $[\![[\varphi]\psi]\!]_{\mathfrak{M}} \in \tau_i(w)$. Since $[\![\psi]\!]_{\mathfrak{M}^{\cap\varphi}} \cap [\![[\varphi]\psi]\!]_{\mathfrak{M}} = [\![[\varphi]\psi]\!]_{\mathfrak{M}}$, we can conclude $[\![[\varphi]\psi]\!]_{\mathfrak{M}} \in \tau_i^{\cap\varphi}(w)$. $\qquad\Box$

Definition 5 (Subset Semantics). *Let $\mathfrak{M} = (W, \{\tau_i\}_{i \in G}, V)$ be a neighborhood model. The notion of truth of an $\mathcal{L}_{\mathbf{PAL}}$-formula φ (notation: $\mathfrak{M}, w \models_\subseteq \varphi$) is defined recursively as usual, except the following semantic clause for $[\varphi]\psi$:*

$$\mathfrak{M}, w \models_\subseteq [\varphi]\psi \text{ iff } \mathfrak{M}, w \models_\subseteq \varphi \text{ implies } \mathfrak{M}^{\subseteq\varphi}, w \models_\subseteq \psi$$

where $\mathfrak{M}^{\subseteq\varphi}$ is the subset submodel $\mathfrak{M}^{\subseteq[\![\varphi]\!]_{\mathfrak{M}}}$.

For $\langle\varphi\rangle\psi$, we obtain the following truth clause:

$$\mathfrak{M}, w \models_\subseteq \langle\varphi\rangle\psi \text{ iff } \mathfrak{M}, w \models_\subseteq \varphi \text{ and } \mathfrak{M}^{\subseteq\varphi}, w \models_\subseteq \psi.$$

Theorem 2. *For any complete epistemic logic Λ under monotone neighborhood semantics, the $\mathcal{L}_{\mathbf{PAL}}$-extension under intersection semantics is completely axiomatized by Λ plus* (RAtom), (R$\neg$), (R$\vee$), (R$\Box_i$sub), *and* (R[·]) *of Table 1.*

Proof. Similar to the proof of Theorem 1. Here, let us check the validity of (R$\Box_i$sub) alone: First of all, it is easy to show that

$$[\![\psi]\!]_{\mathfrak{M}^{\subseteq\varphi}} = [\![\varphi \wedge [\varphi]\psi]\!]_{\mathfrak{M}} = [\![\langle\varphi\rangle\psi]\!]_{\mathfrak{M}}. \tag{$*$}$$

Assume $\mathfrak{M}, w \models_\subseteq [\varphi]\Box_i\psi$ and $\mathfrak{M}, w \models_\subseteq \varphi$. We show $[\![\langle\varphi\rangle\psi]\!]_{\mathfrak{M}} \in \tau_i(w)$. By assumption, $\mathfrak{M}^{\subseteq\varphi}, w \models_\subseteq \Box_i\psi$. Then $[\![\psi]\!]_{\mathfrak{M}^{\subseteq\varphi}} \in \tau_i^{\subseteq\varphi}(w)$. We conclude from $(*)$ that $[\![\langle\varphi\rangle\psi]\!]_{\mathfrak{M}} \in \tau_i(w)$, as desired. Conversely, assume $\mathfrak{M}, w \models_\subseteq \varphi \to \Box_i\langle\varphi\rangle\psi$ and $\mathfrak{M}, w \models_\subseteq \varphi$. Then $\mathfrak{M}, w \models_\subseteq \Box_i\langle\varphi\rangle\psi$, which implies $[\![\langle\varphi\rangle\psi]\!]_{\mathfrak{M}} \in \tau_i(w)$. By $(*)$ and $[\![\psi]\!]_{\mathfrak{M}^{\subseteq\varphi}} \subseteq [\![\varphi]\!]_{\mathfrak{M}}$, we conclude $[\![\psi]\!]_{\mathfrak{M}^{\subseteq\varphi}} \in \tau_i^{\subseteq\varphi}(w)$. $\qquad\Box$

Remark that the axioms (R□$_i$int) and (R□$_i$sub) are different on the right-hand side within the scope of the knowledge operator □$_i$. This syntactic difference shows their semantic difference.

As is shown in Proposition 1, if $[\![\varphi]\!]_{\mathfrak{M}} = \Box_{\tau_i}[\![\varphi]\!]_{\mathfrak{M}}$, then the subset model is the same as the intersection model. It is well-known that the modal logic **S4** is complete with respect to the class of all topological spaces, where we can derive $\Box\varphi \leftrightarrow \Box\Box\varphi$ as a theorem of **S4**.

Proposition 4. *For any topological model* $\mathfrak{M}$ *and formula* φ, $\mathfrak{M}^{\cap\Box_i\varphi} = \mathfrak{M}^{\subseteq\Box_i\varphi}$.

Proof. By reduction axioms, every $\mathcal{L}_{\textbf{PAL}}$-formula is logically equivalent to an $\mathcal{L}_{\textbf{EL}}$-formula. We assume without loss of generality that $\varphi \in \mathcal{L}_{\textbf{EL}}$. Then the domains, and hence valuations, of the two updated models are the same $[\![\Box_i\varphi]\!]_{\mathfrak{M}}$. Since $\mathfrak{M}$ is a topological model, $[\![\Box_i\varphi]\!]_{\mathfrak{M}} = [\![\Box_i\Box_i\varphi]\!]_{\mathfrak{M}} = \Box_{\tau_i}[\![\Box_i\varphi]\!]_{\mathfrak{M}}$. By Proposition 1, $\tau_i^{\cap\Box_i\varphi}(w) = \tau_i^{\subseteq\Box_i\varphi}(w)$ for each w in the domain. Therefore, $\mathfrak{M}^{\cap\Box_i\varphi} = \mathfrak{M}^{\subseteq\Box_i\varphi}$.   □

We note that the proof of Proposition 4 does not require the non-emptiness of frames.

Every Kripke model $\mathfrak{M} = (W, \{R_i\}_{i\in G}, V)$ can be transformed into an equivalent neighborhood model by defining: $\tau_{R_i}(w) = \{X \subseteq W \mid R_i(w) \subseteq X\}$. Let Nbhd($\mathfrak{M}$) = $(W, \{\tau_{R_i}\}_{i\in G}, V)$ be the equivalent neighborhood model.

Proposition 5. *Given any Kripke model* $\mathfrak{M} = (W, \{R_i\}_{i\in G}, V)$ *and any* $X \subseteq W$, *we have* (Nbhd($\mathfrak{M}$))$^{\cap X}$ = Nbhd($\mathfrak{M}^X$), *where* $\mathfrak{M}^X = (X, (R_i^X)_{i\in G}, V^X)$ *where* R_i^X *and* V^X *are restrictions to* X.

Proof. Consider the submodel $\mathfrak{M}^X = (X, (R_i^X)_{i\in G}, V^X)$. For $w \in X$, it suffices to show $\tau_{R_i^X}(x) = \tau_{R_i}^{\cap X}(x)$. Let $P \in \tau_{R_i^X}(x)$. Then $R_i(x) \cap X \subseteq P$. By $R_i(x) \in \tau_{R_i}(x)$, $P \in \tau_{R_i}^{\cap X}(x)$. Conversely, assume $Q \in \tau_{R_i}^{\cap X}(x)$. Then, there exists $P \in \tau_{R_i}(x)$ such that $Q = P \cap X$. Then $R_i(x) \subseteq P$ and so $R_i(x) \cap X \subseteq Q \subseteq X$. Thus $Q \in \tau_{R_i^X}(x)$.   □

This proposition implies that we can also analyze various examples of **PAL** under Kripke semantics in terms of neighborhood semantics with the intersection-update (for instance, backward induction, a procedure for solving extensive games, which was analyzed by van Benthem in [11], and muddy children puzzle in [15]).

Example 1. Let us consider the single agent case. Consider the following Kripke model for **S4**: $\mathfrak{M} = (W, R, V)$ where $W = \{a, b\}$, $R = W^2 \setminus \{(b, a)\}$ and $V(p) = \{b\}$. Then, the equivalent neighborhood model is defined by: $\tau_R(a) = \{\{a, b\}\}$ and $\tau_R(b) = \{\{b\}, \{a, b\}\}$. Note that $\mathfrak{M}, a \models \neg p$ and $\mathfrak{M}, b \models \Box p$ but $\mathfrak{M}, a \not\models \Box p$. Thus, the domain of the updated model of $\mathfrak{M}$ by announcing $\Box p$ in either way is the single point of a, while the domain by announcing $\neg p$ consists of the single point b. Then, $\tau_R^{\cap\neg p}(a) = \{\{a\}\}$ and $\tau_R^{\subseteq\neg p}(a) = \varnothing$, which are not the same. However, $\tau^{\cap\Box p}(b) = \tau^{\subseteq\Box p}(b) = \{\{b\}\}$, which coincides with the result of Proposition 4.

The following provides us with a non-Kripke semantic but topological example.

Example 2. Let $(\mathbb{R}, \tau_{\mathbb{R}})$ be the real line with the ordinary Euclidean topology, i.e., $\tau_{\mathbb{R}}(x)$ is the set of open sets containing $x \in \mathbb{R}$. Define $V(p) = [0, +\infty)$ and write $\mathfrak{M} = (\mathbb{R}, \tau_{\mathbb{R}}, V)$. Then, $[\![\Box p]\!]_{\mathfrak{M}} = (0, +\infty)$. By Proposition 4, we obtain, e.g.,

$$\tau_R^{\subseteq \Box p}(1) = \tau_R^{\cap \Box p}(1) = \{ O \subseteq (0, +\infty) \mid O \text{ is open with respect to } \tau_{\mathbb{R}} \text{ and } 1 \in O \}.$$

Next, let us consider subset- and intersection-updates by p. Then,

$$\tau_{\mathbb{R}}^{\cap p}(1) = \{ O \cap [0, +\infty) \mid O \in \tau_{\mathbb{R}}(1) \}, \quad \tau_{\mathbb{R}}^{\subseteq p}(1) = \{ O \in \tau_{\mathbb{R}}(1) \mid O \subseteq [0, +\infty) \}.$$

It is easy to see that $\tau_{\mathbb{R}}^{\subseteq p}(1) \subseteq \tau_{\mathbb{R}}^{\cap p}(1)$. However, the converse inclusion does not hold, since $[0, 2) \in \tau_{\mathbb{R}}^{\cap p}(1)$ but $[0, 2) \notin \tau_{\mathbb{R}}^{\subseteq p}(1)$. If we regard open sets as evidences (cf. [13, sec.2.2]), the intersection update may give us the new evidences. That is, $[0, 2)$ is not open with respect to the original $\tau_{\mathbb{R}}$ but it is open with respect to $\tau_{\mathbb{R}}^{\cap p}$. On the other hand, the subset-update selects from $\tau_{\mathbb{R}}(x)$ all the evidences that can deduce the announced formula.

Then, what is a possible merit of the subset semantics for $[\varphi]$? Intuitively, the public announcement of φ in subset semantics deletes all propositions in the range of w which cannot entail φ. We could use this procedure to interpret some reasoning patterns in scientific inquiry.

Example 3. Let us introduce a pattern in scientific inquiry called *finite identification* [9]. Suppose that a scientist makes assumptions $\alpha_0, \ldots, \alpha_n$, one of which is actual. Then, one data φ occurs to the scientist. The announcement of φ can eliminate all assumptions which cannot interpret φ. Given an assumption α, let $\alpha \le \varphi$ denote that the assumption α can interpret φ. With the help of the subset semantics, we can explain this pattern. Fix a state w and $\tau(w)$ in a neighborhood model $\mathfrak{M}$. Let $\alpha_0, \ldots, \alpha_n$ be propositions which serve as our assumptions. Let $X_0, \ldots, X_n \in \tau(w)$ such that each $X_i = [\![\alpha_i]\!]_{\mathfrak{M}}$ for all $i \le n$. We interpret the expression $\alpha \le \varphi$ over $\mathfrak{M}$ as follows: $\alpha_i \le \varphi$ iff $X_i \subseteq [\![\varphi]\!]_{\mathfrak{M}}$. Thus all assumptions which cannot interpret φ are eliminated after the announcement of φ in subset semantics. After when finitely many data occur (i.e., are announced), we can identify the actual assumption. By this process, we can also capture the reasoning of abduction in scientific inquiry. Let us consider one simple inference pattern of abduction: from φ and $\psi \vdash \varphi$, we may infer abductively ψ. That is, ψ is a possible reason of φ. In the subset semantics for an announcement of φ, we keep all 'possible reasons' of φ. We hope that the subset semantics could serve as a sort of 'semantics' for abductive reasoning.

2.2 Some Model Theory for Neighborhood PAL

We show some model-theoretic results about the neighborhood semantics for $\mathcal{L}_{\textbf{PAL}}$. We first introduce the notion of neighborhood bisimulation [4].

Definition 6 (Neighborhood Bisimulation). *Let* $\mathfrak{M} = (W, \{\tau_i\}_{i \in G}, V)$ *and* $\mathfrak{M}' = (W', \{\tau_i'\}_{i \in G}, V')$ *be monotone neighborhood models. A non-empty relation* $Z \subseteq W \times W'$ *is called a bisimulation between* $\mathfrak{M}$ *and* $\mathfrak{M}'$ *(notation:* $Z : \mathfrak{M} \rightleftharpoons \mathfrak{M}'$*), if for all* $i \in G$ *and* $(w, w') \in Z$, *the following conditions hold:*

- (Atomic) $w \in V(p)$ *iff* $w' \in V'(p)$ *for each propositional letter* p;
- (Forth) $Z[X] = \{ y \in W' \mid \exists x \in X. xZy \} \in \tau_i'(w)$ *for all* $X \in \tau_i(w)$;
- (Back) $Z^{-1}[X'] = \{ x \in W \mid \exists y \in X'. xZy \} \in \tau_i(w)$ *for all* $X' \in \tau_i'(w)$.

By $\mathfrak{M}, w \rightleftharpoons \mathfrak{M}', w'$ we mean that there is a bisimulation $Z : \mathfrak{M} \rightleftharpoons \mathfrak{M}'$ with wZw'.

Remark 1. In the above definition, the forth condition holds iff for all $X \in \tau_i(w)$, there exists $X' \in \tau_i'(w')$ with $\forall u' \in X' \exists u \in X(uZu')$. The back condition holds iff for all $X' \in \tau_i'(w')$, there exists $X \in \tau_i(w)$ with $\forall u \in X \exists u' \in X'(uZu')$.

We say that two states w and w' in two models $\mathfrak{M}$ and $\mathfrak{M}'$ are $\mathcal{L}_{\mathrm{PAL}}$-equivalent with respect to $*$-semantics, for $* \in \{\cap, \subseteq\}$ (notation: $\mathfrak{M}, w \equiv_{\mathcal{L}_{\mathrm{PAL}}}^* \mathfrak{M}', w'$), if for all $\mathcal{L}_{\mathrm{PAL}}$-formulas φ, $\mathfrak{M}, w \models_* \varphi$ iff $\mathfrak{M}', w' \models_* \varphi$.

Proposition 6. *For any monotone neighborhood models $\mathfrak{M}$ and $\mathfrak{M}'$, If $\mathfrak{M}, w \rightleftharpoons \mathfrak{M}', w$, then $\mathfrak{M}, w \equiv_{\mathcal{L}_{\mathrm{PAL}}}^{\cap} \mathfrak{M}', w'$ and $\mathfrak{M}, w \equiv_{\mathcal{L}_{\mathrm{PAL}}}^{\subseteq} \mathfrak{M}', w'$.*

Proof. By reduction axioms in both semantics, each $\mathcal{L}_{\mathrm{PAL}}$-formula is equivalent to an $\mathcal{L}_{\mathrm{EL}}$-formula and hence invariant under neighborhood bisimulation.   □

As the submodel operation $(\cdot)^{\varphi}$ under Kripke semantics respects bisimulation, here in neighborhood setting, we first obtain the following respecting result.

Proposition 7. *Let $\mathfrak{M}$ and $\mathfrak{M}'$ be monotone neighborhood models. If $\mathfrak{M} \rightleftharpoons \mathfrak{M}'$, then for any $\mathcal{L}_{\mathrm{PAL}}$-formula φ, (i) $\mathfrak{M}^{\cap\varphi} \rightleftharpoons \mathfrak{M}'^{\cap\varphi}$, and (ii) $\mathfrak{M}^{\subseteq\varphi} \rightleftharpoons \mathfrak{M}'^{\subseteq\varphi}$.*

Proof. The item (ii) is similar to (i). For (i), assume $Z : \mathfrak{M} \rightleftharpoons \mathfrak{M}'$. Define $Z^{\cap\varphi} \subseteq [\![\varphi]\!]_{\mathfrak{M}} \times [\![\varphi]\!]_{\mathfrak{M}'}$ by: $Z|_{\varphi} = \{(w, w') \mid wZw' \ \& \ w \in [\![\varphi]\!]_{\mathfrak{M}} \ \& \ w' \in [\![\varphi]\!]_{\mathfrak{M}'}\}$. Then we show $Z|_{\varphi} : \mathfrak{M}^{\cap\varphi} \rightleftharpoons \mathfrak{M}'^{\cap\varphi}$. Only show the forth condition. Assume $wZ|_{\varphi}w'$. Then wZw'. Suppose $X \cap [\![\varphi]\!]_{\mathfrak{M}} \in \tau_i^{\cap\varphi}(w)$ for some $X \in \tau_i(w)$. Then by wZw' and $X \in \tau_i(w)$, we have $Z^{-1}[X] \in \tau_i'(w)$. Now consider the set $Z^{-1}[X] \cap [\![\varphi]\!]_{\mathfrak{M}'}$. For each $v' \in Z^{-1}[X] \cap [\![\varphi]\!]_{\mathfrak{M}'}$, there exists $v \in X$ such that vZv'. Since $\mathcal{L}_{\mathrm{PAL}}$-formulas are invariant under bisimulation, $v \in X \cap [\![\varphi]\!]_{\mathfrak{M}}$, and $vZ|_{\varphi}v'$.   □

We show above that announcement operations $(\cdot)^{\cap\varphi}$ and $(\cdot)^{\subseteq\varphi}$ respect bisimulation. Then, a natural question is whether they respect themselves.

Proposition 8. *For any $\mathcal{L}_{\mathrm{PAL}}$-formula φ, $\models_* [\varphi]\varphi$ iff $(\cdot)^{*\varphi}$ respects itself, i.e., $(\mathfrak{M}^{*\varphi})^{*\varphi} = \mathfrak{M}^{*\varphi}$ for all monotone neighborhood models $\mathfrak{M}$ and $* \in \{\cap, \subseteq\}$.*

Definition 7. *An $\mathcal{L}_{\mathrm{PAL}}$-formula φ is $*$-successful announcement, if $\models_* [\varphi]\varphi$, for $* \in \{\cap, \subseteq\}$. Fix a single agent $i \in G$. Let $\mathcal{L}_i$ be the set of all formulas α defined as:*

$$\alpha ::= p \mid \neg p \mid \alpha \wedge \beta \mid \alpha \vee \beta \mid \Box_i \alpha, \qquad (p \in \mathsf{Prop}).$$

Thus we concentrate on the agent i alone in $\mathcal{L}_i$. Now we are going to establish a result on successful formulas of $\mathcal{L}_i$.

Lemma 1. *Let $\mathfrak{M} = (W, \{\tau_i\}_{i \in G}, V)$ be a regular model satisfying the condition (T) of topological space. $\Box_{\tau_i}(X) \cap [\![\alpha]\!]_{\mathfrak{M}} \subseteq [\![\alpha]\!]_{\mathfrak{M}^{*X}}$, for all $\alpha \in \mathcal{L}_i$, $X \subseteq W$.*

Proof. By induction on $\alpha \in \mathcal{L}_i$. We only show the case where α is $\Box_i\beta$ (we need the condition (T) in atomic case). Assume $w \in \Box_{\tau_i}(X) \cap [\![\Box_i\beta]\!]_{\mathfrak{M}}$. Our goal is to show: $w \in [\![\Box_i\beta]\!]_{\mathfrak{M}^{*X}}$, i.e., $[\![\beta]\!]_{\mathfrak{M}^{*X}} \in \tau_i^{*X}(w)$. Since our τ_i is regular, we obtain $w \in \Box_{\tau_i}(X \cap [\![\beta]\!]_{\mathfrak{M}})$ and so $X \cap [\![\beta]\!]_{\mathfrak{M}} \in \tau_i(w)$. By induction hypothesis: $X \cap [\![\beta]\!]_{\mathfrak{M}} \subseteq [\![\beta]\!]_{\mathfrak{M}^{*X}}$, we get

$[\![\beta]\!]_{\mathfrak{M}^{*X}} \in \tau_i(w)$. When $*$ is $\cap$, we proceed as follows. By $X \in \tau_i(w)$, $[\![\beta]\!]_{\mathfrak{M}^{\cap X}} \cap X \in \tau_i(w)$, which implies our goal of $[\![\beta]\!]_{\mathfrak{M}^{\cap X}} \in \tau_i^{\cap X}(w)$. On the other hand, if $*$ is $\subseteq$, we can establish our goal as follows. Since $[\![\beta]\!]_{\mathfrak{M}^{*X}} \subseteq X$, $[\![\beta]\!]_{\mathfrak{M}^{*X}} \in \tau_i(w)$ tells us that $[\![\beta]\!]_{\mathfrak{M}^{\subseteq X}} \in \tau_i^{\subseteq X}(w)$.   □

Theorem 3. *For every $\alpha \in \mathcal{L}_i$, $[\Box_i\alpha]\Box_i\alpha$ is valid in all topological models under both intersection and subset semantics.*

Proof. Given $\alpha \in \mathcal{L}_i$ and a topological $\mathfrak{M} = (W, \{\tau_i\}_{i \in G}, V)$, for any $w \in W$, assume $\mathfrak{M}, w \models_* \Box_i\alpha$. We need to show $\mathfrak{M}^{*\Box_i\alpha}, w \models_* \Box_i\alpha$. By Lemma 1, $\Box_{\tau_i}([\![\Box_i\alpha]\!]_{\mathfrak{M}}) \cap [\![\Box_i\alpha]\!]_{\mathfrak{M}} \subseteq [\![\Box_i\alpha]\!]_{\mathfrak{M}^{*\Box_i\alpha}}$ (we put $X := [\![\Box_i\alpha]\!]_{\mathfrak{M}}$ and $\alpha := \Box_i\alpha$). Since $\mathfrak{M}$ is a topological model, $[\![\Box_i\alpha]\!]_{\mathfrak{M}} \subseteq \Box_{\tau_i}([\![\Box_i\alpha]\!]_{\mathfrak{M}})$. Therefore, $[\![\Box_i\alpha]\!]_{\mathfrak{M}} \subseteq [\![\Box_i\alpha]\!]_{\mathfrak{M}^{*\Box_i\alpha}}$. By assumption of $w \in [\![\Box_i\alpha]\!]_{\mathfrak{M}}$, we conclude $w \in [\![\Box_i\alpha]\!]_{\mathfrak{M}^{*\Box_i\alpha}}$.   □

Remark 2. For Kripke semantics, a stronger result [15, Proposition 4.36] holds on the common knowledge for all agents than Theorem 3. It is a famous open question to characterize successful announcements of **PAL** in Kripke semantics. Recently, the single agent case is solved in [6].

3   Neighborhood Semantics for DEL

This section explores the general dynamic logic **DEL** of product update. As we do for $\mathcal{L}_{\textbf{PAL}}$, we introduce two kinds of updating mechanism: intersection semantics and subset semantics for the syntax $\mathcal{L}_{\textbf{DEL}}$. These enable us to consider private announcements to a particular single agent as we can see in 'Two Envelopes' example in [11] for Kripke semantics.

Let us define an epistemic neighborhood event model as a tuple $\mathfrak{E} = (E, \{\sigma_i\}_{i \in G}, Pre)$ where E is a non-empty finite set of events, each σ_i is a monotone neighborhood mapping $E \rightarrow \wp(\wp(E))$ which represents the agent i's uncertainty relation, and Pre is a function such that it assigns each event e with a formula of $\mathcal{L}_{\textbf{EL}}$.

Given any two event neighborhood models $\mathfrak{E}_1 = (E_1, \{\sigma_i\}_{i \in G}, Pre_1)$ and $\mathfrak{E}_2 = (E_2, \{\rho_i\}_{i \in G}, Pre_2)$, we define the composition event model $\mathfrak{E}_1 \circ \mathfrak{E}_2 = (E_1 \times E_2, (\sigma_i \times \rho_i)_{i \in G}, Pre)$ as follows:

- $(\sigma_i \times \rho_i)(e_1, e_2) = \{ P \subseteq E_1 \times E_2 \mid X \times Y \subseteq P \ \& \ X \in \sigma_i(e_1) \ \& \ Y \in \rho_i(e_2) \}$.
- $Pre(e_1, e_2) := \langle E_1, e_1 \rangle Pre_2(e_2)$.

The map $\sigma_i \times \rho_i$ in composition of event model is easily seen to be monotone.

3.1   Subset and Intersection Semantics for $\mathcal{L}_{\textbf{DEL}}$

Definition 8 (Intersection Product Update). *Let $\mathfrak{M} = (W, \{\tau_i\}_{i \in G}, V)$ be a monotone neighborhood model, and $\mathfrak{E} = (E, \{\sigma_i\}_{i \in G}, Pre)$ an event model. The product model $\mathfrak{M} \otimes \mathfrak{E} = (W \otimes E, \{\tau_i \otimes \sigma_i\}_{i \in G}, U)$ is defined as follows:*

- $W \otimes E = \{(w, e) \mid \mathfrak{M}, w \models Pre(e)\} \subseteq W \times E$.
- $(\tau_i \otimes \sigma_i)(w, e) = \{P \cap (W \otimes E) \mid X \times Y \subseteq P \ \& \ X \in \tau_i(w) \ \& \ Y \in \sigma_i(e)\}$.
- $U(p) = \{(w, e) \mid w \in V(p)\}$ *for each propositional letter p.*

The definition of $(\tau_i \otimes \sigma_i)(w, e)$ can be written into: $(\tau_i \otimes \sigma_i)(w, e) = \{Q \mid \exists X \in \tau_i(w) \ \& \ \exists Y \in \sigma_i(e)(X \times Y) \cap (W \otimes E) \subseteq Q\}$. Moreover, the map $\tau_i \otimes \sigma_i$ in intersection product update is easily shown to be monotone.

The language $\mathcal{L}_{\mathbf{DEL}}$ is defined by adding operators $[\mathfrak{E}, e]$ to $\mathcal{L}_{\mathbf{EL}}$. We define $\langle \mathfrak{E}, e \rangle$ as an abbreviation of $\neg[\mathfrak{E}, e]\neg$. Given any monotone neighborhood model $\mathfrak{M} = (W, \{\tau_i\}_{i \in G}, V)$ and $w \in W$, the **EL**-part is interpreted as usual. For a formula $[\mathfrak{E}, e]\varphi$, we define:

$$\mathfrak{M}, w \models_\cap [\mathfrak{E}, e]\varphi \text{ iff } \mathfrak{M}, w \models_\cap Pre(e) \text{ implies } \mathfrak{M} \otimes \mathfrak{E}, (w, e) \models_\cap \varphi.$$

An $\mathcal{L}_{\mathbf{DEL}}$-formula φ is valid in intersection semantics (notation: $\models_\cap \varphi$), if it is true at every state of every model.

Theorem 4. *The set of all validities of $\mathcal{L}_{\mathbf{DEL}}$-formulas in intersection semantics is axiomatized by a complete epistemic logic over monotone neighborhood models plus* (DRAtom)*,* (DR¬)*,* (DR∨)*,* (DR□$_i$int)*, and* (DR[·]) *of Table 2.*

Proof. The completeness is reduced to completeness of static epistemic logic. As for the soundness, we only show the validity of (DR□$_i$int). Validity of (DR□$_i$int) is shown as follows: It suffices to show the equivalence between $\mathfrak{M} \otimes \mathfrak{E}, (w, e) \models_\cap \Box_i\varphi$ and $\mathfrak{M}, (w, e) \models_\cap \bigvee_{Y \in \sigma_i(e)} \Box_i \bigwedge_{y \in Y} [\mathfrak{E}, y]\varphi$. It proceeds as follows: $\mathfrak{M} \otimes \mathfrak{E}, (w, e) \models_\cap \Box_i\varphi$ iff there exist some $X \in \tau_i(w)$ and some $Y \in \sigma_i(e)$ such that $(X \times Y) \cap W \otimes E \subseteq [\![\varphi]\!]_{\mathfrak{M} \otimes \mathfrak{E}}$ iff there exist some $X \in \tau_i(w)$ and some $Y \in \sigma_i(e)$ such that, for all $x \in X$ and for all $y \in Y$, $(x, y) \in W \otimes E$ implies $(x, y) \in [\![\varphi]\!]_{\mathfrak{M} \otimes \mathfrak{E}}$ iff there exist some $Y \in \sigma_i(e)$ and some $X \in \tau_i(w)$ such that, for all $x \in X$ and for all $y \in Y$, $\mathfrak{M}, (x, y) \models_\cap [\mathfrak{E}, y]\varphi$ iff $\mathfrak{M}, (x, y) \models_\cap \bigvee_{Y \in \sigma_i(e)} \Box_i \bigwedge_{y \in Y} [\mathfrak{E}, y]\varphi$. $\square$

Table 2. List of Axioms for $[\mathfrak{E}, e]$ for Intersection and Subset Semantics

(DRAtom)	$[\mathfrak{E}, e]p \leftrightarrow (Pre(e) \rightarrow p)$
(DR¬)	$[\mathfrak{E}, e]\neg\varphi \leftrightarrow (Pre(e) \rightarrow \neg[\mathfrak{E}, e]\varphi)$
(DR∨)	$[\mathfrak{E}, e](\varphi \vee \psi) \leftrightarrow ([\mathfrak{E}, e]\varphi \vee [\mathfrak{E}, e]\psi)$
(DR□$_i$int)	$[\mathfrak{E}, e]\Box_i\varphi \leftrightarrow (Pre(e) \rightarrow \bigvee_{Y \in \sigma_i(e)} \Box_i \bigwedge_{y \in Y} [\mathfrak{E}, y]\varphi)$
(DR[·])	$[\mathfrak{E}_1, e_1][\mathfrak{E}_2, e_2]\varphi \leftrightarrow [\mathfrak{E}_1 \circ \mathfrak{E}_2, (e_1, e_2)]\varphi$
(DR□$_i$sub)	$[\mathfrak{E}, e]\Box_i\varphi \leftrightarrow (Pre(e) \rightarrow \bigvee_{Y \in \sigma_i(e)} \Box_i \bigwedge_{y \in Y} \langle \mathfrak{E}, y \rangle\varphi)$

Definition 9 (Subset Update Product). *Let $\mathfrak{M} = (W, \{\tau_i\}_{i \in G}, V)$ be a monotone neighborhood model, and $\mathfrak{E} = (E, \{\sigma_i\}_{i \in G}, Pre)$ an event model. The product model $\mathfrak{M} \odot \mathfrak{E} = (W \odot E, \{\tau_i \odot \sigma_i\}_{i \in G}, U)$ is defined as follows:*

- $W \odot E = \{(w, e) \mid \mathfrak{M}, w \models Pre(e)\} \subseteq W \times E$.
- $\tau_i \odot \sigma_i(w, e) = \{P \subseteq W \odot E \mid X \times Y \subseteq P \text{ for some } (X, Y) \in \tau_i(w) \times \sigma_i(e)\}$.
- $U(p) = \{(w, e) \mid w \in V(p)\}$ *for each propositional letter p.*

The subset semantics for $\mathcal{L}_{\mathbf{DEL}}$ is similar to intersection semantics, but we note that

$$\mathfrak{M}, w \models_{\subseteq} \langle \mathfrak{E}, e \rangle \varphi \text{ iff } \mathfrak{M}, w \models_{\subseteq} Pre(e) \text{ and } \mathfrak{M} \odot \mathfrak{E}, (w, e) \models_{\subseteq} \varphi.$$

Theorem 5. *The set of all validities of $\mathcal{L}_{\mathbf{DEL}}$-formulas in subset semantics is axiomatized by a complete epistemic logic over monotone neighborhood models plus*(DRAtom), (DR¬), (DR∨), (DR□$_i$sub), *and* (DR[·]) *of Table 2.*

Proof. The proof is almost the same as that for theorem 4. Here, we only show the validity of (DR□$_i$sub): We can show that $\mathfrak{M} \odot \mathfrak{E}, (w, e) \models_{\subseteq} \Box_i \varphi$ iff $\mathfrak{M}, w \models_{\subseteq} \bigvee_{Y \in \sigma_i(e)} \Box_i \bigwedge_{y \in Y} \langle \mathfrak{E}, y \rangle \varphi$ as follows: $\mathfrak{M} \odot \mathfrak{E}, (w, e) \models_{\subseteq} \Box_i \varphi$ iff there exist some $X \in \tau_i(w)$ and some $Y \in \sigma_i(e)$ such that $X \times Y \subseteq W \odot E$ and $X \times Y \subseteq [\![\varphi]\!]_{\mathfrak{M} \odot \mathfrak{E}}$ iff there exist some $X \in \tau_i(w)$ and some $Y \in \sigma_i(e)$ such that $X \times Y \subseteq (W \odot E) \cap [\![\varphi]\!]_{\mathfrak{M} \odot \mathfrak{E}}$ iff there exist some $X \in \tau_i(w)$ and some $Y \in \sigma_i(e)$ such that, for all $x \in X$ and for all $y \in Y$, $(x, y) \in (W \odot E) \cap [\![\varphi]\!]_{\mathfrak{M} \odot \mathfrak{E}}$ iff there exist some $Y \in \sigma_i(e)$ and some $X \in \tau_i(w)$ such that, for all $x \in X$ and for all $y \in Y$, $\mathfrak{M}, (x, y) \models_{\subseteq} \langle \mathfrak{E}, y \rangle \varphi$ iff $\mathfrak{M}, (x, y) \models_{\subseteq} \bigvee_{Y \in \sigma_i(e)} \Box_i \bigwedge_{y \in Y} \langle \mathfrak{E}, y \rangle \varphi$. □

Finally, we show that intersection and subset updates by φ can be regarded as special examples of intersection and subset semantics for $\mathcal{L}_{\mathbf{DEL}}$. Let us say that $\mathfrak{M} = (W, \{\tau_i\}_{i \in G}, V)$ and $\mathfrak{M}' = (W', \{\tau_i'\}_{i \in G}, V')$ are *isomorphic* (notation: $\mathfrak{M} \cong \mathfrak{M}'$) if there is a mapping $f : W \to W'$ such that f is a bounded morphism (recall Definition 1), f is one-to-one and onto and $V(p) = f^{-1}[V'(p)]$ for all $p \in \mathsf{Prop}$.

Definition 10. *Fix any φ of $\mathcal{L}_{\mathbf{EL}}$. Define $\mathfrak{E}_\varphi = (E, (\sigma_i)_{i \in G}, pre)$, where $E = \{*\}$, and $\sigma_i(*) = \{\{*\}\}$ for all $i \in G$, and $pre(*) = \varphi$.*

Proposition 9. *Given any monotone neighborhood model $\mathfrak{M}$ and a formula φ of $\mathcal{L}_{\mathbf{EL}}$, $\mathfrak{M}^{\cap\varphi} \cong \mathfrak{M} \otimes \mathfrak{E}_\varphi$ and $\mathfrak{M}^{\subseteq\varphi} \cong \mathfrak{M} \odot \mathfrak{E}_\varphi$.*

Proof. Clearly, $W \odot E = W \otimes E = [\![\varphi]\!]_{\mathfrak{M}} \times \{*\}$. Then, the mapping sending $w \in [\![\varphi]\!]_{\mathfrak{M}}$ to $(w, *) \in [\![\varphi]\!]_{\mathfrak{M}} \times \{*\}$, is a witness for $\mathfrak{M}^{\cap\varphi} \cong \mathfrak{M} \otimes \mathfrak{E}_\varphi$ and $\mathfrak{M}^{\subseteq\varphi} \cong \mathfrak{M} \odot \mathfrak{E}_\varphi$. □

3.2   Some Model Theory for Neighborhood DEL

As we have done for **PAL**, we also give some model-theoretic results for $\mathcal{L}_{\mathbf{DEL}}$ in neighborhood semantics. We consider only neighborhood bisimulation and product update. The question here is whether the operation $(\cdot) \circledast \mathfrak{E}$ (where $\circledast \in \{\otimes, \odot\}$) for any given event model $\mathfrak{E}$ over the class of all monotone neighborhood epistemic models respects bisimulation. The answer is yes.

Proposition 10. *For any monotone neighborhood models $\mathfrak{M} = (W, \{\tau_i\}_{i \in G}, V)$ and $\mathfrak{M}' = (W', \{\tau_i'\}_{i \in G}, V')$, given any event model $\mathfrak{E} = (E, \{\sigma_i\}_{i \in G}, Pre)$, if $\mathfrak{M} \rightleftharpoons \mathfrak{M}'$, then $\mathfrak{M} \circledast \mathfrak{E} \rightleftharpoons \mathfrak{M}' \circledast \mathfrak{E}$.*

Proof. Assume that $Z : \mathfrak{M} \rightleftharpoons \mathfrak{M}'$. We show only that the following relation $Z_{\otimes \mathfrak{E}} = \{((w, e), (w', e)) \mid wZw' \ \& \ \mathfrak{M}, w \models_{\cap} Pre(e) \ \& \ \mathfrak{M}', w' \models_{\cap} Pre(e)\}$ is a bisimulation between intersection updated models. For the forth condition, assume that $X \in \tau_i \otimes \sigma_i(w, e)$ and $(w, e)Z_{\otimes \mathfrak{E}}(w', e)$. Then wZw' and there exist $A \in \tau_i(w)$ and $B \in \sigma_i(e)$ such that $(A \times B) \cap (W \otimes E) \subseteq X$. By wZw' and $A \in \tau_i(w)$, we have $Z[A] \in \tau_i'(w')$. Then $Z_{\otimes \mathfrak{E}}[X] \supseteq Z_{\otimes \mathfrak{E}}[(A \times B) \cap (W \otimes E)] = (Z[A] \times B) \cap (W' \otimes E)$. By $Z[A] \in \tau_i'(w')$ and $B \in \sigma_i(e)$, we obtain $Z_{\otimes \mathfrak{E}}[X] \in \tau_i' \otimes \sigma_i(w', e)$. The back condition is similar. □

We say that two pointed neighborhood models $(\mathfrak{M}, w)$ and $(\mathfrak{N}, v)$ are $\mathcal{L}_{\mathbf{DEL}}$-equivalent with respect to $*$-semantics for $* \in \{\cap, \subseteq\}$ (notation: $\mathfrak{M}, w \equiv^{*}_{\mathcal{L}_{\mathbf{DEL}}} \mathfrak{N}, v$), if for all $\mathcal{L}_{\mathbf{DEL}}$-formulas φ, $\mathfrak{M}, w \models_{*} \varphi$ iff $\mathfrak{N}, v \models_{*} \varphi$. Then we have the following.

Proposition 11. *If* $Z : \mathfrak{M} \rightleftharpoons \mathfrak{N}$ *and* wZv, *then* $\mathfrak{M}, w \equiv^{*}_{\mathcal{L}_{\mathbf{DEL}}} \mathfrak{N}, v$.

Proof. By reduction axioms in both semantics, every $\mathcal{L}_{\mathbf{DEL}}$-formula is equivalent to an $\mathcal{L}_{\mathbf{EL}}$-formula and hence invariant under neighborhood bisimulation. $\qquad\square$

4   Various Notions of Group Knowledge

Here we consider extensions of $\mathcal{L}_{\mathbf{PAL}}$ with various notions of group knowledge over neighborhood models. The following right diagram is useful to explain our notions of group knowledge (cf. [13]). Given two maps τ_1 and τ_2, there are three basic ways for combining them: intersection,  composition and sum. For interpreting notions of knowledge, they correspond to three different notions: general knowledge, common knowledge and distributed knowledge, respectively.

General Knowledge: The general knowledge operator $E_G\varphi$ is interpreted in neighborhood models as follows: for model $\mathfrak{M} = (W, \{\tau_i\}_{i \in G}, V)$, for $* \in \{\cap, \subseteq\}$,

$$\mathfrak{M}, w \models_{*} E_G\varphi \text{ iff } \mathfrak{M}, w \models_{*} \square_i\varphi \text{ for all } i \in G.$$

From this semantic clause, it is easy to see that φ is a general knowledge of some given group G at the current state w, iff every agent in the group knows it, i.e., the truth set $[\![\varphi]\!]_{\mathfrak{M}} \in \bigcap_{i \in G} \tau_i(w)$. Moreover, the following reduction axioms for the operator of general knowledge are easy to check:

(i) $\models_{\cap} [\varphi]E_G\psi \leftrightarrow (\varphi \to E_G(\varphi \to [\varphi]\psi))$, $\models_{\subseteq} [\varphi]E_G\psi \leftrightarrow (\varphi \to E_G(\varphi \wedge [\varphi]\psi))$.

(ii) $\models_{*} [\mathfrak{E}, e]E_G\varphi \leftrightarrow (Pre(e) \to \bigwedge_{i \in G} \bigvee_{Y_i \in \sigma_i(e)} \square_i \bigwedge_{e_i \in Y_i} [\mathfrak{E}, e_i]\varphi)$.

Common Knowledge: Common knowledge is a more complex notion ([11]). Here we provide an interpretation of common knowledge over *regular* neighborhood models.

We use the notion of composition given in [3] to provide the neighborhood intersection semantics with common knowledge. Let τ_1 and τ_2 be neighborhood maps on a non-empty set W. We define the composition of them as follows: for all $w \in W$,

$$\tau_1 \circ \tau_2(w) = \{X \mid \exists Y \in \tau_1(w) \forall v \in Y.X \in \tau_2(v)\}.$$

Then, we define truth condition of the operator of common knowledge as follows:

$$\mathfrak{M}, w \models C_G\varphi \text{ iff for all } i_1 \ldots i_n \in G \text{ with } n \geq 1, [\![\varphi]\!]_{\mathfrak{M}} \in \tau_{i_1} \circ \cdots \circ \tau_{i_n}(w).$$

It is easy to show that $\mathfrak{M}, w \models C_G\varphi$ iff $\mathfrak{M}, w \models_{\cap} \square_{i_1} \cdots \square_{i_n}\varphi$ for all $i_1, \ldots, i_n \in G$ and all $n \in \omega$, over the class of all regular neighborhood models. It is also easy to check that common knowledge is a greatest fixed point in the following sense: (i) $\models C_G\varphi \leftrightarrow E_G C_G\varphi$, and (ii) $\models C_G(\varphi \to E_G\varphi) \to C_G\varphi$.

Now for dynamic operators over common knowledge, it seems hard to give reduction axioms. For instance, the announcement formula $[\alpha]C_G\varphi$ has not been given a reduction axiom yet. However, van Benthem, et.al. gave the notion of conditional common knowledge $C_G^\varphi\psi$ and show a reduction axiom for $[\alpha]C_G^\varphi\psi$ under Kripke semantics ([14]). Now, in the neighborhood setting, we can provide the following neighborhood semantics with the conditional common knowledge: for $* \in \{\cap, \subseteq\}$,

$$\mathfrak{M}, w \models_* C_G^\varphi\psi \text{ iff for all } i_1 \ldots i_n \in G \text{ with } n \geq 1, [\![\psi]\!]_{\mathfrak{M}} \in \tau_{i_n}^{*\varphi} \circ \ldots \circ \tau_{i_1}^{*\varphi}(w).$$

Then, we can easily obtain the reduction axioms, e.g., $\models_\cap [\alpha]C_G^\varphi\psi \leftrightarrow (\alpha \to C_G^{\alpha \wedge [\alpha]\varphi}[\alpha]\psi)$.

Remark 3. Van Benthem and Sarenac [13] studied the logic of common knowledge over product of epistemic topological models. Our approach is different from this topological approach to knowledge in the following two respects. First, we employ the composition of neighborhood maps to treat the interaction between agents in a group directly and define the notion of common knowledge. Second, we could also define the notion of common knowledge in a *weaker* semantic setting than topological spaces, though we need to assume that neighborhood frames are regular. For the common knowledge and/or the belief operator possibly over non-regular neighborhood frames, the reader is refered to [7], [8] and [5].

Distributed Knowledge: Finally, we make some observations on the sum of neighborhood maps and distributed knowledge.

Definition 11. *Let* $\mathfrak{F} = (W, \{\tau_i\}_{i\in G})$ *be a monotone neighborhood frame. Define the sum of* $\{\tau_i\}_{i\in G}$ *by: for every* $w \in W$,

$$\Sigma_{i\in G}\tau_i(w) = \{ \bigcap_{i\in G}X_i \mid (X_i)_{i\in G} \in \prod_{i\in G}\tau_i(w) \}$$

where $\prod_{i\in G}\tau_i(w)$ *is the Cartesian product of* $\{\tau_i(w) \mid i \in G\}$.

Note that $\Sigma_{i\in G}\tau_i$ is monotone: suppose that $\bigcap_{i\in G}X_i \subseteq Z$ and $\bigcap_{i\in G}X_i \in \Sigma_{i\in G}\tau_i(w)$. By assumption, we obtain $X_i \in \tau_i(w)$ for each $i \in G$. Since τ_i is monotone, $X_i \cup Z \in \tau_i(w)$. Then, $Z = (\bigcap_{i\in G} X_i) \cup Z = \bigcap_{i\in G}(X_i \cup Z) \in \Sigma_{i\in G}\tau_i(w)$, as desired.

Given a model $\mathfrak{M} = (W, \{\tau_i\}_{i\in G}, V)$ and a state w, we define

$$\mathfrak{M}, w \models_* D_G\psi \text{ iff } [\![\psi]\!]_{\mathfrak{M}} \in \Sigma_{i\in G} \tau_i(w).$$

Then, we can show the following (the converse of (iii) does not hold):

(i) $\models_\cap [\varphi]D_G\psi \leftrightarrow (\varphi \to D_G(\varphi \to [\varphi]\psi))$, $\models_\subseteq [\varphi]D_G\psi \to (\varphi \to D_G(\varphi \wedge [\varphi]\psi))$.

(ii) $\models_\cap [\mathfrak{E}, e]D_G\varphi \leftrightarrow (Pre(e) \to \bigvee_{(Y_i)_{i\in G}\in\prod_{i\in G}\sigma_i(e)} D_G(\bigwedge_{y\in\bigcap_{i\in G} Y_i}[\mathfrak{E}, y]\varphi))$.

(iii) $\models_\subseteq [\mathfrak{E}, e]D_G\varphi \to (Pre(e) \to \bigvee_{(Y_i)_{i\in G}\in\prod_{i\in G}\sigma_i(e)} D_G(\bigwedge_{y\in\bigcap_{i\in G} Y_i}\langle\mathfrak{E}, y\rangle\varphi))$.

5 Conclusion and Open Questions

We have introduced two ways of updating monotone neighborhood models: taking the intersection submodel and subset submodel, which correspond to two different semantics for both the syntax of **PAL** and **DEL**. We emphasize that these two ways of updating neighborhood models are contributions of this paper. The most important point is that the subset semantics is different from the intersection semantics. The intersection semantics can be regarded as a natural generalization of public announcement logic over Kripke semantics, while the subset semantics gives a totally new perspective on dynamic operators. Finally, let us list some open questions here.

(i) Characterize the fragment of successful announcements under two neighborhood semantics we provided in this paper.
(ii) Characterizing the 'substitution core' of neighborhood **PAL**.
(iii) Extend Theorem 3 on successful formulas to multi-agent case.
(iv) Give an axiomatization of epistemic logic with distributed knowledge in monotone neighborhood semantics.
(v) Show the axiomatization of epistemic logic with common knowledge in regular neighborhood semantics.

References

1. Balbiani, P., van Ditmarsch, H., Kudinov, A.: Subset space logic with arbitrary announcements. In: Lodaya, K. (ed.) ICLA 2013. LNCS (LNAI), vol. 7750, pp. 233–244. Springer, Heidelberg (2013)
2. Baltag, A., Moss, L., Solecki, S.: The logic of public announcements, common knowledge and private suspicions. In: Proceedings of TARK, pp. 43–56. Morgan Kaufmann Publishers, Los Altos (1989)
3. Gasquet, O.: Completeness results in neighborhood semantics for multi-modal monotonic and regualr logics. The Journal of IGPL 4(3), 417–426 (1996)
4. Hansen, H.H.: Monotonic modal logic. Master's thesis. ILLC, University of Amsterdam (2003)
5. Heifetz, A.: Iterative and fixed point common belief. Journal of Philosophical Logic 28, 61–79 (1999)
6. Holliday, W., Icard III, T.: Moorean phenomena in epistemic logic. In: Advances in Modal Logic 2010, pp. 178–199. College Publications (2010)
7. Lismont, L., Mongin, P.: On the logic of common belief and common knowledge. Theory and Decision 37, 75–106 (1994a)
8. Lismont, L., Mongin, P.: Strong completeness theorems for weak logics of common belief. Journal of Philosophical Logic 32, 115–137 (2003)
9. Osherson, de Jongh, D., Martin, D.E., Weinstein, S.: Formal learning theory. In: Van Benthem, J., ter Meulen, A. (eds.) Handbook of Logic and Language, pp. 737–776. Elsevier, Amsterdam (1997)
10. Plaza, J.: Logics of public communication. In: Proceedings of the 4th International Symposium on Methodologies for Intelligent Systems, pp. 201–206 (1989)
11. Van Benthem, J.: Logical Dynamics of Information and Interaction. Cambridge University Press, Cambridge (2011)
12. Van Benthem, J., Pacuit, E.: Dynamic logic of evidence-based belief. Studia Logica 99, 61–92 (2011)
13. Van Benthem, J., Sarenac, D.: The geometry of knowledge. In: Béziau, J.-Y., Leite, A., Facchini, A. (eds.) Aspects of Universal Logic, pp. 1–30. Centre de Recherches Sémiologiques, Université de Neuchatel (2004)
14. Van Benthem, J., Van Eijck, J., Kooi, B.: Logics of communication and change. Information and Computation 204, 1620–1662 (2006)
15. Van Ditmarsch, H., Van der Hoek, W., Kooi, B.: Dynamic Epistemic Logic. Springer (2008)

The Logic of *a Priori* and *a Posteriori* Rationality in Strategic Games

Meiyun Guo[1] and Jeremy Seligman[2]

[1] South West University, Chongqing, China
guomy007@swu.edu.cn
[2] South West University, Chongqing, China, The University of Auckland,
Auckland, New Zealand
j.seligman@auckland.ac.nz

Abstract. We propose a logic for describing the interaction between knowledge, preference, and the freedom to act, and their interactions with the norms of *a Priori* and *a Posteriori* rationality, which we have argued for in previous work [3]. We then apply it to strategic games to characterise weak dominance and Nash equilibrium.

In [3] we proposed a model for rational decision making in which the facts about knowledge, preference and freedom to act are clearly separated from the norms of reasoning. Even the transitivity of the preference relation is considered normative, in our approach. The factual basis for decision making is modelled using what we call 'decision frames' and their multi-agent extensions, 'social decision frames'. We proposed two norms for decision-making, called '*a Priori* rationality' and '*a Posteriori* rationality', which apply to reasoning before making the decision, and after. Before making a decision, one is concerned with making the best, or at least an optimal, decision in ignorance of the effect of contextual factors, especially, in the social setting, the actions of other agents. After making a decision, one is more interested in which the decision was optimal given the conditions that actually applied. We went on to show that these two general norms specialise to the familiar norms of game theory: avoiding (weakly) dominated strategies (*a Priori*) and wanting to have made a best response (*a Posteriori*). The level of abstraction allowed us to provide a uniform account of both pure-strategy and mixed-strategy games.

Here, we propose a language for describing and reasoning with these norms, and show how it formalises and so justifies some of the processes of reasoning that we use to make decisions.

In Section 1 we introduce decision frames and the corresponding concepts of *a Priori* and *a Posteriori* rationality. In Section 2 we propose a language for describing these, and show how it can be embedded in a more powerful language in which an axiomatisation can be given. We go on in Section 3 to show how this is applied to strategic game theory.

D. Grossi, O. Roy, and H. Huang (Eds.): LORI 2013, LNCS 8196, pp. 218–227, 2013.
© Springer-Verlag Berlin Heidelberg 2013

1   The Facts and Norms of Decision-Making

The perspective of our analysis is that of your evaluating a decision that you have just made, to determine whether or not it was a good one. To model this, we propose in [3] the following structures:

Definition 1. *A decision frame* $F = \langle W, \sim, \approx, \leqslant \rangle$ *consists of a non-empty set W of* possible decision situations, *with binary relations* $\sim$, $\approx$ *and* $\leqslant$ *on W, where*

$\sim$ *is an equivalence relation*	$u \sim v$ *represents that v would have been possible in u had you acted differently, given the contingencies of u that are beyond your control (freedom).*
$\approx$ *is an equivalence relation*	$u \approx v$ *represents that in situation u you would not know that you weren't in situation v (epistemic indistinguishability).*
$\leqslant$ *is a relation*	$u \leqslant v$ *represents that you regard situation v as at least as good as situation u.*

Importantly, the structures only represent the facts related to your decision, not the norms. In particular, the relation $\leqslant$ is not required to have any special properties (such as transitivity). Nonetheless, certain norms are definable on the basis of these facts. In [3] we argue that the fundamental norm of decision-making is that you should avoid situations in which you know that a strictly better alternative was possible had you chosen differently. This is ambiguous between two readings of the counterfactual. On the first, *a Priori* reading, you consider only what was known to you at the time of making the decision. On the second, *a Posteriori* reading, you also consider what is known to you after making the decision, specifically those contingent factors such as the actual actions of other agents and the actual circumstances relevant to your decision that you could not have known in advance.

The first of these norms, is formalised using a generalisation of the $\leqslant$ relation, called '*a Priori* free preference' that factors in the contribution of knowledge and freedom.

Definition 2. *The relation $\leqslant_F$ of* a Priori *free preference is defined by*

$$u \leqslant_F v \text{ iff } u' \leqslant v' \text{ for all } u' \approx u \text{ and all } v' \approx v \text{ such that } u' \sim v'.$$

Decision situation u is a Priori *rational iff there is no $v \sim u$ such that $v >_F u$.*

Various specific cases of the *a Priori* free preference relation are worthy of mention. Firstly, assuming that your freedom to choose is unlimited ($\sim$ is the universal relation) and your knowledge unbounded ($\approx$ is the identity relation),

$\leq_F = \leq$, so the free preference relation is a generalisation of the ordinary preference relation. Keeping knowledge unbounded, but allowing for limitations on your freedom to choose, we get that $\leq_F = \sim \cap \leq$. In other words, you only compares your current situation with one that you may have been in had you chosen differently. This is a special case of a more familiar *ceteris paribus* restriction on preference comparisons, which requires the compared situations to be equivalent *ceteris paribus*. In the present setting, two situations are equivalent *ceteris paribus* iff they are free alternatives.[1] When your freedom to choose is unlimited ($\sim$ is universal) but there are some limitations to your knowledge, $u \leq_F v$ iff $u' \leq v'$ for all $u' \approx u$ and all $v' \approx v$. In other words, since you do not know that you are in situation u, only that you are in one of the situations in $[u]$ (the $\approx$-equivalence class of u) and, likewise, were you in situation v, you would know only that you were in one of the situations in $[v]$, to judge that v is (as far as you know) at least as good as u, there should be no $u' \in [u]$ and $v' \in [v]$ for which $u' \nleq v'$. Further discussion of the justification of these various preference relations, including our assuming neither reflexivity nor transitivity of $\leq$ are contained in [3] (p.186). *A Priori* rationality is closely related to the norm of avoiding weakly dominated strategies in game theory, to be discussed in Section 3.

The second norm of *a Posteriori* rationality is formalised using the relation of 'a *Posteriori* free preference' which is the restriction of *a Priori* free preference to alternatives that were in fact possible had you chosen differently, even if you didn't know this at the time.

Definition 3. *The relation $\leq_{F'}$ of* a Posteriori *free preference is defined by*

$$u \leq_{F'} v \text{ iff } u' \leq v' \text{ for all } u' \approx u \text{ and all } v' \approx v \text{ such that } u \sim u' \sim v' \sim v.$$

Decision situation u is a Posteriori *rational iff there is no $v \sim u$ such that $v >_{F'} u$.*

A posteriori rationality is closely related to the norm of achieving a best response in game theory, to be discussed in Section 3.

2 A Logic of Rational Decisions

To describe decision frames and the corresponding norms we will use the following hybrid modal language.

---

[1] Dealing with the *ceteris paribus* aspect of preference comparisons is a matter of degree. Assume that $\leq$ already involves all those *ceteris paribus* considerations that are not concerned with freedom of choice. So, for example, if v is a situation that is the same as your current situation u in all such relevant respects except that you have one million dollars (more?) in your bank account, You may judge $u \leq v$ (and probably that $u < v$) but, unhappily, you are unlikely to be free to choose between u and v and so $u \nleq_F v$. We could have started with a basic preference order $\lesssim$, modelled *ceteris paribus* equivalence (not including considerations of achievability) by an equivalence relation CP and then defined $\leq = CP \cap \lesssim$. Adding freedom as a *ceteris paribus* condition, we would still have $\leq_F = \sim \cap CP \cap \lesssim = \sim \cap \leq$.

Definition 4. *Given disjoint countably infinite sets* PROP *of propositional vari-ables and* NOM *of* nominals, *the language L consists of the following* formulas

$$\varphi ::= p \mid i \mid R \mid R' \mid \neg\varphi \mid (\varphi \wedge \varphi) \mid G\varphi \mid K\varphi \mid C\varphi \mid @_i\varphi$$

for $p \in$ PROP, $i \in$ NOM.

We interpret $G\varphi$ to mean that φ holds in all situations that would have been at least as good for you as the present situation. They may, of course, be no longer possible, as a result of your decision. $K\varphi$ means, as usual, that you know that φ in the present situation, or, more precisely, that φ holds in all situations that you could be in, given your knowledge. We do not, of course, assume that you know precisely which situation you are in. $C\varphi$ means that φ holds in all situations in which you could have been, had you acted differently. The sense of 'could have' here takes into account all those factors that are beyond your control, including the actual actions of other agents and other contingent factors. Finally, R and R' mean that present situation is *a Priori* or *a Posteriori* rational, respectively.

The semantics of L is the standard semantics for hybrid logic, taking G, K and C to be the normal modal operators for the relations $\leqslant$, $\approx$ and $\sim$. R and R' are zero-ary operators that hold in the *a Priori* and *a Posteriori*rational situations, respectively. That R and R' cannot be given an explicit definition in terms of the other operators is easy to shown by a bisimulation argument. This makes the derivation of logical principles relating them somewhat difficult and to solve this problem we will embed the language L in the following, much more powerful language.

Definition 5. *The language of* CPDL[2] *over a sets* PROP *of propositional vari-ables,* NOM *of nominals and* ATPROG *of atomic programs consists of the sets* FORM *of formulas and* PROG *of programs given by*

$$\varphi \in \text{FORM} ::= p \mid i \mid [\pi]\varphi \mid \neg\varphi \mid (\varphi \wedge \varphi)$$
$$\pi \in \text{PROG} ::= a \mid \varphi? \mid \overline{\pi} \mid \pi^\circ \mid \pi^* \mid (\pi;\pi) \mid (\varphi \cup \varphi)$$

for $i \in$ NOM, $p \in$ PROP *and* $\alpha \in$ ATPROG.
Abbreviations: $\top = p \vee \neg p$, $U = \alpha \cup \overline{\alpha}$ *(universal),* $I = \top?$ *(identity),* $\langle \pi \rangle = \neg[\pi]\neg$, $(\pi \cap \rho) = (\overline{\pi} \cup \overline{\rho})$, $(\pi \subset \rho) = \overline{\pi \cap \overline{\rho}}$ *plus the usual Booleans:* $\bot$, $\vee$, $\rightarrow$, $\leftrightarrow$. *Also, where no confusion can arise, especially in the case of atomic programs, we further abbreviate* $(\pi \cap \rho)$ *as* $\pi\rho$. *A formula* φ *is* pure *iff it contains no propositional variables.*

Definition 6. *A structure* $F = \langle W, R \rangle$ *is a* CPDL *frame if* $R(\alpha) \subseteq W^2$ *for each* $\alpha \in$ ATPROG. *A structure* $M = \langle W, R, V \rangle$ *is a* CPDL *model if* $\langle W, R \rangle$ *is a* CPDL *frame,* $V(i) \in W$ *for each* $i \in$ NOM, *and* $V(p) \subseteq W$ *for each* $p \in$ PROP.

---

[2] What we are calling CPDL (Combinatory PDL)is also known as 'full-CPDL'. The lan-guage of CPDL without $^-$ and $^\circ$ is also known as 'hybrid PDL' [1].

Definition 7. *Given a* CPDL *model* $M = \langle W, R, V \rangle$, *and a state* $u \in W$, *we define* $[\![\varphi]\!]^M \subseteq W$ *for each* $\varphi \in$ FORM *and* $[\![\pi]\!]^M \subseteq W^2$ *for each* $\pi \in$ PROG *as follows:*

$$
\begin{aligned}
[\![i]\!]^M &= \{V(i)\} \\
[\![p]\!]^M &= V(p) \\
[\![[\pi]\varphi]\!]^M &= \{u \in W \mid v \in [\![\varphi]\!]^M \text{ for each } v \in W \text{ such that } \langle u,v \rangle \in [\![\pi]\!]^M\} \\
[\![\neg\varphi]\!]^M &= W \backslash [\![\varphi]\!]^M \\
[\![(\varphi \wedge \psi)]\!]^M &= [\![\varphi]\!]^M \cap [\![\psi]\!]^M \\
[\![\alpha]\!]^M &= R(\alpha) \\
[\![\varphi?]\!]^M &= \{\langle u,u \rangle \mid u \in [\![\varphi]\!]^M\} \\
[\![\overline{\pi}]\!]^M &= \{\langle u,v \rangle \mid \langle u,v \rangle \notin [\![\pi]\!]^M\} \\
[\![\pi^\circ]\!]^M &= \{\langle u,v \rangle \mid \langle v,u \rangle \in [\![\pi]\!]^M\} \\
[\![\pi^*]\!]^M &= \text{the smallest transitive, reflexive relation containing } [\![\pi]\!]^M \\
[\![\pi;\rho]\!]^M &= \{\langle u,v \rangle \mid \langle v,w \rangle \in [\![\pi]\!]^M \text{ and } \langle w,v \rangle \in [\![\rho]\!]^M \text{ for some } w \in W\} \\
[\![(\pi \cup \rho)]\!]^M &= [\![\pi]\!]^M \cup [\![\rho]\!]^M
\end{aligned}
$$

When M is clear from the context, we write $[\![\varphi]\!]^M$ as $[\![\varphi]\!]$. Note that $[\![(\pi \cap \rho)]\!]^M = [\![\pi]\!]^M \cap [\![\rho]\!]^M$ and $[\![(\pi \subset \rho)]\!]^M = W$ if $[\![\pi]\!]^M \subseteq [\![\rho]\!]^M$, otherwise $\varnothing$.

As usual, a formula is valid on a model if $M, u \models \varphi$ for all u, valid on a frame F if it is valid on all models $\langle F, V \rangle$ and simply valid if it is valid on all frames.

Theorem 1. *[1] There is an axiomatisation* K *of* CPDL *which is sound and such that for every extension* KΓ *of* K *with pure formulas* Γ *as axioms, if a formula is consistent in* KΓ *then it has a countable model on a frame in which all the formulas in* Γ *are valid. The system* KΓ *is therefore also complete for that class of frames.*

Comment on Axiomatisation and Complexity. Although the validity problem for CPDL is known to be highly undecidable [8], it has a number of well-known decidable fragments, include PDL itself and its extension to allow $^\circ$ and either $\cap$ or $\overline{a}$, i.e., $^-$ restricted to atomic programs, but not both [2] [5]. Hybrid PDL namely PDL with nominals is also decidable [7] but even non-hybrid PDL with unrestricted $^-$ is not. Frame consequence is undecidable even for PDL [8] with premises restricted to pure formulas, and so the decidability of validity for these fragments of CPDL cannot be automatically extended to specific classes of frames defined by pure formulas, despite the existence of a complete axiomatisation.

In order to describe a social decision frame as a frame $F = \langle W, R \rangle$ we take ATPROG $= \{g, k, c\}$ with $R(g) = \leqslant$, $R(k) = \approx$ and $R(c) = \sim$. Then the relations of *a Priori* and *a Posteriori* free preference can be defined as

$$
\begin{aligned}
\mathsf{F} &= \overline{\mathsf{k}; c\overline{\mathsf{g}}; \mathsf{k}} \quad (a\ Priori\ \text{free preference}) \\
\mathsf{F}' &= \overline{\mathsf{ck}; c\overline{\mathsf{g}}; \mathsf{ck}} \quad (a\ Posteriori\ \text{free preference})
\end{aligned}
$$

This enables us to embed L in CPDL as follows:

$$G := [\mathsf{g}] \qquad R := [\mathsf{cF}\overline{\mathsf{F}^\circ}]\bot$$
$$K := [\mathsf{k}] \qquad R' := [\mathsf{cF}'\overline{\mathsf{F}'^\circ}]\bot$$
$$C := [\mathsf{c}]$$

Theorem 2. *A frame is a decision frame iff the following pure formulas* D *are valid on* F:

$$\neg@_i K\neg i \qquad\qquad \neg@_i C\neg i \qquad\qquad (reflexivity\ of \approx and \sim)$$
$$\neg(\neg KK\neg i \wedge K\neg i) \quad \neg(\neg CC\neg i \wedge C\neg i) \quad (transitivity\ of \approx and \sim)$$
$$@_i K\neg j \rightarrow @j K\neg i \quad @_i C\neg j \rightarrow @j C\neg i \quad (symmetry\ of \approx and \sim)$$

Proof. It is enough to check that R and R' are satisfied by precisely the *a Priori* and *a Posteriori* rational situations.

Corollary 1. *The system* $\mathsf{K}D$ *is a complete axiomatisation of the formulas valid in decision frames.*

Proof. It follows from Theorem 1 and 2.

In the full paper we will consider larger fragments of CPDL extending L but which are self-contained in terms of axiomatisation, i.e. to identify exactly which auxiliary operators are needed.

Our language and its CPDL-extension can easily be extended to the multi-agent setting. For a given finite set A of agents, we define the language $L(A)$ to have operators G_a, K_a, C_a, R_a and R'_a and interpret the resulting formulas in 'social decision frames'.

Definition 8. *A social decision frame* $F = \langle W, \approx, \sim, \leqslant \rangle$ *for* A *consists of a decision frame* $F_a = \langle W, \approx_a, \sim_a, \leqslant_a \rangle$ *for each* $a \in A$.

Theorems 2 and Corollary 1 can then be extended to social decision frames, using the corresponding embedding into CPDL with $\text{ATPROG} = \{\mathsf{g}_a, \mathsf{k}_a, \mathsf{c}_a\}_{a \in A}$ and the corresponding $\mathsf{K}D_A$.

3   Games

Our primary examples of social decision frames are taken from the concept of a strategic game in Game Theory.

Definition 9. *Given a set* A *of agents, sets* D_a *of strategies (for each* $a \in A$*), and utility functions*

$$U_a : \prod_{a \in A} D_a \rightarrow \mathbb{R}$$

the strategic game frame $G(A, D, U)$ *is the social decision frame* $\langle W, \sim, \approx, \leqslant \rangle$ *given by*

$W = \prod_{a \in A} D_a$

$w \sim_a v$ *iff* $w_b = v_b$ *for all* $b \neq a$ *in* A

$w \approx_a v$ *iff* $w_a = v_a$

$w \leqslant_a v$ *iff* $U_a(w) \leqslant U_a(v)$

For example, consider the game between players a and b, whose possible strategies are $\{A, B, C\}$ and $\{X, Y, Z\}$ respectively, with utilities given by the table on the left of Figure 1, with (x, y) representing a utility of x to a and y to b for the corresponding outcome. This determines the strategic game frame shown on the right, with a's relations shown with solid lines, and b's with dotted lines. We assume reflexivity and transitivity without displaying the additional links explicitly. The $\approx$ relations are also not shown, since they can be calculated from the capacity relations in a strategic game frame. So, for example, $AY \approx_a AZ$, $AZ \sim_a BZ$, $AY <_a AZ$, $AY \leqslant_b AZ \geqslant AY$.

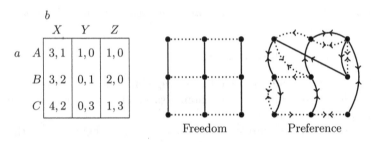

		X	Y	Z
a	A	3,1	1,0	1,0
	B	3,2	0,1	2,0
	C	4,2	0,3	1,3

Fig. 1. A two-player game and its representation as a strategic game frame

Theorem 3. *The extension* $\mathsf{K}G$ *of* $\mathsf{K}DA$ *with the following axioms is complete for the class of strategic game frames:*

$$
\begin{array}{lll}
\mathbf{G_1} & \vdash [U \cap \overline{c_A}]\bot & \textit{(connected)} \\
\mathbf{G_2} & \vdash [c_{\bar{a}} \cap \overline{k_a}]\bot & \textit{(isolated)} \\
\mathbf{G_3} & \vdash [(c_a; c_b) \cap \overline{(c_b; c_a)}]\bot & \textit{(unordered)} \\
\mathbf{G_4} & \vdash [c_a \cap k_a \cap \overline{\top?}]\bot & \textit{(deterministic)} \\
\mathbf{G_5} & \vdash [g_a^* \cap \overline{g_a}]\bot \wedge [U \cap \overline{(g_a \cup g_a^\circ)}]\bot & \textit{(linear)}
\end{array}
$$

Proof. The formulas correspond to the following frame conditions:

$\mathbf{G_1}$ is valid iff $(u)_A = W$ for every $u \in W$ (connected)

$\mathbf{G_2}$ is valid iff every $a \in A$ is isolated   (isolated)

$\mathbf{G_3}$ is valid iff $\sim_a; \sim_b = \sim_b; \sim_a$ for every $a, b \in A$ (unordered)

$\mathbf{G_4}$ is valid iff $(u)[u] = \{u\}$ for every $u \in W$ (deterministic)

$\mathbf{G_5}$ is valid iff $\leqslant$ is reflexive, transitive, and total (linear)

Each formulas are all pure, so by Theorem 1 KG is complete for the class of frames satisfying these conditions. By Theorem 1 of [3] (p. 194) à frame is isomorphic to a strategic game frame iff each of these conditions holds and in addition, it has a 'small' value-size, which means that the number of sets of indifferent situations (i.e. equivalence classes under the equivalence relation $u \leqslant v$ and $v \leqslant u$) is of cardinality $\leqslant 2^{\aleph_0}$. Thus KG is a sound for strategic frames, and complete if every formula satisfiable in such a frame is also satisfiable on a frame with small value-size. But this is guaranteed by the existence of countable models in Theorem 1.

Again, the full version of the paper will contain an exploration of which of these axioms can be stated in fragments extending L with auxiliary operators.

Standard game theoretic concepts such as 'best response', 'Nash equilibrium', 'dominated strategy,' etc. all lift to the slightly more abstract setting of strategic game frames, as shown in [3].

Definition 10. *Given a strategic game frame $G(A, D, U)$, agent a's strategy $d \in D_a$ is (weakly) dominated by another strategy $d' \in D_a$ iff*

1. *d' is sure to be at least as good as d:* $w[^a_{d'}] \geqslant_a w[^a_d]$ *for all $w \in W(A, D)$,[3] and*
2. *d' may be better than d:* $w[^a_{d'}] >_a w[^a_d]$ *for some $w \in W(A, D)$.*

For example, b's strategy Z is dominated by Y because (1) $AY \geqslant_b AZ$, $BY \geqslant_b BZ$, $CY \geqslant_b CZ$, and (2) $BY >_v BZ$. In fact $AY \geqslant_{bF} AZ$ but $AY \not\leqslant_{bF} AZ$, so $AY >_{bF} AZ$ and since $AY \sim_b AZ$, the decision situation AZ is not *a Priori* rational for b. This connection between *a Priori* rationality and domination is quite general.

Theorem 4. *In a model M based on a strategic game frame $G(D, U, A)$, a strategy w_a is dominated iff $M, w \models \neg R_a$.*

Proof. By [3], Theorem 2, p.199, w_a is dominated iff w is *a Priori* rational for a.

Definition 11. *A decision situation w in a strategic game frame $G(A, D, U)$ is a best response for agent a iff there is no strategy $d \in D_a$ such that $w <_a w[^a_d]$ It is a Nash equilibrium iff it is a best response for all agents.*

For example, AY is a best response for a because neither $AY <_a BY$ nor $AY <_a CY$. But it is not a best response for b because $AY <_b AX$. This game has no Nash equilibrium. Since $(AY)_a$ is $\{AY, BY, CY\}$ and $(AY)_b$ is $\{AX, AY, AZ\}$, we can also check that $AY <_{F'b} AX$ but there is no w such that $AY <_{aF'} w$, so AY is a *Posteriori* rational for a but not for b. Again, this connection is general.

---

[3] $w[^a_d]$ is the strategy profile obtained by replacing a's strategy in w by d, i.e.

$$w[^a_d](b) = \begin{cases} d & \text{if } b = a \\ w_b & \text{otherwise} \end{cases}$$

Theorem 5. *In a model M based on a strategic game frame $G(D, U, A)$, a situation w is Nash equilibrium iff $M, w \models \bigwedge_{a \in A} R'_a$.*

Proof. By [3], Theorem 3, p.200. a Nash equilibrium is a situation that is *a Posteriori* rational for all agents.

We can extend this analysis from 'pure strategy' games to 'mixed strategy' games, in which the players randomise their choice of strategy.

Definition 12. *Given a strategic game frame $G(A, D, U)$ with finite[4] D, the mixed-strategy extension of G is the strategic game frame $G^*(A, D^*, U^*)$ in which D^*_a is the set of probability functions $\delta : D_a \to [0, 1]$ and for each $\delta \in D^*$,*

$$U^*_a(\delta) = \sum_{s \in \prod_{b \in A} D_b} u_a(s) \prod_{b \in A} \delta_b(s_b)$$

A frame is a mixed-strategy game frame *iff it is isomorphic to G^* for some strategic game frame G.*

There are, of course, formulas that are valid in every mixed-strategy game frame that are not valid in every strategic game frame and so cannot be derived from **G**. A central example is the following.

Theorem 6. $\mathsf{K}G \nvdash \langle U \rangle \bigwedge_{a \in A} R'_a$ *but this formula is valid on all mixed-strategy strategic game frames.*

Proof. Let M be any model based on a strategic game frame G^*. By [6], every mixed-strategy game has a Nash equilibrium w, and so by Theorem 5, $M, w \models \bigwedge_{a \in A} R'_a$. So the formula is valid on all mixed-strategy strategic game frames. Yet the the frame in Figure 1 does not validate this formula since it lacks a Nash equilibrium, so by Theorem 3, it is not derivable in $\mathsf{K}G$.

4   Concluding Remarks

We have presented a logic that formalises the approach to rational decision-making adopted in [3]. Many salient features of games can be modelled using strategic game frames, which conveniently generalise over pure and mixed strategy games. Our logical investigations, however, are far from complete. In particular, we would like to investigate other fragments of CPDL that are sufficient for use in game theory. A particularly interesting open problem is the axiomatisation of the class of mixed-strategy games. Theorem 6 gives one example of a formula valid over these frames that is not valid in, for example, pure strategy game frames. But approaches to the computation of Nash equilibria (e.g.[4]) suggest many structural features that could be analysed logically.

---

[4] The restriction to finite D is essential, because U^* is calculated as a finite sum. This restriction prevent us from forming G^{**} because D^* is always infinite.

Acknowledgement. Guo is supported by the National Social Science Foundation of China (09CZX033), the Ministry of Education of China (08JC72040002), the Fundamental Research Funds for the Central Universities (SWU1309380) and the key Project of Chongqing key Research base in Humanities and Social Sciences titled with A Study on the Extensions of Dynamic Epistemic Logic (11SKB16).

The research for this paper was conducted during Seligman's Academic Leave from the University of Auckland, New Zealand, in 2010 and a subsequent visit to South West University during May 2012-2013. He is grateful to the University of Auckland for giving him time away from New Zealand to pursue this research and to South West University for their kind invitation and support.

References

1. Areces, C., ten Cate, B.: Hybrid logics. In: van Benthem, J., Blackburn, P., Wolter, F. (eds.) Handbook of Modal Logic. Elsevier, Amsterdam (2006)
2. Danecki, S.: Nondeterministic propositional dynamic logic with intersection is decidable. In: Skowron, A. (ed.) Proceedings of the Fifth Symposium on Computation Theory, vol. 208, pp. 34–53. Springer (1984)
3. Guo, M., Seligman, J.: Making choices in social situations,
   http://www.illc.uva.nl/dg/wp-content/uploads/2012/07/yearbook.20111.pdf
4. Lemke, C.E., Howson, J.T.: Equilibrium points of bimatrix games. Journal of the Society for Industrial and Applied Mathematics 12, 413–423 (1964)
5. Lutz, C., Walther, D.: Pdl with negation of atomic programs. Journal of Applied Non-Classical Logics 15(2), 259–2732 (2005)
6. Nash, J.F.: Non-cooperative games. Annals of Mathematics 54, 286–295 (1951)
7. Passy, S., Tinchev, T.: Pdl with data constants. Information Processing Letters 20(1), 35–41 (1985)
8. Passy, S., Tinchev, T.: An essay in combinatory dynamic logic. Information and Computation 93(2), 263–332 (1991)

Proof Theory, Semantics and Algebra for Normative Systems

Xin Sun

Individual and Collective Reasoning Group, University of Luxembourg
`xin.sun@uni.lu`

Abstract. This paper reports correspondence results between input/ output logic and the theory of joining-systems. The results have the form: every norm (a, x) is logically derivable from a set of norms G if and only if it is in the space of norms algebraically generated by G.

1 Introduction

In their influential book *Normative Systems* [1], Alchourroon and Bulygin conceive a normative system as a deductive mechanism, like black boxes which produce normative statement as output, given we feed them descriptive statements as input. To this tradition belongs as well the input/output logic (I/O logic) of David Makinson and Leon van der Torre in [6–8] and the theory of joining-systems(TJS) proposed by Lars Lindahl and Jan Odelsad in e.g. [4, 5].

Although sharing the same ancestor, I/O logic and TJS have evolved quite separately, and lool very different. I/O logic has a proof theory and a well defined semantics of propositional logic. TJS uses algebra as a tool for modeling normative systems. In this paper, I will show that, nevertheless the two accounts essentially give the same results, and can be seen as "two sides of one and the same coin." The results will illustrate that proof theory, semantics and algebra, as three tools to model normative systems, have their own advantage and disadvantage. Proof theory is neat and easy to be tracked by computers, but hard to be manipulated by human beings. Semantics is intuitive but hard to give us the holistic view of normative systems. Algebra, although it's neither as neat as proof theory nor as intuitive as semantics, it can give rise to holistic results to normative systems in the sense that we can build isomorphisms between the algebraic representation of normative systems. It is their different features that motivate us to use all of them.

The layout of this paper is as follows. In section 2 I will give a brief introduction to I/O logic and TJS. Then, in section 3 I will present two correspondence results between I/O logic and TJS. Section 4 is the section for application of the algebraic tools, illustrating those holistic views we gain by the algebraic representation of normative systems. Section 5 will present some issues for future research.

D. Grossi, O. Roy, and H. Huang (Eds.): LORI 2013, LNCS 8196, pp. 228–238, 2013.

2   Background

2.1   Input/Output Logic

In a series of papers [7–9], Makinson and van der Torre developed a class of deontic logic called input/output logic. A gentle and comprehensive introduction can be found in [10] and [14]. In general, the matured version of I/O logic is the constrained version from [8]. For simplicity's sake, the latter one will be put aside, and only two unconstrained I/O logics will be considered in this paper. I start by describing them.

Let $\mathbb{P} = \{p_0, p_1, \ldots\}$ be a countable set of propositional letters and L be the propositional logic built upon $\mathbb{P}$. Throughout this paper L will be the only logic language we talk about. Let G be a set of ordered pairs of formulas of L. A pair $(a, x) \in G$, call it a norm, is read as "given a, it ought to be x". G can be viewed as a function from 2^L to 2^L such that for a set A of formulas, $G(A) = \{x : (a, x) \in G$ for some $a \in A\}$.

Makison and van der Torre define the operations out_1 and out_2 as following:

- $out_1(G, A) = Cn(G(Cn(A)))$
- $out_2(G, A) = \bigcap\{Cn(G(V)) : A \subset V, V$ is complete$\}$

Here Cn is the classical consequence operator from propositional logic, and a complete set is a set of formulas that is either maxi-consistent or equal to L.

$out_1(G, A)$ and $out_2(G, A)$ are called *simple-minded output* and *basic output* respectively. In [7], simple-minded reusable output and basic reusable output are also defined. I leave them as a topic for future research.

out_1 and out_2 can be fiben a proof theoretic characterization. We say that an ordered pair of formulas is derivable from a set G iff (a, x) is in the least set that includes G, contains the pair (t, t), where t is a tautology, and is closed under a number of rules. The following are the rules we will use:

- SI (strengthening the input): from (a, x) to (b, x) whenever $b \vdash a$
- WO (weakening the output): from (a, x) to (a, y) whenever $x \vdash y$
- AND (conjunction of output): from $(a, x), (a, y)$ to $(a, x \wedge y)$
- OR (disjunction of input): from $(a, x), (b, x)$ to $(a \vee b, x)$

The derivation system based on the rules SI, WO and AND is called $deriv_1$. Adding OR to $deriv_1$ gives $deriv_2$. We use $(a, x) \in deriv_i(G)$, or equivalently $x \in deriv_i(G, a)$, to denote the norm (a, x) is derivable from G using rules of derivation system $deriv_i$. Moreover, for a set A of formulas, we use $(A, x) \in deriv_i(G)$, or equivalently $x \in deriv_i(G, A)$ to denote the fact that there exist $a_1 \ldots a_n \in A$ such that $(a_1 \wedge \ldots \wedge a_n, x) \in deriv_i(G)$. In [7], the following completeness theorems for out_1 and out_2 are given:

Theorem 1 ([7]). *Given an arbitrary normative system G and a set A of formulas,*

1. $x \in out_1(G, A)$ iff $x \in deriv_1(G, A)$
2. $x \in out_2(G, A)$ iff $x \in deriv_2(G, A)$

2.2 Theory of Joining-Systems

An algebraic framework for analyzing normative systems was introduced by Lars Lindahl and Jan Odelstad in their papers [3–5, 12, 13]. The most general form of the theory is called theory of joining-systems(TJS) in [5, 12]. A theory of joining-systems is a triple (B_1, B_2, J) where B_1, B_2 are two ordered algebraic structures and J a relation between B_1 and B_2 satisfying some special conditions. In Lindahl and Odelstad's work, the algebraic structure is usually a Boolean quasi-ordering. In this paper I will work with a Boolean algebra.

Definition 1 (Boolean algebra). *A structure* $\mathfrak{A} = (A, +, \cdot, -, 0, 1)$ *is a Boolean algebra iff it satisfies the following identities:*

(1) $x + y = y + x$, $x \cdot y = y \cdot x$
(2) $x + (y + z) = (x + y) + z$, $x \cdot (y \cdot z) = (x \cdot y) \cdot z$
(3) $x + 0 = x$, $x \cdot 1 = x$
(4) $x + (-x) = 1$, $x \cdot (-x) = 0$
(5) $x + (y \cdot z) = (x + y) \cdot (x + z)$, $x \cdot (y + z) = (x \cdot y) + (x \cdot z)$

We can order the elements of a Boolean algebra by defining $a \leq b$ if $a \cdot b = a$. Here $+$ can be considered as a disjunction, $\cdot$ as a conjunction and $\leq$ as a implication. With this order relation in hand, the *narrowness*($\preceq$) relation between two ordered pairs can be naturally defined as $(a, x) \preceq (b, y)$ iff $b \leq a$ and $x \leq y$. Based on ordered structures, Lindahl and Odelstad define joining-systems as follows:

Definition 2 (Joining-systems, Lindahl and Odelstad's version [12]). *A triple* $(\mathbb{A}, \mathbb{B}, S)$, *where* $\mathbb{A} = (A, \leq)$ *and* $\mathbb{B} = (B, \leq)$ *are ordered structures and* $S \subseteq A \times B$, *is called a joining-system if S satisfies the following conditions:*

1. *If* $(a, x) \in S$ *and* $(a, x) \preceq (b, y)$, *then* $(b, y) \in S$.
2. *For all* $X \subseteq B$, *if for all* $x \in X$, $(a, x) \in S$, *then* $(a, y) \in S$ *for all* $y \in glb(X)$.[1]
3. *For all* $X \subseteq A$, *if for all* $x \in X$, $(x, b) \in S$, *then* $(y, b) \in S$ *for all* $y \in lub(X)$.[2]

In this paper, the major mathematical tool is the joining-systems of Boolean algebra, which is a modified version of Lindahl and Odelstad's joining-systems in the following sense: we let $(1, 1) \in S$ and require the set X in item 2 and 3 above to be finite, of Lindahl and Odelstad's joining-systems.

Definition 3 (Joining-systems of Boolean algebras). *A joining-systems of Boolean algebras is a structure* $\mathbb{S} = (\mathfrak{A}, \mathfrak{B}, S)$ *such that* $\mathfrak{A}, \mathfrak{B}$ *are Boolean algebras and* $S \subseteq A \times B$ *meets the following conditions:*

---

[1] Here glb is the abbreviation of greatest lower bound. Formally, $glb(X) = \{b : \forall x \in X, b \leq x$ and $\forall a$, if $\forall x \in X$, $a \leq x$, then $a \leq b\}$
[2] lub is the abbreviation of least up bound. Formally, $lub(X) = \{a : \forall x \in X, x \leq a$ and $\forall b$, if $\forall x \in X$, $x \leq b$, then $a \leq b\}$

1. $(1,1) \in S$
2. If $(a,x) \in S$ and $(a,x) \preceq (b,y)$, then $(b,y) \in S$.
3. For all finite $X \subseteq B$, if for all $x \in X, (a,x) \in S$, then $(a,y) \in S$ for all $y \in glb(X)$
4. For all finite $X \subseteq A$, if for all $x \in X$, $(x,b) \in S$, then $(y,b) \in S$ for all $y \in lub(X)$

Here we call S a joining-space as Lindahl and Odelstad did in [5]. We can equivalently replace condition (3) and (4) by the following respectively:

3′ If $(a,x) \in S$ and $(a,y) \in S$, then $(a, x \cdot y) \in S$
4′ If $(a,x) \in S$ and $(b,x) \in S$, then $(a+b, x) \in S$

Moreover, we can define joining-space using the standard algebraic terminology of ideal and filter:

Definition 4 (Ideal). Let $\mathfrak{A} = (A, +, \cdot, -, 0, 1)$ be a Boolean algebra and I a subset of A. For I to be an ideal of $\mathfrak{A}$, it is necessary and sufficient that the following three conditions be satisfied:

(1) $0 \in I$
(2) for all $x, y \in I$, $x + y \in I$
(3) for all $x \in I$ and $y \in A$, if $y \leq x$ then $y \in I$

Definition 5 (Filter). Let $\mathfrak{A} = (A, +, \cdot, -, 0, 1)$ be a Boolean algebra and F a subset of A. For F to be a filter of $\mathfrak{A}$, it is necessary and sufficient that the following three conditions are satisfied:

(1) $1 \in F$
(2) for all $x, y \in F$, $x \cdot y \in F$
(3) for all $x \in F$ and $y \in A$, if $x \leq y$ then $y \in F$

Let $F_\uparrow(X)$ be the filter generated by X and $I_\downarrow(X)$ be the ideal generated by X. Then $I_\downarrow(X)(F_\uparrow(X))$ is the smallest ideal(filter) containing X, and we have the following proposition:

Proposition 1. Given a structure $\mathbb{S} = (\mathfrak{A}, \mathfrak{B}, S)$, S is a joining space in $\mathbb{S}$ if and only if it satisfies the following conditions:

1. $(1,1) \in S$
2. For every finite set $X \subseteq A$, if $\forall x \in X$, $(x,b) \in S$, then $\forall y \in I_\downarrow(X)$, $(y,b) \in S$.
3. For every finite set $X \subseteq B$, if $\forall x \in X$, $(a,x) \in S$, then $\forall y \in F_\uparrow(X)$, $(a,y) \in S$.

Proof:
Assume S is a joining space in $\mathbb{S}$. Then trivially we have $(1,1) \in S$.
For the second condition, let X be an arbitrary finite subset of B. Without loss of generality, we can let $X = \{x_1, \ldots, x_n\}$. Suppose $\forall x \in X$, $(a,x) \in S$. Then by

applying clause 3' of Definition 3 finitely many times we have $(a, x_1 \cdot \ldots \cdot x_n) \in S$. Since for all $y \in F_\uparrow(X)$, $x_1 \cdot \ldots \cdot x_n \leq y$, therefore $(a, y) \in S$.

Similarly we can prove that the third condition is satisfied.

Now assume S satisfies the three conditions in this proposition. Then obviously $(1, 1) \in S$.

Assume $(a, x) \in S$ and $(a, x) \preceq (b, y)$, then $x \leq y$ and $y \in F_\uparrow(x)$, hence $(a, y) \in S$. Moreover we have $b \leq a$ and $b \in I_\downarrow(a)$, so we have $(b, y) \in S$.

Assume $(a, x) \in S$ and $(a, y) \in S$. Since $x \cdot y \in F_\uparrow(\{x, y\})$, we know $(a, x \cdot y) \in S$.

Similarly we can prove if $(a, x) \in S$ and $(b, x) \in S$, then $(a + b, x) \in S$. Therefore S is a joining space.   ⊣

Up to now, we have clearly defined what a joining-system and joining space are. But does a joining space always exist? The answer is positive. As the following proposition shows, the largest and the smallest joining space always exists.

Proposition 2. *Given two boolean algebra* $\mathfrak{A}, \mathfrak{B}$,

1 $A \times B$ *is the largest joining space of* $\mathfrak{A} \times \mathfrak{B}$.
2 *If* $\{S_i | i \in I\}$ *is a collection of joining spaces of* $\mathfrak{A} \times \mathfrak{B}$, *then* $S^* = \cap_{i \in I} S_i$ *is a joining space of* $\mathfrak{A} \times \mathfrak{B}$.

Proof:

1 It is easy to check that $A \times B$ satisfies the definition of joining space and it is the largest one.
2 For every S_i, we have $(1, 1) \in S_i$, therefore $(1, 1) \in S^*$.
   For every finite set $X \subseteq A$, if for every $x \in X$, $(x, b) \in S^*$, then $(x, b) \in S_i$ for every $i \in I$. Therefore $\forall y \in I_\downarrow(X)$, $(y, b) \in S_i$. So we must have $(y, b) \in S^*$.
   Similarly we can prove the third statement of Proposition 1 is true. Therefore $S^* = \cap_{i \in I} S_i$ is a joining space of $\mathfrak{A} \times \mathfrak{B}$.

3   Correspondence between I/O Logic and TJS

3.1   Basic I/O Logic and TJS

In this section, I will prove that for a set of norms G, a norm (a, x) is entailed by G in basic I/O logic, if and only if it is in the joining space generated by G. To show this, we need to introduce a special Boolean algebra named Lindenbaum-Tarski algebra.

Let $\equiv$ be the provable equivalence relation on L, i.e. for every formula $\phi, \psi \in L$, $\phi \equiv \psi$ iff $\vdash_L \phi \leftrightarrow \psi$. Let $L/_\equiv$ be the equivalence classes that $\equiv$ induces on L. For any formula $\phi \in L$, let $[\phi]$ denote the equivalence class contains ϕ.

Definition 6 (Lindenbaum-Tarski algebra). *The Lindenbaum-Tarski algebra for a logic* L *is a structure* $\mathfrak{L} = (L/_\equiv, +, \cdot, -, 0, 1)$ *where* $[\phi] + [\psi] = [\phi \vee \psi]$, $[\phi] \cdot [\psi] = [\phi \wedge \psi]$, $-[\phi] = [\neg\phi]$, $0 = [\bot]$ *and* $1 = [\top]$.

For more details of Lindenbaum-Tarski algebra, readers can consult chapter 5 of [2]. It is not hard to check that every Lindenbaum-Tarski algebra is a Boolean algebra. Let G be a set of ordered pairs of formulas of L. Let $G^{\equiv} = \{([a], [x])|(a, x) \in G\}$. Let $\mathbb{S} = (\mathfrak{L}, \mathfrak{L}, S)$ be a joining-systems such that $G^{\equiv} \subseteq S$. By Proposition 2 we know such joining system always exist. Moreover, there must be a smallest joining space G^* such that $G^{\equiv} \subseteq G^*$ and for every joining space S that extends $G^{\equiv}$, $G^* \subseteq S$. Such a G^* is the joining space generated by $G^{\equiv}$, and it satisfies following property:

Proposition 3. *For every $([a], [x]) \in G^*$, at least one of the following holds:*

(1) $([a], [x])$ is $([1], [1])$
(2) for some $([b], [y]) \in G^*$, $([b], [y]) \preceq ([a], [x])$
(3) there exist $([a], [y]), ([a], [z]) \in G^*$ such that $[x] = [y] \cdot [z]$
(4) there exist $([b], [x]), ([c], [x]) \in G^*$ such that $[a] = [b] + [c]$

Proof: Suppose $([a], [x]) \in G^*$ and satisfies none of the above four clause, then we can prove $G' = G^* - \{([a], [x])\}$ is a joining system. This contradicts the fact that G^* is the smallest joining system.

With proposition 3 in hand, we can now prove one main correspondence result:

Theorem 2. *The following three propositions are equivalent:*

1 $(a, x) \in deriv_2(G)$
2 $([a], [x]) \in G^*$
3 $x \in out_2(G, a)$

Proof:
$1 \Rightarrow 2$: This can be proved simply by induction one the length of derivation.
$2 \Rightarrow 3$: Assume $([a], [x]) \in G^*$. By proposition 3 we need to deal with four cases.
(i) If $([a], [x])$ is $([1], [1])$, we need to prove $\top \in \cap\{Cn(G^*(V)) : a \in V, V$ is complete$\}$,which is obviously true.
(ii) If for some $([b], [y]) \in G^*$, $([b], [y]) \preceq ([a], [x])$. Then by induction hypotheses we know $y \in \cap\{Cn(G^*(V)) : b \in V, V$ is complete$\}$. Since $[a] \leq [b]$ and $[y] \leq [x]$ we know $x \in Cn(y)$. Hence $x \in \cap\{Cn(G^*(V)) : b \in V, V$ is complete$\}$. Moreover, every complete set V contains a must contain b, therefore $\cap\{Cn(G^*(V)) : b \in V, V$ is complete$\} \subseteq \cap\{Cn(G^*(V)) : a \in V, V$ is complete$\}$. Therefore $x \in \cap\{Cn(G^*(V)) : a \in V, V$ is complete$\}$, $x \in out_2(G, a)$.
(iii) If there exist $([a], [y]), ([a], [z]) \in G^*$ such that $[x] = [y] \cdot [z]$. Then by induction hypotheses we know $y \in \cap\{Cn(G^*(V)) : a \in V, V$ is complete$\}$ and $z \in \cap\{Cn(G^*(V)) : a \in V, V$ is complete$\}$. Therefore $y \wedge z \in \cap\{Cn(G^*(V)) : a \in V, V$ is complete$\}$. That is, $y \wedge z \in out_2(G, a)$, $x \in out_2(G, a)$.
(iv) If there exist $([b], [x]), ([c], [x]) \in G^*$ such that $[a] = [b] + [c]$. Then by induction hypotheses we know $x \in \cap\{Cn(G^*(V)) : b \in V, V$ is complete$\}$ and $x \in \cap\{Cn(G^*(V)) : c \in V, V$ is complete$\}$. For every complete set V such that $b \vee c \in V$, it must be that either $b \in V$ or $c \in V$. Therefore, for every complete set V that contains $b \vee c$, $x \in Cn(V)$, which means $x \in \cap\{Cn(G^*(V)) : b \vee c \in V, V$ is complete$\}$, i.e. $x \in out_2(G, b \vee c)$, $x \in out_2(G, a)$.
$3 \Rightarrow 1$: This is a special case of observation 2 of [7].

3.2   Simple-Minded I/O Logic and TJS

The previous section proved a correspondence result between basic I/O logic and TJS. In fact, we can prove a similar result between simple-minded I/O logic and a weaker version of TJS.

Definition 7 (Weak joining-systems). *A weak joining-systems of Boolean algebras is a structure* $\mathbb{S} = (\mathfrak{A}, \mathfrak{B}, S^-)$ *such that* $\mathfrak{A}, \mathfrak{B}$ *are boolean algebras and* $S^- \subseteq A \times B$ *satisfies the first three conditions of a joining space. Here we call* S^- *the weak joining space of* $\mathbb{S}$.

Similar to Proposition 2, we can prove the existence of the largest and the smallest weak joining space.

Proposition 4. *Given two Boolean algebra* $\mathfrak{A}, \mathfrak{B}$,

1. $A \times B$ *is the largest weak joining space of* $\mathfrak{A}$ *and* $\mathfrak{B}$.
2. *If* $\{S_i | i \in I\}$ *is a collection of weak joining spaces of* $\mathfrak{A}$ *and* $\mathfrak{B}$, *then* $S^* = \cap_{i \in I} S_i$ *is a weak joining space of* $\mathfrak{A}$ *and* $\mathfrak{B}$.

Let G be a set of ordered pairs of formulas of L and $\mathfrak{L}(\Phi)$ be the Lindenbaum-Tarski algebra of L. Let $G^\equiv = \{([a], [x]) | (a, x) \in G\}$ where $[a], [x]$ are the equivalence classes in $\mathfrak{L}(\Phi)$ respective contains a and x. By Proposition 4 we know that there exists a unique smallest weak joining-systems extends $G^\equiv$. If we denote it as G^+, then we have the following:

Proposition 5. *For every* $([a], [x]) \in G^+$, *at least one of the following holds:*

(1) $([a], [x])$ *is* $([1], [1])$
(2) for some $([b], [y]) \in G^+$, $([b], [y]) \preceq ([a], [x])$
(3) there exists $([a], [y]), ([a], [z]) \in G^+$ *such that* $[x] = [y] \cdot [z]$

Proof: Similar to the proof of proposition 2.

With Proposition 5 in hand, we can prove the following correspondence result:

Theorem 3. *The following three proposition is equivalent:*

1 $(a, x) \in deriv_1(G)$
2 $([a], [x]) \in G^+$
3 $x \in out_1(G, a)$

Proof:
$1 \Rightarrow 2$: This can be proved simply by induction one the length of derivation.
$2 \Rightarrow 3$: Assume $([a], [x]) \in G^+$.
(i) If $([a], [x])$ is $([1], [1])$, we need to prove $\top \in \cap\{Cn(G(T))\}$, which is obviously true.
(ii) If for some $([b], [y]) \in G^+$, $([b], [y]) \preceq ([a], [x])$. Then by induction hypothesis we know $y \in Cn(G(b))$. Since $[a] \leq [b]$ and $[y] \leq [x]$ we know $x \in Cn(y)$. Hence $x \in Cn(G(b))$.
(iii) If there exists $([a], [y]), ([a], [z]) \in G^+$ such that $[x] = [y] \cdot [z]$. Then by induction hypotheses we know $y \in Cn(G(a))\}$ and $z \in Cn(G(a))\}$. Therefore $y \wedge z \in Cn(G(a))$. That is, $y \wedge z \in out_1(G, a), x \in out_1(G, a)$.
$3 \Rightarrow 1$: This is a special case of observation 1 of [7].

4   Application

In this section, we discuss some of the insights obtained from the algebraic approach to normative systems.

4.1   The Core of a Normative System

In section 2.2 the narrowness relation $\preceq$ is defined as $(a, x) \preceq (b, y)$ iff $b \le a$ and $x \le y$. We can further define the strict narrowness relation $\prec$ as $(a, x) \prec (b, y)$ iff $(a, x) \preceq (b, y)$ and not $(b, y) \preceq (a, x)$. A norm (a, x) is minimal in a joining-systems, or normative system, $\mathbb{S}$ iff there is no $(b, y) \in \mathbb{S}$ such that $(b, y) \prec (a, x)$. In [11], such a minimal norm is called a connection from A to B.

As noticed by [11], the set of all minimal elements of a joining-systems can be viewed as the core of the system. If the joining space is finite, then the whole joining-systems is uniquely determined by its minimal norms. If we know the core of the system, we can logically deduce the whole system. Let for a joining-systems $\mathbb{S}$, let $core(\mathbb{S}) = \{(a, x) \in S | (a, x)$ is minimal in $\mathbb{S}\}$ denote the set of all its minimal norms. The following are formal statements about the properties of the core of finite joining-systems.

Observation 1. *For all joining-systems* $\mathbb{S} = (A, B, S)$. *If S is finite, then* $core(\mathbb{S}) \ne \emptyset$

Proof: The proof is trivial. Due to the fact that S is finite, there is no infinite descending chain on $\prec$.

Observation 2. *For all joining-systems* $\mathbb{S} = (A, B, S)$, *if S is finite, then for any $(a, x) \in S$, there exists $(b, y) \in core(\mathbb{S})$ such that $(b, y) \preceq (a, x)$.*

Proof: Let (a, x) be an arbitrary norm in S. If $(a, x) \in core(\mathbb{S})$, then $(a, x) \preceq (a, x)$ and we are done. If $(a, x) \notin core(\mathbb{S})$, then (a, x) is not a minimal norm. Hence there exist some (b, y) such that $(b, y) \prec (a, x)$. If $(b, y) \in core(\mathbb{S})$ then we are done. If not, then there exist some (c, z) such that $(c, z) \prec (b, y)$. Since S is finite, this procedure will stop at some point. Then by transitivity of $\preceq$, there must exist some $(a', x') \in core(\mathbb{S})$ such that $(a, x) \preceq (a', x')$.

Observation 3. *For any joining-systems* $\mathbb{S} = (A, B, S)$ *and* $\mathbb{S}' = (A, B, S')$, *if both S and S' are finite, then $core(\mathbb{S}) = core(\mathbb{S}')$ iff $S = S'$.*

Proof: The right to left direction is trivial. For the left to right direction. Assume $core(\mathbb{S}) = core(\mathbb{S}')$. For any $(a, x) \in S$, by Observation 2 there exist $(b, y) \in core(\mathbb{S})$ such that $(b, y \prec (a, x))$. By assumption we know $(b, y) \in core(\mathbb{S}')$. Then by the definition of joining space we know $(a, x) \in S'$. Therefore $S \subseteq S'$. Similarly we can prove $S \supseteq S'$.

4.2   Harshness of Normative Systems

Suppose there are two norms (a, x) and $(a, x \wedge y)$, it is reasonable to say that the latter is harsher than the former because the latter demand us to do more than the former under the same situation. For illustration we can let a represent "you are invited to a dinner", x represent "you dress your suit" and y represent "you wash your hair". For similar reasons we can consider $(a \vee b, x)$ to be harsher than (a, x).

In general, $(a, x) \preceq (b, y)$ can intuitively be read as (a, x) is "harsher" than (b, y). Moreover, we can lift this harshness concept to the level of normative system as long as we use joining-systems to represent them.

Definition 8 (Harshness). *Let* $\mathbb{S} = (A, B, S)$ *and* $\mathbb{S}' = (A, B, S')$ *be two joining-systems,* $\mathbb{S}$ *is harsher than* $\mathbb{S}'$, *denote as* $\mathbb{S} \precsim \mathbb{S}'$, *iff for all* $(a, x) \in core(\mathbb{S})$ *there exist* $(b, y) \in core(\mathbb{S}')$ *such that* $(a, x) \preceq (b, y)$ *and for all* $(b, y) \in core(\mathbb{S}')$ *there exist* $(a, x) \in core(\mathbb{S})$ *such that* $(a, x) \preceq (b, y)$.

Observation 4. *For any joining-systems* $\mathbb{S} = (A, B, S)$ *and* $\mathbb{S}' = (A, B, S')$, *if* $\mathbb{S} \precsim \mathbb{S}'$, *then* $S' \subseteq S$.

Prove: Assume $(a, x) \in S'$, then there exist $(b, y) \in core(\mathbb{S}')$ such that $(b, y) \preceq (a, x)$. By the definition of harshness there exist $(c, z) \in core(\mathbb{S})$ such that $(c, z) \preceq (b, y)$. There fore $(c, z) \preceq (a, x)$ and $(a, x) \in S$.

This observation shows the more obligation a normative system contains, the harsher it is. Such a result coincides with our intuition quite well.

4.3   Structural Similarity of Normative Systems

For two algebraic structures A and B, if they are isomorphic then they are essentially the same. We can extend the isomorphism of Boolean algebra to joining-systems. But before we do this, we first review the isomorphism of Boolean algebra.

Definition 9 (Isomorphism of Boolean algebra). *For two Boolean algebras* $\mathfrak{A} = (A, +, \cdot, -, 0, 1)$ *and* $\mathfrak{A}' = (A', +, \cdot, -, 0, 1)$ *and* h *a map from* A *to* A'. *We say that* h *is an isomorphism from* $\mathfrak{A}$ *to* $\mathfrak{A}'$ *iff for any* $x, y \in A$, h *satisfies the following conditions:*

1. *h is bijective*
2. *$h(x + y) = h(x) + h(y)$*
3. *$h(x \cdot y) = h(x) \cdot h(y)$*
4. *$h(1) = 1$*

Given an isoporphism h from $\mathfrak{A}$ to $\mathfrak{A}'$, it is easy to check that for all $x, y \in A$ and $x', y' \in A'$, if $h(x) = x'$ and $h(y) = y'$, then $x \leq y$ iff $x' \leq y'$.

Now we extend isomorphism to joining-systems.

Definition 10 (Isomorphism of joining-systems). *For two joining-systems* $\mathbb{S} = (A, B, S)$ *and* $\mathbb{S}' = (A', B', S')$ *and* h *a map from* $A \cup B$ *to* $A' \cup B'$. *We say that* h *is an isomorphism from* $\mathbb{S}$ *to* $\mathbb{S}'$ *iff* h *satisfies the following conditions:*

1. *h is bijective*
2. *the restriction of h on A is an isomorphism from A to A'*
3. *the restriction of h on B is an isomorphism from B to B'*
4. *$(a, x) \in core(\mathbb{S})$ iff $(h(a), h(x)) \in core(\mathbb{S}')$*

If there exist some isomorphism form $\mathbb{S}$ to $\mathbb{S}'$, then we say $\mathbb{S}$ and $\mathbb{S}'$ are isomorphic. Two isomorphic joining-systems can naturally be understood as structurally the same. Although in the last item of the above definition we restrict ourselves to the core of a joining-systems, the correspondence in fact covers the whole system. That is, we have the following observation:

Observation 5. *For any joining-systems* $\mathbb{S} = (A, B, S)$ *and* $\mathbb{S}' = (A, B, S')$, *if* h *is an isomorphism from* $\mathbb{S}$ *to* $\mathbb{S}'$, *then for any* $(a, x) \in A \times B$, *$(a, x) \in S$ iff* $(h(a), h(x)) \in S'$.

5   Conclusion and Future Work

The main contribution of this paper is a correspondence result between input/output logic and the theory of joining-system. These results illustrate that normative systems can be equivalently analyzed using three different tools, proof theory, semantics and algebra. Each tool will give us some special insights of normative systems.

There are a lot of future workto be done. A natural direction is to build a correspondence result between constrained I/O logic and TJS. Another direction is to use more advanced logic and algebra to relate I/O logic and TJS. For example, temporal logic can serve as the basis of I/O logic and Boolean algebra with temporal operator can be the underlying algebra of TJS. Then we can build another correspondence result between the new I/O logic and TJS.

Acknowledgments. The author thanks Leon van der Torre, Xavier Parent and three anonymous referees of the LORI committee for their valuable suggestions and comments.

References

1. Alchourron, C.E., Bulygin, E.: Normative Systems. Springer (1971)
2. Blackburn, P., De Rijke, M., Venema, Y.: Modal logic. Cambridge University Press (2001)
3. Lindahl, L., Odelstad, J.: An algebraic analysis of normative systems. Ratio Juris 13, 261–278 (2000)
4. Lindahl, L., Odelstad, J.: Intermediaries and intervenients in normative systems. Journal of Applied Logic, 229–250 (2008)

5. Lindahl, L., Odelstad, J.: TJS. a formal framework for normative systems with intermediaries. In: Horty, J., Gabbay, D., Parent, X., van der Meyden, R., van der Torre, L. (eds.) Handbook of Deontic Logic and Normative Systems. College Publications (2013)
6. Makinson, D.: On a fundamental problem of deontic logic. In: Mc-Namara, P., Prakken, H. (eds.) Norms, Logics and Information Systems, pp. 29–53. IOS Press, Amsterdam (1999)
7. Makinson, D., van der Torre, L.: Input-output logics. Journal of Philosophical Logic 29, 383–408 (2000)
8. Makinson, D., van der Torre, L.: Constraints for input/output logics. Journal of Philosophical Logic 30(2), 155–185 (2001)
9. Makinson, D., van der Torre, L.: Permission from an input/output perspective. Journal of Philosophical Logic 32, 391–416 (2003)
10. Makinson, D., van der Torre, L.: What is input/output logic? In: Lowe, B., Malzkorn, W., Rasch, T. (eds.) Foundations of the Formal Sciences II: Applications of Mathematical Logic in Philosophy and Linguistics, pp. 163–174 (2003)
11. Odelstad, J., Boman, M.: The role of connections as minimal norms in normative systems. In: Bench-Capon, T., Daskalopulu, A., Winkels, R. (eds.) Legal Knowledge and Information Systems. IOS Press, Amsterdam (2002)
12. Odelstad, J., Boman, M.: Algebras for agent norm-regulation. Annals of Mathematics and Artificial Intelligence 42, 141–166 (2004)
13. Odelstad, J., Lindahl, L.: Normative systems represented by boolean quasi-orderings. Nordic Journal of Philosophical Logic 5, 161–174 (2000)
14. Parent, X., van der Torre, L.: I/O logic. In: Horty, J., Gabbay, D., Parent, X., van der Meyden, R., van der Torre, L. (eds.) Handbook of Deontic Logic and Normative Systems. College Publications (2013)

Explicit and Implicit Knowledge
in Neighbourhood Models

Fernando R. Velázquez-Quesada

Grupo de Lógica, Lenguaje e Información, Universidad de Sevilla
FRVelazquezQuesada@us.es

Abstract. Under relational models, epistemic logic agents are logically omniscient. A common strategy to avoid this has been to distinguish between *implicit* and *explicit* knowledge, and approaches based on relational models have used implicit knowledge as a primitive, defining explicit knowledge as implicit knowledge that satisfies some additional requirement. In this work we follow the opposite direction: using neighbourhood models, we take explicit knowledge as a primitive, then defining implicit knowledge as what the agent will know explicitly in an 'ideal' state. This approach, though natural, does not satisfy two 'intuitive' properties: explicit knowledge does not need to be implicit, and the consequent of an explicitly known implication with explicitly known antecedent does not need to be implicitly known; we discuss why this is the case. Then a modus ponens operation is defined, and it is shown how it satisfies a third 'intuitive' property: if the agent knows explicitly an implication and its antecedent, then after a modus ponens step she will know explicitly the consequent.

1 Introduction

Under relational (Kripke) models, epistemic logic (*EL*; [1,2]) agents are logically omniscient: they know every tautology and their knowledge is closed under modus ponens, so their knowledge is closed under logical consequence. This property has been criticised because the knowledge of 'real' agents does not need to have such properties (not even that of computational ones, who might lack of time and/or space to deal with every logical consequence of what they have).

One of the most prominent ideas for solving this *logical omniscience problem* has been to distinguish between an agent's *explicit* knowledge, what she actually has, and her *implicit* knowledge, what she can eventually obtain (e.g., [3,4,5,6,7]). In particular, within approaches based on relational models, the modal universal operator □ has been understood as describing the agent's *implicit* knowledge, and then *explicit* knowledge has been defined as implicit knowledge that satisfies some extra condition (e.g., *awareness of* the involved formula [5]; *awareness that* the formula is true [6]; the existence of a *justification* for the formula [7]).

But relational models are not the unique semantic model for the epistemic logic language, and recently there have been approaches that, using the so called *minimal* or *neighbourhood* models ([8,9]; see [10] or Chapter 7 of [11] for detailed

D. Grossi, O. Roy, and H. Huang (Eds.): LORI 2013, LNCS 8196, pp. 239–252, 2013.

presentations), have studied not only epistemic phenomena but also their dynamics (prominently, the approach to dynamics of evidence of [12]). The present work uses neighbourhood models to deal with the logical omniscience problem by defining the notions of implicit and explicit knowledge within it. Differently from relational approaches (and given the properties of the operator □ in this setting), we use the notion of explicit knowledge as a primitive, reading formulas of the form □ φ as "the agent knows φ *explicitly*". Then, by using the established relation between neighbourhood and relational models, we define the notion of implicit knowledge in terms of its explicit counterpart: implicit knowledge is what the agent will know explicitly in some 'ideal' final state that can be reached through deductive inference steps.

Interestingly this approach, though natural, does not satisfy two properties found in most (if not all) approaches dealing with explicit and implicit knowledge: here explicit knowledge does not need to be implicit, and the consequent of an explicitly known implication with explicitly known antecedent does not need to be implicitly known. Nevertheless, our approach does have a third 'dynamic' property: if the agent knows explicitly an implication and its antecedent, then after a modus ponens step she will know explicitly the consequent.

Our work is organised as follows. In Section 2 we recall the neighbourhood models framework, in particular, the class of models in which knowledge implies truth (i.e., those where □ φ → φ is valid). Then in Section 3 we recall the connection between neighbourhood and relational models, and also the model operations that can take us from the first to the second. With these tools we move to Section 4 where, using a modality standing for the sequence of model operations that take a neighbourhood model to this 'ideal' state, we define the notions of explicit and implicit knowledge in the lines mentioned above. We also study properties of these definitions, comparing them with what relational approaches typically provide. Then in Section 5 we define a modus ponens operation and we discuss its most important property. Section 6 presents a summary of the work and further research questions.

On *Implicit* Knowledge. The concept of explicit knowledge is, intuitively, easy to grasp: it simply corresponds to what an agent has 'at hand' at some give moment. The concept of implicit knowledge, on the other hand, might differ from work to work; thus, a brief discussion is worthwhile.

In works where explicit knowledge is the primitive concept, the understanding of implicit knowledge is clear from its definition in terms of explicit knowledge (typically, what is implicit is the closure under logical consequence of what is explicit). Even in approaches that use implicit knowledge as the primitive concept, implicit knowledge can be understood as what the agent would know explicitly if she satisfied some 'ideal' property. For example, one can say that while the *awareness of* approach of [5] understands implicit knowledge as what would be explicit if the agent were aware of every formula, the *acknowledgement* approach of [6] understands it as what would be explicit if the agent had acknowledged as true every formula that is so, and the *justification* approach of [7] understands it as what would be explicit if the agent had a justification for every true formula.

The present work understand implicit knowledge as the closure under logical consequence of explicit knowledge or, from a dynamic epistemic perspective, as what the agent will know explicitly in an ideal state reachable via deductive inference steps. Other variations following the lines of (or even combining) the mentioned approaches (implicit is what will be explicit in an ideal state reachable by 'becoming aware' of every formula, or/and by 'building' justifications for all true formulas) are also possible, but will be left for further work.

2   Neighbourhood Models

Let P be a countable non-empty set of atomic propositions.

Definition 1 (Language $\mathcal{L}$). *Formulas φ, ψ of the language $\mathcal{L}$ are given by*

$$\varphi ::= p \mid \neg\varphi \mid \varphi \wedge \psi \mid \Box\,\varphi$$

with $p \in$ P. Formulas of the form $\Box\,\varphi$ can be read (for now) simply as "the agent knows φ". Logical constants $(\top, \bot)$, other boolean operators $(\vee, \rightarrow, \leftrightarrow)$ and the modal operator $\Diamond$ are defined as usual $(\Diamond\,\varphi := \neg\Box\,\neg\varphi$ for the latter).

Definition 2 (Knowledge neighbourhood model). *A* knowledge neighbourhood (KN) model *is a tuple $M = \langle W, N, V \rangle$ where (i) $W \neq \varnothing$ is a finite set of possible worlds; (ii) $N : W \rightarrow \wp(\wp(W))$ is the neighbourhood function, assigning a set of sets of possible worlds to each $w \in W$, and satisfying the following property (the* truth *property): for every $w \in W$, if $U \in N(w)$ then $w \in U$; (iii) $V : $ P $\rightarrow \wp(W)$ is a valuation function.*

Definition 3 (Semantic interpretation). *Take a KN model $M = \langle W, N, V \rangle$. Given a formula $\varphi \in \mathcal{L}$, the set $[\![\varphi]\!]^M$, containing the possible worlds in M where φ is true (the truth set of φ), is recursively defined as follows.*

$$[\![p]\!]^M := V(p) \qquad\qquad [\![\varphi \wedge \psi]\!]^M := [\![\varphi]\!]^M \cap [\![\psi]\!]^M$$

$$[\![\neg\varphi]\!]^M := W \setminus [\![\varphi]\!]^M \qquad [\![\Box\,\varphi]\!]^M := \left\{ w \in W \mid [\![\varphi]\!]^M \in N(w) \right\}$$

The relevant clause is the one for $\Box\,\varphi$: a world w is in $[\![\Box\,\varphi]\!]^M$ (i.e., $\Box\,\varphi$ is true at w) if and only if the set of worlds where φ is true, $[\![\varphi]\!]^M$, is in the neighbourhood of w, $N(w)$, that is,

$$w \in [\![\Box\,\varphi]\!]^M \quad \text{iff} \quad [\![\varphi]\!]^M \in N(w)$$

By unfolding the definition of $\Diamond$ we obtain $[\![\Diamond\,\varphi]\!]^M = \left\{ w \in W \mid [\![\neg\varphi]\!]^M \notin N(w) \right\}$, that is, $w \in [\![\Diamond\,\varphi]\!]^M$ if and only if $[\![\neg\varphi]\!]^M \notin N(w)$.

When reading $\Box\,\varphi$ as "the agent knows φ", the neighbourhood function can be seen simply as a function that indicates the set of formulas the agent knows at each possible world: "the agent knows φ" at w $(w \in [\![\Box\,\varphi]\!]^M)$ if and only if $[\![\varphi]\!]^M \in N(w)$. This is very similar to what is done in syntactic approaches, where an agent's knowledge is given by a set of formulas; the only difference is that in a neighbourhood function each formula is represented not by a string of symbols, but rather by a set of worlds: those in which the formula is true.

Note also how, different from [12], we use the 'weakest' semantic interpretation in neighbourhood models. A stronger alternative establishes that

$$\llbracket \Box\, \varphi \rrbracket^M := \left\{ w \in W \mid \text{there is } U \in N(w) \text{ s. t. } U \subseteq \llbracket \varphi \rrbracket^M \right\}$$

Under this alternative, w does not need to have $\llbracket \varphi \rrbracket^M$ in its neighbourhood in order to satisfy $\Box\, \varphi$; a subset of it is enough.[1] Under this alternative, if $\varphi \to \psi$ is a valid formula, then so is $\Box\, \varphi \to \Box\, \psi$, a property that does not hold under the semantic interpretation of Definition 3. In later sections we will understand $\Box$ as indicating the agent's *explicit* knowledge, so a system in which this operator has less closure properties is better for our purposes since, from our perspective, an agent's *explicit* knowledge does not need to have any minimal amount of information; what matters are the epistemic actions the agent can perform.

3   From Neighbourhood to Relational Models

When epistemic logic formulas are interpreted in relational models we obtain omniscient agents: their knowledge contains every validity (the validity of φ implies the validity of $\Box\, \varphi$), and is closed under modus ponens (the K formula $\Box\, (\varphi \to \psi) \to (\Box\, \varphi \to \Box\, \psi)$ is valid), thus making it also closed under logical consequence. It is precisely because of these properties that, when working in relational models, the operator $\Box$ is better understood as describing the agent's *implicit semantic information*. Then, typically, explicit knowledge is defined as implicit knowledge that satisfies certain additional properties.

In neighbourhood models, the operator $\Box$ does not have the mentioned properties: $\Box\, \varphi$ does not need to be valid, even if φ is, and $\Box\, \psi$ does not need to be true, even if $\Box\, (\varphi \to \psi)$ and $\Box\, \varphi$ are. The only closure property that $\Box$ has is closure under substitution of equivalents: if φ and ψ are true in the same worlds of every model (if $\varphi \leftrightarrow \psi$ is valid), then the agent knows the former if and only if she knows the latter ($\Box\, \varphi \leftrightarrow \Box\, \psi$ is valid).[2] Thus, under this semantic model, it makes more sense to understand $\Box$ as indicating the agent's *explicit* knowledge. One natural question is, then, can we define the agent's *implicit* knowledge within this framework? And, if so, how?

For this we can follow the idea of syntactic approaches where explicit knowledge is the primitive concept (usually given by an arbitrary set of formulas without any closure property) and implicit knowledge is defined as the closure

---

[1] These two semantic interpretations are compared in [13].

[2] This property, still an important idealisation, is unavoidable when the truth of a formula depends only on the set of possible words where it holds. It does not need to hold when a formula's semantic interpretation involves additional components, as its syntactic form or its justification.

under logical consequence of its explicit counterpart (see, e.g., the *AGM* approach to belief revision [14] or the foundationalist discussion in [15]). Thus, we can define implicit knowledge in terms of what can be derived from its explicit counterpart or, in a more 'dynamic' way, in terms of *the epistemic actions* that make the agent's explicit knowledge closed under logical consequence. Within neighbourhood models there is a very natural way to do this, and the key is the known connection between the logically omniscient relational models and the 'more realistic' neighbourhood models: a relational model can be seen as a neighbourhood model in which the neighbourhood function satisfies certain closure properties. Then we can define implicit knowledge as what the agent knows explicitly when she reaches such 'ideal' state.

Let us make the connection precise (see [11] and [10] for details and proofs).

Definition 4. *Let N be a neighbourhood function N over a domain W. We say that N (i) is* closed under supersets *at w whenever $U \in N(w)$ and $U \subseteq T$ imply $T \in N(w)$; (ii) is* closed under binary intersections *at w whenever $U \in N(w)$ and $T \in N(w)$ imply $(U \cap T) \in N(w)$; (iii)* contains the unit *at w whenever $W \in N(w)$. We say that the function N is* closed under supersets/is closed under binary intersections/contains the unit *in case it has the respective property for every element of its domain. Similarly, we will say that a KN model M has one of this properties in case its neighbourhood function has it.*

What makes the three properties in Definition 4 interesting is that a *finite* neighbourhood model that satisfies them corresponds to a relational model. First, we state formally what "correspond" means.

Definition 5 (Pointwise equivalence). *Let $M^K = \langle W, R, V \rangle$ be a relational model (the 'K' stands for Kripke) with $R \subseteq W \times W$ its the epistemic indistinguishability relation; define the satisfiability relation $\Vdash$ between (M^K, w) and a formula φ in $\mathcal{L}$ in the standard way. Let $M = \langle W, N, V \rangle$ be a neighbourhood model with the same domain and the same atomic valuation. We say that M^K and M are* pointwise equivalent *whenever for every $w \in W$ and for every formula φ in $\mathcal{L}$,*

$$(M^K, w) \Vdash \varphi \qquad \text{if and only if} \qquad w \in [\![\varphi]\!]^M$$

Then the theorem in our 'knowledge' ($\Box\, \varphi \to \varphi$) setting.

Theorem 1. *Let $M^K = \langle W, R, V \rangle$ be a finite and reflexive relational model; then M^K is pointwise equivalent to some KN model that is closed under supersets and binary intersections and contains the unit. Likewise (the direction relevant for us), let $M = \langle W, N, V \rangle$ be a KN model that is closed under supersets and binary intersections and contains the unit; then M is pointwise equivalent to some finite and reflexive relational model.*

Then, given any *KN* model, we can apply the appropriate operations to make it satisfy the mentioned properties, and thus make it pointwise equivalent to a reflexive relational model.

Definition 6. *Let $M = \langle W, N, V \rangle$ be a KN model.*

(i) *The model $M_{\subseteq^*} = \langle W, N_{\subseteq^*}, V \rangle$ differs from M only in the neighbourhood function, given for every $w \in W$ by*

$$T \in N_{\subseteq^*}(w) \text{ iff } U \subseteq T \text{ for some } U \in N(w)$$

(ii) *The model $M_{\cap^*} = \langle W, N_{\cap^*}, V \rangle$ differs from M only in the neighbourhood function, given for every $w \in W$ by*

$$T \in N_{\cap^*}(w) \text{ iff } T = T_1 \cap \cdots \cap T_n \text{ with } n > 0 \text{ and } T_i \in N(w) \ (1 \leq i \leq n)$$

(iii) *The model $M_{+W} = \langle W, N_{+W}, V \rangle$ differs from M only in the neighbourhood function, given for every $w \in W$ by*

$$T \in N_{+W}(w) \text{ iff } T \in (N(w) \cup \{W\})$$

Each operation extends the original model, yielding one with the expected properties.

Proposition 1. *Let $M = \langle W, N, V \rangle$ be a KN model.*

(i) *Take the model $M_{\subseteq^*} = \langle W, N_{\subseteq^*}, V \rangle$. Then, (a) $M_{\subseteq^*}$ is a KN model, (b) for every $w \in W$, $N(w) \subseteq N_{\subseteq^*}(w)$, and (c) $M_{\subseteq^*}$ is closed under supersets.*

(ii) *Take the model $M_{\cap^*} = \langle W, N_{\cap^*}, V \rangle$. Then, (a) $M_{\cap^*}$ is a KN model, (b) for every $w \in W$, $N(w) \subseteq N_{\cap^*}(w)$, and (c) $M_{\cap^*}$ is closed under binary intersections.*

(iii) *Take the model $M_{+W} = \langle W, N_{+W}, V \rangle$. Then, (a) M_{+W} is a KN model, (b) for every $w \in W$, $N(w) \subseteq N_{+W}(w)$, and (c) M_{+W} contains the unit.*

Since each operation preserves *KN* models, so does their composition. This has two readings. First, technically, it allows us to introduce a modality for describing the model that results from such composition (it is a *KN* model, so we can evaluate $\mathcal{L}$ formulas in it). Second, the *truth* property is what makes $\Box\varphi \rightarrow \varphi$ valid, and therefore what makes the agent's information *true*. The fact that the operations (each one of them and their composition) preserve the property tells us that they *preserve truth*, that is, each one of them can be seen as a *deductive* inference operation. There are, of course, other operations that preserve this property, like closure under union, but there are also operations that do not preserve it, like closure under subsets.

4 Explicit and Implicit Knowledge

We now introduce a modality for describing the behaviour of the composition of these operations.

Definition 7 (Full closure operation). *Take a KN model M. The KN model $M*$ is given by the application of closure under supersets, then closure under binary intersections and finally adding the unit, that is,*

$$M* := \left((M_{\subseteq*})_{\cap*} \right)_{+W}{}^{3}$$

Definition 8 (Full closure modality). *Take a KN model M, and let φ be a formula in $\mathcal{L}$; then $\langle * \rangle \varphi$ is also in $\mathcal{L}$. The truth-set of formulas with $\langle * \rangle$ (the full closure modality) is given by*

$$[\![\langle * \rangle \varphi]\!]^M := [\![\varphi]\!]^{M*}$$

Its dual *formula $[*]\varphi$ is defined as $[*]\varphi := \neg \langle * \rangle \neg \varphi$.*

As mentioned, the neighbourhood function can be seen as returning the set of formulas the agent knows at each possible world, close to the idea of explicit knowledge in syntactic approaches. Then it is natural to use formulas that allows us to look into such set for defining our notion of explicit knowledge:

$$K_{\mathrm{Ex}}\varphi := \Box\,\varphi$$

In words, the agent knows *explicitly* a given φ at a world w in a *KN* model M if and only if $[\![\varphi]\!]^M \in N(w)$, that is, if and only if the truth set of φ at M is in the neighbourhood of w at M.

The notion of implicit knowledge, on the other hand, is not defined as what the agent has in the neighbourhood of w *at M*, but rather as what she will have in the neighbourhood of that world after the model reaches the 'ideal' relational-model-pointwise-equivalent state $M*$:

$$K_{\mathrm{Im}}\varphi := \langle * \rangle \Box\,\varphi$$

Thus, the agent knows *implicitly* a given φ at a world w in a *KN* model M if and only if $[\![\varphi]\!]^{M*} \in N*(w)$, that is, if and only if the truth set of φ at $M*$ is in the neighbourhood of w at $M*$. This states that the agent knows φ *implicitly* if and only if she knows it explicitly after her knowledge set (given by the neighbourhood function) has reached certain 'ideal' closure properties.

It is worthwhile to emphasise some relevant features of these definitions. First, we are using explicit knowledge as the primitive, and then defining implicit knowledge in terms of it. This is the typical idea in syntactic approaches where explicit knowledge is given as a set of formulas and implicit knowledge is given as its closure under logical consequence. But in semantic approaches, like our current one, what is typically done is the opposite: implicit knowledge is taken as a primitive (in relational models it corresponds to the modal universal operator) and then explicit knowledge is defined in terms of it, usually as implicit knowledge that satisfy some extra property(ies).

---

[3] In fact, the three operations commute pairwise when working with finite models, so we can apply them in any order.

Second, we have defined implicit knowledge in M as what the agent knows explicitly in a KN model that extends the current one (Proposition 1) and that behaves like a relational model (Theorem 1). Thus, implicit knowledge has the omniscient properties.

Fact 1. *Let φ and ψ be formulas in $\mathcal{L}$. Then*

(i) *if φ is valid then so is $K_{\text{Im}}\varphi$ (if φ is valid, then it is implicitly known),*

(ii) *the formula $K_{\text{Im}}(\varphi \to \psi) \to (K_{\text{Im}}\varphi \to K_{\text{Im}}\psi)$ is valid (implicit knowledge is closed under modus ponens).*

Proof. For the first, given any KM model $M = \langle W, N, V \rangle$, $M*$ is also a KM model (Proposition 1) and therefore, since φ is valid, $[\![\varphi]\!]^{M*} = W$. By Proposition 1 the neighbourhood of every $w \in W$ at $M*$ contains the unit, so $W \in N*(w)$, that is, $[\![\varphi]\!]^{M*} \in N*(w)$. Then $w \in [\![\Box\,\varphi]\!]^{M*}$, that is, $w \in [\![\langle * \rangle\,\Box\,\varphi]\!]^{M}$.

For the second, observe that the full closure operation produces a unique result. Then, for any KN model $M = \langle W, N, V \rangle$ and any world $w \in W$, $w \in [\![\langle * \rangle\,\Box\,(\varphi \to \psi) \land \langle * \rangle\,\Box\,\varphi]\!]^{M}$ gives us $w \in [\![\langle * \rangle\,(\Box\,(\varphi \to \psi) \land \Box\,\varphi)]\!]^{M}$ (if the operation produced more than one result, we could have $\Box\,(\varphi \to \psi)$ and $\Box\,\varphi$ true in different models, not in the unique $M*$). Then we have $w \in [\![\Box\,(\varphi \to \psi) \land \Box\,\varphi]\!]^{M*}$, that is, w is in both $[\![\Box\,(\varphi \to \psi)]\!]^{M*}$ and $[\![\Box\,\varphi]\!]^{M*}$. But since $M*$ is pointwise equivalent to a relational model (in which the K formula $\Box\,(\varphi \to \psi) \to (\Box\,\varphi \to \Box\,\psi)$ is valid), we also have $w \in [\![\Box\,\psi]\!]^{M*}$ and hence $w \in [\![\langle * \rangle\,\Box\,\psi]\!]^{M}$.

When proving part *(ii)* we used the fact that implicit knowledge is defined as what the agent knows explicitly in the *single* 'ideal' (and final, from a deductive inference point of view) model $M*$. We could have defined this notion as what the agent knows explicitly in some intermediate point, stating in this way that implicit knowledge is what the agent can eventually derive, but we did not follow that idea because, as we will see and discuss (Subsection 4.1), the agent might know explicitly a given φ at some KN model M and yet not know it explicitly in the 'ideal' model $M*$. Thus, we have chosen to define implicit knowledge not as what might be explicitly known in some intermediate step, but rather as what will survive all the reasoning process and therefore not only will become explicitly known at some point, but also will still be explicitly known in the 'ideal' final state. This highlights that our implicit knowledge corresponds to what will be explicit 'in the limit' of the deductive inference reasoning process, that is, it is a form of *stable* explicit knowledge that will not be affected by further *deductive* inferences.[4]

It is also interesting to observe how the model operations that define $M*$ are *closure operations*, that is, operations that make the neighbourhood function of the model closed under certain operations between sets. Thus, it is not only that we are defining implicit knowledge as what agent knows explicitly after

---

[4] In this sense, our approach is close to the idea of *identification in the limit* [16] proposed in learning theory, which has been already studied in a dynamic epistemic logic framework [17].

some operations: we are defining implicit knowledge as what the agent knows explicitly after *a certain fixed point is reached*, the fixed point $M*$ where the neighbourhood function is closed under supersets and binary intersections and contains the unit. From this perspective, the omniscient properties the agent's knowledge has in the 'ideal' relational-model-pointwise-equivalent state are then not seen as 'static' properties that the agent's knowledge either has or does not have; instead, they are understood as the final result of the iterative application of the adequate epistemic actions, in this case, truth-preserving reasoning steps.

4.1 Properties of the Defined Notions

Most approaches that deal with both explicit and implicit knowledge [3,4,5,6] satisfy two intuitive properties. The first, a property that has been considered almost mandatory in any approach for these notions, states that explicit knowledge is also implicit: if the agent knows explicitly any formula, then she will still know it once her knowledge is closed under logical consequence. The second indicates that the consequent of an explicitly known implication with explicitly known antecedent is already implicitly known (what we might call an implicit/explicit version of the K formula): if the agent knows explicitly and implication and its antecedent, then surely the consequent is something she can derive.

These two properties are not satisfied in our framework. First, explicit knowledge is not always implicit.

Fact 2. *The formula $K_{Ex}\varphi \to K_{Im}\varphi$ is not valid in KN models.*

Proof. We just need to provide a formula φ together with a KN model and a world in it where $K_{Ex}\varphi \wedge \neg K_{Im}\varphi$ (that is, $\Box\,\varphi \wedge \neg\langle*\rangle\,\Box\,\varphi$) holds. Take φ as $\neg\Box\,q$, and consider a four-worlds KN model $M = \langle W = \{w_1, w_2, w_3, w_4\}, N, V\rangle$ over the set of atomic propositions $\{p, q\}$ where $V(p) = \{w_1, w_2\}$ and $V(q) = \{w_1, w_3\}$. Moreover, suppose that the neighbourhood function is given by

$$N(w_1) := \{\{w_1, w_2\}, \{w_1, w_3, w_4\}, W\}, \quad N(w_2) = N(w_3) = N(w_4) := \varnothing$$

Observe how $[\![\neg\Box\,q]\!]^M = W$ is in $N(w_1)$ so $w_1 \in [\![\Box\,\neg\Box\,q]\!]^M$: the agent knows explicitly $\neg\Box\,q$ at w_1 in M. Now, this is the neighbourhood function of $M*$:

$$N*(w_1) = \{\{w_1\}, \{w_1, w_2\}, \{w_1, w_3\}, \{w_1, w_4\},$$
$$\{w_1, w_2, w_3\}, \{w_1, w_2, w_4\}, \{w_1, w_3, w_4\}, W\}$$
$$N*(w_2) = N*(w_3) = N*(w_4) = \{W\}$$

Note how $[\![\neg\Box\,q]\!]^{M*} = \{w_2, w_3, w_4\}$ is not in $N*(w_1)$ so $w_1 \notin [\![\Box\,\neg\Box\,q]\!]^{M*}$, that is, $w_1 \notin [\![\langle*\rangle\,\Box\,\neg\Box\,q]\!]^M$: the agent does not know implicitly $\neg\Box\,q$ at w_1 in M.

Second, the consequent of an explicitly known implication whose antecedent is also explicitly known does not need to be implicitly known.

Fact 3. *The formula $K_{Ex}(\varphi \to \psi) \to (K_{Ex}\varphi \to K_{Im}\psi)$ is not valid in KN models.*

Proof. Consider the KN model M in the proof of Fact 2 and take φ as $\top$ and ψ as $\neg\Box\, q$; we will show that, at w_1 in M, although $\top \to \neg\Box\, q$ and $\top$ are explicitly known, $w_1 \in [\![\Box\,(\top \to \neg\Box\, q) \wedge \Box\, \top]\!]^M$, $\neg\Box\, q$ is not implicitly known, $w_1 \notin [\![\langle * \rangle\, \Box\, \neg\Box\, q]\!]^M$.

We have $[\![\top]\!]^M = [\![\top \to \neg\Box\, q]\!]^M = W$ in $N(w_1)$, so $w_1 \in [\![\Box\,(\top \to \neg\Box\, q) \wedge \Box\, \top]\!]^M$: the agent knows explicitly both $\top \to \neg\Box\, q$ and $\top$ at w_1 in M. Still, as before, $[\![\neg\Box\, q]\!]^{M*} = \{w_2, w_3, w_4\}$ is not in $N*(w_1)$ so $w_1 \notin [\![\Box\,\neg\Box\, q]\!]^{M*}$, that is, $w_1 \notin [\![\langle * \rangle\, \Box\, \neg\Box\, q]\!]^M$: the agent does not know implicitly $\neg\Box\, q$ at w_1 in M.

The reason for the failure of these properties is that our agent has knowledge not only about propositional facts but also about her own (and eventually other agents') knowledge. Since the agent's knowledge (semantically, the neighbourhood function) changes through the operations, she might know something at some point and not know it afterwards (semantically, we might have $U \in N(w)$ with $U = [\![\varphi]\!]^M$ for some φ but, though Proposition 1 guarantees $U \in N*(w)$, nothing guarantees $U = [\![\varphi]\!]^{M*}$). This is nothing but an instance of the so called 'Moorean phenomena', which occurs when an epistemic action invalidates itself. In its best known incarnation it appears as formulas that, after being publicly announced [18,19], are not known [20,21]; in our case it appears as situations in which some logical consequences of what is explicitly known at some stage are not explicitly known after our closure-under-logical-consequence operation.

But this does not mean that neighbourhood models are not adequate for representing explicit and implicit knowledge. They behave as expected when we restrict our attention to formulas whose truth-values are not affected by the operations (this includes every purely propositional formula).

Proposition 2. *Let $\varphi \to \psi$ and φ be formulas whose truth-value is preserved by the full closure operation, that is, assume*

$$\varphi \to \langle * \rangle\, \varphi, \qquad \text{and} \qquad (\varphi \to \psi) \to \langle * \rangle\, (\varphi \to \psi)$$

Then the following formulas are valid

$$K_{\mathrm{Ex}}\varphi \to K_{\mathrm{Im}}\varphi, \qquad \text{and} \qquad K_{\mathrm{Ex}}(\varphi \to \psi) \to (K_{\mathrm{Ex}}\varphi \to K_{\mathrm{Im}}\psi)$$

Proof. Take any KN model $M = \langle W, N, V \rangle$ and a world $w \in W$. The validity of $\varphi \to \langle * \rangle\, \varphi$ says that $w \in [\![\varphi]\!]^M$ implies $w \in [\![\langle * \rangle\, \varphi]\!]^M$, that is, $[\![\varphi]\!]^M \subseteq [\![\langle * \rangle\, \varphi]\!]^M$, i.e., $[\![\varphi]\!]^M \subseteq [\![\varphi]\!]^{M*}$. Likewise, the validity of $(\varphi \to \psi) \to \langle * \rangle\, (\varphi \to \psi)$ gives us $[\![\varphi \to \psi]\!]^M \subseteq [\![\varphi \to \psi]\!]^{M*}$. Recall that $K_{\mathrm{Ex}}\varphi := \Box\, \varphi$ and $K_{\mathrm{Im}}\varphi := \langle * \rangle\, \Box\, \varphi$.

For the first, suppose $w \in [\![\Box\, \varphi]\!]^M$; then $[\![\varphi]\!]^M \in N(w)$. Since $N*(w)$ extends $N(w)$ we have $[\![\varphi]\!]^M \in N*(w)$; since $[\![\varphi]\!]^M \subseteq [\![\varphi]\!]^{M*}$ and $N*(w)$ is closed under supersets we have $[\![\varphi]\!]^{M*} \in N*(w)$, that is, $w \in [\![\Box\, \varphi]\!]^{M*}$ so $w \in [\![\langle * \rangle\, \Box\, \varphi]\!]^M$.

For the second, suppose $w \in [\![\Box\, (\varphi \to \psi) \wedge \Box\, \varphi]\!]^M$; then we have $[\![\varphi \to \psi]\!]^M$ and $[\![\varphi]\!]^M$ in $N(w)$. Since $N*(w)$ extends $N(w)$ we have $[\![\varphi \to \psi]\!]^M$ and $[\![\varphi]\!]^M$ in $N*(w)$; since $[\![\varphi]\!]^M \subseteq [\![\varphi]\!]^{M*}$, $[\![\varphi \to \psi]\!]^M \subseteq [\![\varphi \to \psi]\!]^{M*}$ and $N*(w)$ is closed under supersets, we have $[\![\varphi \to \psi]\!]^{M*}$ and $[\![\varphi]\!]^{M*}$ in $N*(w)$, that is, w is in both $[\![\Box\, (\varphi \to \psi)]\!]^{M*}$ and $[\![\Box\, \varphi]\!]^{M*}$, i.e., in both $[\![\langle * \rangle\, \Box\, (\varphi \to \psi)]\!]^M$ and $[\![\langle * \rangle\, \Box\, \varphi]\!]^M$. Then by Fact 1 we have w in $[\![\langle * \rangle\, \Box\, \psi]\!]^M$.

Note how we have not required for φ's (and $(\varphi \to \psi)$'s) truth-value to be invariant under the full closure operation (syntactically, $\varphi \leftrightarrow \langle * \rangle \varphi$; semantically, $\llbracket \varphi \rrbracket^M = \llbracket \varphi \rrbracket^{M*}$). We have only asked for φ's (and $(\varphi \to \psi)$'s) truth to be preserved by it. Thus, in order to obtain the expected properties we do not need to restrict ourselves to formulas φ whose truth set does not change after the full closure operation, $\llbracket \varphi \rrbracket^M = \llbracket \varphi \rrbracket^{M*}$; we just need to restrict ourselves to formulas φ whose truth set *does not shrink*, $\llbracket \varphi \rrbracket^M \subseteq \llbracket \varphi \rrbracket^{M*}$. More precisely, an explicitly know formula is also implicitly known and the consequent of an explicitly known implication whose antecedent is also explicitly known is implicitly known *as long as we are dealing with formulas whose truth set will not become smaller as a consequence of our deductive inference operations.*

5 A Modus Ponens Action

We have defined implicit knowledge in terms of an operation that makes the agent's knowledge closed under logical consequence (technically, an operation that makes the neighbourhood model pointwise equivalent to a relational model). The operation represents the result of several deductive inference steps, but it is also interesting to see how the agent's knowledge changes after just one of them.

Definition 9. *Let $M = \langle W, N, V \rangle$ be a KN model, and let $\eta \to \chi$ be a formula in $\mathcal{L}$. The modus ponens operation with $\eta \to \chi$ produces the model $M_{\underrightarrow{\eta \to \chi}} = \langle W, N_{\underrightarrow{\eta \to \chi}}, V \rangle$ in which, for every $w \in W$, the set $N_{\underrightarrow{\eta \to \chi}}(w)$ is given by*

$$N_{\underrightarrow{\eta \to \chi}}(w) := \begin{cases} N(w) \cup \{\llbracket \chi \rrbracket^M\} & \text{if } \{\llbracket \eta \rrbracket^M, \llbracket \eta \to \chi \rrbracket^M\} \subseteq N(w) \\ N(w) & \text{otherwise} \end{cases}$$

In words, a modus ponens operation adds the truth set of the implication's consequent to the neighbourhood of every world in which the agent already has the truth set of both the implication and its antecedent.

Proposition 3. *Let $M = \langle W, N, V \rangle$ be a KN model. Then so is $M_{\underrightarrow{\eta \to \chi}}$.*

The previous proposition tells us that our operation represents a truth-preserving (i.e., deductive) inference step. Now, for the language.

Definition 10. *The modality $\langle \overset{\eta \to \chi}{\longrightarrow} \rangle$ is semantically defined as follows:*

$$\llbracket \langle \overset{\eta \to \chi}{\longrightarrow} \rangle \varphi \rrbracket^M := \llbracket \varphi \rrbracket^{M_{\underrightarrow{\eta \to \chi}}}$$

There is no precondition for this action, but this does not affect its spirit. Instead of understanding it as a partial function (defined only when the implication and its antecedent are present), we understand it as a total one that in some cases (the implication or its antecedent or both are not present) will have no effect.

Interestingly, if the agent knows explicitly an implication and its antecedent, then after a modus ponens step she will know explicitly the consequent.

Proposition 4. *The following is valid in KN models:*

$$K_{\text{Ex}}(\eta \to \chi) \to (K_{\text{Ex}}\eta \to \langle \overset{\eta \to \chi}{\longrightarrow} \rangle K_{\text{Ex}}\chi)$$

In words, if the agent knows explicitly η and $\eta \to \chi$, then after a modus ponens step she will know η explicitly.

Proof. Take any KN model $M = \langle W, N, V \rangle$, any world $w \in W$, and suppose $\{ [\![\eta]\!]^M, [\![\eta \to \chi]\!]^M \} \subseteq N(w)$; then $[\![\chi]\!]^M \in N_{\eta \to \chi}(w)$. Now consider the models M and $M_{\overrightarrow{\eta \to \chi}}$: the only difference between them is that, in the second, the neighbourhood of each world might have been extended with one single set: $[\![\chi]\!]^M$. But then, the only formulas whose truth set can be affected are those that state something about the knowledge of χ, that is, formulas with $\Box \chi$ as a subformula. Thus, the truth set of χ itself cannot change, so $[\![\chi]\!]^M = [\![\chi]\!]^{M_{\overrightarrow{\eta \to \chi}}}$. Therefore, $[\![\chi]\!]^{M_{\overrightarrow{\eta \to \chi}}} \in N_{\eta \to \chi}(w)$.

Note how it is essential that the operation adds only one set. Operations that add two or more sets, even truth-preserving ones, do not guarantee that the agent will know explicitly the just inferred formulas: one of them might state lack of knowledge about the other. Consider a conjunction elimination *operation for $\eta \wedge \chi$* that adds the truth-set of η and the truth-set of χ: we might have $\chi := \neg \Box \eta$ so χ might not be true after the operation, and hence cannot be known (recall that in our models $\Box \varphi \to \varphi$ is valid).[5]

Though at first sight this proposition might seem to contradict Fact 3, this is not the case. The difference is that, while implicit knowledge is defined as what will be explicit when truth-preserving inference steps have reached a fixed point, our modus ponens operation represents one single such step that adds at most one single formula to the agent's knowledge. Together, these results tell us that a single modus ponens step cannot invalidate itself (Proposition 4), but it might be invalidated by *further* deductive steps (Fact 3). This is indeed what happens in the proof of Fact 3: the agent knows explicitly both $\top$ and $\top \to \neg \Box q$ (at w_1 in M) so after a modus ponens step she reaches a stage where she knows explicitly $\neg \Box q$. But she also knows explicitly both p and $p \to q$, so further modus ponens steps will make q explicitly known, thus invalidating the previous outcome. This is yet another proof that one should always know 'where to stop'.

6   Conclusions and Further Work

We have recalled the neighbourhood models framework, arguing that from an epistemic point of view, in these models formulas of the form $\Box \varphi$ are better read as describing the agent's *explicit knowledge*. Then, using the known connection between neighbourhood and relational models, we have defined a model operation that makes a neighbourhood model pointwise equivalent to a relational model, and we have defined a modality for it. With this modality we have

---

[5] The simplest instance of this, $\eta := p$, gives us exactly the paradigmatic form of Moorean formulas, since $\eta \wedge \chi$ becomes $p \wedge \neg \Box p$ (see [21]).

defined the notion of *implicit knowledge* as what the agent will know explicitly in this 'ideal' final state, and we have discussed some properties of these two notions, comparing them with some properties typically obtained in other frameworks. We have also introduced and briefly discussed an operation and a modality representing a single-step modus ponens action.[6]

There are several ways to extend the ideas presented in this work; here we mention the ones we find more appealing. (i) As Proposition 1 shows, the defined model operations preserve the truth property, and hence they can be seen as *deductive* inference operations, but it would be also interesting to deal with *non-deductive* inference steps. This is related to the following point since in the present work (ii) we have worked only with the notion of knowledge, but it also makes sense to work with other notions, like *beliefs*. Given the properties of $\Box$, this would correspond to a notion of *explicit* beliefs, so we can also define a notion of *implicit* beliefs in terms of the explicit ones. Finally, (iii) though studying different attitudes and their dynamics separately is interesting, it is even more interesting to study them together and, in particular, to explore the epistemic actions that arise from their combination (e.g., [22] understand abductive reasoning as a process that changes the agent's beliefs according to what she knows). It would be interesting to work with both an agent's knowledge and beliefs in a neighbourhood model setting, and also to explore inference steps that combine these two notions (e.g., in the style of [6]).

References

1. Hintikka, J.: Knowledge and Belief: An Introduction to the Logic of the Two Notions. Cornell University Press, Ithaca (1962)
2. Fagin, R., Halpern, J.Y., Moses, Y., Vardi, M.Y.: Reasoning about knowledge. The MIT Press, Cambridge (1995)
3. Levesque, H.J.: A logic of implicit and explicit belief. In: Proc. of AAAI 1984, Austin, TX, pp. 198–202 (1984)
4. Vardi, M.Y.: On epistemic logic and logical omniscience. In: Halpern, J.Y. (ed.) TARK, pp. 293–305. Morgan Kaufmann Publishers Inc., San Francisco (1986)
5. Fagin, R., Halpern, J.Y.: Belief, awareness, and limited reasoning. Artificial Intelligence 34(1), 39–76 (1988)
6. Velázquez-Quesada, F.R.: Dynamic epistemic logic for implicit and explicit beliefs (2013), http://personal.us.es/frvelazquezquesada/docs/delieb-01-05.pdf (accepted for publication)
7. Renne, B.: Multi-agent justification logic: Communication and evidence elimination. Synthese 185(suppl. 1), 43–82 (2012)
8. Scott, D.: Advice in modal logic. In: Lambert, K. (ed.) Philosophical Problems in Logic, Reidel, Dordrecht, The Netherlands, pp. 143–173 (1970)
9. Montague, R.: Universal grammar. Theoria 36(3), 373–398 (1970)
10. Pacuit, E.: Neighborhood semantics for modal logic. an introduction. In: Lecture Notes for the ESSLLI Course A Course on Neighborhood Structures for Modal Logic (2007)

---

[6] Derivation systems for the introduced modalities are not presented here for space reasons.

11. Chellas, B.F.: Modal Logic: An Introduction. Cambridge University Press, Cambridge (1980)
12. van Benthem, J., Pacuit, E.: Dynamic logics of evidence-based beliefs. Studia Logica 99(1), 61–92 (2011)
13. Areces, C., Figueira, D.: Which semantics for neighbourhood semantics? In: Boutilier, C. (ed.) IJCAI 2009, pp. 671–676. Morgan Kaufmann Publishers Inc., San Francisco (2009)
14. Alchourrón, C.E., Gärdenfors, P., Makinson, D.: On the logic of theory change: Partial meet contraction and revision functions. The Journal of Symbolic Logic 50(2), 510–530 (1985)
15. Rott, H.: Change, Choice and Inference: a Study of Belief Revision and Nonmonotonic Reasoning. Oxford Logic Guides, vol. 42. Oxford Science Publications (2001)
16. Gold, E.M.: Language identification in the limit. Information and Control 10(5), 447–474 (1967)
17. Gierasimczuk, N.: Bridging learning theory and dynamic epistemic logic. Synthese (Knowledge, Rationality and Action) 169(2), 371–384 (2009)
18. Plaza, J.A.: Logics of public communications. In: Emrich, M.L., Pfeifer, M.S., Hadzikadic, M., Ras, Z.W. (eds.) Proceedings of the 4th International Symposium on Methodologies for Intelligent Systems, pp. 201–216. Oak Ridge National Laboratory, ORNL/DSRD-24, Tennessee (1989)
19. Gerbrandy, J., Groeneveld, W.: Reasoning about information change. Journal of Logic, Language, and Information 6(2), 147–196 (1997)
20. van Ditmarsch, H., Kooi, B.P.: The secret of my success. Synthese 151(2), 201–232 (2006)
21. Holliday, W.H., Icard, T.F.: Moorean phenomena in epistemic logic. In: Beklemishev, L., Goranko, V., Shehtman, V. (eds.) Advances in Modal Logic, pp. 178–199. College Publications (2010)
22. Velázquez-Quesada, F.R., Nepomuceno-Fernández, Á., Soler-Toscano, F.: An epistemic and dynamic approach to abductive reasoning: Abductive problem and abductive solution. To appear in Journal of Applied Logic (2013), doi:10.1016/j.jal.2013.07.002

Expressivity Hierarchy of Languages for Epistemic Awareness Models

Fernando R. Velázquez-Quesada

Grupo de Lógica, Lenguaje e Información, Universidad de Sevilla
FRVelazquezQuesada@us.es

Abstract. We study the expressivity hierarchy of languages for epistemic awareness models based on the operators for implicit and explicit knowledge, implicit and explicit possibility and awareness, providing in each case an expressivity characterisation in terms of bisimulation.

1 Introduction

One of the most influential approaches to deal with the logical omniscience problem within epistemic logic (*EL*; [1,2]) has been *awareness logic* [3]. This approach is based on the idea of distinguishing between the agent's potential *implicit* knowledge, what she can eventually get, and her actual *explicit* knowledge, what she currently has (see, e.g., [4,5,6]). In particular, the key observation in awareness logic is that, in order for an implicitly known φ to be explicitly known, the agent needs to be *aware of* it.

Of particular interest not only in logic but also in computer science and economics has been the case of *awareness based on atomic propositions*, in which an agent's awareness is generated by a set of atoms. In such case, and within the *EL* framework, the agent's knowledge is semantically represented with an *epistemic awareness model*: a relational model extended with a function $\mathcal{A}$ that assigns a set of atomic propositions to the agent at each possible world. Then it is said that the agent is aware of a given φ at some world w if and only if every atom in φ is in $\mathcal{A}(w)$ (in symbols, $atm(\varphi) \subseteq \mathcal{A}(w)$).

Typically, the language used for describing epistemic awareness models has been the standard modal language extended with an operator that verifies whether the atoms of a given formula belong to the $\mathcal{A}$-set of a given possible world, that is, a language whose primitive epistemic operators are that of implicit knowledge (the standard universal modal operator) and that of awareness (the just described one). Yet, there are several other possibilities: one might consider a language in which the basic operators are those for *explicit* knowledge and awareness, or another based only on explicit knowledge, or even one that uses only explicit possibility (the 'dual' of explicit knowledge). The choice of the language is crucial: different languages might have different expressivity, and hence might represent different intuitions about which kind situations (i.e., models) can or cannot be distinguished *from the agent's perspective*.

D. Grossi, O. Roy, and H. Huang (Eds.): LORI 2013, LNCS 8196, pp. 253–266, 2013.

The present work explores the expressivity hierarchy of different languages over epistemic awareness models. We take primitive operators for the notions of implicit knowledge, implicit possibility, explicit knowledge, explicit possibility and awareness, and then we compare the expressivity of the different languages that arise from these operator's combinations.[1] Besides the expressivity hierarchy, the main results of our work are the characterisation of the discussed languages' expressivity in terms of their appropriate concepts of bisimulation.

Notation. Let $\mathcal{L}_1$ and $\mathcal{L}_2$ be languages for the same semantic model.

- $\mathcal{L}_1 \preccurlyeq \mathcal{L}_2$, "$\mathcal{L}_2$ *is at least as expressive as* $\mathcal{L}_1$", indicates that every formula in $\mathcal{L}_1$ is semantically equivalent to some formula in $\mathcal{L}_2$. A typical proof for such a statement is a translation $tr : \mathcal{L}_1 \to \mathcal{L}_2$ from any formula φ in $\mathcal{L}_1$ to a formula $tr(\varphi)$ in $\mathcal{L}_2$ such that, in any given semantic model, φ holds if and only if $tr(\varphi)$ holds. Note how $\preccurlyeq$ is reflexive and transitive.
- $\mathcal{L}_1 \not\preccurlyeq \mathcal{L}_2$, "$\mathcal{L}_2$ *is not as expressive as* $\mathcal{L}_1$", indicates that there is a formula in $\mathcal{L}_1$ that does not correspond semantically to any formula in $\mathcal{L}_2$. A typical proof for such a statement is to provide two semantic models that satisfy exactly the same formulas in $\mathcal{L}_2$, and then provide a formula in $\mathcal{L}_1$ that holds in one model but fails in the other.
- $\mathcal{L}_1 \prec \mathcal{L}_2$, "$\mathcal{L}_2$ *is more expressive than* $\mathcal{L}_1$" (or, alternatively, "$\mathcal{L}_1$ *is less expressive than* $\mathcal{L}_2$"), is defined as $\prec := \preccurlyeq \cap \not\succcurlyeq$.
- $\mathcal{L}_1 \approx \mathcal{L}_2$, "$\mathcal{L}_1$ *and* $\mathcal{L}_1$ *are equally expressive*", is defined as $\approx := \preccurlyeq \cap \succcurlyeq$.
- $\mathcal{L}_1 \asymp \mathcal{L}_2$, "$\mathcal{L}_1$ *and* $\mathcal{L}_2$ *are incomparable*", is defined as $\asymp := \not\preccurlyeq \cap \not\succcurlyeq$.

2   Basic Definitions

Let P be a countable non-empty set of atomic propositions.

Definition 1 (Epistemic awareness model/state). *An* epistemic awareness model *(EAM)* M *is a tuple* $\langle W, R, \mathcal{A}, V \rangle$ *where* W *is a non-empty set of* possible worlds, $R \subseteq (W \times W)$ *is an* indistinguishability relation *(we define* $R[w] := \{u \in W \mid Rwu\}$*),* $\mathcal{A} : W \to \wp(P)$ *is an* awareness function, *and* $V : P \to \wp(W)$ *is an* atomic valuation. *A pair* (M, w) *with* $w \in W$ *is called an* epistemic awareness state *(EAS) and* w *is called its* evaluation point. *An epistemic awareness model* M *is* image-finite *if and only if* $R[w]$ *is finite for every world* w *in it.*

Epistemic awareness models can be described with different languages, from propositional ones to different variations of modal languages and even first-order ones. Here we focus on extensions of the propositional language that include a subset of the following modal operators:

$\square\,\varphi$ - The agent knows φ implicitly	$\Diamond\,\varphi$ - The agent considers φ possible implicitly
$K\,\varphi$ - The agent knows φ explicitly	$L\,\varphi$ - The agent considers φ possible explicitly
$A\,\varphi$ - The agent is aware of φ	

---

[1] We work in the single agent case, leaving group-knowledge operators like "everybody knows" and "it is common knowledge that" for further work.

Considering $\Box$ and $\Diamond$ is natural for any extension of a modal framework. For A (awareness) and K (explicit knowledge), they are standard operators for languages over epistemic awareness models. Finally, L is the dual of K.

Definition 2 (Languages).

- $\mathcal{L}(\ )$ *is the propositional language, crucially including the always true constant* $\top$.
- $\mathcal{L}(O_1, \ldots, O_n)$ *is the language extending* $\mathcal{L}(\ )$ *with the operators* $O_1, \ldots, O_n$.
- *For any language* $\mathcal{L}$ *and any* $\mathsf{Q} \subseteq \mathsf{P}$, $\mathcal{L}|\mathsf{Q}$ *is the set of formulas in* $\mathcal{L}$ *built by using only atoms in* Q, *that is,*

$$\mathcal{L}|\mathsf{Q} := \{\varphi \in \mathcal{L} \mid atm(\varphi) \subseteq \mathsf{Q}\}^2$$

Here is the semantic interpretation for each one of our relevant operators.

Definition 3 (Semantic interpretation). *Let* (M, w) *be an epistemic awareness state with* $M = \langle W, R, \mathcal{A}, V \rangle$. *Then we have*

$(M, w) \Vdash \Box \varphi$	iff	for all $u \in R[w]$ we have $(M, u) \Vdash \varphi$
$(M, w) \Vdash \Diamond \varphi$	iff	there is a $u \in R[w]$ such that $(M, u) \Vdash \varphi$
$(M, w) \Vdash K \varphi$	iff	$atm(\varphi) \subseteq \mathcal{A}(w)$ and for all $u \in R[w]$ we have $(M, u) \Vdash \varphi$
$(M, w) \Vdash L \varphi$	iff	$atm(\varphi) \subseteq \mathcal{A}(w)$ and there is a $u \in R[w]$ such that $(M, u) \Vdash \varphi$
$(M, w) \Vdash A \varphi$	iff	$atm(\varphi) \subseteq \mathcal{A}(w)$

where the function atm, *returning the set of atoms appearing in a given formula* φ, *is formally defined as follows:*

$$atm(\top) := \varnothing \qquad atm(\Box \varphi) := atm(\varphi)$$
$$atm(p) := \{p\} \qquad atm(\Diamond \varphi) := atm(\varphi)$$
$$atm(\neg\varphi) := atm(\varphi) \qquad atm(K \varphi) := atm(\varphi)$$
$$atm(\varphi \wedge \psi) := atm(\varphi) \cup atm(\psi) \qquad atm(L \varphi) := atm(\varphi)$$
$$atm(A \varphi) := atm(\varphi)$$

When negation is present, the implicit possibility operator $\Diamond$ and the implicit knowledge operator $\Box$ can be defined in terms of each other ($\Diamond \varphi \leftrightarrow \neg\Box\neg\varphi$ and $\Box \varphi \leftrightarrow \neg\Diamond\neg\varphi$ are valid), and hence one of them can be dropped from any language (here we will drop $\Diamond$) without losing expressivity. This is not the case with their explicit counterparts. Explicit possibility L is indeed definable in terms of explicit knowledge K and propositional operators ($L \varphi \leftrightarrow \neg K \neg \varphi \wedge K (\varphi \vee \neg\varphi)$), but the converse is not the case (this statement will be proved formally later on). Thus, here we will work with the operators in $\{\Box, K, L, A\}$.

One last definition before starting our work.

Definition 4 (Modal equivalence). *Let* $\mathcal{L}$ *be a language, and take any* $\mathsf{Q} \subseteq \mathsf{P}$. *Two awareness epistemic states* (M, w) *and* (M', w') *are* $\mathcal{L}$-*modally equivalent up to* Q, *notation* $(M, w) \equiv_{\mathsf{Q}}^{\mathcal{L}} (M', w')$, *whenever for all* $\varphi \in \mathcal{L}|\mathsf{Q}$

$$(M, w) \Vdash \varphi \qquad \text{if and only if} \qquad (M', w') \Vdash \varphi$$

---

[2] The function *atm* is defined formally in Definition 3.

3   Expressivity Hierarchy

3.1   Equivalence Classes for Languages $\mathcal{L}(\Box, A)$ and $\mathcal{L}(K, L)$

The following results are from [7].

Proposition 1 ([7]). *The languages* $\mathcal{L}(\Box, A)$, $\mathcal{L}(\Box, K)$, $\mathcal{L}(\Box, A, K)$, $\mathcal{L}(\Box, A, L)$, $\mathcal{L}(\Box, K, L)$ *and* $\mathcal{L}(\Box, A, K, L)$ *are all equally expressive.*

Proof. *The proposition follows from the following equivalences:*

$$A\,\varphi \leftrightarrow K\,(\varphi \vee \neg\varphi), \qquad K\,\varphi \leftrightarrow (\Box\,\varphi \wedge A\,\varphi), \qquad L\,\varphi \leftrightarrow (\neg\Box\,\neg\varphi \wedge A\,\varphi)$$

Proposition 2 ([7]). *The languages* $\mathcal{L}(K, L)$, $\mathcal{L}(K, A)$, $\mathcal{L}(L, A)$, $\mathcal{L}(K)$ *and* $\mathcal{L}(K, L, A)$ *are all equally expressive.*

Proof. *The proposition follows from the following equivalences:*

$$L\,\varphi \leftrightarrow (\neg K\,\neg\varphi \wedge A\,\varphi), \qquad K\,\varphi \leftrightarrow (\neg L\,\neg\varphi \wedge A\,\varphi), \qquad A\,\varphi \leftrightarrow K\,(\varphi \vee \neg\varphi)$$

Each one of these equivalence classes can be characterised semantically.

Definition 5 (Extended bisimulation [7]). *Take any* $\mathsf{Q} \subseteq \mathsf{P}$. *An* extended Q-bisimulation *between epistemic awareness models* $M = \langle W, R, \mathcal{A}, V \rangle$ *and* $M' = \langle W', R', \mathcal{A}', V' \rangle$ *is a relation* $Z_\mathsf{Q} \subseteq (W \times W')$ *such that, for every* (w, w') *in* Z_Q:

- **atoms:** *for every* $p \in \mathsf{Q}$, $w \in V(p)$ *iff* $w' \in V'(p)$;
- **aware:** $\mathsf{Q} \cap \mathcal{A}(w) = \mathsf{Q} \cap \mathcal{A}'(w')$;
- **forth:** *if* $u \in R[w]$ *then there is a* $u' \in R'[w']$ *such that* $Z_\mathsf{Q}uu'$;
- **back:** *if* $u' \in R'[w']$ *then there is a* $u \in R[w]$ *such that* $Z_\mathsf{Q}uu'$.

(M, w) *and* (M', w') *are* extended Q-bisimilar *states*, $(M, w) \simeq_\mathsf{Q}^{\mathcal{A}} (M', w')$, *when there is an extended* Q-bisimulation between M and M' that contains (w, w').

Theorem 1 ([7]). *Let* (M, w) *and* (M', w') *be epistemic awareness states with* M *and* M' *image-finite models, and take* $\mathsf{Q} \subseteq \mathsf{P}$. *Then,*

$$(M, w) \simeq_\mathsf{Q}^{\mathcal{A}} (M', w') \qquad iff \qquad (M, w) \equiv_\mathsf{Q}^{\mathcal{L}(\Box, A)} (M', w')$$

Definition 6 (Awareness bisimulation [7]). *Take any* $\mathsf{Q} \subseteq \mathsf{P}$. *An* awareness Q-bisimulation *between epistemic awareness models* $M = \langle W, R, \mathcal{A}, V \rangle$ *and* $M' = \langle W', R', \mathcal{A}', V' \rangle$ *is a relation* $Z_\mathsf{Q} \subseteq (W \times W')$ *such that every* (w, w') *in* Z_Q *satisfies the clauses* **atoms** *and* **aware** *of Definition 5, plus:*

- **a-forth:** *if* $u \in R[w]$ *then there is a* $u' \in R'[w']$ *such that* $Z_{\mathsf{Q}\cap\mathcal{A}(w)}uu'$;
- **a-back:** *if* $u' \in R'[w']$ *then there is a* $u \in R[w]$ *such that* $Z_{\mathsf{Q}\cap\mathcal{A}'(w')}uu'$.

where $Z_{\mathsf{Q}\cap\mathcal{A}(w)}$ *is an awareness* $(\mathsf{Q} \cap \mathcal{A}(w))$*-bisimulation and* $Z_{\mathsf{Q}\cap\mathcal{A}'(w')}$ *is an awareness* $(\mathsf{Q}\cap\mathcal{A}'(w'))$*-bisimulation, both between* M *and* M'. *Two states* (M, w) *and* (M', w') *are* awareness Q-bisimilar, $(M, w) \leftrightarrow_\mathsf{Q} (M', w')$, *when there is an awareness* Q*-bisimulation between* M *and* M' *that contains* (w, w').

Theorem 2 ([7]). *Let (M, w) and (M', w') be epistemic awareness states with M and M' image-finite models, and take $Q \subseteq P$. Then,*

$$(M, w) \mathbin{\underline{\leftrightarrow}}_Q (M', w') \qquad \textit{iff} \qquad (M, w) \equiv_Q^{\mathcal{L}(K,L)} (M', w')$$

The following relation between these equivalence classes can be established.

Proposition 3. $\mathcal{L}(K, L)$ *is not as expressive as* $\mathcal{L}(\Box, A)$ *($\mathcal{L}(\Box, A) \not\preceq \mathcal{L}(K, L)$).*

Proof. Consider the following epistemic awareness states N_1 and N_1' (evaluation point doubled circled), in which each possible world contains the atoms true at it (absence represents falsity) and its respective $\mathcal{A}$-set appears next to it.

$$N_1 \qquad\qquad\qquad\qquad N_1'$$

These states are awareness $\{p\}$-bisimilar, $(N_1, w) \mathbin{\underline{\leftrightarrow}}_{\{p\}} (N_1', w')$: the evaluation points coincide, modulo p, in atomic valuation and in $\mathcal{A}$-sets, and their successors are ($\{p\} \cap \{\ \}$)- (i.e., $\varnothing$-) bisimilar. Thus, by Theorem 2, these states are modally equivalent in $\mathcal{L}(K, L)|\{p\}$. Still, $\Box p \in \mathcal{L}(\Box, A)|\{p\}$ distinguishes them: it holds at (N_1, w) but fails at (N_1', w')). Hence, $\mathcal{L}(\Box, A) \not\preceq \mathcal{L}(K, L)$.

Given that the operators in $\mathcal{L}(K, L)$ can be defined in terms of those in $\mathcal{L}(\Box, A)$, in fact we have the following:

Corollary 1 ([7]). $\mathcal{L}(\Box, A)$ *is more expressive than* $\mathcal{L}(K, L)$ *($\mathcal{L}(K, L) \prec \mathcal{L}(\Box, A)$).*

Thus, we have the following relation between the languages in the equivalence classes of $\mathcal{L}(\Box, A)$ and $\mathcal{L}(K, L)$ (an arrow from $\mathcal{L}_1$ to $\mathcal{L}_2$ indicates $\mathcal{L}_1 \prec \mathcal{L}_2$):

$$\boxed{\mathcal{L}(K, L)} \longrightarrow \boxed{\mathcal{L}(\Box, A)}$$

We are left with the task of finding the position of the following five languages: $\mathcal{L}(\ )$, $\mathcal{L}(\Box)$, $\mathcal{L}(A)$, $\mathcal{L}(L)$ and $\mathcal{L}(\Box, L)$.

3.2 Language $\mathcal{L}(\ )$

We clearly have $\mathcal{L}(\ ) \prec \mathcal{L}(K, L)$ and, since $\mathcal{L}(K, L) \prec \mathcal{L}(\Box, A)$ and $\prec$ is transitive, we also have $\mathcal{L}(\ ) \prec \mathcal{L}(\Box, A)$.

3.3 Language $\mathcal{L}(\Box)$

Clearly, $\mathcal{L}(\Box)$ is less expressive than $\mathcal{L}(\Box, A)$, that is, $\mathcal{L}(\Box) \prec \mathcal{L}(\Box, A)$. Also clearly, $\mathcal{L}(\Box)$ is more expressive than $\mathcal{L}(\ )$, that is, $\mathcal{L}(\ ) \prec \mathcal{L}(\Box)$.

In order to compare $\mathcal{L}(\Box)$ with the equivalence class of $\mathcal{L}(K, L)$, we recall the definition of a standard bisimulation and its known relation with the language $\mathcal{L}(\Box)$ (see, e.g., Chapter 2 of [8] for details).

Definition 7 (Standard bisimulation). *Let* $Q \subseteq P$. *A standard* Q-*bisimulation between epistemic awareness models* $M = \langle W, R, \mathcal{A}, V \rangle$ *and* $M' = \langle W', R', \mathcal{A}', V' \rangle$ *is a relation* $Z_Q \subseteq (W \times W')$ *such that every* (w, w') *in* Z_Q *satisfies* **atoms**, **forth** *and* **back** *of Definition 5. The notation for standard* Q-*bisimilarity between epistemic awareness states is* $(M, w) \simeq_Q (M', w')$.

Theorem 3. *Let* (M, w) *and* (M', w') *be epistemic awareness states with* M *and* M' *image-finite models, and take* $Q \subseteq P$. *Then,*

$$(M, w) \simeq_Q (M', w') \qquad \textit{iff} \qquad (M, w) \equiv_Q^{\mathcal{L}(\Box)} (M', w')$$

Proposition 4. $\mathcal{L}(\Box)$ *and* $\mathcal{L}(K, L)$ *are incomparable* $(\mathcal{L}(\Box) \asymp \mathcal{L}(K, L))$.

Proof. For showing $\mathcal{L}(K, L) \npreceq \mathcal{L}(\Box)$, consider the following EASs:

$$N_2 \qquad\qquad\qquad\qquad N_2'$$

We have $(N_2, w) \simeq_{\{p\}} (N_2', w')$ *and by Theorem 3, the states are modally equivalent in* $\mathcal{L}(\Box)|\{p\}$. *Still,* $K p$ *in* $\mathcal{L}(K, L)|\{p\}$ *holds at* (N_2, w) *but fails at* (N_2', w').

For showing $\mathcal{L}(\Box) \npreceq \mathcal{L}(K, L)$, consider the EASs N_1 and N_1' in the proof of Proposition 3 (page 257). As mentioned, we have $(N_1, w) \leftrightarroweq_{\{p\}} (N_1', w')$ and hence, by Theorem 2, the two states are modally equivalent in $\mathcal{L}(K, L)|\{p\}$. Still, the states can be distinguished by the formula $\Box p$ in $\mathcal{L}(\Box)|\{p\}$.

3.4 Language $\mathcal{L}(A)$

Clearly, $\mathcal{L}(\ ) \prec \mathcal{L}(A)$. Moreover, $\mathcal{L}(A) \prec \mathcal{L}(K, L)$: A can be defined in terms of K (so $\mathcal{L}(A) \preccurlyeq \mathcal{L}(K, L)$) and, since L allows us to look at other worlds, there are states indistinguishable with $\mathcal{L}(A)$ but distinguishable with $\mathcal{L}(K, L)$ (so $\mathcal{L}(K, L) \npreceq \mathcal{L}(A)$). For comparing $\mathcal{L}(A)$ and $\mathcal{L}(\Box)$ we use the following.

Definition 8 (Local bisimulation). *Let* $Q \subseteq P$. *A local* Q-*bisimulation between epistemic awareness models* $M = \langle W, R, \mathcal{A}, V \rangle$ *and* $M' = \langle W', R', \mathcal{A}', V' \rangle$ *is a relation* $Z_Q \subseteq (W \times W')$ *such that every* (w, w') *in* Z_Q *satisfies both* **atoms** *and* **aware** *of Definition 5. The notation for local* Q-*bisimilarity between epistemic awareness states is* $(M, w) \simeq_Q^l (M', w')$.

Theorem 4. *Let* (M, w) *and* (M', w') *be EASs, and take* $Q \subseteq P$. *Then,*

$$(M, w) \simeq_Q^l (M', w') \qquad \textit{iff} \qquad (M, w) \equiv_Q^{\mathcal{L}(A)} (M', w')$$

Proposition 5. $\mathcal{L}(A)$ *and* $\mathcal{L}(\Box)$ *are incomparable* $(\mathcal{L}(A) \asymp \mathcal{L}(\Box))$.

Proof. For $\mathcal{L}(\Box) \not\preccurlyeq \mathcal{L}(A)$, consider (N_1, w) and (N_1', w') from Proposition 3 (page 257). We have $(N_1, w) \simeq_{\{p\}}^l (N_1', w')$ and, by Theorem 4, the states are modally equivalent in $\mathcal{L}(A)|\{p\}$. Still, $\Box p$ in $\mathcal{L}(\Box)|\{p\}$ distinguishes between them.

For $\mathcal{L}(A) \not\preccurlyeq \mathcal{L}(\Box))$, consider (N_2, w) and (N_2', w') from Proposition 4 (page 258). We have $(N_2, w) \simeq_{\{p\}} (N_2', w')$ and, by Theorem 3, the states are modally equivalent in $\mathcal{L}(\Box)|\{p\}$. Still, $A p$ in $\mathcal{L}(A)|\{p\}$ distinguishes between them.

The expressivity picture so far is the following, with transitive arrows omitted and unreachability indicating incomparability.

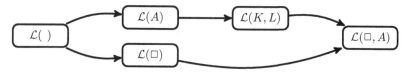

Up to now the results have been straightforward; the ones for the two remaining languages, $\mathcal{L}(L)$ and $\mathcal{L}(\Box, L)$, constitute the core of the present work.

3.5 Language $\mathcal{L}(L)$

Clearly, $\mathcal{L}(\ ) \prec \mathcal{L}(L)$. Moreover, $\mathcal{L}(L) \preccurlyeq \mathcal{L}(K, L)$; in order to establish $\mathcal{L}(K, L) \not\preccurlyeq \mathcal{L}(L)$ (and hence $\mathcal{L}(L) \prec \mathcal{L}(K, L)$), we use the following definition and results.

Definition 9 (Restricted awareness bisimulation). *Let* $\mathsf{Q} \subseteq \mathsf{P}$. *A restricted awareness* Q-*bisimulation between epistemic awareness models* $M = \langle W, R, \mathcal{A}, V \rangle$ *and* $M' = \langle W', R', \mathcal{A}', V' \rangle$ *is a relation* $Z_\mathsf{Q} \subseteq (W \times W')$ *such that every* (w, w') *in* Z_Q *satisfies* **atoms** *of Definition 5 plus the following two clauses:*

- **a-forth-bis:** *if* $u \in R[w]$ *then* (I) $(\mathsf{Q} \cap \mathcal{A}(w)) \subseteq (\mathsf{Q} \cap \mathcal{A}'(w'))$ *and* (II) *there is a* $u' \in R'[w']$ *such that* $Z_{\mathsf{Q} \cap \mathcal{A}(w)} uu'$;
- **a-back-bis:** *if* $u' \in R'[w']$ *then* (I) $(\mathsf{Q} \cap \mathcal{A}'(w')) \subseteq (\mathsf{Q} \cap \mathcal{A}(w))$ *and* (II) *there is a* $u \in R[w]$ *such that* $Z_{\mathsf{Q} \cap \mathcal{A}'(w')} uu'$.

The notation for restricted awareness Q-*bisimilar states is* $(M, w) \underline{\leftrightarrow}_\mathsf{Q}^r (M', w')$.

Note how a restricted awareness bisimulation does not ask for every pair in the relation to satisfy the **aware** clause. Instead, this requirement is distributed as an additional consequent within the **a-forth-bis** and **a-back-bis** clauses (their respective item (I)).

Our first important characterisation result is that, in image-finite models, restricted awareness bisimulation characterises modal equivalence in $\mathcal{L}(L)$.

Proposition 6. *Let* (M, w) *and* (M', w') *be epistemic awareness states, and take* $\mathsf{Q} \subseteq \mathsf{P}$. *Then*

$$(M, w) \underline{\leftrightarrow}_\mathsf{Q}^r (M', w') \qquad \text{implies} \qquad (M, w) \equiv_\mathsf{Q}^{\mathcal{L}(L)} (M', w')$$

Proof. See appendix A.

Proposition 7. *Let (M, w) and (M', w') be epistemic awareness states with M and M' image-finite models, and take $\mathbb{Q} \subseteq \mathbb{P}$. Then*

$$(M, w) \equiv_{\mathbb{Q}}^{\mathcal{L}(L)} (M', w') \quad implies \quad (M, w) \leftrightarrow_{\mathbb{Q}}^r (M', w')$$

Proof. See appendix A.

Theorem 5. *Let (M, w) and (M', w') be epistemic awareness states with M and M' image-finite models, and take $\mathbb{Q} \subseteq \mathbb{P}$. Then,*

$$(M, w) \leftrightarrow_{\mathbb{Q}}^r (M', w') \quad iff \quad (M, w) \equiv_{\mathbb{Q}}^{\mathcal{L}(L)} (M', w')$$

Now we can proceed.

Proposition 8. $\mathcal{L}(L)$ *is not as expressive as* $\mathcal{L}(K, L)$ *($\mathcal{L}(K, L) \not\preceq \mathcal{L}(L)$).*

Proof. Consider (N_2, w) and (N_2', w') from Proposition 4 *(page 258)*. Note how $(N_2, w) \leftrightarrow_{\{p\}}^r (N_2', w')$ and, by Theorem 5, the states are modally equivalent in $\mathcal{L}(L)|\{p\}$. Still, $K(p \vee \neg p)$ in $\mathcal{L}(K, L)|\{p\}$ distinguishes between them.

Incidentally, this also proves an earlier claim: K is not definable in $\mathcal{L}(L)$, that is, it is not definable in terms of L and boolean operators.

Corollary 2. $\mathcal{L}(K, L)$ *is more expressive than* $\mathcal{L}(L)$ *($\mathcal{L}(L) \prec \mathcal{L}(K, L)$).*

We also have the following results.

Proposition 9. $\mathcal{L}(L)$ *and* $\mathcal{L}(\square)$ *are incomparable ($\mathcal{L}(L) \asymp \mathcal{L}(\square)$).*

Proof. For $\mathcal{L}(\square) \not\preceq \mathcal{L}(L)$ we use EASs (N_1, w) and (N_1', w') from Proposition 3 *(page 257)*. We have $(N_1, w) \leftrightarrow_{\{p\}}^r (N_1', w')$ and, by Theorem 5, the states are modally equivalent in $\mathcal{L}(L)|\{p\}$. Still, $\square p$ in $\mathcal{L}(\square)|\{p\}$ distinguishes them.

For $\mathcal{L}(L) \not\preceq \mathcal{L}(\square)$ consider the states below: we have $(N_3, w) \simeq_{\{p\}} (N_3', w')$ and, by Theorem 3, the states are modally equivalent in $\mathcal{L}(\square)|\{p\}$. Still, Lp in $\mathcal{L}(L)|\{p\}$ holds at (N_3, w) but fails at (N_3', w').

Proposition 10. $\mathcal{L}(L)$ *and* $\mathcal{L}(A)$ *are incomparable ($\mathcal{L}(L) \asymp \mathcal{L}(A)$).*

Proof. For $\mathcal{L}(A) \not\preceq \mathcal{L}(L)$ consider (N_2, w') and (N_2', w') from Proposition 4, page 258: we have $(N_2, w) \leftrightarrow_{\{p\}}^r (N_2', w')$ and, by Theorem 5, the states are modally equivalent in $\mathcal{L}(L)|\{p\}$. Still, Ap in $\mathcal{L}(A)|\{p\}$ distinguishes between them.

For $\mathcal{L}(L) \not\preceq \mathcal{L}(A)$ consider the states below: we have $(N_4, w) \simeq_{\{p\}}^l (N_4', w')$ and, by Theorem 4, the states are modally equivalent in $\mathcal{L}(A)|\{p\}$. Yet, Lp in $\mathcal{L}(L)|\{p\}$ distinguishes between them.

3.6 Language $\mathcal{L}(\Box, L)$

Proposition 11. $\mathcal{L}(\Box, L)$ *is more expressive than* $\mathcal{L}(L)$ *(*$\mathcal{L}(L) \prec \mathcal{L}(\Box, L)$*).*

Proof. *Clearly* $\mathcal{L}(\Box, L)$ *is at least as expressive as* $\mathcal{L}(L)$ *so* $\mathcal{L}(L) \preccurlyeq \mathcal{L}(\Box, L)$. *Moreover, states* (N_1, w) *and* (N_1', w') *(the proof of the first part of Proposition 4, page 258) are such that* $(N_1, w) \underline{\leftrightarrow}^r_{\{p\}} (N_1', w')$ *and hence (Theorem 5) modally equivalent in* $\mathcal{L}(L)|\{p\}$. *Nevertheless, the formula* $\Box p$ *in* $\mathcal{L}(\Box)|\{p\}$ *distinguishes between them, and therefore* $\mathcal{L}(\Box, L) \not\preccurlyeq \mathcal{L}(L)$.

For comparing $\mathcal{L}(\Box, L)$ with $\mathcal{L}(A)$ and $\mathcal{L}(K, L)$ we use the following.

Definition 10 (Restricted extended bisimulation). *Let* $\mathsf{Q} \subseteq \mathsf{P}$. *A restricted extended* Q-*bisimulation between epistemic awareness models* $M = \langle W, R, \mathcal{A}, V \rangle$ *and* $M' = \langle W', R', \mathcal{A}', V' \rangle$ *is a relation* $Z_{\mathsf{Q}} \subseteq (W \times W')$ *such that every* (w, w') *in* Z_{Q} *satisfies* **atoms** *of Definition 5 plus the following two clauses:*

- **forth-bis:** *if* $u \in R[w]$ *then* (I) $(\mathsf{Q} \cap \mathcal{A}(w)) \subseteq (\mathsf{Q} \cap \mathcal{A}'(w'))$ *and* (II) *there is a* $u' \in R'[w']$ *such that* $Z_{\mathsf{Q}} u u'$;
- **back-bis:** *if* $u' \in R'[w']$ *then* (I) $(\mathsf{Q} \cap \mathcal{A}'(w')) \subseteq (\mathsf{Q} \cap \mathcal{A}(w))$ *and* (II) *there is a* $u \in R[w]$ *such that* $Z_{\mathsf{Q}} u u'$.

The notation for restricted extended Q-*bisimilar states is* $(M, w) \simeq^r_{\mathsf{Q}} (M', w')$.

Again, note how a restricted extended bisimulation does not ask for every pair in the relation to satisfy the **aware** clause. Instead, this requirement is distributed as an additional consequent within the **forth-bis** and **back-bis** clauses (their respective item (I)).

Our second important characterisation result is that, in image-finite models, restricted extended bisimulation characterises modal equivalence in $\mathcal{L}(\Box, L)$.

Proposition 12. *Let* (M, w) *and* (M', w') *be epistemic awareness states, and take* $\mathsf{Q} \subseteq \mathsf{P}$. *Then*

$$(M, w) \simeq^r_{\mathsf{Q}} (M', w') \qquad \text{implies} \qquad (M, w) \equiv^{\mathcal{L}(\Box, L)}_{\mathsf{Q}} (M', w')$$

Proof. See appendix A.

Proposition 13. *Let* (M, w) *and* (M', w') *be epistemic awareness states with* M *and* M' *image-finite models, and take* $\mathsf{Q} \subseteq \mathsf{P}$. *Then*

$$(M, w) \equiv^{\mathcal{L}(\Box, L)}_{\mathsf{Q}} (M', w') \qquad \text{implies} \qquad (M, w) \simeq^r_{\mathsf{Q}} (M', w')$$

Proof. See appendix A.

Theorem 6. *Let* (M, w) *and* (M', w') *be epistemic awareness states with* M *and* M' *image-finite models, and take* $\mathsf{Q} \subseteq \mathsf{P}$. *Then,*

$$(M, w) \simeq^r_{\mathsf{Q}} (M', w') \qquad \text{iff} \qquad (M, w) \equiv^{\mathcal{L}(\Box, L)}_{\mathsf{Q}} (M', w')$$

Now we can proceed.

Proposition 14. $\mathcal{L}(\Box, L)$ and $\mathcal{L}(A)$ are incomparable ($\mathcal{L}(\Box, L) \asymp \mathcal{L}(A)$).

Proof. For $\mathcal{L}(A) \npreceq \mathcal{L}(\Box, L)$, the states (N_2, w) and (N_2', w') from Proposition 4 (page 258) are such that $(N_2, w) \simeq^r_{\{p\}} (N_2', w')$ and, by Theorem 6, they are modally equivalent in $\mathcal{L}(\Box, L)|\{p\}$. Still, they can be distinguished by $A p$ in $\mathcal{L}(A)|\{p\}$.

For $\mathcal{L}(\Box, L) \npreceq \mathcal{L}(A)$, the states (N_4, w) and (N_4', w') from Proposition 10 (page 260) are such that $(N_4, w) \simeq^l_{\{\}} (N_4', w')$ and, by Theorem 4, they are modally equivalent in $\mathcal{L}(A)|\{\}$. Still, they can be distinguished by $\Diamond \top$ in $\mathcal{L}(\Box, L)|\{\}$.

Proposition 15. $\mathcal{L}(\Box, L)$ and $\mathcal{L}(K, L)$ are incomparable ($\mathcal{L}(\Box, L) \asymp \mathcal{L}(K, L)$).

Proof. For $\mathcal{L}(K, L) \npreceq \mathcal{L}(\Box, L)$, the states (N_2, w) and (N_2', w') (Proposition 4, page 258) satisfy $(N_2, w) \simeq^r_{\{p\}} (N_2', w')$ and, by Theorem 6, they are modally equivalent in $\mathcal{L}(\Box, L)|\{p\}$. Still, they can be distinguished by $K (p \lor \neg p)$ in $\mathcal{L}(K, L)|\{p\}$.

For $\mathcal{L}(\Box, L) \npreceq \mathcal{L}(K, L)$, the states (N_1, w) and (N_1', w') (Proposition 4, page 258) are awareness $\{p\}$-bisimilar (Definition 6) and, by Theorem 2, also modally equivalent in $\mathcal{L}(K, L)|\{p\}$. Still, they can be distinguished by $\Box p$ in $\mathcal{L}(\Box, L)|\{p\}$.

To complete the picture, we also have the following results.

Proposition 16. $\mathcal{L}(\Box)$ is less expressive than $\mathcal{L}(\Box, L)$ ($\mathcal{L}(\Box) \prec \mathcal{L}(\Box, L)$).

Proposition 17. $\mathcal{L}(\Box, L)$ is less expressive than $\mathcal{L}(\Box, A)$ ($\mathcal{L}(\Box, L) \prec \mathcal{L}(\Box, A)$).

Proof. For $\mathcal{L}(\Box, L) \prec \mathcal{L}(\Box, A)$, L can be defined in terms of $\Box$ and A (Proposition 1). For $\mathcal{L}(\Box, A) \npreceq \mathcal{L}(\Box, L)$, states (N_2, w) and (N_2', w) (page 258) are $\mathcal{L}(\Box)$-equivalent (Proposition 4) and also $\mathcal{L}(L)$-equivalent (first part of Proposition 10), thus $\mathcal{L}(\Box, L)$-equivalent, yet distinguishable by $A p$ in $\mathcal{L}(A)$.

The full expressivity picture is shown in Figure 1 (arrows represent $\prec$, transitive arrows are omitted and unreachability indicates incomparability).

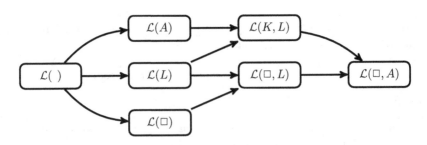

Fig. 1. Expressivity hierarchy for languages over epistemic awareness models

3.7 Discussion

The bisimulations results (Theorems 1 and 2) show that the difference between the two main expressivity classes, $\mathcal{L}(\Box, A)$ and $\mathcal{L}(K, L)$, is that the latter is restricted to the atoms the agent is aware of every time an epistemically indistinguishable world is visited. It can be argued, then, that $\mathcal{L}(K, L)$ is a more adequate language to work with epistemic awareness models, since it gives us exactly what the agent can see (just like $\mathcal{L}(\Box, \Diamond)$, equally expressive as $\mathcal{L}(\Box)$, is the adequate language for an agent with full awareness).[3]

One relevant result is that, different from the implicit knowledge case, the languages with only explicit knowledge ($\mathcal{L}(K)$, in the equivalence class of $\mathcal{L}(K, L)$) and that with only explicit possibility ($\mathcal{L}(L)$) are not equally expressive. Their respective bisimulation characterisations (awareness bisimulation, Theorem 2; restricted awareness bisimulation, Theorem 5) tell us precisely which one is the difference: different from $\mathcal{L}(K)$, the language $\mathcal{L}(L)$ can talk about the agent's awareness only when there are epistemic possibilities, that is, only when the indistinguishability relation is serial.[4] In more epistemic terms, in $\mathcal{L}(L)$ the agent can talk about her own awareness only when her implicit knowledge is consistent.

Another relevant result is the position in the expressivity hierarchy of the 'hybrid' language $\mathcal{L}(\Box, L)$ that combines implicit knowledge (and thus implicit possibility) with explicit possibility. The language lies between $\mathcal{L}(L)$ and $\mathcal{L}(\Box, A)$, and jumps to the latter when the indistinguishability relation is serial, just like the case of $\mathcal{L}(L)$. Interestingly, $\mathcal{L}(\Box, L)$ is incomparable to the language of explicit knowledge and explicit possibility, $\mathcal{L}(K, L)$, the differences being that while the former can see differences in implicit knowledge that the latter cannot, the latter can see differences in awareness, something the former can do only when the agent's implicit knowledge is consistent.

4 Summary and Further Work

We have presented the expressivity hierarchy of languages over epistemic awareness models based on the operators of implicit knowledge ($\Box$), implicit possibility ($\Diamond$), explicit knowledge (K), explicit possibility (L) and awareness (A). Besides stating their position within the expressivity hierarchy, we have characterised (in the class of image-finite epistemic awareness models) each one of these languages in terms of a (semantic) notion of bisimulation.

There is still further work to do. First, there are other useful operators for working with epistemic awareness models; in particular, we have the "speculative knowledge" of [9,10]. This and other operators allow us to build languages whose position within the expressivity hierarchy is also relevant.

---

[3] Note how $\mathcal{L}(K, L)$ and $\mathcal{L}(K, L, A)$ are equally expressive, so in $\mathcal{L}(K, L)$ the agent can talk about her own awareness.

[4] In such cases we have the validity $A\varphi \leftrightarrow L(\varphi \vee \neg\varphi)$, and thus $\mathcal{L}(L)$ jumps to the equivalence class where $\mathcal{L}(L, A)$ lies: that of $\mathcal{L}(K, L)$.

Maybe more interesting are the languages that allows us to talk not only about an agent's (implicit and explicit) knowledge and awareness, but also about the way these notions change. Previous works have introduced the so-called epistemic awareness models [11,12], semantic structures that allow us to represent actions that can change an agent's knowledge and awareness. It is, thus, also relevant to find if (and how) the expressivity of the languages studied in this paper changes when we add modalities for representing the effect of such actions to each one of them.

A Appendix

Proof of Proposition 6 The proof is by induction on formulas in $\mathcal{L}(L)|\mathsf{Q}$. The case for atomic propositions follows from the **atoms** clause in the definition of a restricted awareness bisimulation (Definition 9) and the cases for propositional connectives follow from the inductive hypothesis. For the remaining case:

- **Case L.** Suppose $atm(\varphi) \subseteq \mathsf{Q}$ and $(M, w) \Vdash L\varphi$. Then $atm(\varphi) \subseteq \mathcal{A}(w)$ and there is a $u \in R[w]$ such that $(M, u) \Vdash \varphi$.

  By (II) of the **a-forth-bis** clause (Definition 9), there is a $u' \in R'[w']$ such that $(M, u) \underleftrightarrow{r}_{\mathsf{Q} \cap \mathcal{A}(w)} (M', u')$, and since $atm(\varphi) \subseteq \mathsf{Q}$ and $atm(\varphi) \subseteq \mathcal{A}(w)$ we get $atm(\varphi) \subseteq \mathsf{Q} \cap \mathcal{A}(w)$; hence, since $(M, u) \Vdash \varphi$, inductive hypothesis gives us $(M', u') \Vdash \varphi$. But by (I) of the **a-forth-bis** clause, we also have $\mathsf{Q} \cap \mathcal{A}(w) \subseteq \mathsf{Q} \cap \mathcal{A}'(w')$ and hence $atm(\varphi) \subseteq \mathsf{Q} \cap \mathcal{A}'(w')$, that is, $atm(\varphi) \subseteq \mathcal{A}'(w')$; then $(M', w') \Vdash L\varphi$. The other direction (from $(M', w') \Vdash L\varphi$ to $(M, w) \Vdash L\varphi$) is similar, using **a-back-bis** instead of **a-forth-bis**.

Proof of Proposition 7. The proof consists in showing that $\equiv_{\mathsf{Q}}^{\mathcal{L}(L)}$ is a restricted awareness Q-bisimulation, that is, that it satisfies **atoms**, **a-forth-bis** and **a-back-bis**.

- **atoms.** Take any $p \in \mathsf{Q}$ with $w \in V(p)$; then $(M, w) \Vdash p$ and hence by $\equiv_{\mathsf{Q}}^{\mathcal{L}(L)}$ we get $(M', w') \Vdash p$, that is, $w' \in V'(p)$. The other direction is similar.
- **a-forth-bis.** Suppose there is a $u \in R[w]$; we will show that (I) $\mathsf{Q} \cap \mathcal{A}(w) \subseteq \mathsf{Q} \cap \mathcal{A}'(w')$ and (II) there is a $u' \in R'[w']$ such that $(M, u) \equiv_{\mathsf{Q} \cap \mathcal{A}(w)}^{\mathcal{L}(L)} (M', u')$. For (II) we work by contradiction. Suppose no element of $R'[w']$ is $\mathcal{L}(L)|\mathsf{Q} \cap \mathcal{A}(w)$-modally equivalent to u. Observe that $R'[w']$ is finite and non-empty: finite by image-finiteness, and non-empty because the existence of $u \in R[w]$ gives us $(M, w) \Vdash L\top$, which by $(M, w) \equiv_{\mathsf{Q}}^{\mathcal{L}(L)} (M', w')$ implies $(M', w') \Vdash L\top$, that is, there is at least an element in $R'[w']$.

  Now, since no element of $R'[w']$ is $\mathcal{L}(L)|\mathsf{Q} \cap \mathcal{A}(w)$-modally equivalent to u, then each $u'_i \in R'[w']$ is distinguishable from u by some formula φ'_i in $\mathcal{L}(L)|\mathsf{Q} \cap \mathcal{A}(w)$ and thus, since negation is in the language, for each $u'_i \in R'[w']$ there is a φ_i in $\mathcal{L}(L)|\mathsf{Q} \cap \mathcal{A}(w)$ that holds at u but fails at u'_i. Define $\varphi := \varphi_1 \wedge \cdots \wedge \varphi_n$ (with n the cardinality of $R'[w']$); then $(M, u) \Vdash \varphi$ (u satisfies every φ_i) but $(M', u'_i) \not\Vdash \varphi$ (each u'_i fails at least in φ_i). Now, since each φ_i is in $\mathcal{L}(L)|\mathsf{Q} \cap \mathcal{A}(w)$, φ is also in $\mathcal{L}(L)|\mathsf{Q} \cap \mathcal{A}(w)$ and thus $atm(\varphi) \subseteq \mathsf{Q} \cap \mathcal{A}(w)$ so $atm(\varphi) \subseteq \mathcal{A}(w)$. Then, we get $(M, w) \Vdash L\varphi$. However, $(M', w') \not\Vdash L\varphi$ because no u'_i satisfies φ. But $L\varphi$ is in $\mathcal{L}(L)|\mathsf{Q}$ so this contradicts our starting point $(M, w) \equiv^{\mathcal{L}(L)}_{\mathsf{Q}} (M', w')$. Hence, there should be a $u' \in R'[w']$ such that $(M, u) \equiv^{\mathcal{L}(L)}_{\mathsf{Q} \cap \mathcal{A}(w)} (M', u')$.

In order to show (I), take any p in $\mathsf{Q} \cap \mathcal{A}(w)$. Then, $p \in \mathsf{Q}$ and $p \in \mathcal{A}(w)$, and since $R[w]$ is non-empty (this clause starts by assuming there is a $u \in R[w]$), we get $(M, w) \Vdash L(p \vee \neg p)$. But $L(p \vee \neg p)$ is in $\mathcal{L}(L)|\mathsf{Q}$ and hence $(M, w) \equiv^{\mathcal{L}(L)}_{\mathsf{Q}} (M', w')$ gives us $(M', w') \Vdash L(p \vee \neg p)$, which implies $p \in \mathcal{A}'(w')$ and hence $p \in \mathsf{Q} \cap \mathcal{A}'(w')$. Thus, $\mathsf{Q} \cap \mathcal{A}(w) \subseteq \mathsf{Q} \cap \mathcal{A}'(w')$.

– **a-back-bis.** Similar to the previous one.

Proof of Proposition 12. The proof is by induction on formulas in $\mathcal{L}(\Box, L)|\mathsf{Q}$. The case for atomic propositions follows from the **atoms** clause of Definition 10 and the cases for propositional connectives follow from inductive hypothesis. For the remaining,

– **Case L.** Suppose $atm(\varphi) \subseteq \mathsf{Q}$ and $(M, w) \Vdash L\varphi$. Then $atm(\varphi) \subseteq \mathcal{A}(w)$ and there is a $u \in R[w]$ such that $(M, u) \Vdash \varphi$. By (II) of the **forth-bis** clause of Definition 10, there is a $u' \in R'[w']$ such that $(M, u) \simeq^r_{\mathsf{Q}} (M', u')$; hence, since $atm(\varphi) \subseteq \mathsf{Q}$, inductive hypothesis gives us $(M', u') \Vdash \varphi$. But by (I) of the **forth-bis** clause, we also have $\mathsf{Q} \cap \mathcal{A}(w) \subseteq \mathsf{Q} \cap \mathcal{A}'(w')$; since $atm(\varphi) \subseteq \mathsf{Q}$ and $atm(\varphi) \subseteq \mathcal{A}(w)$, we have $atm(\varphi) \subseteq \mathsf{Q} \cap \mathcal{A}(w)$ and hence $atm(\varphi) \subseteq \mathsf{Q} \cap \mathcal{A}'(w')$, that is, $atm(\varphi) \subseteq \mathcal{A}'(w')$. Therefore we have $(M', w') \Vdash L\varphi$. The other direction is similar, using **back-bis** instead of **forth-bis**.[5]

– **Case $\Box$.** Suppose $atm(\varphi) \subseteq \mathsf{Q}$ and $(M, w) \Vdash \Box\varphi$. Then for all $u \in R[w]$ we have $(M, u) \Vdash \varphi$. Now take any $u' \in R'[w']$; by (II) of **back-bis** there is a $u \in R[w]$ such that $(M, u) \simeq^r_{\mathsf{Q}} (M', u')$ but then, since $atm(\varphi) \subseteq \mathsf{Q}$, induction hypothesis gives us $(M', u') \Vdash \varphi$. Given that u' is an arbitrary element of $R'[w']$, we get $(M', w') \Vdash \Box\varphi$, as required. The other direction is similar, using (II) of **forth-bis** instead.

Proof of Proposition 13. The proof consists in showing that the relation $\equiv^{\mathcal{L}(\Box, L)}_{\mathsf{Q}}$ is a restricted extended Q-bisimulation, that is, that it satisfies **atoms**, **forth-bis** and **back-bis**. The proof of these clauses is very similar to the proof of clauses **atoms**, **a-forth-bis** and **a-back-bis** (respectively) of Proposition 7.

---

[5] This case is slightly different from the one for the same operator in Proposition 6 as in the latter, clause (II) gives us (restricted awareness) bisimilarity up to $\mathsf{Q} \cap \mathcal{A}(w)$ and our current case gives us (restricted extended) bisimilarity up to Q.

References

1. Hintikka, J.: Knowledge and Belief. Cornell University Press, Ithaca (1962)
2. Fagin, R., Halpern, J.Y., Moses, Y., Vardi, M.Y.: Reasoning about knowledge. The MIT Press, Cambridge (1995)
3. Fagin, R., Halpern, J.Y.: Belief, awareness, and limited reasoning. Artificial Intelligence 34(1), 39–76 (1988)
4. Konolige, K.: Belief and incompleteness. T.R. 319. SRI International (1984)
5. Levesque, H.J.: A logic of implicit and explicit belief. In: Proc. of AAAI 1984, Austin, TX, pp. 198–202 (1984)
6. Vardi, M.Y.: On epistemic logic and logical omniscience. In: Halpern, J.Y. (ed.) TARK, pp. 293–305. Morgan Kaufmann Publishers Inc., San Francisco (1986)
7. van Ditmarsch, H., French, T., Velázquez-Quesada, F.R., Wang, Y.: Knowledge, awareness, and bisimulation. In: Schipper, B.C. (ed.) 14th TARK, pp. 61–70 (2013)
8. Blackburn, P., de Rijke, M., Venema, Y.: Modal logic. CUP, New York (2001)
9. van Ditmarsch, H., French, T.: Awareness and forgetting of facts and agents. In: Web Intelligence/IAT Workshops, pp. 478–483. IEEE (2009)
10. van Ditmarsch, H., French, T.: Becoming aware of propositional variables. In: Banerjee, M., Seth, A. (eds.) ICLA 2011. LNCS, vol. 6521, pp. 204–218. Springer, Heidelberg (2011)
11. van Benthem, J., Velázquez-Quesada, F.R.: The dynamics of awareness. Synthese (Knowledge, Rationality and Action) 177(suppl. 1), 5–27
12. van Ditmarsch, H., French, T., Velázquez-Quesada, F.R.: Action models for knowledge and awareness. In: van der Hoek, W., Padgham, L., Conitzer, V., Winikoff, M. (eds.) AAMAS, IFAAMAS, pp. 1091–1098 (2012)

Public Announcements, Private Actions and Common Knowledge in S5 Structures

Yì N. Wáng[1,*] and Thomas Ågotnes[2,3]

[1] Center for the Study of Language and Cognition
Zhejiang University, China
[2] Department of Information Science and Media Studies
University of Bergen, Norway
[3] Institute of Logic and Intelligence
Southwest University, China
wonease@gmail.com, thomas.agotnes@infomedia.uib.no

Abstract. In this paper we take the S5 definition of knowledge, and have a new look at logics combining modalities for public announcements, action models, common knowledge and relativized common knowledge. In particular, we prove two expressivity results which previously have only been shown for the case where knowledge is represented using arbitrary Kripke models but have remained open for the case of S5 models: public announcement logic with relativized common knowledge is strictly more expressive than public announcement logic with common knowledge, and action model logic with common knowledge is strictly more expressive than public announcement logic with common knowledge. We also propose and study a definition of relativized common knowledge for action model logic.

1 Introduction

Dynamic epistemic logics extend traditional ("static") epistemic logic [1,2] in order to be able to express epistemic pre- and post conditions of actions and other events. The two probably most prominent examples are *public announcement logic* [3] in which actions are assumed to be truthful public announcements, and *action model logic* [4,5] which can be used to reason about a very general class of events.

A key consideration in the study of dynamic epistemic logics is relative expressive power. For example, it is well known that public announcement logic is not more expressive than (static) epistemic logic, but public announcement logic *with common knowledge* is strictly more expressive than epistemic logic with common knowledge. However, the latter is not true again when common

* The author gratefully acknowledges funding support from the Major Project of National Social Science Foundation of China (No. 11&ZD088). The author was affiliated with Bergen University College, Norway, when this paper was submitted.

D. Grossi, O. Roy, and H. Huang (Eds.): LORI 2013, LNCS 8196, pp. 267–281, 2013.

knowledge is replaced by *relativized common knowledge* [6,7]: public announcement logic with relativized common knowledge can be reduced to epistemic logic with relativized common knowledge.

In this paper we settle two open problems. We show that:

(i) public announcement logic with relativized common knowledge is strictly more expressive than public announcement logic with common knowledge;

(ii) action model logic with common knowledge is strictly more expressive than public announcement logic with common knowledge;

under the S5 assumptions about knowledge. While the S5 assumption is very common, in fact almost standard, in epistemic logic [1,2], these results have so far only been proved, in [8] and [7] respectively, under the assumption that knowledge is represented by general Kripke structures. van Ditmarsch *et al.* [9, p242] says that "Whether [(i)] is true for S5 is still an open problem". In general, it is not trivial to transfer results for general Kripke models to S5. In order to prove these expressivity results for S5, we introduce a class of models called *canyon models*. (ii) was proved for general Kripke models in [8] by the use of *private actions* (which were not S5 actions). We show that private actions also exist in the S5 case, and that they are not expressible in public announcement logic.

Relativized common knowledge was "designed" to provide reduction axioms for public announcement logic. There is no corresponding notion of relativized common knowledge for action model logic in the literature. In this paper we also introduce and study a relativized common knowledge operator for action model logic.

The paper is organized as follows. In the next section we briefly review background definitions and results, before proving the two expressivity results using canyon models in Section 3. Relativized common knowledge for action model logic is discussed in Section 4 and we end with a discussion in Section 5.

2   Background

We briefly review the key definition and results we build on from (dynamic) epistemic logic. The presentation is terse due to lack of space; we refer the reader to, e.g., [9] for details.

2.1   Epistemic Logic, Action Model Logic, and Public Announcement Logic

Let PROP be a countable set of propositional variables and AG a finite set of agent symbols.

Definition 1 (Action models). *Let $\mathcal{L}$ be a language. $\mathfrak{A}^{\mathcal{L}} = (A, \simeq, \mathsf{pre})$ is called an action model for $\mathcal{L}$ (or simply an action for $\mathcal{L}$), if the following hold:*

− *A is a non-empty finite set of action states, called the domain of $\mathfrak{A}^{\mathcal{L}}$;*

− *$\simeq$: AG $\to \wp(A \times A)$ maps every agent a to an equivalence relation $\simeq_a$ on A;*

− *$\mathsf{pre} : A \to \mathcal{L}$ is a precondition function.*

$(\mathfrak{A}^{\mathcal{L}}, a)$ *is called a* pointed action (model) *for* $\mathcal{L}$, *if* $\mathfrak{A}^{\mathcal{L}}$ *is an action for* $\mathcal{L}$ *and* a *is an action state in the domain of* $\mathfrak{A}^{\mathcal{L}}$.

Example 2 (Public announcement actions). A *public announcement action* for a formula ψ is an action $\mathsf{Pub}(\psi) = (\{\mathsf{pub}\}, \backsimeq, \mathsf{pre})$ such that $\mathsf{pre}(\mathsf{pub}) = \psi$ and for all $a \in \mathrm{AG}$, $\backsimeq_a = \{(\mathsf{pub}, \mathsf{pub})\}$.

Definition 3 (Languages). *The languages of epistemic logics and action model logics are given as follows:*

$$
\begin{aligned}
(\mathcal{EL}) \quad & \varphi ::= p \mid \neg\varphi \mid \varphi \wedge \varphi \mid K_a\varphi \\
(\mathcal{ELC}) \quad & \varphi ::= p \mid \neg\varphi \mid \varphi \wedge \varphi \mid K_a\varphi \mid C_A\varphi \\
(\mathcal{AML}) \quad & \varphi ::= p \mid \neg\varphi \mid \varphi \wedge \varphi \mid K_a\varphi \mid [\mathfrak{A}, a]\varphi \\
(\mathcal{AMC}) \quad & \varphi ::= p \mid \neg\varphi \mid \varphi \wedge \varphi \mid K_a\varphi \mid C_A\varphi \mid [\mathfrak{B}, b]\varphi
\end{aligned}
$$

where $p \in \mathrm{PROP}$, $i \in \mathrm{AG}$, $A \subseteq \mathrm{AG}$, *and* $(\mathfrak{A}, a)$ *(resp.* $(\mathfrak{B}, b)$*) is a pointed action for* $\mathcal{AML}$ *(resp.* $\mathcal{AMC}$*).*[1] *The parentheses surrounding a pointed action is often omitted, as we did in the above. We write* $\hat{K}_a\varphi$ *as a shorthand for* $\neg K_a\neg\varphi$, $\hat{C}_A\varphi$ *for* $\neg C_A\neg\varphi$, *and* $\langle\mathfrak{A}, a\rangle\psi$ *for* $\neg[\mathfrak{A}, a]\neg\psi$, *in addition to the standard derived propositional connectives.*

Let $\mathfrak{A} = (A, \backsimeq, \mathsf{pre}^{\mathfrak{A}})$ and $\mathfrak{B} = (B, \backsimeq, \mathsf{pre}^{\mathfrak{B}})$ be two $\mathcal{L}$-actions, where $\mathcal{L}$ is one of the languages above. Their *composition*, denoted $\mathfrak{A} \circ \mathfrak{B}$, is the action $(A \times B, =$, $\mathsf{pre})$ such that $\mathsf{pre}((a, b)) = \langle\mathfrak{A}, a\rangle\mathsf{pre}^{\mathfrak{B}}(b)$, and for all $a \in \mathrm{AG}$, $(a, b) =_a (a', b')$ iff $a \backsimeq_a a'$ & $b \backsimeq_a b'$. In this definition $\langle\mathfrak{A}, a\rangle\mathsf{pre}^{\mathfrak{B}}(b)$ is a formula starting with the action operator $\langle\mathfrak{A}, a\rangle$ (see Definition 3). The *composition* of two pointed actions, $(\mathfrak{A}, a) \circ (\mathfrak{B}, b)$, is defined as the pointed action $((\mathfrak{A} \circ \mathfrak{B}), (a, b))$.

Our definition of action model logic only allows *atomic* actions. Standard definitions [9] also allow compound actions to be constructed using a *nondeterministic choice* operator $\cup$, but that operator does not increase the expressive power and can be defined as a derived operator.[2] Allowing only atomic actions as primitives makes the technical treatment somewhat simpler. We emphasize that our definition of the languages is equivalent to the standard definition in [9], and that all the results in this paper trivially extend to the versions of the languages with $\cup$ (and in fact *test* and *sequential composition* as well) as primitives.

Public announcement logic (PAL) [3] extends classical (static) epistemic logic (EL) with an operator which can be used to express public announcements. It is one of the simplest dynamic epistemic logics, and has been investigated extensively in the past few decades.

The language $\mathcal{PAL}$ of public announcement logic is a sublanguage of $\mathcal{AML}$ where only public announcement actions (Example 2) are allowed. In a similar

---

[1] To show that $\mathcal{AML}$ is well defined, we can start with $\mathcal{AML}_0$ where only actions for $\mathcal{EL}$ are allowed. Then, $\mathcal{AML}_n$ is defined such that only actions for $\mathcal{AML}_{n-1}$ are allowed. Finally, $\mathcal{AML}$ is defined as $\bigcup_{i \in \omega} \mathcal{AML}_i$. Similarly for $\mathcal{AMC}$.

[2] $[\alpha \cup \alpha']\varphi \leftrightarrow ([\alpha]\varphi \wedge [\alpha']\varphi)$ holds (also for all the logics with action model operators considered in this paper).

fashion, the language $\mathcal{PAC}$ of public announcement logic with common knowledge is a sublanguage of $\mathcal{AMC}$. In public announcement logics, we often use $[\psi]\varphi$ as a shorthand for $[\mathsf{Pub}(\psi), \mathsf{pub}]\varphi$. In other words, the languages $\mathcal{PAL}$ and $\mathcal{PAC}$ look as follows:

$$(\mathcal{PAL}) \quad \varphi ::= p \mid \neg\varphi \mid \varphi \wedge \varphi \mid K_a\varphi \mid [\varphi]\varphi$$
$$(\mathcal{PAC}) \quad \varphi ::= p \mid \neg\varphi \mid \varphi \wedge \varphi \mid K_a\varphi \mid C_A\varphi \mid [\varphi]\varphi.$$

Interpretation of all these languages is defined in terms of *epistemic models* (a.k.a. *S5 models*) $\mathfrak{M} = (M, \sim, V)$ consisting of a set M of *states*, for each agent a an *indistiguishability relation* $\sim_a$ which is an equivalence relation on M, and an *evaluation function* $V : \mathrm{PROP} \to \wp(M)$. What we call a (general) *Kripke model* is exactly like an epistemic model except that the indistinguishability relation is not required to be an equivalence relation.

Definition 4 (Semantics). *Given an epistemic model* $\mathfrak{M} = (M, \sim, V)$ *and a point* $m \in M$, *the satisfaction relation*, $\models$, *is defined as follows:*

$$
\begin{array}{llll}
\mathfrak{M}, m \models p & \textit{iff} & m \in V(p) \\
\mathfrak{M}, m \models \neg\varphi & \textit{iff} & \mathfrak{M}, m \not\models \varphi \\
\mathfrak{M}, m \models \varphi \wedge \psi & \textit{iff} & \mathfrak{M}, m \models \varphi \ \& \ \mathfrak{M}, m \models \psi \\
\mathfrak{M}, m \models K_a\varphi & \textit{iff} & \forall n \in M. \ (m \sim_a n \Rightarrow \mathfrak{M}, n \models \varphi) \\
\mathfrak{M}, m \models C_A\varphi & \textit{iff} & \forall n \in M. \ (m \sim_A n \Rightarrow \mathfrak{M}, n \models \varphi) \\
\mathfrak{M}, m \models [\mathfrak{A}, \mathsf{a}]\varphi & \textit{iff} & \mathfrak{M}, m \models \mathsf{pre}(\mathsf{a}) \Rightarrow \mathfrak{M} \otimes \mathfrak{A}, (m, \mathsf{a}) \models \varphi \\
\textit{in particular,} \quad \mathfrak{M}, m \models [\psi]\varphi & \textit{iff} & \mathfrak{M}, m \models \psi \Rightarrow \mathfrak{M}|\psi, m \models \varphi.
\end{array}
$$

In the above, $\sim_A$ *is the transitive closure of* $\bigcup_{a \in A} \sim_a$ *and* $\mathfrak{M} \otimes \mathfrak{A} = (N, \approx, \nu)$ *is an epistemic model such that*

$$
\begin{array}{llll}
N & = & \{(m, \mathsf{a}) \in M \times A \mid \mathfrak{M}, m \models \mathsf{pre}(\mathsf{a})\} \\
(m, \mathsf{a}) \approx_a (n, \mathsf{b}) & \textit{iff} & m \sim_a n \ \& \ \mathsf{a} \eqsim_a \mathsf{b}, & \textit{for all } a \in \mathrm{AG} \\
(m, \mathsf{a}) \in \nu(p) & \textit{iff} & m \in V(p), & \textit{for every } p \in \mathrm{PROP}.
\end{array}
$$

$\mathfrak{M}|\psi$ *is the submodel of* $\mathfrak{M}$ *restricted to* $\{m \in M \mid \mathfrak{M}, m \models \psi\}$. *To understand the interpretation of the public announcement operator, observe that i) by definition* $\psi = \mathsf{pre}(\mathsf{pub})$, *ii)* $\mathfrak{M}|\psi$ *is the submodel of* $\mathfrak{M}$ *restricted to* $\{m \in M \mid \mathfrak{M}, m \models \mathsf{pre}(\mathsf{pub})\}$, *and iii) the pointed epistemic models* $(\mathfrak{M} \otimes \mathfrak{A}, (m, \mathsf{pub}))$ *and* $(\mathfrak{M}|\psi, m)$ *are bisimilar. Validity is defined as usual.*

We use capital letters (in roman font), e.g., EL, AML and AMC, to name the logics induced by the interpretation of the different languages.

2.2 Relativized Common Knowledge

Relativized common knowledge is a variant of common knowledge proposed for public announcement logic by treating knowledge update as *relativization* [6,7]. Public announcement logic with common knowledge cannot be reduced to epistemic logic with common knowledge, but it becomes reducible when we replace common knowledge with relativized common knowledge.

We consider the following languages with relativized common knowledge:

$$(\mathcal{ELRC}) \quad \varphi ::= p \mid \neg\varphi \mid \varphi \wedge \varphi \mid K_a\varphi \mid C_A^\varphi \varphi$$
$$(\mathcal{PARC}) \quad \varphi ::= p \mid \neg\varphi \mid \varphi \wedge \varphi \mid K_a\varphi \mid C_A^\varphi \varphi \mid [\varphi]\varphi.$$

As usual, $\hat{C}_A^\psi \varphi$ is a shorthand for $\neg C_A^\psi \neg\varphi$. The clause $C_A^\psi \varphi$ is interpreted as follows:

$$\mathfrak{M}, m \models C_A^\psi \varphi \quad \text{iff} \quad \forall n \in M. \, (m \sim_{A,\psi} n \Rightarrow \mathfrak{M}, n \models \varphi),$$

where $\sim_{A,\psi}$ is the transitive closure of $(\bigcup_{a \in A} \sim_a) \cap (M \times \{m \in M \mid \mathfrak{M}, m \models \psi\})$.

2.3 Expressivity and Axiomatizations

Bisimulation is a well-known notion which can be used to compare expressivity. Here we refer to standard modal logic textbooks (e.g., [10]) for details. The PAC$_\mathcal{P}$-game, as defined below, is a tool introduced in [11,7] for studying the relative expressivity of the logic PAC$_\mathcal{P}$; PAC restricted to propositions from the set $\mathcal{P}$.

Definition 5 (PAC$_\mathcal{P}$-game). *Let $\mathcal{P}$ be a set of propositions. Let pointed epistemic models $(\mathfrak{M}, m) = (M, \sim, V, m)$ and $(\mathfrak{N}, n) = (N, \approx, \nu, n)$ be given. The r-round PAC$_\mathcal{P}$-game between spoiler and duplicator on $(\mathfrak{M}, m)$ and $(\mathfrak{N}, n)$ is the following:*

- *If $r = 0$, spoiler wins iff $V(p) \neq \nu(p)$ for some $p \in \mathcal{P}$.*
- *If $r > 0$, spoiler can initiate one of the following scenarios (unless specified, the rest of the game is the $(r-1)$-round PAC$_\mathcal{P}$-game on $(\mathfrak{M}, m')$ and $(\mathfrak{N}, n')$):*

  (K-forth) Spoiler chooses an agent a and an $m' \in M$ such that $m \sim_a m'$. Duplicator responds by choosing an $n' \in N$ such that $n \approx_a n'$.

  (K-back) Spoiler chooses an agent a and an $n' \in N$ such that $n \approx_a n'$. Duplicator responds by choosing an $m' \in M$ such that $m \sim_a m'$.

  (C-forth) Spoiler chooses a group A and an $m' \in M$ such that $m \sim_A m'$. Duplicator responds by choosing an $n' \in N$ such that $n \approx_A n'$.

  (C-back) Spoiler chooses a group A and an $n' \in N$ such that $n \approx_A n'$. Duplicator responds by choosing an $m' \in M$ such that $m \sim_A m'$.

  ([φ]-move) Spoiler chooses a number $s < r$, and sets $M' \subseteq M$ and $N' \subseteq N$ such that $m \in M'$ and $n \in N'$.

    ⟨Stage 1⟩ Duplicator chooses states $x \in M' \cup N'$ and $y \in (M - M') \cup (N - N')$. Then the players play the s-round PAC$_\mathcal{P}$-game on x and y. Duplicator wins the r-round game if she wins this subgame.

    ⟨Stage 2⟩ Otherwise, the players continue with the $(r - s - 1)$-round PAC$_\mathcal{P}$-game on $(\mathfrak{M}|M', m)$ and $(\mathfrak{N}|N', n)$.

If either player cannot perform an action prescribed above, that player loses.

Lemma 6 (Invariance of games). *Let $(\mathfrak{M}, m)$ and $(\mathfrak{N}, n)$ be two pointed epistemic models and $\mathcal{P}$ be a finite set of propositions. For all $r \in \mathbb{N}$, duplicator has a* *winning strategy for the r-round PAC$_\mathcal{P}$-game on* $(\mathfrak{M}, m)$ *and* $(\mathfrak{N}, n)$, *if and only if* $(\mathfrak{M}, m)$ *and* $(\mathfrak{N}, n)$ *satisfy exactly the same set of PAC$_\mathcal{P}$-formulas of degree at most* r.

Proof. This lemma is in fact [9, Theorem 8.53]. A proof can be found there.

Definition 7 (Expressivity relations). *Let $\mathcal{L}$ and $\mathcal{L}'$ be languages. $\mathcal{L}$ is at least as expressive as $\mathcal{L}'$ (notation: $\mathcal{L}' \preceq \mathcal{L}$) iff for every $\mathcal{L}'$-formula there is an $\mathcal{L}$-formula equivalent to it (satisfied in exactly the same pointed models) on a given class of models. Unless we say otherwise, the class of models is in the following implicitly taken to be the class of all epistemic models (sometimes we will consider the more general class of Kripke models). $\mathcal{L}$ and $\mathcal{L}'$ are equally expressive (notation: $\mathcal{L} \equiv \mathcal{L}'$) iff $\mathcal{L} \preceq \mathcal{L}'$ and $\mathcal{L}' \preceq \mathcal{L}$. $\mathcal{L}$ is (strictly) more expressive than $\mathcal{L}'$ (notation: $\mathcal{L}' \prec \mathcal{L}$) iff $\mathcal{L}' \preceq \mathcal{L}$ but $\mathcal{L}' \not\equiv \mathcal{L}$. We also say $\mathcal{L}'$ is less expressive than $\mathcal{L}$ in this case. $\mathcal{L}$ and $\mathcal{L}'$ are not comparable iff neither $\mathcal{L} \preceq \mathcal{L}'$ nor $\mathcal{L}' \preceq \mathcal{L}$.*

It is easy to observe that all the introduced languages are equally expressive when we allow only a single agent. We present the known results for the case that $|\mathrm{AG}| > 1$ in Fig. 1.

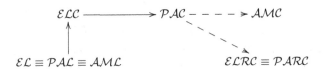

Fig. 1. Overview of expressivity results in the literature. $\equiv$ stands for the relationship of being equally expressive. A solid arrow going from one logic to another means that the first is strictly less expressive. A dashed arrow going from one logic to another means that the first is not more expressive (over S5 models), and is known to be strictly less expressive over arbitrary Kripke models.

It is known that $\mathcal{AML}$ is equally expressive as $\mathcal{EL}$, while $\mathcal{AMC}$ is more expressive than $\mathcal{ELC}$ [4] (in particular, $\mathcal{PAL}$ is as expressive as $\mathcal{EL}$ [3]), but whether $\mathcal{AMC}$ and/or $\mathcal{PARC}$ is more expressive than $\mathcal{PAC}$ has been shown only for arbitrary Kripke models [8,7] (see also [9, Chapter 8]).

A sound and complete axiomatization for EL is the well-known Hilbert system **S5** (Fig. 2). The axiomatization **AML** for AML (Fig. 2) is obtained by adding to **S5** reduction axioms for the action operators [4,9,11,12]. The axiomatization **AMC** for AMC (Fig. 3) is obtained by adding to **AML** extra axioms and rules for characterizing common knowledge [4,9]. Note that there is no reduction axiom for common knowledge, and this closely relates to the fact that $\mathcal{ELC} \prec \mathcal{AMC}$.

(PC) Instances of tautologies	(MP) $\vdash \varphi \ \& \ \vdash \varphi \to \psi \Rightarrow \vdash \psi$
(N) $\vdash \varphi \Rightarrow \vdash K_a\varphi$	(K) $K_a(\varphi \to \psi) \to K_a\varphi \to K_a\psi$
(T) $K_a\varphi \to \varphi$	(5) $\neg K_a\varphi \to K_a\neg K_a\varphi$
(AP) $[\mathfrak{A}, a]p \leftrightarrow (\text{pre}(a) \to p)$	(AC) $[\mathfrak{A}, a](\varphi \wedge \psi) \leftrightarrow ([\mathfrak{A}, a]\varphi \wedge [\mathfrak{A}, a]\psi)$
(AN) $[\mathfrak{A}, a]\neg\varphi \leftrightarrow (\text{pre}(a) \to \neg[\mathfrak{A}, a]\varphi)$	(AA) $[\mathfrak{A}, a][\mathfrak{B}, b]\varphi \leftrightarrow [(\mathfrak{A}, a) \circ (\mathfrak{B}, b)]\varphi$
(AK) $[\mathfrak{A}, a]K_a\varphi \leftrightarrow (\text{pre}(a) \to \bigwedge_{a \simeq_a b} K_a[\mathfrak{A}, b]\varphi)$	

Fig. 2. The axiomatization **AML** of action model logic, and the sub-system **S5** consisting of (PC), (MP), (N), (K), (T) and (5). The 4 axiom, i.e., $K_a\varphi \to K_aK_a\varphi$, meaning *positive introspection*, is often also included, but technically redundant — it can be derived in **S5**.

All axioms and rules of **AML**	$(N_C) \vdash \varphi \Rightarrow \vdash C_A\varphi$
$(K_C) \ C_A(\varphi \to \psi) \to (C_A\varphi \to C_A\psi)$	$(T_C) \ C_A\varphi \to \varphi$
(C1) $C_A\varphi \to E_AC_A\varphi$	(C2) $C_A(\varphi \to E_A\varphi) \to (\varphi \to C_A\varphi)$
$(N_A) \vdash \varphi \Rightarrow \vdash [\mathfrak{A}, a]\varphi$	
(RA) $\vdash \psi_b \to [\mathfrak{A}, b]\varphi \ \& \ \vdash \psi_b \wedge \text{pre}(b) \to E_A\psi_c \Rightarrow \vdash \psi_a \to [\mathfrak{A}, a]C_A\varphi$	

Fig. 3. The axiomatization **AMC** of action model logic with common knowledge, where $\mathfrak{A} = (A, \simeq, \text{pre})$, $a, b, c \in A$, and $a \simeq_A b$. For any $x, y \in A$, by writing φ_x and φ_y we mean that $\varphi_x[y/x] = \varphi_y$ and $\varphi_y[x/y] = \varphi_x$.

3 New Expressivity Results

Expressivity results for public announcement logics and action model logics have been studied extensively in the literature (for a survey see [9, Chapter 8]). However, some results are based on interpretations over arbitrary Kripke models, instead of epistemic models (S5 models). In particular we are interested in the following open questions:

1. Is $\mathcal{PARC}$ more expressive than $\mathcal{PAC}$?

2. Is $\mathcal{AMC}$ more expressive than $\mathcal{PAC}$?

In the literature on action model logic [4,13,14,8,5], an important concept that can be characterized in this logic is a *private action*. Private actions were used in particular to show that $\mathcal{AMC}$ is more expressive than $\mathcal{PAC}$ (on arbitrary Kripke models; cf. [8]). Since in S5 modeling, an action model must also be "S5" (see Definition 1), we are interested in these questions:

3. Is there a private action under S5 modeling? And furthermore,

4. is there a private action under S5 modeling which is not equivalent to any formula of public announcement logic?

The answers to all the above four questions are yes, and when showing $\mathcal{PAC} \prec \mathcal{AMC}$ (the answer to question 2) we make use of a private action (which answers questions 3 and 4).

We now introduce and discuss *canyon models* before using them to prove the two mentioned expressivity results.

3.1   Canyon Models

We first illustrate canyon models in Fig. 4. Formal definitions of these models will be given afterwards. Intuitively, a canyon model is composed of two "peeks" which only differ by a •-state. A †-state links the two peeks, so that the •-state is $\sim_{\{a,b\},p}$-reachable from states in the right peek, but not from any state in the left peek, while it is $\sim_{\{a,b\}}$-reachable from all states (including those from the left peek). Such models make a difference when interpreting common knowledge operators and relativized common knowledge operators. Public announcements update models by eliminating states; in regards to this we introduce *weathering* of canyon models. The canyon models are designed to discern between common knowledge and relativized common knowledge operators, even when public announcement operators are involved.

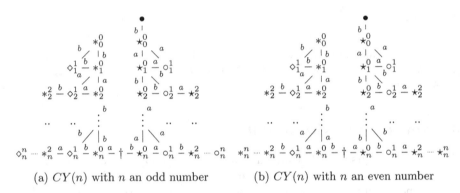

(a) $CY(n)$ with n an odd number   (b) $CY(n)$ with n an even number

Fig. 4. Illustration of Canyon Models. The dotted edges in the bottom of each diagram stand for a path of a similar pattern.

Definition 8 (Canyon models). *For any natural number x, we introduce a canyon model (canyon for short) $CY(x)$ which is an epistemic model $(M, \sim, V)$ such that*

- $M = \{\star_i^{2j}, *_i^{2j}, \circ_i^{2k+1}, \diamond_i^{2k+1} \mid i \leq x, 2j \leq i, 2k+1 \leq i\} \cup \{\dagger, \bullet\}$

- $\sim_a = [\{(\star_{2i}^0, *_{2i+1}^0) \mid 2i+1 \leq x\} \cup \{(\circ_{2i'+1}^0, *_{2i'+2}^0) \mid 2i'+2 \leq x\} \cup$
  $\{(\star_{2i+1}^{2j}, \circ_{2i+1}^{2j+1}) \mid 2i+1 \leq x, j \leq i\} \cup \{(*_{2i'+2}^{2j'}, \diamond_{2i'+2}^{2j'+1}) \mid 2i'+2 \leq x, j' \leq$
  $i'+1\} \cup \{(*_x^0, \dagger) \mid x \text{ is even}\} \cup \{(*_x^0, \dagger) \mid x \text{ is odd}\}]^{\#}$

- $\sim_b = [\{(*_{2i}^0, *_{2i+1}^0) \mid 2i+1 \leq x\} \cup \{(\star_{2i'+1}^0, *_{2i'+2}^0) \mid 2i'+2 \leq x\} \cup \{(\star_0^0, \bullet)\} \cup$
  $\{(*_{2i+1}^{2j}, \diamond_{2i+1}^{2j+1}) \mid 2i+1 \leq x, j \leq i\} \cup \{(*_{2i'+2}^{2j'}, \circ_{2i'+2}^{2j'+1}) \mid 2i'+2 \leq x, j' \leq$
  $i'+1\} \cup \{(*_x^0, \dagger) \mid x \text{ is even}\} \cup \{(*_x^0, \dagger) \mid x \text{ is odd}\}]^{\#}$

- $V(p)$ is the set of all $\star$-, $*$-, and $\bullet$-states,

where all i, i', j, j' and k are natural numbers, and the superscript $^{\#}$ in the definition of $\sim$ stands for the reflexive symmetric transitive closure of a relation.

Every state of a canyon $CY(x) = (M, \sim, V)$ has a depth, *which is defined by the function $d : M \to \mathbb{N}$ such that*

$$
\begin{aligned}
d(\bullet) &= 0, \\
d(\dagger) &= x + 1, \\
d(m_i^j) &= i - j, \quad \text{where } m \in \{\star, *, \circ, \diamond\}
\end{aligned}
$$

A positive state *of a canyon is a state that is different from $\bullet$ and at which p is true, i.e., a $\star$- or $*$-state. Analogously, a* negative state *of a canyon is a state that is different from $\bullet$ and at which p is false, i.e., a $\dagger$-, $\circ$-, or $\diamond$-state.*

We define the *weathering* of a canyon, to be used in the proof of Lemma 10 below.

Definition 9 (Weathering of a canyon). *Let $CY(x)$ be a canyon. For any $y \in \mathbb{N}$ such that $y \leq x$, a* y-weathering *of $CY(x)$ is a submodel of $CY(x)$ over a set N of states such that N includes all states of depth no less than y. Such a set N is called the* core *of a y-weathering of $CY(x)$. A weathering of a canyon is also called a* weathered canyon. *Intuitively, a y-weathered canyon is a canyon with its surface weathered up to depth y.*

We write $CY_y(x)$ for the set of all y-weatherings of $CY(x)$. Clearly, $CY(x) \in CY_y(x)$, $CY_{y-1}(x) \subseteq CY_y(x)$, and $CY_0(x) = \{CY(x)\}$.

For a state m of a weathered canyon in $CY_y(x)$, the real depth *of m is defined by the function rd such that $rd(m) = d(m) - y$. Clearly, The real depth of a canyon is equivalent to its depth, i.e., for any $m \in CY(x)$, $rd(m) = d(m)$.*

3.2 PARC Is More Expressive Than PAC

We show that the $\mathcal{PARC}$-formula $C_{ab}^p \neg (K_a p \wedge K_b p)$ is not equivalent to any $\mathcal{PAC}$-formula.

Lemma 10. *Let $\mathcal{P}$ be a finite set of propositions. Given a canyon model $CY(x)$, and $\mathfrak{M} \in CY_y(x)$ such that $y \in \mathbb{N}$ with $y \leq x$,*

1. *For any positive states m and n in the core of $\mathfrak{M}$, any $r \leq \min(rd(m), rd(n))$, duplicator has a winning strategy for the r-round $PAC_{\mathcal{P}}$-game on $(\mathfrak{M}, m)$ and $(\mathfrak{M}, n)$;*

2. *For any negative states m and n in the core of $\mathfrak{M}$, any $r \leq \min(rd(m), rd(n))$, duplicator has a winning strategy for the r-round $PAC_{\mathcal{P}}$-game on $(\mathfrak{M}, m)$ and $(\mathfrak{M}, n)$.*

Proof. We show items 1 and 2 simultaneously by mutual induction on the number of rounds. If the number of rounds is 0, then the two states only have to agree on propositions. They do agree, as being both positive states or both negative states.

Suppose the lemma holds for the number of rounds r. We show that it holds also for $r + 1$. Note that we have the assumption $r + 1 \leq \min(rd(m), rd(n))$. We first look at clause 1, and explore the three different kinds of moves spoiler can take.

- K-move. We consider K-forth here (K-back is similar). If spoiler chooses a K-forth-move, he chooses i) a K_a-, K_b-, or K_c-move ($c \notin \{a,b\}$) to a positive state of real depth at least r, or ii) a K_a-, K_b-, or K_c-move to a negative state of real depth at least r. Duplicator can always respond by choosing a K_a-, K_b-, or K_c-move to a positive or negative state of real depth at least r, respectively. They will then play the r-round game on the new states. By the induction hypothesis, duplicator has a winning strategy.

- C-move. We consider C-forth here (C-back is similar). If spoiler choose a C-forth-move, he chooses i) a $C_{\{a\}\cup A}$-, $C_{\{b\}\cup A}$-, or C_A-move, where $a,b \notin A$, or ii) a $C_{\{a,b\}\cup A}$-move. Case (i) is just like a K-move, as it can only reach a state of real depth at least r. Duplicator can respond likewise by choosing a $C_{\{a\}\cup A}$-, $C_{\{b\}\cup A}$-, or C_A-move. For case (ii), spoiler can reach any state. So does duplicator. She responds by moving to that state as well. Since for any state m of $\mathfrak{M}$, $(\mathfrak{M}, m)$ is bisimilar to itself, duplicator has a winning strategy by simply following spoiler's move.

- $[\varphi]$-move. Let $\mathfrak{M} = (M, \sim, V)$. Spoiler chooses a number of rounds $s < r + 1$ and two sets $M' \subseteq M$ and $M'' \subseteq M$ such that $m \in M'$ and $n \in M''$. It must be the case that $M' = M''$, since otherwise duplicator can choose the same state twice in the s-round stage-1 subgame, and has a winning strategy for it. Moreover, all states of real depth no less than s must be in M', since otherwise duplicator has a winning strategy in the s-round stage-1 subgame by choosing the state m and an $m' \in M \setminus M'$ with $rd(m') \geq s$, and then the induction hypothesis applies.

  In the stage-2 subgame the generated submodel over M' is an s-weathering of $\mathfrak{M}$, i.e., an $(s+y)$-weathering of $CY(x)$.[3] Let it be denoted by $\mathfrak{M}'$. The real depth of m and n in $\mathfrak{M}'$ are $rd(m) - s$ and $rd(n) - s$ respectively. Moreover, the new game is $(r + 1 - s - 1)$-round PAC$_{\mathcal{P}}$ game on $(\mathfrak{M}', m)$ and $(\mathfrak{M}', n)$. Clearly, $r + 1 - s - 1 = r - s < \min(rd(m) - s, rd(n) - s)$, since by assumption we have $r < \min(rd(m), rd(n))$. Therefore the induction hypothesis applies, and duplicator has a winning strategy.

The cases for clause 2 are similar.

We get the following immediately by Lemmas 10 and 6.

Lemma 11. *For all $x \in \mathbb{N}$, $(CY(x), \star_x^0) \equiv_{PAC}^x (CY(x), \star_x^0)$.*

Theorem 12. $\mathcal{PAC} \prec \mathcal{ELRC} \equiv \mathcal{PARC}$

Proof. It is known that $\mathcal{PAC} \preceq \mathcal{ELRC} \equiv \mathcal{PARC}$. What remains is to show that $\mathcal{PARC}$ is strictly more expressive that $\mathcal{PAC}$. Kooi and van Benthem [11] proved this for Kripke models, but not for epistemic models. Here we show it by the canyon models introduced above. Suppose towards a contradiction that $\mathcal{ELRC}$ is as expressive as $\mathcal{PAC}$. Then there is a $\mathcal{PAC}$-formula φ equivalent to $C_{ab}^p \neg (K_a p \wedge K_b p)$. It is not hard to see that for all $x \in \mathbb{N}$, $(CY(x), \star_x^0) \not\models$

---

[3] Note that $s + y \leq x$, because $s < r + 1 \leq \min(rd(m), rd(n)) = \min(d(m) - y, d(n) - y) \leq x - y$, where $d(m)$ and $d(n)$ are the depth of m and n in $CY(x)$.

$C_{ab}^p \neg (K_a p \wedge K_b p)$ and $(CY(x), *_x^0) \models C_{ab}^p \neg (K_a p \wedge K_b p)$, since $\bullet$ is the only state where $K_a p \wedge K_b p$ is true. Suppose $d(\varphi) = n$. Therefore $(CY(n), *_n^0) \not\models \varphi$ and $(CY(n), *_n^0) \models \varphi$, which contradicts Lemma 11.

3.3 AMC Is More Expressive Than PAC

We now define a *private (S5) action* and use it, together with the canyon models introduced previously, to show the expressivity result claimed in the heading.

Definition 13 (Private actions). *A private knowledge update of φ for agent a is an action* $\mathsf{Priv}^a(\varphi) = (\{\mathsf{p}, \mathsf{np}\}, \simeq, \mathsf{pre})$ *where:*

- *p and np are two action states;*
- *$\simeq$ is defined such that $\simeq_a = \{(\mathsf{p}, \mathsf{p}), (\mathsf{np}, \mathsf{np})\}$ and for all agent x different from a, $\simeq_x = \{(\mathsf{p}, \mathsf{p}), (\mathsf{np}, \mathsf{np}), (\mathsf{p}, \mathsf{np})\}$;*
- *$\mathsf{pre}(\mathsf{p}) = \varphi$ and $\mathsf{pre}(\mathsf{np}) = \neg\varphi$.*

A private action for φ is illustrated in Figure 5.

$$\mathsf{p}^\varphi \xrightarrow{\quad \text{AG}\setminus\{a\} \quad} \mathsf{np}^{\neg\varphi}$$

Fig. 5. $\mathsf{Priv}^a(\varphi)$, a private action for φ. The precondition of an action state is written as a superscript. A line labeled with a set of agents from an action state to another means none of those agents can distinguish between the two action states.

Example 14 (Private actions in effect). Let $\mathfrak{M} = (\{m_0, m_1, m_2\}, \sim, V)$ be an epistemic model illustrated as follows:

$$m_2^{\neg p} \xrightarrow{\quad \sim_a \quad} m_0^p \xrightarrow{\quad \sim_b \quad} m_1^{\neg p}$$

where a formula near a state means the formula is true at the state by the definition of V. It is not hard to verify that $K_a p$ and $K_b p$ are both false at the state m_0. Now the epistemic model $\mathfrak{M} \otimes \mathsf{Priv}^a(p) = (N, \approx, \nu)$ is as follows:

$$(m_2, \mathsf{np})^{\neg p} \qquad (m_0, \mathsf{p})^p \xrightarrow{\quad \approx_b \quad} (m_1, \mathsf{np})^{\neg p}$$

Clearly $K_a p$ is true at (m_0, p) while $K_b p$ not. Intuitively, the action $\mathsf{Priv}^a(p)$ behaves like a private announcement of p to the agent a.

We now show that the $\mathcal{AMC}$-formula $[\mathsf{Priv}^a(p), \mathsf{p}]\hat{C}_{ab}(K_a p \wedge K_b p)$ is not expressible by any $\mathcal{PAC}$-formula.

Lemma 15. *For all $x \in \mathbb{N}$, $(CY(x), *_x^0) \not\equiv_{AMC}^x (CY(x), *_x^0)$.*

Proof. Given an $n \in \mathbb{N}$, consider the epistemic model $CY(n) \otimes \mathsf{Priv}^a(p)$ (cf. Fig. 6). Note that $(\bullet, \mathsf{p})$ and $(*_0^0, \mathsf{p})$ are the only states reachable from $*_n^0$ and $*_n^0$ at which $K_a p \wedge K_b p$ is true. Therefore the $\mathcal{AMC}$-formula $[\mathsf{Priv}^a(p), \mathsf{p}]\hat{C}_{ab}(K_a p \wedge K_b p)$ makes a difference between $*_n^0$ and $*_n^0$, i.e., $CY(x) \otimes \mathsf{Priv}^a(p), *_n^0 \not\models [\mathsf{Priv}^a(p), \mathsf{p}]\hat{C}_{ab}(K_a p \wedge K_b p)$ and $CY(x) \otimes \mathsf{Priv}^a(p), *_n^0 \models [\mathsf{Priv}^a(p), \mathsf{p}]\hat{C}_{ab}(K_a p \wedge K_b p)$.

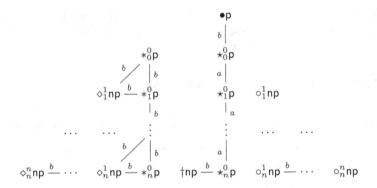

Fig. 6. Epistemic models $CY(n) \otimes \mathsf{Priv}^a(p)$ with n an odd number. It is analogous when n is an even number. Parentheses and a comma are omitted in the notation of a state, e.g., $*_n^0 p$ stands for $(*_n^0, p)$.

Theorem 16. $\mathcal{PAC} \prec \mathcal{AMC}$

Proof. It is known that $\mathcal{PAC} \preceq \mathcal{AMC}$. That $\mathcal{AMC}$ is more expressive follows from Lemmas 11 and 15.

4   Relativized Common Knowledge for Action Model Logic

We now introduce a relativized common knowledge operator for action model logic, in the spirit of the relativized common knowledge operator for public announcement logic (enabling reduction axioms).

Definition 17 (A-$\mathfrak{A}aa'$-paths). *Let A be a set of agents. Let $\mathfrak{M} = (M, \sim, V)$ be an epistemic model, and $\mathfrak{A} = (\mathsf{A}, \eqsim, \mathsf{pre})$ be an action model. Let m_0 and m_n be two states of $\mathfrak{M}$. Let a and a' be two action states of $\mathfrak{A}$.*

An A-$\mathfrak{A}aa'$-path from m_0 to m_n is a path $\langle m_0 \sim_{a_0} \cdots \sim_{a_{n-1}} m_n \rangle$ of $\mathfrak{M}$ such that $\{a_0, \ldots, a_{n-1}\} \subseteq A$ and there is a sequence $\langle \mathsf{a}_0 \cdots \mathsf{a}_n \rangle$ satisfying i) $\mathsf{a}_0 = \mathsf{a}$, ii) $\mathsf{a}_n = \mathsf{a}'$, iii) $\langle \mathsf{a}_0 \eqsim_{a_0} \cdots \eqsim_{a_{n-1}} \mathsf{a}_n \rangle$ forms an path of $\mathfrak{A}$, and iv) for all $i \leq n$, $\mathfrak{M}, m_i \models \mathsf{pre}(\mathsf{a}_i)$.

We write $m \sim_{C_A^{\mathfrak{A}aa'}} n$ iff there is an A-$\mathfrak{A}aa'$-path from m to n.

Intuitively, an A-$\mathfrak{A}aa'$-path is still a path in the updated epistemic model $\mathfrak{M} \otimes \mathfrak{A}$. We now define a formula of the form $C_A^{\mathfrak{A}aa'} \varphi$ to be true iff φ is true at all states reachable via an A-$\mathfrak{A}aa'$-path.

Formally, the language and semantics are defined as follows.

Definition 18 (Languages). *The following grammar rules define the language for epistemic logics with relativized common knowledge.*

$$(\mathcal{ELRC'}) \quad \varphi ::= p \mid \neg\varphi \mid \varphi \wedge \varphi \mid K_a\varphi \mid C_A^{\mathfrak{A}aa'} \varphi$$
$$(\mathcal{AMRC}) \quad \varphi ::= p \mid \neg\varphi \mid \varphi \wedge \varphi \mid K_a\varphi \mid C_A^{\mathfrak{A}aa'} \varphi \mid [\mathfrak{A}, \mathsf{a}]\varphi$$

where the superscript $\cdot^{\mathfrak{A}aa'}$ is such that $\mathfrak{A}$ is an action for the corresponding language, and a and a' are two action states of $\mathfrak{A}$. In case of possible confusion, we use commas to separate the components $(\cdot^{\mathfrak{A},a,a'})$. Other standard conventions are used.

Definition 19 (Satisfaction). *Let $\mathfrak{M} = (M, \sim, V)$ be an epistemic model, and m be a state in M. Satisfaction at $(\mathfrak{M}, m)$ is defined as follows:*

$$
\begin{aligned}
&\mathfrak{M}, m \models p && \text{iff } m \in V(p) \\
&\mathfrak{M}, m \models \neg\varphi && \text{iff } \mathfrak{M}, m \not\models \varphi \\
&\mathfrak{M}, m \models \varphi \wedge \psi && \text{iff } \mathfrak{M}, m \models \varphi \text{ \& } \mathfrak{M}, m \models \psi \\
&\mathfrak{M}, m \models K_a \varphi && \text{iff } (\forall n \in M)(m \sim_a n \Rightarrow \mathfrak{M}, n \models \varphi) \\
&\mathfrak{M}, m \models C_A^{\mathfrak{A}aa'} \varphi && \text{iff } (\forall n \in M)(m \sim_{C_A^{\mathfrak{A}aa'}} n \Rightarrow \mathfrak{M}, n \models \varphi) \\
&\mathfrak{M}, m \models [\mathfrak{A}, a]\varphi && \text{iff } \mathfrak{M}, m \models \mathsf{pre}(a) \Rightarrow \mathfrak{M} \otimes \mathfrak{A}, (m, a) \models \varphi
\end{aligned}
$$

Satisfaction in an epistemic model $\mathfrak{M}$ *(denoted by $\mathfrak{M} \models \varphi$) is defined as usual. The notation $\models \varphi$ denotes validity with respect to all epistemic models, i.e., $\mathfrak{M}, m \models \varphi$ for any $(\mathfrak{M}, m)$.*

Proposition 20. *The following $\mathcal{ELRC}'$-formulas are valid:*

1. $C_A^{\mathfrak{A}aa'} \varphi \leftrightarrow E_A \bigwedge_{b \simeq_{E_A} a} (\mathsf{pre}(b) \rightarrow C_A^{\mathfrak{A}ba'} \varphi)$

2. $\bigwedge_{a'} C_A^{\mathfrak{A}aa'} (\varphi_{a'} \rightarrow E_A(\mathsf{pre}(b) \rightarrow \varphi_b)) \rightarrow ((\mathsf{pre}(a) \rightarrow \varphi_a) \rightarrow C_A^{\mathfrak{A}ab} \varphi_b)$

3. $[\mathfrak{A}, a] C_A^{\mathfrak{B}bb'} \varphi \leftrightarrow (\mathsf{pre}^{\mathfrak{A}}(a) \rightarrow \bigwedge_{a \simeq_A a'} C_A^{\mathfrak{A} \circ \mathfrak{B}, (a,b), (a',b')} [\mathfrak{A} \circ \mathfrak{B}, (a', b')] \varphi)$.

In the above proposition, Formula 1 is a fix-point characterization of the $C_A^{\mathfrak{A}aa'}$-operator. Formula 2 characterizes the induction principle of the $C_A^{\mathfrak{A}aa'}$-operator. Formula 1 tells the recursive construction of an A-$\mathfrak{A}aa'$ from the tail, while Formula 2 tells the construction from the head. Formula 3 is a reduction principle for formulas of the type $[\mathfrak{A}, a] C_A^{\mathfrak{B}bb'} \varphi$. Such a principle does not exist when using the C_A-operator instead.

We easily get the following theorem by Proposition 20(3).

Theorem 21. $\mathcal{ELRC}' \equiv \mathcal{AMRC}$

We now introduce the axiomatization **S5RC'** of the logic ELRC'. **S5RC'** is composed of all axioms and rules of **S5**, and several additional axioms and rules for the $C_A^{\mathfrak{A}aa'}$-operator. Details are given in Figure 7. In the remainder of this section we will only be concerned with the logic ELRC', and implicitly mean "$\mathcal{ELRC}'$-formula" when we say "formula", and so on.

Theorem 22 (Soundness). *$S5RC'$ is sound. I.e., for all formulas φ, if $\vdash_{S5RC'} \varphi$, then $\models \varphi$.*

Proof. It suffices to show that all axioms of **S5RC'** are valid and all rules of **S5RC'** keep validity. The validity of RC'1 and RC'2 are stated in Proposition 20. The validity of other axioms and the validity preservation of the rules are easy to verify.

All axioms and rules of **S5**

$(\mathrm{K}_{RC'})$ $C_A^{\mathfrak{A}aa'}(\varphi \to \psi) \to (C_A^{\mathfrak{A}aa'}\varphi \to C_A^{\mathfrak{A}aa'}\psi)$

$(\mathrm{T}_{RC'})$ $C_A^{\mathfrak{A}aa'}\varphi \to (\mathrm{pre}(a) \to \varphi)$

$(\mathrm{RC'}1)$ $C_A^{\mathfrak{A}aa'}\varphi \leftrightarrow E_A \bigwedge_{b \simeq_{E_A} a}(\mathrm{pre}(b) \to C_A^{\mathfrak{A}ba'}\varphi)$

$(\mathrm{RC'}2)$ $\bigwedge_{a'} C_A^{\mathfrak{A}aa'}(\varphi_{a'} \to E_A(\mathrm{pre}(c) \to \varphi_c)) \to ((\mathrm{pre}(a) \to \varphi_a) \to C_A^{\mathfrak{A}ac}\varphi_c)$

$(\mathrm{N}_{RC'})$ $\vdash \varphi \Rightarrow \vdash C_A^{\mathfrak{A}aa'}\varphi$

Fig. 7. The proof system **S5RC'**, where $\mathfrak{A} = (A, \simeq, \mathrm{pre})$ and $a, a', b, c \in A$ such that $a' \simeq_{E_A} c$. For any $x, y \in A$, by writing φ_x and φ_y we mean that $\varphi_x[y/x] = \varphi_y$ and $\varphi_y[x/y] = \varphi_x$.

The following can be proved using a finitary canonical model method (the proof is left out here due to lack of space).

Theorem 23 (Completeness). ***S5RC'*** *is weakly complete: for all* $\mathcal{ELRC'}$-*formulas* φ, $\models \varphi$ *implies* $\vdash_{S5RC'} \varphi$.

The axiomatization **AMRC** (Fig. 8) for AMRC is achieved by adding to **S5RC'** reduction axioms for the $[\mathfrak{A}, a]$-operator. Compared with the reduction axioms used in **AML**, we need an extra one for the relativized common knowledge operator.

All axioms and rules of **S5RC'** and **AML**

(ARC) $[\mathfrak{A}, a]C_A^{\mathfrak{B}bb'}\varphi \leftrightarrow (\mathrm{pre}^{\mathfrak{A}}(a) \to \bigwedge_{a \simeq_A a'} C_A^{\mathfrak{A}\circ\mathfrak{B},(a,b),(a',b')}[\mathfrak{A} \circ \mathfrak{B}, (a', b')]\varphi)$

Fig. 8. The proof system **AMRC**

It is easy to show the completeness of **AMRC**, based on the completeness result of **S5RC'** (Theorem 23).

5   Discussion

In this paper we proved two expressivity results for S5-based logics that previously have been shown only for the case of arbitrary Kripke models, solving what [9] refers to as an open problem. As usual, the proofs for the latter case cannot be trivially extended to the former. We showed how canyon models can be used to compare expressivity of S5-based logics. Another key concept we discussed was the existence of non-reducible private actions in S5-based logics. We believe that the technical machinery can be useful also for future work on S5-based logics. A final contribution of the paper was a definition of a relativized common knowledge operator for action model logic, in the spirit of the relativized common knowledge for public announcement logic, i.e., allowing reduction axioms.

There are many opportunities for future work. When it comes to expressive power, it has not been shown whether $\mathcal{ELRC} \prec \mathcal{ELRC}'$ holds. Also, while the expressivity of public announcement logics with *distributed knowledge* has recently been studied [15], relative expressivity results for many of the languages discussed in the current paper extended with distributed knowledge do not exist. In Section 2 we mentioned some action model composition operators that do not change the expressive power of the language. One such operator that *does*, however, is the Kleene star. Relative expressivity results for variants of the logics considered here with Kleene star is interesting future work, as is the relationship between AMRC and PDL. Other opportunities for future work include characterizations of computational complexity for most of the logics considered in this paper.

References

1. Fagin, R., Halpern, J.Y., Moses, Y., Vardi, M.Y.: Reasoning about Knowledge. MIT (1995)
2. Meyer, J.J.C., van der Hoek, W.: Epistemic Logic for AI and Computer Science. Cambridge Tracts in Theoretical Computer Science, vol. 41. Cambridge University Press (1995)
3. Plaza, J.A.: Logics of public communications. In: Proceedings of ISMIS, pp. 201–216 (1989)
4. Baltag, A., Moss, L.S., Solecki, S.: The logic of public announcements, common knowledge, and private suspicions. In: Gilboa, I. (ed.) Proceedings of TARK, pp. 43–56 (1998)
5. Baltag, A., Moss, L.S.: Logics for epistemic programs. Synthese 139, 165–224 (2004)
6. van Benthem, J.F.A.K.: Information update as relativisation. Technical report, ILLC, University of Amsterdam (2000)
7. van Benthem, J.F.A.K., van Eijck, J., Kooi, B.: Logics of communication and change. Information and Computation 204, 1620–1662 (2006)
8. Baltag, A., Moss, L.S., Solecki, S.: Logics for epistemic actions: Completeness, decidability, expressivity (2003) (manuscript)
9. van Ditmarsch, H., van der Hoek, W., Kooi, B.: Dynamic Epistemic Logic. Springer (2007)
10. Blackburn, P., de Rijke, M., Venema, Y.: Modal Logic. Cambridge University Press (2001)
11. Kooi, B., van Benthem, J.: Reduction axioms for epistemic actions. In: Schmidt, R., Pratt-Hartmann, I., Reynolds, M., Wansing, H. (eds.) Advances in Modal Logic, pp. 197–211 (2004)
12. Kooi, B.: Expressivity and completeness for public update logics via reduction axioms. Journal of Applied Non-Classical Logics 17, 231–253 (2007)
13. Baltag, A.: A logic of epistemic actions. In: van der Hoek, W., Meyer, J.J.C., Witteveen, C. (eds.) Proceedings of FACAS 1999, Utrecht University (1999)
14. Baltag, A., Moss, L.S., Solecki, S.: The logic of public announcements, common knowledge, and private suspicions. Technical report, SEN-R9922, CWI (1999)
15. Wang, Y., Ågotnes, T.: Public announcement logic with distributed knowledge: expressivity, completeness and complexity. Synthese (2013), doi:10.1007/s11229-012-0243-3

Logic Aggregation

Xuefeng Wen and Hu Liu

Institute of Logic and Cognition & Department of Philosophy,
Sun Yat-sen University, Guangzhou, 510275, China
wxflogic@gmail.com, liuhu2@mail.sysu.edu.cn

Abstract. We study the possibility and impossibility of aggregating logics, which may come from different sources (individuals, agents, groups, societies, cultures). A logic is treated as a binary relation between sets of formulas and formulas (or a set of accepted arguments). Logic aggregation is treated as argument-wise. We prove that certain logical properties can be preserved by some desired aggregation functions, while some other logical properties cannot be preserved together under nondegenerate aggregation functions, as long as some natural conditions for the aggregation function are satisfied. We compare our framework of logic aggregation with other aggregation frameworks, including preference aggregation and judgment aggregation.

1 Motivations

Judgment aggregation [15] is to study how to aggregate individual judgments on logically correlated propositions to collective judgments. Since it is both a generalization of preference aggregation in social choice theory, and closely related to deliberative democracy in political science, as well as belief merging in informatics, it has been quickly developed in the past decade. For an up-to-date survey of it, refer to [14] and [11] (more technical). This paper is a further development of judgment aggregation, by setting up a framework called logic aggregation, or aggregation of logics. The research issue of logic aggregation is: given a set of logics, which may come from different sources (individuals, agents, groups, societies, cultures, etc.), how to aggregate these logics into one by some generally acceptable methods. Why is this problem interesting? Here are several motivations for studying logic aggregation.

Firstly, when a pluralistic view on logic is taken, the problem of logic aggregation arises naturally. All present research in judgment aggregation presumes a unified underlying logic, though it need not be the classical logic (see [2]). Not only different individuals have the same logic, but also the logic underlying the collective judgments is the same as those of individuals. But different sources may use different logics in judgments. This could be true for aggregating information from distributed systems, which may come from different domains and use different logics in representing their knowledge. It could also be the case for judgment aggregation in a situation of cross-cultural communication, where individuals or groups from different cultures may have different reasoning patterns.

D. Grossi, O. Roy, and H. Huang (Eds.): LORI 2013, LNCS 8196, pp. 282–295, 2013.
© Springer-Verlag Berlin Heidelberg 2013

Moreover, even if different sources use a unified logic, it is not necessary for the collective to use the same logic. In other words, it is unjustified to presume the collective rationality to be the same as the normal rationality for individuals.

Secondly, logic aggregation provides a framework that may avoid some philosophical difficulties in judgment aggregation. Most research in judgment aggregation adopts a proposition-wise approach, i.e., aggregation of sets of judgments is reduced to aggregation of propositions. When the standard independence condition is assumed, the collective judgment of a proposition does not depend on individual judgments on other propositions. Though independence is natural in preference aggregation, it is controversial in judgment aggregation, because different propositions may have relevance in content apart from pure logical correlations. In particular, some propositions may be premises or reasons for others. From a deliberative point of view, proposition-wise aggregation with independence is undesirable. There are several approaches to this problem. One is to keep proposition-wise aggregation but weaken or generalize the notion of independence [17,4]. The other is to give up the proposition-wise aggregation completely and adopt a holistic method, like distance-based aggregation [18]. Logic aggregation is a compromised approach, going from proposition-wise aggregation to argument-wise aggregation. To realize it, we treat a logic as a set of (accepted) arguments. Then logic aggregation boils down to aggregation of sets of arguments, where the aggregation is argument-wise – a proposition is considered together with its premises (reasons) in aggregation.

Thirdly, logic aggregation opens the door for exploring more notions of rationality and collective rationality. In judgment aggregation, consistency is often required to be preserved from individual judgment sets to the collective one. But consistency is only one property in logic. There are other interesting properties in logics that can be considered, for example, transitivity (a.k.a. cut). In other words, going from judgment aggregation to logic aggregation, we are able to consider more notions of rationality and collective rationality. Unlike [10], which studies different rationality constraints for aggregation in different languages, we explore different rationality in the same language.

Last but not the least, logic aggregation is more or less an application of graph aggregation proposed in [5]. Graph aggregation is in effect the aggregation of arbitrary binary relations on a given set. Though it generalizes preference aggregation, where the binary relation is an ordering, it is too abstract to illustrate interesting applications. Since a logic can also be treated as a binary relation (between sets of formulas and formulas), logic aggregation can be roughly embedded in graph aggregation and thus provides an interesting instantiation of the latter. See Section 5 for more on this.

The problem of aggregating logics has been touched in judgment aggregation before [1,16]. But it was discussed in particular cases. In this paper, we propose the problem explicitly and study it generally in the framework of logic aggregation. The rest of the paper is organized as follows. In Section 2, we introduce the general notion of logic and some typical properties for logic. Section 3 presents the framework of logic aggregation, for which we prove some possibility and impossibility results in Section 4. Section 5 compares logic aggregation with other aggregation frameworks, including preference aggregation and judgment aggregation. The conclusion section indicates some future works.

2   Logics

A logic used to be treated as a set of formulas (which are valid in the logic). This view has been proved to be too narrow since the emergence of numerous non-classical logics, in particular, logics without valid formulas at all, such as Kleene's three-valued logic. Now a logic is usually considered to be a consequence relation (either syntactically or semantically defined), which we will adopt in this paper.

Let $\mathcal{L}$ be a fixed language, namely a set of sentences, which has at least three elements. It can be either finite or infinite. Generally, a *logic* for $\mathcal{L}$ is a binary relation $\vdash$ between $\wp(\mathcal{L})$ and $\mathcal{L}$, where $\wp(\mathcal{L})$ is the power set of $\mathcal{L}$. A pair $(\Sigma, \varphi) \in \wp(\mathcal{L}) \times \mathcal{L}$ is called an *argument* in $\mathcal{L}$, with Σ the set of premises and φ the conclusion. An argument with empty premise is called a *judgment*. An argument (Σ, φ) is called *valid* (or *accepted*) in a logic $\vdash$, if $(\Sigma, \varphi) \in \vdash$, which is often denoted by $\Sigma \vdash \varphi$ instead. Thus, a logic is treated as the set of all valid (or accepted) arguments (rather than formulas) in it. As we do not specify how validity is syntactically or semantically defined, as in standard logic textbooks, we do not distinguish between validity and acceptance of an argument. We assume that any binary relation between $\wp(\mathcal{L})$ and $\mathcal{L}$ is a logic. Instead of $\emptyset \vdash \varphi$, we write $\vdash \varphi$. By $\Sigma \vdash \Delta$, we mean $\Sigma \vdash \varphi$ for all $\varphi \in \Delta$.[1] By Σ, φ (or φ, Σ) and Σ, Σ' occurring on the left hand side of $\vdash$ or $\subseteq$, we mean $\Sigma \cup \{\varphi\}$ and $\Sigma \cup \Sigma'$, respectively. The following are typical properties of logics considered in the literature (see [8] for example).

- Non-triviality: A logic $\vdash$ for $\mathcal{L}$ is *non-trivial* if $\vdash \neq \wp(\mathcal{L}) \times \mathcal{L}$, i.e., a logic is non-trivial, if it does not accept all arguments in the language.
- Consistency: For $\mathcal{L}$ with negation $\neg$, a logic $\vdash$ for $\mathcal{L}$ is *consistent* if there is no $\varphi \in \mathcal{L}$ such that $\vdash \varphi$ and $\vdash \neg\varphi$.

For classical logic, non-triviality and consistency boil down to the same notion. But for non-classical logics, they are not the same. It is well known that para-consistent logics can be inconsistent but non-trivial.

- Reflexivity: A logic $\vdash$ for $\mathcal{L}$ is *reflexive* if for all $\varphi \in \mathcal{L}$, $\varphi \vdash \varphi$.

This is also known as *restricted reflexivity*. A stronger version of reflexivity is as follows.

- Strong reflexivity: A logic $\vdash$ for $\mathcal{L}$ is *strongly reflexive* if for all $\Sigma, \varphi \subseteq \mathcal{L}$, $\varphi \in \Sigma$ implies $\Sigma \vdash \varphi$.

---

[1] Note that this is different from the standard multi-conclusion consequence, where $\Sigma \vdash \Delta$ means $\Sigma \vdash \varphi$ for *some* $\varphi \in \Delta$.

It is easily seen that strong reflexivity can be derived from reflexivity plus monotonicity given below.

- Monotonicity: A logic $\vdash$ for $\mathcal{L}$ is *monotonic* if for all $\Sigma, \Sigma', \varphi \subseteq \mathcal{L}$, $\Sigma \vdash \varphi$ implies $\Sigma, \Sigma' \vdash \varphi$.

Monotonicity is not as uncontroversial as reflexivity. In common sense reasoning, monotonicity is not obeyed, which motivates the branch of nonmonotonic reasoning. The following restricted version of monotonicity is weaker but less controversial.

- Cautious monotonicity: A logic $\vdash$ for $\mathcal{L}$ is *cautiously monotonic* if for all $\Sigma, \varphi, \varphi' \subseteq \mathcal{L}$, $\Sigma \vdash \varphi$ together with $\Sigma \vdash \varphi'$ imply $\Sigma, \varphi' \vdash \varphi$.
- Transitivity: A logic $\vdash$ for $\mathcal{L}$ is *transitive* if for all $\Sigma, \Sigma', \varphi, \varphi' \subseteq \mathcal{L}$, $\Sigma \vdash \varphi$ and $\varphi, \Sigma' \vdash \varphi'$ imply $\Sigma, \Sigma' \vdash \varphi'$.

Transitivity is also known as *cut* in proof theory. It is crucial in composing a valid argument (proof) from other valid arguments (proofs). Many logics (such as relevant logic, linear logic, nonmonotonic logic) lack monotonicity, while preserving transitivity.

- Compactness: A logic $\vdash$ for $\mathcal{L}$ is *compact* if for all $\Sigma, \varphi \subseteq \mathcal{L}$, if $\Sigma \vdash \varphi$ then there is a finite $\Sigma' \subseteq \Sigma$ such that $\Sigma' \vdash \varphi$.

In some literature, compactness refers to the following stronger property, which we call m-compactness.

- M-compactness: A logic $\vdash$ for $\mathcal{L}$ is *m-compact* if for all $\Sigma, \varphi \subseteq \mathcal{L}$, $\Sigma \vdash \varphi$ iff there is a finite $\Sigma' \subseteq \Sigma$ such that $\Sigma' \vdash \varphi$.

It is easily seen that m-compactness is actually the conjunction of compactness and monotonicity.[2]

- Formality: A logic $\vdash$ for $\mathcal{L}$ is *(universally) formal* (a.k.a. *structural*) if for all $\Sigma, \varphi \subseteq \mathcal{L}$, for all substitution σ, $\Sigma \vdash \varphi$ implies $\Sigma^\sigma \vdash \varphi^\sigma$, where φ^σ is the substitution of φ by σ and $\Sigma^\sigma = \{\psi^\sigma \mid \psi \in \Sigma\}$.[3]

Formality, which is realized by the substitution rule (or in effect by axiom schemes), is included in most logics, since it is the mechanism for a logic to be characterized by a finite set of arguments (or, axioms and rules). Universal formality, however, also restricts the power of logic to cover many valid arguments in natural language. For instance, "all bachelors are unmarried" is not valid in standard logic, since it is not true by its form but rather by the meanings of the expressions in it, unless we formalize 'bachelors' as a compound predicate. For this reason, defining a relativized formality as follows is reasonable.

---

[2] In abstract algebraic logic theory [12], pioneered by Tarski, only a binary relation $\vdash$ that satisfies reflexivity, monotonicity, and transitivity can be called a logic. In Tarski's original theory of logical consequence, the three minimal properties of a logic are reflexivity, transitivity, and m-compactness.

[3] As we do not specify the language $\mathcal{L}$ in our general framework, substitution here is underspecified. It can be defined precisely as long as the language $\mathcal{L}$ is specified.

Definition 1 (Formality). *Given $\mathcal{A} \subseteq \mathcal{L}$, a logic $\vdash$ for $\mathcal{L}$ is $\mathcal{A}$-formal if for all $\Sigma, \varphi \subseteq \mathcal{A}$, for all substitution σ, $\Sigma \vdash \varphi$ implies $\Sigma^\sigma \vdash \varphi^\sigma$, where $\Sigma^\sigma = \{\psi^\sigma \mid \psi \in \Sigma\}$. A logic for $\mathcal{L}$ is formal if it is $\mathcal{L}$-formal.*

For similar reasons, we propose relativized completeness and disjunctiveness as follows.

Definition 2 (Syntactical completeness). *For $\mathcal{L}$ with negation[4] $\neg$ and nonempty $\mathcal{A} \subseteq \mathcal{L}$, a logic $\vdash$ for $\mathcal{L}$ is $\mathcal{A}$-complete if for all $\varphi \in \mathcal{A}$, either $\vdash \varphi$ or $\vdash \neg\varphi$. A logic $\vdash$ for $\mathcal{L}$ is (syntactically) complete (a.k.a. negation complete) if it is $\mathcal{L}$-complete.*

Definition 3 (Disjunction property). *For $\mathcal{L}$ with disjunction[5] $\vee$ and $\mathcal{A} \subseteq \mathcal{L}$ including at least one formula of the form $\varphi \vee \psi$, a logic $\vdash$ for $\mathcal{L}$ is $\mathcal{A}$-disjunctive if for all $\varphi \vee \psi \in \mathcal{A}$, $\vdash \varphi \vee \psi$ implies $\vdash \varphi$ or $\vdash \psi$. A logic $\vdash$ for $\mathcal{L}$ has the disjunction property if it is $\mathcal{L}$-disjunctive.*

It is well known that intuitionistic logic has the disjunction property. The following two properties are usually assumed for logics.

Definition 4 (Conjunction property). *For $\mathcal{L}$ with conjunction $\wedge$, a logic $\vdash$ for $\mathcal{L}$ is conjunctive if for all $\Sigma, \varphi, \psi \subseteq \mathcal{L}$, $\Sigma \vdash \varphi \wedge \psi$ iff $\Sigma \vdash \varphi$ and $\Sigma \vdash \psi$.*

Definition 5 (Confluency). *A logic $\vdash$ for $\mathcal{L}$ is confluent if for all $\Sigma, \Sigma', \varphi \subseteq \mathcal{L}$, $\Sigma \vdash \varphi$ and $\Sigma' \vdash \varphi$ imply $\Sigma, \Sigma' \vdash \varphi$.*

Note that monotonicity implies confluency but not vice versa. Finally, we introduce a property which is more familiar in informal logic than in formal logic.

Definition 6 (Non-tautologicity). *A logic $\vdash$ for $\mathcal{L}$ is non-tautological if for all $\Sigma, \varphi \subseteq \mathcal{L}$, $\Sigma \vdash \varphi$ implies $\varphi \notin \Sigma$.*

Non-tautologicity is usually not required in formal logic. Indeed, if a logic is reflexive or monotonic, then it can not be non-tautological. But non-tautologicity is rather plausible for natural language arguments, where begging the question is not allowed. In other words, a good argument should not contain its conclusion as one of its premises. To save monotonicity, we could slightly restrict it as: $\Sigma, \varphi' \vdash \varphi$ whenever $\Sigma \vdash \varphi$ and $\varphi' \neq \varphi$.

For brevity, we use the following notations in the sequel:

- **L**: the set of all logics for $\mathcal{L}$, namely, $\mathbf{L} = \wp(\wp(\mathcal{L}) \times \mathcal{L})$.
- $\mathbf{L}_{cc}$: the set of all consistent and complete logics for $\mathcal{L}$.
- $\mathbf{L}_{cj}$: the set of all conjunctive logics for $\mathcal{L}$.
- $\mathbf{L}_{nt}$: the set of all non-tautological logics for $\mathcal{L}$.

---

[4] For our purpose, it need not be interpreted as the standard negation. It can be any unary connective or operator.

[5] For our purpose, it need not be interpreted as the standard disjunction. It can be any binary connective or operator.

3   Social Logic Function

Let $N = \{1, \ldots, n\}$ be a finite set of agents (groups, societies, cultures) with at least three members. A profile $\Vdash = (\vdash_1, \ldots, \vdash_n)$ is a vector of logics in $\mathbf{L}$. Analogously, we write $\Vdash'$ for the profile $(\vdash_1', \ldots, \vdash_n')$. Let $N_{\Sigma,\varphi}^{\Vdash}$ be the set of agents who accept the argument (Σ, φ) in the profile $\Vdash$, namely, $N_{\Sigma,\varphi}^{\Vdash} = \{i \in N \mid \Sigma \vdash_i \varphi\}$. We write $N_\varphi^{\Vdash}$ for $N_{\emptyset,\varphi}^{\Vdash}$ and $N_{\varphi,\psi}^{\Vdash}$ for $N_{\{\varphi\},\psi}^{\Vdash}$, respectively. Let $N_{\Sigma,\Delta}^{\Vdash} = \bigcap_{\varphi \in \Delta} N_{\Sigma,\varphi}^{\Vdash}$, namely, the set of agents who accept the arguments (Σ, φ) for all $\varphi \in \Delta$ in $\Vdash$. Instead of $F(\Vdash)$, we write $\vdash_F$, and analogously, we write $\vdash_F'$ for $F(\Vdash')$.

Definition 7 (Social logic function). *A social logic function (SLF) (for n logics) for $\mathcal{L}$ is a map $F : \mathbf{L}^n \to \mathbf{L}$.*

Some natural desiderata for SLFs borrowed from social choice theory are listed below.

Definition 8 (Unanimity). *An SLF $F : \mathbf{L}^n \to \mathbf{L}$ is unanimous, if for all $\Sigma, \varphi \subseteq \mathcal{L}$, for all profiles $\Vdash$ in $\mathbf{L}^n$, $\Sigma \vdash_i \varphi$ for all $i \in N$ implies $\Sigma \vdash_F \varphi$, i.e., if an argument is accepted by all individuals, then it is collectively accepted.*

The following is the counterpart of groundedness proposed in [20].

Definition 9 (Groundedness). *An SLF $F : \mathbf{L}^n \to \mathbf{L}$ is grounded, if for all $\Sigma, \varphi \subseteq \mathcal{L}$, for all profiles $\Vdash$ in $\mathbf{L}^n$, $\Sigma \vdash_F \varphi$ implies $\Sigma \vdash_i \varphi$ for some $i \in N$, i.e., if an argument is collectively accepted, then it must be accepted by one of the individuals. An SLF is ungrounded if it is not grounded.*

Unanimity and groundedness of F determine the lower and upper bound of $\vdash_F$, respectively. More precisely, if F is unanimous and grounded then for all profiles $\Vdash$, $\bigcap_{i \in N} \vdash_i \subseteq \vdash_F \subseteq \bigcup_{i \in N} \vdash_i$. We call an SLF *bounded* if it is both unanimous and grounded. If we restrict arguments to judgments, we get weak unanimity and weak groundedness, respectively.

Definition 10 (Weak unanimity). *An SLF $F : \mathbf{L}^n \to \mathbf{L}$ is weakly unanimous, if for all $\varphi \in \mathcal{L}$, for all profiles $\Vdash$ in $\mathbf{L}^n$, $\vdash_i \varphi$ for all $i \in N$ implies $\vdash_F \varphi$, i.e., if a judgment is accepted by all individuals, then it is collectively accepted.*

Definition 11 (Weak groundedness). *An SLF $F : \mathbf{L}^n \to \mathbf{L}$ is weakly grounded, if for all $\varphi \in \mathcal{L}$, for all profiles $\Vdash$ in $\mathbf{L}^n$, $\vdash_F \varphi$ implies $\vdash_i \varphi$ for some $i \in N$, i.e., if a judgment is collectively accepted, then it must be accepted by one of the individuals.*

The following fact should be easily verified. Recall that $\mathbf{L}_{cc}$ is the set of all consistent and complete logics.

Proposition 1. *$F : \mathbf{L}_{cc}^n \to \mathbf{L}_{cc}$ is weakly unanimous iff it is weakly grounded.*

Definition 12 (IIA). *An SLF F is independent of irrelevant arguments (IIA), if for all $\Sigma, \varphi \subseteq \mathcal{L}$, for all profiles $\Vdash$ and $\Vdash'$, $N_{\Sigma,\varphi}^{\Vdash} = N_{\Sigma,\varphi}^{\Vdash'}$ implies that $\Sigma \vdash_F \varphi$ iff $\Sigma \vdash_F' \varphi$, i.e., the collective acceptance of an argument only depends on the individual acceptance of this argument.*

Independence is the most controversial property in social choice, particularly in judgment aggregation. Since we lift aggregation from proposition-wise to argument-wise, independence in logic aggregation is more justified than that in judgment aggregation. Of course, we can still ask the reasons for an argument, just as we can ask the reasons for a proposition. But we can then take the reasons of an argument to be the premises of the argument and form a meta-argument. Thus, an abstract argument-wise aggregation framework is applicable unless we lift the level of arguments constantly.

Definition 13 (Neutrality). *An SLF F is neutral (for arguments) if for all $\Sigma, \Sigma', \varphi, \varphi' \subseteq \mathcal{L}$, for all profiles $\Vdash$, $N^{\Vdash}_{\Sigma, \varphi} = N^{\Vdash}_{\Sigma', \varphi'}$ implies that $\Sigma \vdash_F \varphi$ iff $\Sigma' \vdash_F \varphi'$, i.e. if two arguments receive the same individual acceptance, their collective acceptances are also the same. In other words, all arguments are treated equal.*

We define two weak versions of neutrality as follows, which seem to have no counterparts in the literature.

Definition 14 (C-Neutrality). *An SLF F is neutral for conclusions if for all $\Sigma, \varphi, \varphi' \subseteq \mathcal{L}$, for all profiles $\Vdash$, $N^{\Vdash}_{\Sigma, \varphi} = N^{\Vdash}_{\Sigma, \varphi'}$ implies that $\Sigma \vdash_F \varphi$ iff $\Sigma \vdash_F \varphi'$.*

Definition 15 (P-Neutrality). *An SLF F is neutral for premises if for all $\Sigma, \Sigma', \varphi \subseteq \mathcal{L}$, for all profile $\Vdash$, $N^{\Vdash}_{\Sigma, \varphi} = N^{\Vdash}_{\Sigma', \varphi}$ implies that $\Sigma \vdash_F \varphi$ iff $\Sigma' \vdash_F \varphi$.*

Proposition 2. *An IIA SLF is neutral iff it is both C-neutral and P-neutral.*

Proof. The direction from left to right is obvious. For the other direction, suppose F is both C-neutral and P-neutral. Given a profile $\Vdash$, suppose $N^{\Vdash}_{\Sigma, \varphi} = N^{\Vdash}_{\Sigma', \varphi'} =_{df} C$ and $\Sigma \vdash_F \varphi$. We need to show that $\Sigma' \vdash_F \varphi'$. Consider a profile $\Vdash'$ such that $N^{\Vdash'}_{\Sigma, \varphi} = N^{\Vdash'}_{\Sigma, \varphi'} = N^{\Vdash'}_{\Sigma', \varphi'} = C$. Since $N^{\Vdash'}_{\Sigma, \varphi} = N^{\Vdash}_{\Sigma, \varphi}$, by IIA it follows from $\Sigma \vdash_F \varphi$ that $\Sigma \vdash'_F \varphi$, which implies $\Sigma \vdash'_F \varphi'$ by C-neutrality of F. It in turn implies $\Sigma' \vdash'_F \varphi'$ by P-neutrality of F. By IIA again, we have $\Sigma' \vdash_F \varphi'$. $\qquad\blacksquare$

We call an SLF *systematic* if it is both IIA and neutral.

Definition 16 (N-monotonicity). *An SLF F is n-monotonic if for all all $\Sigma, \varphi \subseteq \mathcal{L}$, for all profiles $\Vdash$, $N^{\Vdash}_{\Sigma, \varphi} \subseteq N^{\Vdash}_{\Sigma', \varphi'}$ and $\Sigma \vdash_F \varphi$ imply $\Sigma' \vdash_F \varphi'$, i.e., compared to a collectively accepted argument, any argument with the same or additional acceptance will also be collectively accepted.*

It is a bit surprising that this natural property had not been proposed before, until its first presence in [6]. Note that n-monotonicity is different from the standard notion of monotonicity, which involves two profiles rather than one. It is easily seen that n-monotonicity implies neutrality but not vice versa.

Definition 17 (Dictatorship). *An SLF F is dictatorial if there exists an $i \in N$ such that for all profiles $\Vdash$, $\vdash_F = \vdash_i$, i.e. the social logic is always the same as i's logic.*

We say that a profile $\Vdash$ satisfies a property P if $\vdash_i$ satisfies P for all $i \in N$. The following notion is adapted from [10].

Definition 18 (Collective rationality, Robustness). *An SLF F is collectively rational for a property P if for all profiles $\Vdash$, $\vdash_F$ satisfies P whenever $\Vdash$ satisfies P. In this case, we also say that P is robust under F.*

We intentionally give another name for collective rationality. With the name of collective rationality, preserving certain logical properties are considered to be 'rational' (and thus desired) for an aggregation function. But there is no reason to assume that these properties on the individual level should also be satisfied on the social level. If we take a serious *social* view on logic, the fact that some logical properties are not preserved under aggregation does not mean that the aggregation is not rational. It only indicates that the rationality or logic on the social level is different. By using the name of robustness, we can compare different logical properties under the framework of social choice and provide a new perspective on logic.

The main question we are to address is: what logical properties are robust? More precisely, which logical properties can be preserved under the desired social logic functions?

4   Some Possibility and Impossibility Results

First we give some easy possibility results.

Proposition 3. *(Strong) reflexivity is robust under any unanimous SLF F : $\mathbf{L}^n \to \mathbf{L}$.*

Proof. Let $F : \mathbf{L}^n \to \mathbf{L}$ be unanimous and $\Vdash$ satisfy (strong) reflexivity. Then (given $\varphi \in \Sigma$), for every $i \in N$, $\varphi \vdash_i \varphi$ ($\Sigma \vdash_i \varphi$). By unanimity, $\varphi \vdash_F \varphi$ ($\Sigma \vdash_F \varphi$). $\qquad \square$

Proposition 4. *Monotonicity is robust under any n-monotonic SLF $F : \mathbf{L}^n \to \mathbf{L}$.*

Proof. Let $F : \mathbf{L}^n \to \mathbf{L}$ be an n-monotonic SLF and $\Vdash$ satisfy monotonicity. Suppose $\Sigma \vdash_F \varphi$ and $\Sigma \subseteq \Sigma'$. Since $\Vdash$ is monotonic, $N_{\Sigma,\varphi}^{\Vdash} \subseteq N_{\Sigma',\varphi}^{\Vdash}$. Then by the n-monotonicity of F, we have $\Sigma' \vdash_F \varphi$. $\qquad \square$

Proposition 5. *$\mathcal{A}$-formality is robust under any n-monotonic SLF $F : \mathbf{L}^n \to \mathbf{L}$.*

Proof. Let $F : \mathbf{L}^n \to \mathbf{L}$ be an n-monotonic SLF and $\Vdash$ satisfy $\mathcal{A}$-formality. Suppose $\Sigma, \varphi \subseteq \mathcal{A}$ and $\Sigma \vdash_F \varphi$. For any substitution σ, we have $N_{\Sigma,\varphi}^{\Vdash} \subseteq N_{\Sigma^\sigma,\varphi^\sigma}^{\Vdash}$, since $\Vdash$ is $\mathcal{A}$-formal. It follows from the n-monotonicity of F that $\Sigma^\sigma \vdash_F \varphi^\sigma$. $\qquad \square$

Proposition 6. *M-compactness is robust under any n-monotonic SLF $F : \mathbf{L}^n \to \mathbf{L}$.*

Proof. Let $F : \mathbf{L}^n \to \mathbf{L}$ be an n-monotonic and grounded SLF and $\Vdash$ satisfy m-compactness. Since $\Vdash$ is monotonic, by Proposition 4, $\vdash_F$ is also monotonic. Thus it suffices to prove that $\vdash_F$ is compact. Suppose $\Sigma \vdash_F \varphi$. Since $\Vdash$ is compact, for each $i \in N_{\Sigma,\varphi}^{\Vdash}$, there is a finite $\Delta_i \subseteq \Sigma$ such that $\Delta_i \vdash_i \varphi$. Let $\Delta = \bigcup_{i \in N_{\Sigma,\varphi}^{\Vdash}} \Delta_i$. Then $\Delta \subseteq \Sigma$ is also finite. Since $\Vdash$ is monotonic, $\Delta \vdash_i \varphi$ for all $i \in N_{\Sigma,\varphi}^{\Vdash}$. Hence, $N_{\Sigma,\varphi}^{\Vdash} \subseteq N_{\Delta,\varphi}^{\Vdash}$. It follows from n-monotonicity of F that $\Delta \vdash_F \varphi$, as required.

Note that m-compactness can not be replaced by compactness in the above proposition. Actually, compactness alone is not robust even under the majority rule. Consider $\vdash_i = \{(\{p_i\}, p), (\{p_i \mid i \in \mathbb{N}\}, p)\}$ for $i \in N$. Then every $\vdash_i$ is compact. But by the majority rule, $\vdash = \{(\{p_i \mid i \in \mathbb{N}\}, p)\}$, which is not compact.

Proposition 7. *Cautious monotonicity is robust under any n-monotonic SLF $F : \mathbf{L}_{cj}^n \to \mathbf{L}_{cj}$.*

Proof. Recall that $\mathbf{L}_{cj}$ is the set of all conjunctive logics. Let $F : \mathbf{L}_{cj}^n \to \mathbf{L}_{cj}$ be an n-monotonic SLF and $\Vdash$ satisfy cautious monotonicity. Suppose $\Sigma \vdash_F \varphi$ and $\Sigma \vdash_F \varphi'$. Since $\vdash_F$ is conjunctive, we have $\Sigma \vdash_F \varphi \wedge \varphi'$. By the conjunction property of $\Vdash$, $N_{\Sigma,\varphi \wedge \varphi'}^{\Vdash} \subseteq N_{\Sigma,\varphi}^{\Vdash} \cap N_{\Sigma,\varphi'}^{\Vdash}$. By cautious monotonicity of $\Vdash$, $N_{\Sigma,\varphi}^{\Vdash} \cap N_{\Sigma,\varphi'}^{\Vdash} \subseteq N_{\Sigma \cup \{\varphi'\},\varphi}^{\Vdash}$. Hence, $N_{\Sigma,\varphi \wedge \varphi'}^{\Vdash} \subseteq N_{\Sigma \cup \{\varphi'\},\varphi}^{\Vdash}$. Since F is n-monotonic, it follows from $\Sigma \vdash_F \varphi \wedge \varphi'$ that $\Sigma, \varphi' \vdash_F \varphi$.

Now we give some impossibility results. First, an easy one, which says that n-monotonicity for non-trivial logics forces groundedness.

Proposition 8. *There is no n-monotonic and ungrounded SLF $F : \mathbf{L}^n \to \mathbf{L}$ that is collectively rational for non-triviality.*

Proof. Suppose $F : \mathbf{L}^n \to \mathbf{L}$ is n-monotonic and ungrounded that is collectively rational for non-triviality. Then there is a profile $\Vdash$ of non-trivial logics and $\Sigma, \varphi \subseteq \mathcal{L}$ such that no one accepts (Σ, φ) but $\Sigma \vdash_F \varphi$. Thus, for any $\Sigma', \varphi' \subseteq \mathcal{L}$, $N_{\Sigma,\varphi}^{\Vdash} = \emptyset \subseteq N_{\Sigma',\varphi'}^{\Vdash}$, which implies by n-monotonicity that $\Sigma' \vdash_F \varphi'$. Hence, $\vdash_F$ is trivial, contradicting the assumption.

This result is not as pessimistic as the usual impossibility results in social choice theory. It only indicates that n-monotonicity should be applied together with groundedness; otherwise, we may get a trivial logic by aggregation. The following results are more parallel with the usual impossibility results. The proofs are canonical, using the property of ultrafilters, which was first introduced in [7] for an alternative proof of Arrow's theorem, and later adapted and refined for other aggregation frameworks in the literature, including [9], [3], and [13] for judgment aggregation and [5] for graph aggregation.

Definition 19. *A group $C \subseteq N$ is a* winning coalition *of (Σ, φ) (under F), if for all profiles $\Vdash$, $N_{\Sigma,\varphi}^{\Vdash} = C$ implies $\Sigma \vdash_F \varphi$.*

The following lemma is easily verified.

Lemma 1. *Let F be an IIA SLF and $\mathcal{W}^F_{\Sigma,\varphi}$ the set of all winning coalitions of (Σ,φ) under F.*

1. *If there is a profile $\Vdash$ such that $N^{\Vdash}_{\Sigma,\varphi} = C$ and $\Sigma \vdash_F \varphi$, then C is a winning coalition of (Σ,φ).*
2. *For all profiles $\Vdash$, $\Sigma \vdash_F \varphi$ iff $N^{\Vdash}_{\Sigma,\varphi} \in \mathcal{W}^F_{\Sigma,\varphi}$.*

We often omit the superscript F in $\mathcal{W}^F_{\Sigma,\varphi}$ if F is clear from the context. The following lemma is adapted from [5].

Lemma 2. *Suppose F is IIA. Let $\mathcal{W}_{\Sigma,\varphi}$ be defined as above. Then*

1. *F is unanimous iff $N \in \mathcal{W}_{\Sigma,\varphi}$ for all $\Sigma,\varphi \subseteq \mathcal{L}$.*
2. *F is grounded iff $\emptyset \notin \mathcal{W}_{\Sigma,\varphi}$ for all $\Sigma,\varphi \subseteq \mathcal{L}$.*
3. *F is neutral iff $\mathcal{W}_{\Sigma,\varphi} = \mathcal{W}_{\Sigma',\varphi'}$ for all $\Sigma,\Sigma',\varphi,\varphi' \subseteq \mathcal{L}$.*

Proof. The first two clauses are obvious. For (3), suppose F is neutral and $N^{\Vdash}_{\Sigma,\varphi} \in \mathcal{W}_{\Sigma,\varphi}$. Let $\Vdash'$ be a profile such that $N^{\Vdash'}_{\Sigma',\varphi'} = N^{\Vdash'}_{\Sigma,\varphi} = N^{\Vdash}_{\Sigma,\varphi}$. Since $N^{\Vdash}_{\Sigma,\varphi} \in \mathcal{W}_{\Sigma,\varphi}$, we have $\Sigma \vdash_F \varphi$, which implies $\Sigma \vdash'_F \varphi$ by IIA. It in turn implies $\Sigma' \vdash'_F \varphi'$ by neutrality. Thus, $N^{\Vdash}_{\Sigma,\varphi} = N^{\Vdash'}_{\Sigma',\varphi'} \in \mathcal{W}_{\Sigma',\varphi'}$. Hence, $\mathcal{W}_{\Sigma,\varphi} \subseteq \mathcal{W}_{\Sigma',\varphi'}$. Similarly, we have $\mathcal{W}_{\Sigma',\varphi'} \subseteq \mathcal{W}_{\Sigma,\varphi}$. For the other direction of (3), suppose $\mathcal{W}_{\Sigma,\varphi} = \mathcal{W}_{\Sigma',\varphi'}$ for all $\Sigma,\Sigma',\varphi,\varphi' \subseteq \mathcal{L}$ and $N^{\Vdash}_{\Sigma,\varphi} = N^{\Vdash}_{\Sigma',\varphi'}$. Then $\Sigma \vdash_F \varphi$ iff $N^{\Vdash}_{\Sigma,\varphi} \in \mathcal{W}_{\Sigma,\varphi}$ iff $N^{\Vdash}_{\Sigma',\varphi'} \in \mathcal{W}_{\Sigma',\varphi'}$ iff $\Sigma' \vdash_F \varphi'$.

Now let's recall the definition of ultrafilters.

Definition 20. *(Ultrafilter) An ultrafilter $\mathcal{W}$ over N is a set of subsets of N satisfying the following conditions:*

1. *$\mathcal{W}$ is proper, i.e. $\emptyset \notin \mathcal{W}$;*
2. *$\mathcal{W}$ is closed under (finite) intersection, i.e. $C_1, C_2 \in \mathcal{W}$ implies $C_1 \cap C_2 \in \mathcal{W}$;*
3. *$\mathcal{W}$ is maximal, i.e. for all $C \subseteq N$, either $C \in \mathcal{W}$ or $\overline{C} \in \mathcal{W}$, where $\overline{C} = N - C$ is the complement of C.*

An ultrafilter $\mathcal{W}$ over N is principal *if $\mathcal{W} = \{C \subseteq N \mid i \in C\}$ for some $i \in N$.*

The following is a well-known fact of ultrafilters.

Lemma 3. *Any ultrafilter over a finite set is principal.*

Theorem 1. *For all nonempty $\mathcal{A} \subseteq \mathcal{L}$, any bounded and systematic SLF $F : \mathbf{L}^n \to \mathbf{L}$ that is collectively rational for transitivity and $\mathcal{A}$-completeness must be dictatorial.*

Proof. Let $F : \mathbf{L}^n \to \mathbf{L}$ be bounded (unanimous and grounded), systematic (IIA and neutral), and collectively rational for transitivity and $\mathcal{A}$-completeness. By Lemma 2, there is a set $\mathcal{W}$ of winning coalitions such that $\Sigma \vdash_F \varphi$ iff $N^{\Vdash}_{\Sigma,\varphi} \in \mathcal{W}$ for all $\Sigma,\varphi \subseteq \mathcal{L}$. By Lemma 3, it suffices to prove that $\mathcal{W}$ is an ultrafilter. First, $\mathcal{W}$ is proper by Lemma 2(2), since F is grounded. Second, suppose $C_1, C_2 \in \mathcal{W}$. Consider a transitive profile $\Vdash$ such that $C_1 = N^{\Vdash}_{\varphi}$, $C_2 = N^{\Vdash}_{\varphi,\psi}$, and $C_1 \cap C_2 = N^{\Vdash}_{\psi}$.

This is possible since by the transitivity of $\Vdash$, $C_1 \cap C_2 \subseteq N_\psi^{\Vdash}$. As C_1, C_2 are winning coalitions, we have $\vdash_F \varphi$ and $\varphi \vdash_F \psi$. It follows that $\vdash_F \psi$ by the collective rationality of F for transitivity. Hence, $C_1 \cap C_2 \in \mathcal{W}$. Finally, Let $C \subseteq N$. Consider an $\mathcal{A}$-complete profile $\Vdash$ such that $C = N_\varphi^{\Vdash}$ and $\overline{C} = N_{\neg\varphi}^{\Vdash}$, where $\varphi \in \mathcal{A}$. By the collective rationality of F for $\mathcal{A}$-completeness, $\vdash_F \varphi$ or $\vdash_F \neg\varphi$. Hence, $C \in \mathcal{W}$ or $\overline{C} \in \mathcal{W}$.

Theorem 2. *For all $\mathcal{A} \subseteq \mathcal{L}$ including at least one formula of the form $\varphi \vee \psi$ with $\varphi \neq \psi$, any bounded and systematic SLF $F : \mathbf{L}^n \to \mathbf{L}$ that is collectively rational for transitivity and $\mathcal{A}$-disjunctiveness must be dictatorial.*

Proof. The proof is almost the same as above, except the verification of the maximality of $\mathcal{W}$. Let $C \subseteq N$. Consider an $\mathcal{A}$-disjunctive profile $\Vdash$ such that $C = N_\varphi^{\Vdash}$, $\overline{C} = N_\psi^{\Vdash}$, and $N = N_{\varphi\vee\psi}^{\Vdash}$, where $\varphi \vee \psi \in \mathcal{A}$. By unanimity, $\vdash_F \varphi \vee \psi$. Since F is collectively rational for $\mathcal{A}$-disjunctiveness, we have $\vdash_F \varphi$ or $\vdash_F \psi$. Hence, $C \in \mathcal{W}$ or $\overline{C} \in \mathcal{W}$.

Here is a reformulation of the above two theorems: the conjunction of transitivity and $\mathcal{A}$-completeness ($\mathcal{A}$-disjunctiveness) is not robust under any bounded, systematic, and non-dictatorial social logic function.

For non-tautological logic, the above theorems can be strengthened by dropping neutrality in the assumption, due to the following lemma. Recall that $\mathbf{L}_{nt}$ is the set of all non-tautological logics.

Lemma 4. *Every unanimous and IIA SLF $F : \mathbf{L}_{nt}^n \to \mathbf{L}_{nt}$ that is collectively rational for both transitivity must be neutral.*

Proof. Let F be unanimous, IIA, and collectively rational for transitivity. Let C be a winning coalition of (Σ, φ). By Lemma 2(3), it suffices to prove that C is also winning coalition of (Σ', φ'). First, we prove that C is a winning coalition of (Σ, φ'). Let $\Vdash$ be a transitive profile such that $N_{\Sigma,\varphi}^{\Vdash} = N_{\Sigma,\varphi'}^{\Vdash} = C$ and $N_{\varphi,\varphi'}^{\Vdash} = N$. This is possible, since $\varphi' \notin \Sigma$ by non-tautologicity. Since C is a winning coalition of (Σ, φ), we have $\Sigma \vdash_F \varphi$. On the other hand, we have $\varphi \vdash_F' \varphi'$ by unanimity. It follows that $\Sigma \vdash_F \varphi'$ by transitivity. Hence, C is a winning coalition of (Σ, φ'). To prove that C is a winning coalition of (Σ', φ'), let $\Vdash$ be a transitive profile such that $N_{\Sigma,\varphi'}^{\Vdash} = N_{\Sigma',\varphi'}^{\Vdash} = C$ and $N_{\Sigma',\Sigma}^{\Vdash} = N$. This is possible, since $\varphi' \notin \Sigma$ by non-tautologicity. Since C is a winning coalition of (Σ, φ'), we have $\Sigma \vdash_F \varphi'$. On the other hand, we have $\Sigma' \vdash \Sigma$ by unanimity. It follows that $\Sigma' \vdash_F \varphi'$ by transitivity. Hence, C is a winning coalition of (Σ', φ').

Using the above lemma, by slightly modifying the proofs of Theorem 1 and Theorem 2, we obtain the following results.

Theorem 3. *For all nonempty $\mathcal{A} \subseteq \mathcal{L}$, any bounded and IIA SLF $F : \mathbf{L}_{nt}^n \to \mathbf{L}_{nt}$ that is collectively rational for transitivity and $\mathcal{A}$-completeness must be dictatorial.*

Theorem 4. *For all $\mathcal{A} \subseteq \mathcal{L}$ including at least one formula of the form $\varphi \vee \psi$ with $\varphi \neq \psi$, any bounded and IIA SLF $F : \mathbf{L}_{nt}^n \to \mathbf{L}_{nt}$ that is collectively rational for transitivity and $\mathcal{A}$-disjunctiveness must be dictatorial.*

In all the above theorems, transitivity can be replaced by confluency or the conjunction property, proofs of which are almost the same. Actually, all properties of the form $A \wedge B \rightarrow C$ can replace the role of transitivity in the above theorems. A parallel result in aggregation of general binary relations has already been obtained in [5].

5   Relating to Other Aggregation Frameworks

Preference Aggregation. If we treat a logic as a binary relation on formulas, instead of that between sets formulas and formulas, then a logic can be roughly regarded as a preference. The reflexivity and transitivity of preferences can be naturally assumed for logics. But the completeness of preferences are not suitable for logics. In other words, when logic is treated as a binary relation on formulas, a framework of general binary relation (as in [5]) or partial relation (as in [19]) is more suitable for logic aggregation than standard preference aggregation.

Conversely, preference aggregation can be embedded into logic aggregation, since preference aggregation can be embedded into judgment aggregation, and the latter can in turn be embedded into logic aggregation (see the next subsection).

Judgment Aggregation. A logic $\vdash$ can be regarded as a set of judgments $J = \{(\Sigma, \varphi) \mid \Sigma \vdash \varphi\} \cup \{\neg(\Sigma, \varphi) \mid \Sigma \nvdash \varphi\}$. The difference is that J is infinite if the language $\mathcal{L}$ is infinite, whereas in judgment aggregation a judgement set is usually finite. Regardless of this difference, logic aggregation can be turned to judgment aggregation of special propositions, where a proposition expresses whether an argument holds.

On the other hand, each set J of judgments together with the underlying logic $\vdash$ can be regarded as a new logic $\vdash' = \vdash \cup \{(\emptyset, p) \mid p \in J\}$. But notice that usually the judgments in J are not formal, in the sense that the substitution rule is not applicable to them in the new logic. In this way, judgment aggregation under a unified logic can be regarded as logic aggregation, where the individual logics are obtained from the unified logic augmented with the individual judgments as non-logical axioms. In this sense, judgment aggregation can be translated into logic aggregation. The relation between these two frameworks is just like that between object language and metalanguage. We can always turn a metalanguage into an object one and vice versa. The logics underlying the judgments expressed by metalanguage in judgment aggregation are turned into object language in logic aggregation, which helps to understand the logical properties better in judgment aggregation.

Graph Aggregation. Graph aggregation proposed in [5] is the aggregation of arbitrary binary relations on a given set V. If we take V to be the set of formulas, and assume compactness of logic, then a logic can be treated as a binary relation on V. An edge from vertex φ to ψ in a graph G represents an accepted argument from premise φ to conclusion ψ in the logic G. But notice that in graph aggregation V is usually assumed to be finite, whereas in logic aggregation the set of formulas are usually infinite. Besides this difference, vertices in a graph are independent of each other while a formula in a logic can be composed from other formulas. Hence, even for compact logics, graph aggregation is too abstract to express logic aggregation.

On the other hand, if we take a graph as a frame for modal logic, then a set of graphs define a logic. Thus the aggregation of sets of graphs can be transformed to the aggregation of modal logics. Moreover, graph aggregation is a special case of the aggregation of sets of graphs (aggregating singleton sets). In this way, graph aggregation can be transformed to logic aggregation.

6   Conclusion and Future Works

We propose a formal framework for logic aggregation, in which some possibility and impossibility results are proved. We also compare logic aggregation with other aggregation frameworks. Our contribution is mainly conceptual rather than technical. This is only a first step in applying the social choice framework to logic. There are a lot left to be explored.

Firstly, more general possibility and impossibility results can be explored under the framework we proposed. Secondly, the framework of logic aggregation itself can also be generalized. An immediate generalization is to take a substructural view on logic, where a logic is no longer a binary relation between sets of formulas and formulas, but between structures of formulas and formulas. Then more nonclassical logics, such as linear logic and Lambek calculus can be incorporated in logic aggregation. Thirdly, logic can be treated in a more functional or dynamic way, where a logic is a procedure (an algorithm, or a method) rather than reducing it to the input-output data (accepted arguments). This means that even if two algorithms produce the same set of valid arguments, they are different logics. Finally, the method of logic aggregation need not be argument-wise. Just as there are global or holistic methods in judgment aggregation, we can also explore global or holistic methods in logic aggregation, for example, the distance-based approach. If this approach is taken, we have to clarify what it means for two logics to be close, and how to define the distance between logics.

Acknowledgements. The paper was supported by the National Fund of Social Science (No. 13BZX066), Humanity and Social Science Youth foundation of Ministry of Education of China (No.11YJC72040001), and the Fundamental Research Funds for the Central Universities (No. 1309073). We would like to than the anonymous referees for their helpful comments.

References

1. Benamara, F., Kaci, S., Pigozzi, G.: Individual Opinions-Based Judgment Aggregation Procedures. In: Torra, V., Narukawa, Y., Daumas, M. (eds.) MDAI 2010. LNCS (LNAI), vol. 6408, pp. 55–66. Springer, Heidelberg (2010)

2. Dietrich, F.: A generalised model of judgment aggregation. Social Choice and Welfare 28(4), 529–565 (2007)

3. Dietrich, F., List, C.: Arrow's theorem in judgment aggregation. Social Choice and Welfare 29(1), 19–33 (2007)

4. Dietrich, F., Mongin, P.: The premiss-based approach to judgment aggregation. Journal of Economic Theory 145(2), 562–582 (2010)

5. Endriss, U., Grandi, U.: Graph Aggregation. In: Proceedings of the 4th International Workshop on Computational Social Choice (COMSOC 2012) (2012)

6. Endriss, U., Grandi, U., Porello, D.: Complexity of Judgment Aggregation: Safety of the Agenda. In: Proceedings of AAMAS 2010, pp. 359–366 (2010)

7. Fishburn, P.C.: Arrow's impossibility theorem: Concise proof and infinite voters. Journal of Economic Theory 2(1), 103–106 (1970)

8. Gabbay, D.M. (ed.): What is a Logical System? Oxford University Press (1995)

9. Gärdenfors, P.: A Representation Theorem for Voting with Logical Consequences. Economics and Philosophy 22(2), 181 (2006)

10. Grandi, U., Endriss, U.: Lifting Rationality Assumptions in Binary Aggregation. In: Proceedings of AAAI 2010, pp. 780–785 (2010)

11. Grossi, D., Pigozzi, G.: Introduction to Judgment Aggregation. In: Bezhanishvili, N., Goranko, V. (eds.) ESSLLI 2010/2011. LNCS, vol. 7388, pp. 160–209. Springer, Heidelberg (2012)

12. Jansana, R.: Propositional Consequence Relations and Algebraic Logic. In: Zalta, E.N. (ed.) The Stanford Encyclopedia of Philosophy. Spring 2011 edn. (2011), http://plato.stanford.edu/archives/spr2011/entries/consequence-algebraic/

13. Klamler, C., Eckert, D.: A simple ultrafilter proof for an impossibility theorem in judgment aggregation. Economics Bulletin 29(1), 319–327 (2009)

14. List, C.: The theory of judgment aggregation: an introductory review. Synthese 187(1), 179–207 (2012)

15. List, C., Pettit, P.: Aggregating Sets of Judgments: An Impossibility Result. Economics and Philosophy 18(1), 89–110 (2002)

16. Miller, M.K.: Judgment Aggregation and Subjective Decision-Making. Economics and Philosophy 24(2) (2008)

17. Mongin, P.: Factoring out the impossibility of logical aggregation. Journal of Economic Theory 141(1), 100–113 (2008), http://linkinghub.elsevier.com/retrieve/pii/S0022053107001457

18. Pigozzi, G.: Belief merging and the discursive dilemma: an argument-based account to paradoxes of judgment aggregation. Synthese 152(2), 285–298 (2006)

19. Pini, M.S., Rossi, F., Venable, K.B., Walsh, T.: Aggregating Partially Ordered Preferences. Journal of Logic and Computation 19(3), 475–502 (2009)

20. Porello, D., Endriss, U.: Ontology Merging as Social Choice. In: Leite, J., Torroni, P., Ågotnes, T., Boella, G., van der Torre, L. (eds.) CLIMA XII 2011. LNCS, vol. 6814, pp. 157–170. Springer, Heidelberg (2011)

The Task Model of Court Investigation in a Multi-agent System of Argumentation in Court

Qiaoting Zhong[1], Xudong Luo[1,*], Frans H. van Eemeren[2], and Fan Huang[1]

[1] Institute of Logic and Cognition,
Sun Yat-sen University,
Guangzhou, 510275, China
{i.qt.zhong,wong.vanz}@gmail.com, luoxd3@mail.sysu.edu.cn
[2] Department of Speech Communication, Argumentation Theory and Rhetoric,
University of Amsterdam,
Amsterdam, 1012 VB, The Netherlands
F.H.vanEemeren@uva.nl

Abstract. To develop a court argumentation system of three agents: a plaintiff, a defendant, and a judge, this paper constructs task models of courtroom investigation. More specifically, we develop the algorithms of evidence questioning and evaluation, and establish a mechanism for fact finding, which consists of *evidence-claim* networks and evidence aggregation. Finally, we illustrate our system with a real legal scenario.

1 Introduction

Argumentation is a communication process that aims at resolving a difference of opinion with the addressee [1]. By proposing a constellation of propositions, the arguer, who is challenged, can try to make his standpoint acceptable to a rational judge. Argumentation not only is omnipresent every day but also has become an influential approach in Artificial Intelligence [2]. So, many argumentation systems have been developed in domains such as law [3] and medicine [4]. However, although a court argumentation simulation system can help lawyers to analyse their cases, predict judicial decisions, and prepare some strategies for legal processes, few such systems have been developed.

This paper is to address this issue. More specifically, our multi-agent system simulates the process of conducting a civil lawsuit in accordance with Chinese civil procedural law. Normally, a lawsuit consists of four stages: courtroom investigation, courtroom debate, mediation, and finally judgement.[1] In this paper, we focus on dealing with the courtroom investigation stage. The main goal of a courtroom investigation is to come up with facts that can be admitted by law, called legal facts, and thus can be used as the premises of the final judgement. According to [5], a courtroom investigation proceeds as follows: (i) both parties present opening statements; (ii) the plaintiff provides evidence; (iii) the defendant asks questions; (iv) the defendant provides evidence; (v) the

---

* Corresponding author.
[1] The mediation part just indicates whether the judge should put forward a decision or leave the things to a further mediation process.

D. Grossi, O. Roy, and H. Huang (Eds.): LORI 2013, LNCS 8196, pp. 296–310, 2013.

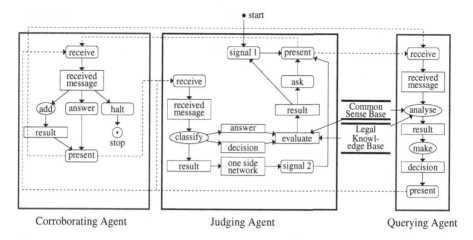

Fig. 1. The interaction process of courtroom investigation

plaintiff asks questions; and (vi) the judge asks questions. In our system, the pieces of evidence are presented one by one and so actually steps (ii) and (iii) are repeated until no more evidence. The same goes for steps (iv) and (v).

The three agents in our model are shown in Fig. 1. During the courtroom investigation process, after the left agent that corroborates (called corroborating agent) has presented some pieces of evidence and propositions, the right agent that addresses inquiries to evidence (called querying agent) will first analyse the evidence and inferences by answering some critical questions, and then decides how to respond accordingly. Each time after steps (ii) to (iii) (or (iv) to (v)), the judging agent (in the middle of Fig. 1) tries to check out any conflicts among all the received messages.[2] If there is, he needs judge which claim is more convincing, or asks questions for more information. During the lawsuit process, each agent has its own actions and beliefs, but all the agents share some premise knowledge, which includes legal rules and interpretations (stored in *legal knowledge base* in Fig. 1) and a certain amount of common sense (stored in *common sense base* in Fig. 1).[3] So, basically our system is a knowledge based system [6].

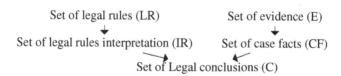

Fig. 2. General structure of judge's litigation reasoning

---

[2] By conflicts, we mean that there is evidence supporting claim A while there is another evidence supporting claim $\neg A$.

[3] The common sense knowledge base is used to support the reasoning on legal knowledge. If there is inconsistency between these two bases, the legal knowledge base will override the common sense knowledge.

To solve the conflicts between the two disputing parties, the most important thing for the judge is to give an acceptable justification of his legal decision. In [7], Xiong presents a general structure of judge's litigation reasoning, as shown in Fig. 2, where the right part relates to fact finding and the left part relates to the debate. In this paper, our focus is on the reasoning process involved in the right part in Fig. 2 (in other papers, we will discuss the left part and the integration of two parts to form a complete system). In this right part, before the judge can draw an inference from evidence to case fact (claim), one agent needs to provide evidence and the other needs to question. Accordingly, our system consists of the task models of evidence questioning, evaluating, and aggregating, and an *evidence-to-claim* network, which is used to keep track of the reasoning process during courtroom investigation.

The rest of the paper is organised as follows. Sections 2 and 3 detail the computational models of evidence questioning and evaluating. Sections 4 and 5 introduce the mechanism of fact finding, which includes an *evidence-claim* network and an evidence aggregation operator. Section 6 illustrates our system in a real civil case. Section 7 discusses related work. Finally, Section 8 concludes the paper with future work.

2   Task Model of Questioning on Evidence

This section presents the algorithm of evidence questioning.

Usually the corroborating agent cannot use indirect evidence e_i to support some case fact cf_j directly. For example, a fingerprint at the scene of a crime cannot directly prove that someone is involved in the case. Instead, the corroborating agent should introduce an intermediate claim E_i to specify the usage of e_i in the case, and then provide an inference from E_i to cf_j.[4]

Fig. 3 shows the algorithm of questioning on evidence. *Firstly*, the querying agent *classifies* received evidence e_i, according to the *legal knowledge base*, into: (i) documentary evidence; (ii) physical evidence; (iii) audio and visual material; (iv) testimony of witnesses; (v) statements of the parties involved; (vi) conclusions of identification experts; and (vii) transcripts of survey.

Secondly, the querying agent *verifies* the evidence of e_i by answering the critical questions about the *legality and genuineness* of evidence e_i. The critical questions are about the following key points [9]: whether it is obtained: (i) by infringing upon the lawful rights and interests of other people, or (ii) by means prohibited by law. If e_i is verified as illegal, the querying agent will inform the judging agent about this to end the cross-examination. The genuineness of evidence is examined in a similar way. All these critical questions are sorted in *legality and genuineness rule base*.

Thirdly, if the querying agent can ensure that e_i is *illegal* or *non-genuine* by above steps, then he can *select* the rule from *legal knowledge base* that supporting his proof and present it to the judging agent. Otherwise, the querying agent is assumed to believe

---

[4] It allows multi-steps in the reasoning process of e_i proving cf_j, *i.e.*, there is reasoning chain $e_i \rightarrow E_{i1} \rightarrow E_{i2} \rightarrow \ldots \rightarrow E_{in} \rightarrow cf_j$, but in this paper we restricted on the form of $e_i \rightarrow E_i \rightarrow cf_j$ where E_i is a key step in the above chain. And the reasoning involved is default reasoning [8], *i.e.*, if e_i is admissible and the inference from e_i to cf_j is made, then cf_j is admissible when inconsistency information is not supplied.

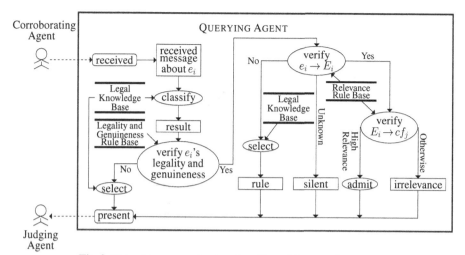

Fig. 3. The evidence questioning algorithm of the querying agent

that e_i is legal and genuine. Then in this case, he will further verify the relevance between e_i and cf_j. Since we introduce a key step E_i to connect the whole inference from e_i to cf_j, the relevance check is started by *verifying* $e_i \to E_i$, followed by *verifying* $E_i \to cf_j$ (if $e_i \to E_i$ is assumed to be acceptable).

- When *verifying* $e_i \to E_i$, the querying agent will answer the following questions (included in relevant rule base) [9,10], depending on types of e_i:[5] (i) Documentary evidence, physical evidence, audio and visual material: *e.g.*, is e_i relevant to E_i? (ii) Testimony of witnesses: *e.g.*, does the witness not have interests in any of the party involved? (iii) Conclusions of expert witnesses: *e.g.*, is their assertion based on evidence? (iv) Statements of the parties involved: *e.g.*, do the parties provide other relevant evidence? The critical questions are stored in the order of the difficulty of answering them, *i.e.*, from the easiest to the hardest. The system checks the questions one by one starting from the top one in the list. If the querying agent cannot acquire enough information to answer any of these questions (*unknown*), he can choose to remain *silent*; if the answer to one of critical questions is *negative*, the querying agent needs to *select* the relevant rule from *legal knowledge base*, and present to the judging agent; otherwise, $e_i \to E_i$ is assumed to be acceptable, and he should go ahead to *verify* the strength of $E_i \to cf_j$.
- When *verifying* $E_i \to cf_j$, the querying agent just answers some questions such as "Are they relevant?" and "How relevant are they?" (these questions are stored in the *relevance rule base*). According to [11], the querying agent can determine the strength of $E_i \to cf_j$. If *highly relevant*, the querying agent *admits* the evidence and the relevant propositions, and then *presents* to the judging agent about his admission; otherwise, he *presents* the result of *irrelevance* to the judging agent.

---

[5] No doubts for transcripts of inspection and examination [9].

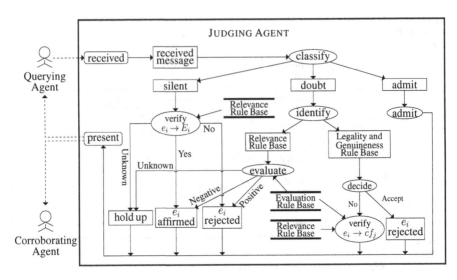

Fig. 4. The evidence evaluation algorithm of the judging agent

3   Task Model of Evidence Evaluation

In this section, we discuss the algorithm of evidence evaluation.

After the querying agent finishes the evidence questioning process shown in Fig. 3, the judging agent will *classify*, as shown in Fig. 4, the received massages from the querying agent to see whether or not the received evidence should be admitted as legal evidence. Firstly, the judging agent *classifies* the received messages into three types: silent, doubt and admit. According to different types, different evaluation process will be carried out. (i) When the querying agent keeps *silent*, the judge needs to *verify* the strength of $e_i \rightarrow E_i$ because the judging agent knows that according to the method of questioning (see Fig. 3), it means that the querying agent does not have sufficient information. According to *relevance rule base*, if the verified result is yes, then the judging agent *affirm* the evidence e_i; if no, he should *reject* e_i. Otherwise, if the judging agent cannot decide the strength of $e_i \rightarrow E_i$, it is reasonable to set his temporary evaluation result as *hold up*. (ii) when the querying agent *doubts* the admissibility of e_i by providing some propositions about rule violations, the judging agent's job is more complicated. That is, first the judging agent needs to *identify* which rule base the querying agent talked about. Here are the following two possibilities:

- *Relevance rule base*, which is about the inference from evidence to case fact. In the previous section, we divide the relevance check into two steps, but since giving a counter-argument is more critical than just presenting doubts, for both steps, the judging agent *evaluates* the response of the querying agent by asking himself questions such as: "Does he have opposite evidence or reasons?", "Are the opposite evidence or reasons powerful enough to rebut the evidence of the other party?", which are stored in the *evaluation rule base*. If the answers to these questions are *negative*, the judging agent should reject the querying agent's attack and *affirm* the

Fig. 5. Independent case   **Fig. 6.** Dependent case

related evidence provided by the corroborating agent (see [9]: Article 72), otherwise he should admit the querying agent's attack and *reject* that evidence. With incomplete information, the judging agent cannot decide to admit the evidence or not. So, in this case it is reasonable to set his temporary evaluation result as *hold up* and do not let such evidence be the basis of his judgement.[6]

- *Legality and genuineness rule base*, which is about the legality and genuineness of evidence, in which all are yes/no questions. So, the judging agent can easily *decide* to accept some doubt or not. If he *accepts* the doubt, he will *reject* the related evidence e_i provided by the corroborating agent. Otherwise, he needs to further *verify* whether or not that evidence supports the case fact by answering some critical questions in the *relevance rule base* and the *evaluation rule base*.

And (iii) when the querying agent *admits* the evidence provided by the corroborating agent, the judging agent needs not check it further but *admits* it (see [9]: Article 72).

4   Evidence-Claim Network

This section presents our evidence-claim network to record the process of courtroom investigation.

The key to the success in a lawsuit is to provide the court sufficient, solid evidence. However, often it is unclear how an inference from evidence to claim (*i.e.*, fact of case) is made. Therefore, the agents concerned, especially the judging agent, needs to figure out the relationship between a piece of evidence and a certain case fact. To solve the problem, we introduce an *evidence-claim* network, which consists of points (labelled by cf, e, and E), links and marks. Its links are the relations between evidence and claims, and also between pieces of evidence. Since evidence could be independent or not, there are two kinds of link as shown in Figs. 5 and 6. The marks in the network are set to indicate the status of evidence and consistency about the links. Therefore, the marks are on the points and links both. In particular, symbol ✓ marks *acceptable*, symbol × marks *unacceptable*, and symbol ? means *doubts* or *need of more information*.

After the corroborating agent provides evidence e_i to support his claim cf_j, each agent uses the following operations to establish a network: $add\_point(e_i)$, $add\_point(E_i)$, and $add\_point(cf_j)$; $add\_link(e_i, E_i)$, and $add\_link(E_i, cf_j)$, where E_i is an intermediate fact of agent that corroborates.

---

[6] In fact, the judging agent can ask the other agents for more information, but then the case involves to decide which questions could be asked and to ask which agent. Thus, the whole problem becomes much more complicated and beyond the scope of this paper. So, we leave the problem to our future work.

After questioning: (i) The corroborating agent changes nothing in his network. (ii) if admitting the other's evidence and claims, the querying agent puts symbol ✓ on the related points and links; if doubts, put symbol ?; and if silent, do nothing. And (iii) for the judging agent, as shown in Fig. 4, if he sorts the received message into the class of:

- Admit: Put symbol ✓ on all the related points and links.
- Doubt: (a) On e_i's legality and genuineness: First put symbol ? on e_i, then call the evaluation method. If it is verified as legal and genuine, replace symbol ? with symbol ✓ and call a further evaluation; otherwise, with symbol ×. (b) On the strength of $e_i \to E_i$ (or $E_i \to cf_j$): First put symbol ? on the link, then call the evaluation algorithm. If the querying agent does not have any opposite evidence or reasons powerful enough to rebut e_i, replace symbol ? with symbol ×; otherwise, with symbol ✓. However, if the judging agent cannot judge whether or not it is powerful enough, then do not replace.
- Silent: First put symbol ? on link $e_i \to E_i$, then call the questioning algorithm (as shown in Fig. 3). If any of the following cases is confirmed, the judging agent will replace symbol ? with symbol ×, otherwise with symbol ✓: (a) Documentary evidence, physical evidence, audio and visual material: e_i is irrelevant to E_i. (b) Testimony of witnesses: the witness has interests in some of the parties involved. (c) Conclusions of expert witnesses: the assertion is not based on evidence. And (d) statements of the parties involved: the party who claims cannot provide other relevant evidence. As to transcripts of inspection and examination, the judging agent just replaces symbol ? with symbol ✓. If the judging agent does not have enough information to answer the above questions, he will not replace.

5 Evidence Aggregation

This section will discuss: (i) how the judging agent aggregates different evidence, which support the same claim to some extents;[7] and (ii) after such an aggregation, how the judging agent judges whether or not the case fact is affirmative.

Now we answer the first question. When agents plaintiff and defendant identify different pieces of evidence to prove their contradicting facts, different types of evidence have different probative forces. According to Articles 76 and 77 in [9], the probative force of physical evidence, the conclusions of identification experts, the transcripts of survey and the document evidence that have been notarised are, as a general rule, much stronger than that of non-notarised document evidence, audio-visual materials and testimonies. Thus, we can set the probative force of each type of evidence, denoted as $F(e_i)$, with numbers in $[0, 1]$ as shown in Table 1.[8] Moreover, depending on supporting the truth of some case fact directly or not, different pieces of evidence could have

---

[7] We will not discuss the reliability of intermediate facts. However, we will investigate the relationship between different intermediate facts (supportive or contradictory) and their effects on the supportive of a claim.

[8] Intuitively, the probative force of evidence e can be viewed as the judge's subjective belief of the statement that e proves by default, unless e is proved not admissible. The numbers in Table 1 could be changed for different judges but remain to reflect the preference order stated in [9].

Table 1. Probative force of evidence of each type

$F(e_i)$	type of evidence e_i
0.7	non-notarised document evidence
0.9	notarised document evidence
0.9	physical evidence
0.7	audio and visual material
0.7	testimony of witnesses
0.6	statements of involving parties
0.9	conclusions of expert witnesses
0.9	transcripts of inspection and examination

different influences, which we call the weight of evidence. The more important a piece of evidence, the heavier its weight; and the stronger the probative force of a piece of evidence, the more important it is. Formally, we have:

Definition 1. *For each evidence e_i ($i = 1, \ldots, n$) that supports the same case fact cf, its weight w_i is given by:*

$$w_i = \frac{F(e_i)}{\sum\limits_{i=1}^{n} F(e_i)}, \tag{1}$$

where $F(e_i)$ is the probative force of evidence e_i.

For the final network, if symbol $\times$ is on point e_i, it means that evidence e_i cannot be admitted as legal evidence, and so the weight of evidence e_i reduces to zero, *i.e.*, $w_i = 0$. Only in this case, the weight of a piece of evidence will change. Thus, we can define the status of evidence as follows:

Definition 2. *Evidence e_i is rejected if $w_i = 0$ or there is a symbol of $\times$ on the link started from e_i; evidence e_i is admissible if symbol of $\checkmark$ is on all the links started from e_i; otherwise, evidence e_i cannot be a basis of a judgement.*[9]

Although e_i is verified as admissible, different admissibility strength of $e_i \to cf_j$ influences the final status of cf_j differently. According to the results of the evaluation algorithm and the final network, for any link $e_i \to E_i \to cf_j$, (i) if symbol $\checkmark$ is on both $e_i \to E_i$ and $E_i \to cf_j$ (*i.e.*, e_i and cf_j are highly relevant), and e_i is admissible, then cf_j will be affirmed by the judging agent surely; (ii) if a symbol of $\times$ is on one of them, then cf_j cannot be affirmed; and (iii) otherwise, e_i cannot be the basis of judgement (namely, e_i cannot affect the final status of cf_j). More specifically, we have:

Definition 3. *For link $e_i \to E_i \to cf_j$, its strength, which reflects the admissibility of evidence e_i to cf_j, is given by:*

$$S(e_i \to cf_j) = \begin{cases} 1 & \text{if symbol } \checkmark \text{ is on both } e_i \to E_i \text{ and } E_i \to cf_j, \\ 0.25 & \text{if symbol } \times \text{ is on one of } e_i \to E_i \text{ and } E_i \to cf_j, \\ 0.5 & \text{otherwise.} \end{cases}$$

---

[9] Especially, a piece of evidence e_i cannot be a basis of case fact cf's final status judgement (affirmed or not) if there is a symbol of ? on the link between e_i and cf.

To calculate the reliability of a case fact, cf, from its weight as well as the strength of $e_i \rightarrow cf$, intuitively: (i) if the most important evidence is admitted and no evidence is against cf, cf should be affirmed absolutely; (ii) if the most important evidence is rejected and others are not strong enough to support cf, cf should be absolutely rejected; (iii) if multiple pieces of evidence to support the same fact of cf and carry an equal importance, the reliability of cf should increased with the amounts of the evidence; (iv) if multiple pieces of evidence are against the same fact of cf and carry an equal importance, the reliability of cf should be decreased with the amounts of the evidence; and (v) if some evidence supports and the others are against the same fact of cf, and all carry an equal importance, the reliability of cf should be inbetween the reliability from support and that from against. Formally, we have:

Definition 4. *The function of $f : ([0,1] \times [0,1])^n \rightarrow [0,1]$ is a reliability function for case fact cf from the admissibilities and weights of evidence $e_1, \ldots, e_n$ if it satisfies:*

(i) *if $w_j = \max\{w_1, \ldots, w_n\}$, $S(e_j \rightarrow cf) = 1$ and $S(e_k \rightarrow cf) > 0.5 (\forall k \neq j)$ then $f((w_1, S(e_i \rightarrow cf)), \ldots, (w_j, S(e_j \rightarrow cf))) = 1$;*

(ii) *if $w_j = \max\{w_1, \ldots, w_n\}$, $S(e_j \rightarrow cf) = 0$ and $S(e_k \rightarrow cf) < 0.5 (\forall k \neq j)$ then $f((w_1, S(e_i \rightarrow cf)), \ldots, (w_j, S(e_j \rightarrow cf))) = 0$;*

(iii) *if $\forall j, k \in \{1, \ldots, n\}$, $w_j = w_k$ and $\forall i \in \{1, \ldots, n\}$, $S(e_i \rightarrow cf) > 0.5$ then $\forall j \in \{1, \ldots, n\}$, $f((w_1, S(e_i \rightarrow cf)), \ldots, (w_j, S(e_j \rightarrow cf))) \leq f((w_1, S(e_i \rightarrow cf)), \ldots, (w_{j+1}, S(e_{j+1} \rightarrow cf)))$;*

(iv) *if $\forall j, k \in \{1, \ldots, n\}$, $w_j = w_k$ and $\forall i \in \{1, \ldots, n\}$, $S(e_i \rightarrow cf) < 0.5$ then $\forall j \in \{1, \ldots, n\}$, $f((w_1, S(e_i \rightarrow cf)), \ldots, (w_j, S(e_j \rightarrow cf))) \geq f((w_1, S(e_i \rightarrow cf)), \ldots, (w_{j+1}, S(e_{j+1} \rightarrow cf)))$;*

(v) *if $\forall j, k \in \{1, \ldots, n\}$, $w_j = w_k$ and $\exists m \in \{1, \ldots, n\}$ such that $\forall i \in \{1, \ldots, m\}$, $S(e_i \rightarrow cf) > 0.5$ and $\forall i \in \{m+1, \ldots, n\}$, $S(e_i \rightarrow cf) < 0.5$, then $f((w_{m+1}, S(e_{m+1} \rightarrow cf)), \ldots, (w_n, S(e_n \rightarrow cf))) \leq f((w_i, S(e_i \rightarrow cf)), \ldots, (w_n, S(e_n \rightarrow cf))) \leq f((w_1, S(e_1 \rightarrow cf)), \ldots, (w_m, S(e_m \rightarrow cf)))$.*

According to [12,13], uninorm operator $\oplus$ has the following properties listed in the following lemma.

Lemma 1. *Let $\oplus$ is a uninorm with identity 0.5.*

(i) *Commutativity: $\oplus(a, b) = \oplus(b, a)$;*

(ii) *Associativity: $\oplus(a, \oplus(b, c)) = \oplus(\oplus(a, b), c)$;*

(iii) *Monotonicity: $\oplus(a, b) \geq \oplus(c, d)$ if $a \geq c$ and $b \geq c$;*

(iv) *Strengthening: if $a_{n+1} > 0.5$, $\oplus(a_1, \ldots, a_n) \leq \oplus(a_1, \ldots, a_n, a_{n+1})$;*

(v) *Weakening: if $a_{n+1} < 0.5$, $\oplus(a_1, \ldots, a_n) \geq \oplus(a_1, \ldots, a_n, a_{n+1})$;*

(vi) *Absorptivity: $\forall a \geq 0.5$, $\oplus(a, 1) = 1$ and $\forall a \leq 0.5$, $\oplus(0, 1) = 0$; and*

(vii) *Comprising: for any $a < 0.5$ and $b > 0.5$, $a \leq \oplus(a, b) \leq b$.*

The following theorem presents our evidence aggregation operator:

Theorem 1. *Given a case fact of cf, let e_i with $w_i \neq 0$ $(i = 1, \ldots, n)$ be the evidence that supports cf. Then the following is the reliability function for cf:*

$$R(cf) = \oplus_{i=1}^n \left(\frac{w_i}{\max\{w_1, \ldots, w_n\}} \times S(e_i \rightarrow cf) \right), \tag{2}$$

where uninorm operator $\oplus$ is given by:

$$\oplus(a, b) = \frac{ab}{ab + (1 - a)(1 - b)}. \tag{3}$$

Proof. (i) Assume that the most important evidence is admitted and others are not strong enough to support cf, then there exists j such that $w_j = \max\{w_1, \ldots, w_n\}$ and $S(e_j \to cf) = 1, \forall k \neq j, S(e_k \to cf) \geq 0.5$, thus according to properties (i) -(iii) and (vi) of Lemma 1, we have $R(cf) = \oplus(1, b) = 1$ where

$$b = \oplus\left(\oplus_{i=1}^{j-1} \left(\frac{w_i \times S(e_i \to cf)}{\max\{w_1, \ldots, w_{j-1}\}}\right), \oplus_{i=j+1}^{n} \left(\frac{w_i \times S(e_i \to cf)}{\max\{w_{j+1}, \ldots, w_n\}}\right)\right).$$

(ii) Assume that the most important evidence is rejected and others are not strong enough to support cf, then the proof is similar to that of case (i) above, but $\forall k \neq j, S(e_k \to cf) < 0.5$.

(iii) Assume that multiple pieces of evidence to support the same fact of cf and carry an equal importance, then $\forall j, k \in \{1, \ldots, n\}, w_j = w_k$ and $\forall i \in \{1, \ldots, n\}, S(e_i \to cf) > 0.5$. Then according to property (iv) of Lemma 1, for any $j \in \{1, \ldots, n\}$,

$$\oplus_{i=1}^{j} \left(\frac{w_i}{\max\{w_1, \ldots, w_j\}} \times S(e_i \to cf)\right) \leq \oplus_{i=1}^{j+1} \left(\frac{w_i}{\max\{w_1, \ldots, w_{j+1}\}} \times S(e_i \to cf)\right).$$

Therefore, $R(cf)$ is increased with the amounts of the evidence.

(iv) Assume that multiple pieces of evidence are against the same fact of cf and carry an equal importance, then the proof is similar to that of case (iii) above, but $\forall i \in \{1, \ldots, n\}, S(e_i \to cf) < 0.5$ and by property (v) of Lemma 1.

(v) Assume that some evidence supports and the others are against the same fact of cf, and all carry an equal importance, then $\forall j, k \in \{1, \ldots, n\}, w_j = w_k$ and $\exists j \in \{1, \ldots, n\}$ such that $\forall i \in \{1, \ldots, j\}, S(e_i \to cf) > 0.5$ and $\forall i \in \{j + 1, \ldots, n\}, S(e_i \to cf) < 0.5$. Then, according to property (iii) of Lemma 1, we get

$$a = \oplus_{i=1}^{j} \left(\frac{w_i}{\max\{w_1, \ldots, w_j\}} \times S(e_i \to cf)\right) \geq 0.5,$$

$$b = \oplus_{i=j+1}^{k} \left(\frac{w_i}{\max\{w_{j+1}, \ldots, w_k\}} \times S(e_i \to cf)\right) \leq 0.5.$$

Therefore, by property (vii) of Lemma 1, $b \leq \oplus(a, b) \leq a, i.e., b \leq R(cf) \leq a$. $\qquad\square$

Formula (3) is a uninorm operator which concept is introduced by Yager in [12]. Finally, we answer the second question as follows:

Definition 5. *Let $\theta \in [0, 1]$ be the priori reliability of case fact cf, then it is affirmed if and only if $R(cf) > \theta$; it is rejected if and only if $R(cf) < \theta$; and it is neutral if and only if $R(cf) = \theta$.*

In the above definition, θ is the priori reliability of cf, which can be viewed as the judging agent's subjective priori probability on the truth of cf. Thus, if the judging agent has no bias, it is unknown to him whether cf is true or not, then it is reasonable to set θ as 0.5 [14]. However, if the judging agent has bias, the case will become complicated, and we will not discuss here since Chinese law does not allow this to happen [15].

Suppose evidence e_1 and e_2 support the same case fact of cf. According to reliability function (2), Fig. 7 shows clearly the relation between evidence probative force F and

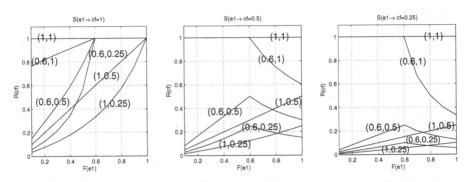

Fig. 7. The relation between probative force and reliability, where (1,1) means that $F(e_2) = 1$ and $S(e_2 \to cf) = 1$, and others are similar

reliability R: (i) when evidence e_2 is more important, if it is 100% admissible, the reliability of cf is always 1 for any probative force of e_1; (ii) for a given probative force of e_2 and a given admissibility of e_2 to cf (not equal to 1), if e_1 is proven to be absolutely relevant to the case (*i.e.*, $S(e_1 \to cf) = 1$), then the reliability of cf increases with the probative force of e_1 (see the leftmost graph); otherwise, from the peak points in the middle and in the rightmost subfigure, it is impossible to make cf affirmed (*i.e.*, $R(cf) > 0.5$); and (iii) the peak points in each subfigure shows that the reliability of cf depends on the less relevant evidence when e_1 and e_2 are equally important. From the dropping lines from those peak points, we can see that irrelevant evidence is not helpful and sometimes harmful to increase the reliability of cf.

6 A Legal Scenario

This section will show how our system simulates a real civil case [16]: Zhiqiang Zhang sued Suning Company in Quanshan District Court for the injury of his rights and interests as a consumer. Zhang alleged that he bought a refrigerator from Suning and had it replaced due to its quality problems, but the replacement is a used one, which is a fraudulent act. So, Zhang claimed against Suning for refunding the purchase price at doubled amount and paying for his loss of income due to missed working time, the relevant traffic expenses and telephone bill, *i.e.*, 3320 Chinese Yuan in total. However, Suning requested the court to reject Zhang's claims because they replaced Zhang with a new one without any quality problem.

Because of the page limit, we omit the procedure of evidence questioning and evaluating, but just show that Zhang alleged some facts of case supported by certain evidence in Table 2, and Suning Company argued that the fact of case in Table 3 is true.

Fig. 8 shows the judge's final *evidence-claim* network with respect to this legal case.[10] By Definition 2, $e_0, \ldots, e_4$, and e_6 are admissible but e_9 is rejected; e_5 cannot be a basis of cf_4's final status judgement; and e_7 and e_8 cannot be a basis of cf_5's

---

[10] Suppose the judge has no bias. The number below each e_i is its probative force in this case according to Table 1.

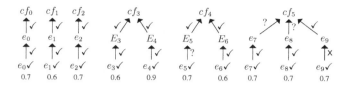

Fig. 8. The judge's final network

Table 2. Case facts provided by the plaintiff

$e_0 \longrightarrow cf_0$	cf_0: Zhang (Z) bought refrigerator R from Suning (S). e_0: The invoices of R (*Documentary*).
$e_1 \longrightarrow cf_1$	cf_1: R has some quality problems because S dropped in and repaired it for twice but failed. e_1: The statements provided by Z.
$e_2 \longrightarrow cf_2$	cf_2: To replace R, S provided Z with a new one, R', of the same brand and the same model on 24 July 2004. e_2: The pick-up list provided by S (*Documentary*).
e_4 e_3 $\downarrow$ $\downarrow$ E_4 E_3 $\searrow$ $\swarrow$ cf_3	cf_3: S had subjective fault when providing commodities to Z. E_3: When S's employees delivered R' at the downstairs of Z's apartment, they unpacked it without Z presented, carried it to upstairs, and then left without having it checked by Z. e_3: The statements provided by Z. E_4: S's employees left R without taking back its voucher, or leaving the voucher of R'. e_4: The voucher of R (*Physical*).
e_6 e_5 $\downarrow$ $\downarrow$ E_6 E_5 $\searrow$ $\swarrow$ cf_4	cf_4: Z found stains and mildew on R', and thought it was a second-hand one. E_5: A videotape manifested that R' had some problems. e_5: A videotape on R' (*Audio and visual material*). E_6: If R' is a new one then there must be a certificate that indicates the date of production. However, the fact is that there is no certificate provided for R'. e_6: The statements provided by Z.

Table 3. Case facts provided by the defendant

e_9 e_8 e_7 $\searrow$ $\downarrow$ $\swarrow$ cf_5	cf_5: S replaced R with a new one R' without any quality problem. e_7: The inventory provided by S (*Documentary*); e_8: The pick-up list provided by S (*Documentary*); and e_9: The deliverer's testimony.

final status judgement, either. Thus, by Definition 1, Definition 3 and Theorem 1, we have:

$$
\begin{aligned}
R(cf_3) &= \oplus \left(\frac{w_3 S(e_3 \rightarrow cf_3)}{\max\{w_3, w_4\}}, \frac{w_4 S(e_4 \rightarrow cf_3)}{\max\{w_3, w_4\}} \right) \\
&= \oplus \left(\frac{0.4 \times 1}{\max\{0.4, 0.6\}}, \frac{0.6 \times 1}{\max\{0.4, 0.6\}} \right) \\
&= \frac{0.67 \times 1}{0.67 \times 1 + (1 - 0.67) \times (1 - 1)} \\
&= 1.
\end{aligned}
$$

Similarly, by formula (2) we can obtain $R(cf_0) = R(cf_1) = R(cf_2) = 1$, $R(cf_4) = 0.86$, and $R(cf_5) = 0.25$. Then, by Definition 5, $cf_0, \ldots, cf_3$ and cf_4 are affirmed but cf_5 is rejected.

7 Related Work

Our work advanced the state-of-art in the following three aspects:

(i) *Evidence admissibility.* In dispute resolution, Prakken [17] modelled a third party to allocate the burden of proof, decide the admissibility of evidence, and adjudicate the conflict in the judge's final decision. Our work is significantly different from theirs as follows: (a) In our model, the judge's belief changes on facts of case claimed by the disputing parties can be recorded and represented clearly through our evidence-claim network. (b) To judge the admissibility of evidence, we introduce a mechanism of fact finding, which employs the fuzzy method of uninorm operation [12,13].

(ii) *Argument diagram.* Wigmore' evidence chart is the most well-known argument diagram, which is a directed graph structure made up of points (representing propositions) and arrows (representing steps of inference) [18,19]. The evidence-claim network in our system is similar to Wigmore's chart because we use nodes to represent statements or evidence and links to represent relation between nodes. However, in our network, we use the same kind of node to represent different type of evidence and indicate their difference by showing their probative force under the node, while Wigmore's chart have different kinds of nodes. And more importantly, we have shown how to aggregate all the probative force of relevant evidence in the network to obtain the overall reliability of a legal fact, while this part is not involved in Wigmore's work. Moreover, although Wigmore's chart is widely used in the legal domain, it cannot be used to predict judicial decisions. Rather, lawyers can use our system, especially our evidence-claim network and fact finding mechanism, to predict and prepare court strategies by guessing the judge's subjective belief on evidence before going to the court. Since the opening statement provided by the plaintiff can be read before the lawsuit, the defence lawyers can use the algorithm of evidence questioning in Fig. 3 to find out some ways to question and then build up a final evidence-claim network by using the algorithm of evidence evaluation in Fig. 4. Through guessing the judge's subjective prior belief on evidence, the defence lawyers can find out facts with evidence aggregation.

(iii) *Evidence aggregation.* The existing judgement aggregation work (*e.g.*, [20]) is to find a collective judgement for a group of judges, but they do not consider how a judgement is made according to amounts of uncertain evidence. On the other hand, in our model, only one judge is involved and his own judgement based on the multiple pieces of uncertain evidence, and thus, in particular, we design an evidence aggregation mechanism under uncertainty. Unfortunately, in their work there are no such ones.

8   Conclusions and Future Work

Based on knowledge engineering methodology for negotiating agent development [6], to developed a multi-agent system of argumentation in court, which simulates the process of a civil lawsuit in accordance with the Chinese civil law procedure, we present a detailed task model to carry out court investigation. In particular, we introduce evidence-claim network and design the evidence aggregation mechanism.

It remains for future work to investigate the computational properties of efficiency of our algorithms and our contribution with respect to the techniques proposed in [12,13], as well as to compare our work with Bayesian networks [21] and Carneades [3]. Moreover, it is worth studying how other factors could determine the weight of evidence. The task model of courtroom debate needs to be developed and integrated with the task model of courtroom investigation to form a completed simulation system of court argumentation.

Acknowledgements. The authors appreciate the anonymous referees for their insightful comments, which have been used to improve the paper. Moreover, this paper is partially supported by Raising Program of Major Project of Sun Yat-sen University (No. 1309089), National Natural Science Foundation of China (No.61173019), Bairen Plan of Sun Yat-sen University, and Major Projects of the Ministry of Education (No. 10JZD0006) China.

References

1. van Eemeren, F.H., Garssen, B., Krabbe, E.C.W., Henkemans, A.F.S., Verheij, B., Wagemans, J.H.M.: Handbook of Argumentation Theory: An Overview of Classical and Neo-Classical Perspectives on Argumentation and Modern Theoretical Approaches to Argumentative Discourse. Springer (2013)
2. Bench-Capon, T.J.M., Dunne, P.E.: Argumentation in artificial intelligence. Artificial Intelligence 171(10-15), 619–641 (2007)
3. Gordon, T., Prakken, H., Walton, D.: The carneades model of argument and burden of proof. Artificial Intelligence 171(10), 875–896 (2007)
4. Bromuri, S., Morge, M.: Multiparty argumentation game for consensual expansion applied to evidence based medicine. In: Proceedings of the 14th Conference on Artificial Intelligence in Medicine, pp. 33–37 (2013)
5. SCNPC No.75: Civil Procedure Law of the People's Republic of China (2007 Amendment) (2007)
6. Luo, X., Miao, C., Jennings, N., He, M., Shen, Z., Zhang, M.: KEMNAD: A knowledge engineering methodology for negotiating agent development. Computational Intelligence 28(1), 51–105 (2012)

7. Xiong, M.: Litigational Argumentation: Logical Perspective of Litigation Games. China University of Political Science and Law Press (2010)

8. Reiter, R.: A logic for default reasoning. Artificial Intelligence 13(1-2), 81–132 (1980)

9. SPC of PRC No.33: Some Provisions of the Supreme People's Court on Evidence in Civil Procedures (come into force in January 2002) (2001)

10. Walton, D.: Fundamentals of Critical Argumentation. Cambridge University Press (2006)

11. Lavery, J., Hughes, W.: Critical Thinking: An Introduction to the Basic Skills, 5th edn. Broadview Press (2008)

12. Yager, R.R., Rybalov, A.: Uninorm aggregation operators. Fuzzy Sets and Systems 80(1), 111–120 (1996)

13. Luo, X., Jennings, N.R.: A spectrum of compromise aggregation operators for multi-attribute decision making. Artificial Intelligence 171(2-3), 161–184 (2007)

14. Smets, P., Kennes, R.: The transferable belief model. Artificial Intelligence 66(2), 191–234 (1994)

15. Zhou, M., Zong, L.: Chinese law and oriental culture. In: Blanpain, R. (ed.) Law in Motion: Recent Developments in Civil Procedure, Constitutional, Contract, Criminal, Environmental, Family & Succession, Intellectual Property, Labour, Medical, Social Security, Transport Law, pp. 163–174. Kluwer Law International (1997)

16. SPC of PRC: Case of dispute over consumer's interests infringement: Zhang Zhiqiang v. Suning Electronic Materials Company in Xuzhou. Gazette of the Supreme People's Court of the People's Republic of China 120(10), 32–36 (2006)

17. Prakken, H.: A formal model of adjudication dialogues. Artificial Intelligence and Law 16(3), 305–328 (2008)

18. Wigmore, J.: Evidence in Trials at Common Law. In: Tillers, P. (ed.), vol. 1a. Little, Brown and Company, Boston (1983)

19. Walton, D.: Argumentation and theory of evidence. In: Breur, C., Kommer, M.M., Nijboer, J.F., Reijntjes, J. (eds.) New Trends in Criminal Investigation and Evidence, vol. 2, pp. 711–732. Intersentia (2000)

20. List, C.: The theory of judgment aggregation: An introductory review. Synthese 187(1), 179–207 (2012)

21. Pearl, J.: Probabilistic reasoning in intelligent systems: networks of plausible inference. Morgan Kaufmann (1988)

A Deontic Action Logic for Complex Actions

Huimin Dong and Xiaowu Li

Institute of Logic and Cognition, Sun Yat-Sen University, Guangzhou, 510275, China
ellutung@gmail.com, lxw121@aliyun.com

Abstract. Here we choose an object-oriented approach to model a deontic action logic. The interpretation of an action, related to its execution circumstance, is a set of events charactered by a structure, named event-base, which satisfies some algebra properties. Different from Modal Action Logic (MAL), this structure is not a Boolean one, but reflects the algebra properties of sequent actions and true concurrent actions. At last, our work includes an axiomatic system for deontic complex actions as well as its completeness.

1 Introduction

This paper is about a Deontic Action Logic for complex action norms in human life. It specifies the properties of normative sentences, which could talk about norms for complex actions or behaviours in human society, and gives a sound and complete axiomatic system for complex action norms.

In the opinion of von Wright [21], norms should be things that govern actions, which are missing in Standard Deontic Logic. However, not until Segerberg's effort did a formal analysis of action norms come up, and Meyer developed a classical deontic action framework, called dynamic deontic logic (DDL), which is an adaptation of the Andersonian-like reduction to character deontic notions over actions [13]. According to the dynamic approach, an action is interpreted as a modality of states transformation, which are treated as the executions of this action. And then an action is permitted if all its executions do not lead to any violated states. Such semantics reduces the deontic notion into all actual possibilities or impossibilities to perform actions with violation constants. Yet Kent, Maibaum, and Quirk proposed a Boolean algebraic semantics on action norms to reject the dynamic modality reduction in MAL [15]. In this logic, the specifications of agents' actions are related to deontic predicates over the actions' executions, where the deontic predicates are different from action modalities. Moreover, this Boolean action algebra works through many paradoxes in dynamic deontic logics, although it contains limited types of actions. However, simple action norms are not enough for the normative formal theory about complicated human-like reasoning.

To model the daily life normative reasoning, this paper will take complex actions into account, and define some algebraic structure which reflects the properties of choice actions, concurrent actions and sequential actions as well as their corresponding norms. With the help of them, we have a new deontic action logic

D. Grossi, O. Roy, and H. Huang (Eds.): LORI 2013, LNCS 8196, pp. 311–315, 2013.

to analyse human normative reasoning. In our daily life, norms are not only about simple actions, such as "permitted to read", but also the complex ones. For example, a fresh is permitted to spend weekend at home after provided his signed himself out of college when he finishes his military training. The trandition deontic action logic could not properly describe these action norms.

The first example is for the concurrent action norms, called true concurrent action norms. These days in China, it is not allowed to smoke in some indoor public places. For instance, in the restaurant, office, store and so on. No one is allowed to smoke inside a restaurant, but it is permitted to smoke if he or she is outside the restaurant [20]. Let α denote "Sit inside", β denote "Smoke". In trandition, $P(\alpha \times \beta) \to P(\alpha)$ is valid. But it contradicts our intuition about concurrent action norms in daily life. We want to find another way to describe concurrent action norms, so that the characteration fits the true concurrent action norms in daily life.

Secondly, we want to get rid of the norms similar to the example below: If after shooting the president it is permitted to remain silent, then it is permitted to shoot the president and remain silent. Denote "shoot the president" as α, "Remain silent" as β. Then it could be formalized as $\langle\alpha\rangle P(\beta) \to P(\alpha\,;\beta)$. This normative sentence is valid but undesired in Meyer's approach. However, it can not be expressed in [5][6][7][8]. So, it is also an important task for this paper to character these sequential action norms.

2   A Deontic Action Logic for Complex Actions

This section gives out a deontic action logic for complex actions, including a deductive system and its semantics, to describe complex action norms, such as "permitted to drink water or whisky", "not permitted to knock the door and then come in the office in the restime", and "not allowed to drink and drive". So, the action norms here are more close to our daily life and capture our intuitive understanding of normative reasoning about human activity.

Let Φ_0 be a countable set of propositional variables, and Δ_0 be a finite set of primitive actions, such that Φ_0 and Δ_0 are mutually disjoint. The primitive symbols include $\mathbf{0}$, $\mathbf{1}$, and the elements of Φ_0 and Δ_0, where $\mathbf{0}$ and $\mathbf{1}$ are not in Φ_0 or Δ_0. Actions in one norm are always finite, so are the primitive actions here. In our language, compound actions are formed by the primitive actions $a \in \Phi_0$, the impossible action $\mathbf{0}$ and the trivial action $\mathbf{1}$ with three kinds of action operators $\cup, ;, \times$. Except for the boolean sentences, $[\alpha]\varphi$ and $P(\alpha)$, we also have $\alpha \equiv \beta$.

In our logic, the range of the execution of actions is always bigger than the range of the permission of actions. For this reason, we describe the properties of the impossible actions and the trivial action as follows: Let $\mathbf{0}$ denote the impossible action and $\mathbf{1}$ denote the trivial action. If something is impossible to be executed, it is always not allowed to be executed, meanwhile if do some action changes noting, it is allowed to be executed since its result makes no harm to the current state. Thus, this logic acesses the power of norms over the utility of executions. The axiomatization below will show us this opinion. The trivial action $\mathbf{1}$ can not be defined as the non-deterministic choice of all the actions here. Moreover, the compound actions $\alpha \cup \beta, \alpha ; \beta, \alpha \times \beta$ are respectively denoted as the choice actions, the sequent actions and the concurrent actions. Here we do not consider the finite repeated actions α^* or the test actions φ? for simplicity, and also the negative actions $\overline{\alpha}$ in [9]. Being of lack of the negative action, it is not possible for us to character negative action as well as the duality, a weak permission, of $P(\alpha)$, a strong permission. According to [9], the obligation is defined by the strong permission and the weak permission. However, these notions are not hard to be added up into our system because of its algebra semantics. $P(\alpha)$ means that the executing of action α is permitted or allowed. $\alpha \equiv \beta$ is read as action α and action β are identical, representing that the utility of do α is identical to the utilityof do β. For example, use a cipher card to open a door has the same effect with the password.

Our deontic action logic for complex actions here contains the axioms in Table 1 and Table 2, meanwhile it contains the Modus Ponens and the Generated Rule. To prove the completeness, we need to add axiom **AM** into the system:

$$\gamma \subseteq \alpha \; O \; \beta \wedge \alpha \; O \; \beta \equiv \sqcup \Delta_{\alpha \, O \, \beta} \to \gamma \equiv \alpha_1 \; O \; \beta_1 \vee \cdots \vee \gamma \equiv \alpha_n \; O \; \beta_m,$$

where $\alpha_i \; O \; \beta_j \in \Delta_{\alpha \, O \, \beta}, 1 \le i \le n, 1 \le j \le m, O \in \{\cup, ;, \times\}$, and $\sqcup \Delta_\alpha$ is closed under $\cup, ;, \times$ in $\{\gamma \in \Delta_o \;|\!\vdash \gamma \subseteq \alpha\}$, where $\alpha \in \Delta_0 \cup \{\mathbf{0}, \mathbf{1}\}$. We define **MDAL = CDAL + AM**.

Table 1. Action axioms for Complex Action Terms

Ac1 $\alpha \equiv \alpha$	**Ac11** $\alpha \times \mathbf{1} \equiv \mathbf{1} \times \alpha \equiv \alpha$
Ac2 $\alpha \cup \alpha \equiv \alpha$	**Ac12** $\alpha \times \beta \equiv \beta \times \alpha$
Ac3 $\alpha \cup \beta \equiv \beta \cup \alpha$	**Ac13** $(\alpha \times \beta) \times \gamma \equiv \alpha \times (\beta \times \gamma)$
Ac4 $(\alpha \cup \beta) \cup \gamma \equiv \alpha \cup (\beta \cup \gamma)$	**Ac14** $\chi \subseteq \alpha \wedge \gamma \subseteq \beta \leftrightarrow \chi \times \gamma \subseteq \alpha \times \beta$
Ac5 $\gamma \subseteq \alpha \cup \beta \to \gamma \subseteq \alpha \vee \gamma \subseteq \beta$	**Ac15** $\alpha \equiv \beta \leftrightarrow \alpha \subseteq \beta \wedge \beta \subseteq \alpha$
Ac6 $\alpha ; \mathbf{0} \equiv \mathbf{0} ; \alpha \equiv \mathbf{0}$	**Ac16** $\neg(\mathbf{0} \equiv \mathbf{1})$
Ac7 $\alpha ; \mathbf{1} \equiv \mathbf{1} ; \alpha \equiv \alpha$	**Ac17** $\gamma \subseteq a \to \gamma \equiv a$, where $a \in \Delta_0$
Ac8 $(\alpha ; \beta) ; \gamma \equiv \alpha ; (\beta ; \gamma)$	**Ac18** $\gamma \subseteq \mathbf{0} \to \gamma \equiv \mathbf{0}$
Ac9 $\chi \subseteq \alpha \wedge \gamma \subseteq \beta \leftrightarrow \chi ; \gamma \subseteq \alpha ; \beta$	**Ac19** $\gamma \subseteq \mathbf{1} \to \gamma \equiv \mathbf{1}$
Ac10 $\alpha \times \mathbf{0} \equiv \mathbf{0} \times \alpha \equiv \mathbf{0}$	**Ac20** $\gamma \; O \; \alpha \cup \beta \equiv \gamma \; O \; \alpha \cup \gamma \; O \; \beta$,
	where $O \in \{\cup, ;, \times\}$

Table 2. Axioms for Action Norms

A1 $\neg P(\mathbf{0})$	**A6** $[\mathbf{0}]\varphi$
A2 $P(\mathbf{1}$	**A7** $\varphi \to [\mathbf{1}]\varphi$
A3 $P(\alpha \cup \beta) \leftrightarrow P(\alpha) \wedge P(\beta)$	**A8** $[\alpha]\varphi \wedge \beta \subseteq \alpha \to [\beta]\varphi$
A4 $P(\alpha ; \beta) \wedge P(\beta ; \gamma) \to P(\alpha ; \beta ; \gamma)$	**A9** $[\alpha](\varphi \to \psi) \to ([\alpha]\varphi \to [\alpha]\psi)$
A5 $P(\alpha) \wedge \beta \subseteq \alpha \to P(\beta)$	**Sub** $\varphi \wedge (\alpha \equiv \beta) \to \varphi(\alpha/\beta)$

In the semantic part, we have to define a event-base $\Sigma = \langle E, \circ, \times, 0, 1 \rangle$, which contain the annihilator elements 0, the identity element 1, the commutative and associative operator $\circ$ as well as the associative operator $\times$, and it is also closed under operatiors $\circ$ and $\times$. This structure gives out the properties of complex events as well as their corresponding actions. Then we have the model defintion:

Definition 1. *Structure* $M = \langle W, \Sigma, V, \theta, I, P \rangle$ *as a deontic action model for* L, *if and only if it satisfies the following conditions:*

1. W *is a non-empty set, in which every element is treated as a possible world;*
2. $\Sigma = \{ E_w \mid E_w$ *is a event-base, for any* $w \in W \};$
3. V *is a proposition evaluation function from* Φ_0 *to* $\wp(W);$
4. θ *is a mapping from* E_w *to* $\wp(W \times W)$, *such that for any* $e \in E_w$, *we have* $\theta(e) \subseteq W \times W$, *denoted as* θ_e, *which is an event relation satisfies the following condition: For any* $e, g \in E_w, 0, 1 \in E_w$, *we have* $\theta_{e \circ g} = \theta_e \circ \theta_g$, $\theta_0 = \emptyset$ *and* $\theta_1 = Id$, *where* Id *is an dentity function from* $W \times W$.
5. I *is a function on* W *such that for any* $w \in W$, $I(w)$ *is a function from* $\Delta_0 \cup \{0, 1\}$ *to* $\wp(E_w)$, *named as* I_w, *satisfies that* $I_w(a_0) \subseteq E_w$ *with* $\mid I_w(a_0) \mid = 1$, $I_w(0) = \{0\}$ *and* $I_w(1) = \{1\}$.
6. $P \subseteq \bigsqcup_{w \in W} \{w\} \times E_w$ *satisfies that: For any* $w \in W, e_1, e_2, e_3 \in E_w$,
   P-com *If* $(w, e_1 \circ e_2) \in P$ *and* $(w, e_2 \circ e_3) \in P$, *then* $(w, e_1 \circ e_2 \circ e_3) \in P;$
   P-E $(w, 0) \notin P;$
   P-T $(w, 1) \in P.$

We use $\mathbf{M}(DA)$ for the class of all the deontic action models defined as above. In this model, the interpretation of an action are a set of events. For every primitive action, its interpretation is only one event, because the primitive action is executed in a specific condition, and then one condition lead out one consequence. However, it seems to be more natural that the execution of one action leads to finite many outcome. We think this considerarion is easy to clarify under our frame work, so we left it to our further work. Moreover, the function I could be inductively extended by the Boolean operation $\cup$ and the event operations $\circ$ and $\times$ within the condition that if $I_w^*(\gamma) \subseteq I_w^*(\alpha \ O \ \beta)$, then there is some $\chi \in \Delta_{\alpha \ O \ \beta}$ such that $I_w^*(\gamma) = I_w^*(\chi)$, where $O \in \{\cup, ;, \times\}$. According to these, we could define the truth of normative sentences. The truth of boolean sentences are defined as usual, while given any $M = \langle W, \Sigma, V, \theta, I, P \rangle \in \mathbf{M}(DA), w \in W$,

$$M, w \vDash \alpha \equiv \beta \Leftrightarrow I_w(\alpha) = I_w(\beta),$$
$$M, w \vDash [\alpha]\varphi \Leftrightarrow e \in I_w(\alpha) \text{ and } w\theta_e u \Rightarrow M, u \vDash \varphi,$$
$$M, w \vDash P(\alpha) \Leftrightarrow e \in I_w(\alpha) \Rightarrow P(w, e).$$

At last, we show the completeness.

This paper not only give a sound and complete axiomazation for complex action norms, but also give a more powerful theoretical tool than the dynamic approach to character normative reasoning. It solves the same paradoxes as dynamic deontic logic does, and make the dynamic deontic undesire but valid sentences invalid.

References

1. Anglberger, A.: Dynamic Deontic Logic and its Paradoxes. Studia Logica: An International Journal for Symbolic Logic 89(3), 427–435 (2008)
2. Blackburn, P., de Rijke, M., de Venema, Y.: Modal Logic. Cambridge Tracts in Theoretical Computer Science 53 (2001)
3. Broersen, J.: Modal Action Logics for Reasoning about Reactive Systems. PhD thesis, Vrije University (2003)
4. Castro, P.F.: A Complete and Compact Propositional Deontic Logic. In: Jones, C.B., Liu, Z., Woodcock, J. (eds.) ICTAC 2007. LNCS, vol. 4711, pp. 109–123. Springer, Heidelberg (2007)
5. Castro, P.F., Maibaum, T.S.E.: A Tableaux System for Deontic Action Logic. In: van der Meyden, R., van der Torre, L. (eds.) DEON 2008. LNCS (LNAI), vol. 5076, pp. 34–48. Springer, Heidelberg (2008)
6. Castro, P., Maibaum, T.: Deontic action logic, Atomic Boolean Algebras and Fault-tolerance. Journal of Applied Logic 7(4), 441–466 (2009)
7. Castro, P., Maibaum, T.: Deontic Logic, Contrary to Duty Reasoning and Fault Tolerance. Electronic Notes in Theoretical Computer Science 258, 17–34 (2009)
8. Castro, P., Maibaum, T.: Towards a First-Order Deontic Action Logic. In: Mossakowski, T., Kreowski, H.-J. (eds.) WADT 2010. LNCS, vol. 7137, pp. 61–75. Springer, Heidelberg (2012)
9. Fiadeiro, J., Maibaum, T.: Temporal theories as modularisation units for concurrent system specification. Formal Aspects of Computing 4(3), 239–272 (1992)
10. Gargov, G., Passy, S.: A Note on Boolean Logic. In: Petkov, P.P. (ed.) Proceedings of the Heyting Summer School. Plenum Press (1990)
11. Kent, S., Maibaum, T., Quirk, W.: Formally Specifying Temporal Constraints and Error Recovery. In: Proceedings of IEEE International Symposium on Requirements Engineering, pp. 208–215 (1993)
12. Kulicki, P., Trypuz, R.: A Deontic Action Logic with Sequential Composition of Actions. In: Ågotnes, T., Broersen, J., Elgesem, D. (eds.) DEON 2012. LNCS, vol. 7393, pp. 184–198. Springer, Heidelberg (2012)
13. McNamara, P.: Deontic Logic. In: Gabbay, D.M., Woods, J. (eds.) Handbook of the History of Logic, vol. 7 (2006)
14. Meyer, J.J.: A Different Approach to Deontic Logic: Deontic Logic Viewed as Variant of Dynamic Logic. Notre Dame Journal of Formal Logic 29 (1988)
15. Meyer, J.J., Wieringa, R.J.: Applications of Deontic Logic in Computer Science: A Concise Overview. In: Deontic Logic in Computer Science: Normative System Specification, pp. 17–40. John Wiley and Sons, Chichester (1993)
16. Meyer, J.J., Wieringa, R.J., Dignum, F.P.M.: The Paradoxes of Deontic Logic Revisited: A Computer Science Perspective. In: Electrical Engineering, Mathematics and Computer Science (EEMCS) (1994)
17. Monk, J.D.: Mathematical logic. Springer (1976)
18. Peleg, D.: Communication in Concurrent Dynamic Logic. Journal of Computer and System Sciences 35, 23–58 (1987)
19. Prisacariu, C.: Synchronous Kleene Algebra. Logic Algebra. Programming 79, 608–635 (2010)
20. Prisacariu, C., Schneider, G.: A Dynamic Deontic Logic for Complex Contracts. Journal of Logic and Algebraic Programming 81, 458–490 (2012)
21. von Wright, G.: An Essay in Modal Logic and Deontic Logic (1951)
22. von Wright, G.: Norm and Action: A Logical Enquiry. Routledge & Kegan Paul. The Humanities Press, London, New York (1963)

Planning Using Dynamic Epistemic Logic: Correspondence and Complexity

Martin Holm Jensen

Technical University of Denmark
mhje@imm.dtu.dk

Abstract. A growing community investigates planning using dynamic epistemic logic. Another framework based on similar ideas is knowledge-based programs as plans. Here we show how actions correspond in the two frameworks. We finally discuss fragments of DEL planning obtained by the restriction of event models. Fragments are separated by virtue of their computational complexity.

1 Introduction

We consider planning where actions may be non-deterministic and the environment partially observable. This requires modeling the agent's knowledge state as actions change it. In some cases we might want to consider planning problems with less strict assumptions (e.g. non-deterministic actions in a fully observable environment), and a desirable property of a planning formalism is that it allows for doing so. In the following we explore the link between the *knowledge-based programs (KBPs) as plans* approach of [5], and planning using dynamic epistemic logic (DEL) (see e.g [2,1]). This link is established by showing how actions correspond in the two approaches.

Example 1. An agent has two dice cups and two coins with a white and red side at her disposal. Her knowledge state is described by the coin configurations she considers possible. She can shuffle a coin using a cup, she can lift a cup concealing a coin, and she can simply toss the coin without using a cup.

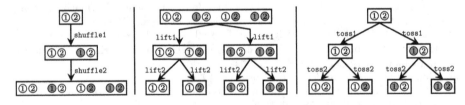

The actions change her knowledge state as depicted above, where a rectangle containing coin configurations corresponds to a knowledge state. On the left we see that shuffling a coin leads to ignorance of which side is up, contrasted by the center illustration where feedback is received upon lifting a cup. The former is an example of an ontic action, and the latter of an epistemic action as we introduce below. On the right is illustrated the non-deterministic outcomes of an *observable* coin toss.

D. Grossi, O. Roy, and H. Huang (Eds.): LORI 2013, LNCS 8196, pp. 316–320, 2013.

1.1   Preliminaries: Knowledge-Based Programs as Plans

We repeat the KBPs as plans approach of [5]. Let $X = \{x_1, \ldots, x_n\}$ be a finite set of propositional symbols, $L_P(X)$ (resp. $L_E(X)$) denote formulas of the propositional (resp. epistemic) language. A *state* s is a valuation of X assigning each x_i true or false (we also identify a state with a subset of X in the obvious manner). A *knowledge state* M is a nonempty set of states. A formula $\phi \in L_E(X)$ is *objective* if it does not contain K, and for an objective ϕ we name $K\phi$ an *epistemic atom*. Formulas formed by the usual propositional connectives and epistemic atoms are called *purely subjective*. Given a knowledge state M and purely subjective formula ϕ, then satisfaction is as usual for combinations of epistemic atoms with propositional connectives. For epistemic atoms we define $M \models K\phi$ if $s \models \phi$ for all $s \in M$, where $s \models \phi$ is standard propositional satisfaction.

Given is a set of ontic action symbols A_O and a set of epistemic action symbols A_E. Let $X' = \{x' \mid x \in X\}$ and $\alpha \in A_O$, then its associated *ontic theory* Σ_α is a formula of $L_P(X \cup X')$. Σ_α is required to be *consistent*, meaning that for all states $s \in 2^X$ the set $\{s' \in 2^{X'} \mid ss' \models \Sigma_\alpha\}$ is nonempty (ss' is the valuation obtained by combining two states — valuations — defined over different symbols). Given a knowledge state M' involving only primed symbols of the form x' ($x \in X$), we write $Plain(M')$ for the knowledge state obtained by replacing x' with x for all $x \in X$. $\gamma_M(\alpha)$ denotes the *progression* of ontic action α in the knowledge state M, defined as the knowledge state $Plain(\{s' \mid s' \in 2^{X'}, s \in M, ss'' \models \Sigma_\alpha\})$.

For $\alpha \in A_E$ its associated *feedback theory* $\Omega_\alpha = (K\phi_1, \ldots, K\phi_k)$ is a list of epistemic atoms, whose elements are named *feedbacks*. We require epistemic actions to be *tautological* meaning that the objective formula $\phi_1 \vee \cdots \vee \phi_k$ is a tautology. The progression of feedback $K\phi_j$ in M is given by $\gamma_M^j(\alpha) = \{s \in M \mid s \models \phi_j\}$, defined if $M \not\models K\neg\phi_j$, and undefined otherwise. The set of immediate successors of α in M is denoted $\Gamma_M(\alpha)$ and given by the set (of defined progressions) $\{\gamma_M^1(\alpha), \ldots, \gamma_M^k(\alpha)\}$.

1.2   Preliminaries: Dynamic Epistemic Logic

A thorough introduction to DEL is [3], however here we use a version due to [4] — particularly important is the inclusion of ontic change. We consider exclusively the single agent version of DEL. Let P be a *countable* set of propositional symbols. An *epistemic model* on P is a tuple $\mathcal{M} = (W, \sim, V)$, where W is a set of *worlds*, $\sim$ is an *indistinguishably relation* on W, and $V : P \to 2^W$ is a *valuation function*. Satisfaction of a formula $\phi \in L_E(P)$ is given as usual for epistemic logic. $\neg K\neg\phi$ is abbreviated $\widehat{K}\phi$, and $\mathcal{M} \models \phi$ is shorthand for $\mathcal{M}, w \models \phi$ for all $w \in W$.

An *event model* of $L_E(P)$ is a tuple $\mathcal{E} = (E, R, pre, post)$, where E is a set of *events*, R is the indistinguishability relation on E. Further, $pre : E \to L_E(P)$ maps each event to a *precondition* $\phi \in L_E(P)$, and $post : E \to (P \to L_E(P))$ assigns to each event a *postcondition* $\phi \in L_E(P)$ for each symbol. We consider both $\sim$ and R to be equivalence relations, and we use $\lfloor \mathcal{M} \rfloor$ to denote the bisimulation contraction of $\mathcal{M}$. $\mathcal{M} \otimes \mathcal{E} = (W', \sim', V')$ denotes the *product update* of $\mathcal{M}$ with $\mathcal{E}$, where $W' = \{(w,e) \in W \times E \mid \mathcal{M}, w \models pre(e)\}$, $\sim' = \{((w,e),(w',e')) \in W' \times W' \mid w \sim w' \text{ and } eRe'\}$, and $V'(p) = \{(w,e) \in W' \mid \mathcal{M}, w \models post(e)(p)\}$ for each $p \in P$.

1.3   Induced Epistemic Models and Progression of Event Models

Given a knowledge state $M = \{s_1, \ldots, s_m\}$, we define the induced epistemic model (of M) as $\mathcal{M} = (W, W \times W, V)$ on X where $W = \{w_1, \ldots, w_m\}$, and for $1 \leq i \leq m$: $x \in V(w_i)$ iff $s_i \models x$ for all $x \in X$. Using this construction it is easy to see that $M \models \phi$ iff $\mathcal{M} \models \phi$ for any purely subjective formula ϕ, and when this is the case we say that M and $\mathcal{M}$ are *X-equivalent*. Observe that $\mathcal{M}$ is finite, connected and bisimulation minimal, as no two worlds of W have the same valuation.

Knowledge states are simply special cases of epistemic models, but this not so when considering KBP actions versus event models. We will define progression in DEL planning using product update, motivated by the following observation. A bisimulation minimal epistemic model represents a *set* of knowledge states; i.e. each connected component is a knowledge state. Therefore each connected component of $\mathcal{M}' = \mathcal{M} \otimes \mathcal{E}$ describes an immediate successor of $\mathcal{E}$ in $\mathcal{M}$, meaning that progressions are induced by the connected components of $\mathcal{E}$.

The *feedbacks* of $\mathcal{E}$ is the set of R-equivalence classes of E denoted $\{E_1, \ldots, E_k\}$. The progression of E_j in $\mathcal{M}$ is given by $\gamma_{\mathcal{M}}^j(\mathcal{E}) = \lfloor \mathcal{M} \otimes (\mathcal{E} \upharpoonright E_j) \rfloor$, defined if $\mathcal{M} \models \hat{K} \bigvee_{e \in E_j} pre(e)$, and otherwise undefined. The set of immediate successors of $\mathcal{E}$ in $\mathcal{M}$ is denoted $\Gamma_{\mathcal{M}}(\mathcal{E})$ and given by the set (of defined progressions) $\{\gamma_{\mathcal{M}}^1(\mathcal{E}), \ldots, \gamma_{\mathcal{M}}^k(\mathcal{E})\}$. We say that $\mathcal{E}$ is *applicable* in $\mathcal{M}$ if some $\gamma_{\mathcal{M}}^j(\mathcal{E})$ is defined, in which case $\Gamma_{\mathcal{M}}(\mathcal{E})$ is non-empty.

2   Correspondence

Consider now an epistemic action $\alpha \in A_E$ with $\Omega_\alpha = (K\phi_1, \ldots, K\phi_k)$. The corresponding event model is denoted $\mathcal{E}_{\Omega_\alpha} = (E, R, pre, post)$, where $E = \{e_1, \ldots, e_k\}$, R is the identity and for $1 \leq j \leq k$: $pre(e_j) = \phi_j$ and $post(e_j)$ is the identity (i.e. $post(e_j)(x) = x$, for each $x \in X$).

Proposition 1. *Given a knowledge state M and $\alpha \in A_E$. Let $\mathcal{M}$ denote the induced epistemic model of M, $\mathcal{E}_{\Omega_\alpha}$ be as above, and for $1 \leq j \leq k$ we further let $M_j = \gamma_M^j(\alpha)$ and $\mathcal{M}_j = \gamma_{\mathcal{M}}^j(\mathcal{E}_{\Omega_\alpha})$. Then M_j and $\mathcal{M}_j$ are X-equivalent.*

Proof. By construction $\gamma_M^j(\alpha)$ is defined exactly when $\gamma_M^j(\mathcal{E}_{\Omega_\alpha})$ is. We have that $s \in M_j \Leftrightarrow s \models \phi_j$ meaning that M_j contains every state satisfying ϕ_j. Further we have that $(w, e_j) \in dom(\mathcal{M}_j) \Leftrightarrow (w, e_j) \in dom(\lfloor \mathcal{M} \otimes (\mathcal{E}_{\Omega_\alpha} \upharpoonright \{e_j\}) \rfloor) \Leftrightarrow \mathcal{M}, w \models pre(e_j) \Leftrightarrow \mathcal{M}, w \models \phi_j$, hence as R is the identity $\mathcal{M}_j$ contains every world of $\mathcal{M}$ satisfying ϕ_j. Consequently with $post(e_j)$ being the identity we see that $\mathcal{M}_j$ is exactly the induced epistemic model of M_j, and so they are *X-equivalent*.

The progression of an ontic action α in a knowledge state M can lead to a knowledge state of size exponential in α and M — consider for instance the ontic theory $\Sigma_\alpha = \top$. This succinctness cannot be achieved with a single event model, as $M \otimes \mathcal{E}$ is polynomial in the size of $\mathcal{M}$ and $\mathcal{E}$. Nonetheless, event models *can* polynomially simulate ontic actions, by adding primer event models that "blow up" an epistemic model.

Let $X = \{x_1, \ldots, x_n\}$ and define $P = X \cup X' \cup \{y_0, \ldots, y_n\}$. In the postconditions below for an event e, we assume $post(e)(p) = p$ for any unmentioned proposition $p \in P$. For $1 \leq i \leq n$ construct the connected event model $\mathcal{E}_{b_i}$ of $L_E(P)$ containing events e_i and f_i, where $pre(e_i) = pre(f_i) = y_{i-1}$. Further, $post(e_i)(y_{i-1}) = post(f_i)(y_{i-1}) = \bot$, $post(e_i)(y_i) = post(f_i)(y_i) = \top$, $post(e_i)(x_i') = \top$, and $post(f_i)(x_i') = \bot$ (observe the difference in the assignment of x_i'). For an ontic action $\alpha \in A_O$ with ontic theory Σ_α, construct $\mathcal{E}_{\Sigma_\alpha}$ of $L_E(P)$ containing a single event e_{Σ_α}, where $pre(e_{\Sigma_\alpha}) = y_n \wedge \Sigma_\alpha$, $post(e_{\Sigma_\alpha})(y_n) = \bot$, $post(e_{\Sigma_\alpha})(y_0) = \top$ and $post(e_{\Sigma_\alpha})(x_i) = x_i'$ for $1 \leq i \leq n$. We say an epistemic model on $L_E(P)$ is *i-ready* when it satisfies y_{i-1} and no other y symbol distinct from y_{i-1}.

Proposition 2. *Given a knowledge state M and $\alpha \in A_O$. Let $\mathcal{M}$ be an epistemic model on $L_E(P)$, s.t. $\mathcal{M}$ is 1-ready and M and $\mathcal{M}$ are X-equivalent. With $\gamma_M(\alpha) = M'$ and $\mathcal{M}'$ denoting the iterated progression of $\mathcal{E}_{b_1}, \ldots, \mathcal{E}_{b_n}, \mathcal{E}_{\Sigma_\alpha}$ in $\mathcal{M}$, we have that M' and $\mathcal{M}'$ are X-equivalent.*

Proof. Let $\mathcal{M}_i$ be i-ready and consider $\Gamma_{\mathcal{M}_i}(\mathcal{E}_{b_i}) = \{\mathcal{M}_{i+1}\}$. Clearly $\mathcal{M}_{i+1}$ is $(i+1)$-ready, and is further double the size of $\mathcal{M}_i$. To see this, for each world w in $\mathcal{M}_i$, $\mathcal{M}_{i+1}$ contains worlds (w, e_i) and (w, f_i), with x_i' being true in the former and false in the latter. As $\mathcal{M}$ is 1-ready we have that the iterated progression of $\mathcal{E}_{b_1}, \ldots, \mathcal{E}_{b_n}$ in $\mathcal{M}$ is the $(n+1)$-ready model $\mathcal{M}_{n+1}$ exponential in the size of $\mathcal{M}_1$, and where each world of $\mathcal{M}$ gives rise to 2^n worlds in $\mathcal{M}_{n+1}$ — one for each valuation in $2^{X'}$. $\mathcal{E}_{\Sigma_\alpha}$ is applicable in $\mathcal{M}_{n+1}$, and for $\Gamma_{\mathcal{M}_{n+1}}(\mathcal{E}_{\Sigma_\alpha}) = \{\mathcal{M}'\}$ we have that $\mathcal{M}'$ contains exactly the worlds of $\mathcal{M}_{n+1}$ satisfying Σ_α and unprimed symbols are assigned the truth value of their primed counterpart. Therefore M' and $\mathcal{M}'$ are X-equivalent.

Given a KBP planning problem with actions A_E and A_O, we can transform this into a DEL planning problem using the event models presented above. A minor modification is needed for $\mathcal{E}_{\Omega\alpha}$, namely $pre(e_j) = \phi_j \wedge y_0$ ensuring $\mathcal{E}_{\Omega\alpha}$ is applicable only in 1-ready epistemic models. We can map a plan for either problem to a plan for the other, and this mapping is polynomial and preserves weak bisimilarity of plans upto X-equivalence. This is shown by an inductive proof on the plan structure. The base cases are slight variations of Propositions 1 and 2. Most pressing is the inductive step from DEL plans to KBP plans. Here applicability guarantees that no $\mathcal{E}_{\Omega\alpha}$ can be interleaved within a sequence of event models that simulate ontic actions, so we achieve the proper interleaving of epistemic actions and ontic actions.

3   DEL Fragments

From [5] we have that plan existence for a KBP planning problem is PSPACE-Complete with only epistemic actions, EXPSPACE-Complete with only ontic actions and 2EXP-Complete with both types. We now consider the corresponding DEL fragments separated by computational complexity. From the previous section we make the following observations. (i) Epistemic actions corresponds to event models, where every event is distinguishable and have identity postconditions. (ii) Ontic actions corresponds to sequences of connected event models with postconditions. (iii) We can combine event models as in (i) and (ii) to represent both epistemic and ontic actions. This reveals three DEL fragments with distinct complexity.

The separation of epistemic actions and ontic actions in the KBPs as plans approach does not lend itself to describing actions such as the observable coin toss action described in Example 1. We can see this as epistemic actions cannot make ontic change and ontic actions always have a single outcome. This is an example of a fully observable and non-deterministic problem, whose corresponding plan existence problem is EXP-Complete [6]. This is the fragment of DEL planning where the event models of (i) is extended to include postconditions, and we remark that postconditions in this case affect the computational complexity of DEL planning. In the multi-agent case DEL planning is undecidable [2]. The recent [7] discuss a decidable fragment obtained by extensively restricting the indistinguishability relation, having propositional preconditions and no postconditions. Further work is to find more decidable fragments.

Acknowledgements. I thank the three anonymous reviewers of LORI-13 for their useful feedback on a longer version of this paper.

References

1. Andersen, M.B., Bolander, T., Jensen, M.H.: Conditional epistemic planning. In: del Cerro, L.F., Herzig, A., Mengin, J. (eds.) JELIA 2012. LNCS, vol. 7519, pp. 94–106. Springer, Heidelberg (2012)
2. Bolander, T., Andersen, M.B.: Epistemic planning for single and multi-agent systems. Journal of Applied Non-Classical Logics 21(1), 9–34 (2011)
3. van Ditmarsch, H., van der Hoek, W., Kooi, B.: Dynamic Epistemic Logic, 1st edn. Springer Publishing Company, Incorporated (2007)
4. van Ditmarsch, H., Kooi, B.: Semantic results for ontic and epistemic change. In: Bonanno, G., van der Hoek, W., Wooldridge, M. (eds.) Logic and the Foundations of Game and Decision Theory (LOFT 7), pp. 87–117. Texts in Logic and Games, Amsterdam University Press (2008)
5. Lang, J., Zanuttini, B.: Knowledge-based programs as plans: Succinctness and the complexity of plan existence. In: Proc. TARK 2013 (2013)
6. Rintanen, J.: Complexity of planning with partial observability. In: Zilberstein, S., Koehler, J., Koenig, S. (eds.) ICAPS, pp. 345–354. AAAI (2004)
7. Yu, Q., Wen, X., Liu, Y.: Multi-agent epistemic explanatory diagnosis via reasoning about actions. In: To appear in Proceedings of the Twenty-Third International Joint Conference on Artificial Intelligence (IJCAI 2013) (2013)

Judgment Aggregation with Abstentions: A Hierarchical Approach

Guifei Jiang[1], Dongmo Zhang[1], and Xiaojia Tang[2]

[1] AIRG, University of Western Sydney, Australia
[2] CSLI, Southwest University, China

Abstract. This paper presents a quasi-lexicographic judgment aggregation rule based on the hierarchy of judges. We do not assume completeness at both individual and collective levels, which means that a judge can abstain from a proposition and the collective judgment on a proposition can be undetermined. We prove that the proposed rule is (weakly) oligarchic. This is by no means a negative result. In fact, our result demonstrates that with abstentions, oligarchic aggregation is not necessarily a single level determination but can be a multiple-level democracy, which partially explains its pervasiveness in the real world.

1 Introduction

Judgment aggregation deals with the problem of how a group judgement on certain issues can be formed based on individuals' judgements on the same issues. List and Pettit presented an impossibility result which says that no aggregation rule can generate consistent collective judgments if we require the rule to satisfy a set of "plausible" conditions [1]. However, such an impossibility result does not discourage the investigation of judgement aggregation. By weakening or varying these conditions, a growing body of literature on judgement aggregation has emerged in recent years [2–6].

In our society the hierarchy is one of the most basic organization forms and a hierarchical group may give individual members or subgroups the priority to determine the collective judgments on certain propositions. However, such kind of expert rights has been rarely investigated in the current literature [7], let alone proposing a specific judgment aggregation rule to formally display how the hierarchical groups generate the collective judgments. In addition, the potential aggregation rule in the hierarchical group is likely to be oligarchic defined in [8], but the non-oligarchs still have chance to make contributions to the collective judgments on some issues. Then one of the challenges is to find the specific condition under which non-oligarchs have such power. To clarify these two questions, it is feasible to explore a plausible judgment aggregation rule for hierarchical groups.

In this paper, we focus on dealing with above questions by introducing a quasi-lexicographic approach for hierarchical groups without requiring the completeness at both the collective and individual level. This proposed aggregate

D. Grossi, O. Roy, and H. Huang (Eds.): LORI 2013, LNCS 8196, pp. 321–325, 2013.

procedure is proved to be oligarchic. This is by no means a negative result. In fact, our result reveals that with abstentions, oligarchic aggregation is no longer a single level determination but can also be a multiple-level democracy, which partially explains its pervasiveness in the real world.

2   The Model

The formal model of judgment aggregation with abstentions is the same as the formalism in [1, 9] except the followings: First, we restrict the agenda to a finite set of *literals* i.e., atomic propositions or their negation in the underlying logic **L**. That is $X = \{p, \neg p : p \in X^*\}$ where $X^* \subseteq \mathbf{L}$ is a set of unnegated atomic propositions. Secondly, we will not assume that each individual's judgment set $\Phi_i \subseteq X$ and the collective judgment set $\Phi \subseteq X$ must be complete. And individual i abstains from making a judgment on p, which is denoted by $p\#\Phi_i$. In other words, $p\#\Phi_i$ if and only if $p \notin \Phi_i$ and $\neg p \notin \Phi_i$. Lastly, we will assume that each individual's judgment is *individual consistence*[1]. That is, for every $p \in X$, if $p \in \Phi_i$, then $\neg p \notin \Phi_i$.

3   Conditions on Aggregation Rules

We now turn to investigating the conditions which are desirable to put on an aggregation rule in terms of abstentions. Let F be an aggregation function.

Consistency (C). *For each consistent profile $(\Phi_i)_{i \in N} \in Dom(F)^2$, $F((\Phi_i)_{i \in N})$ is consistent as well. That is, $p \in F((\Phi_i)_{i \in N})$ implies $\neg p \notin F((\Phi_i)_{i \in N})$ for all $p \in X$.*

Non-dictatorship (D). *There is no $x \in N$ such that for all $\{\Phi_i\}_{i \in N} \in Dom(F)$, $F(\{\Phi_i\}_{i \in N}) = \Phi_x$.* This is a basic democratic requirement: no single individual should always determine the collective judgment set.

Unanimity with Abstentions (U). *For every $p \in X$ and any $\alpha \in \{p, \neg p\}$, if there is some $V \subseteq N$ such that $V \neq \emptyset$, $\forall i \in V.\alpha \in \Phi_i$ and $\forall j \in N\backslash V.\alpha\#\Phi_j$, then $\alpha \in F((\Phi_i)_{i \in N})$.* Intuitively, if a set of individuals agree on a certain judgment on a proposition α while all the others abstain from α, then this condition requires that $F((\Phi_i)_{i \in N})$ also make the same judgment on α.

Systematicity (S). *For every $p, q \in X$ and every profiles $(\Phi_i)_{i \in N}$, $(\Phi'_i)_{i \in N} \in \mathrm{Dom}(F)$, if for every $i \in N$, $p \in \Phi_i$ iff $q \in \Phi'_i$ and $\neg p \in \Phi_i$ iff $\neg q \in \Phi'_i$, then $p \in F((\Phi_i)_{i \in N})$ iff $q \in F((\Phi'_i)_{i \in N})$.* This condition including independency and neutrality parts requires that propositions in the agenda should be treated in an even-handed way by the aggregation function, and the collective judgments on each proposition should depend exclusively on the pattern of individual judgment on that proposition.

---

[1] Given that the agenda is a set of literals, it is equivalent to logical consistency, i.e., there is a valuation v such that $v(p) = T$ for all $p \in \Phi_i$.

[2] $Dom(F)$ denotes the domain of F, i.e., the set of admissible profiles.

Note that comparing to other conditions, *Unanimity with Abstentions* is specially designed for judgments with abstentions, and *non-dictatorship* can be derived from it.

Proposition 1. *Every judgment aggregation rule satisfying unanimity with abstentions is non-dictatorial.*

In the following, we denote *Consistency*, *Unanimity with Abstentions* and *Systematicity* as **CUS** for short.

4   The Quasi-Lexicographic Aggregation Rule

In this section, we define an aggregation rule that satisfies the above-mentioned three fundamental conditions. Firstly we need two auxiliary concepts, which are precisely defined as follows:

Definition 1. *Let N be a set of individuals, $(N, <)$ is called a hierarchy over N if $<$ satisfies transitivity and asymmetry.*

Definition 2. *Let $<$ be a hierarchy over N. For every $p \in X$, p is not collectively rejected by aggregation rule F, denoted by $p \rhd F((\Phi_i)_{i \in N})$, if there is an individual with greater priority accepting it once it is rejected by some individual. That is,*

$$p \rhd F((\Phi_i)_{i \in N}) \text{ iff } \forall i \in N(\neg p \notin \Phi_i \lor \exists j \in N(i < j \land p \in \Phi_j)) \qquad (1)$$

We denote the negation of $p \rhd F((\Phi_i)_{i \in N})$ as $\overline{p \rhd F((\Phi_i)_{i \in N})}$, meaning that p is neither collectively accepted nor collectively undetermined. Based on the concept of "non-rejection" $\rhd$, we can define an aggregate procedure F for that p *is collectively accepted*, denoted by $p \in F((\Phi_i)_{i \in N})$, as follows.

Definition 3. *For all $p \in X$,*

$$p \in F((\Phi_i)_{i \in N}) \text{ iff } p \rhd F((\Phi_i)_{i \in N}) \text{ and } \exists j \in N.p \in \Phi_j \qquad (2)$$

Intuitively, this aggregate procedure says that p is accepted by a group if p is not collectively rejected and there is at least one individual who accepts it.

Obviously, a proposition p is *collectively undetermined* is decided by the following condition:

$$p \# F((\Phi_i)_{i \in N}) \text{ iff } p \notin F((\Phi_i)_{i \in N}) \text{ and } \neg p \notin F((\Phi_i)_{i \in N}). \qquad (3)$$

We will call the above defined rule F *the quasi-lexicographic rule*. And the next proposition shows that the quasi-lexicographic rule F satisfies the desirable conditions **CUS**.

Proposition 2. *The quasi-lexicographic rule F satisfies conditions **CUS**.*

The following theorem has double value: on one hand, it is helpful to understand the quasi-lexicographic rule more intuitively; on the other hand, it is useful to prove the main result.

Theorem 1. *Given that $(N, <)$ is well-prioritized[3], and let $\Phi = F((\Phi_i)_{i \in N})$. Then*

1. $p \triangleright \Phi$ *iff* $\forall i \in N((\forall j > i.p \# \Phi_j) \to \neg p \notin \Phi_i)$.
2. $p \triangleright \Phi$ *iff* $\forall i \in N(\neg p \notin \Phi_i \vee (\exists j > i.p \in \Phi_j \wedge \forall j' > j.p \# \Phi_{j'}))$.

In the last part of this section, we will show that the quasi-lexicographic rule F is weakly oligarchic [4] and displays nicely under what conditions the non-oligarchs can have the power to affect the collective decision-making.

Proposition 3. *The quasi-lexicographic rule F is weakly oligarchic, but not strictly oligarchic.*

In order to reveal how the non-oligarchs can have the power to make collective decisions through F, we need two further definitions.

Definition 4. *A set of judges D is decisive on $p \in X$ for a judgment aggregation function G iff for every profile $(\Phi_i)_{i \in N} \in Dom(G)$, if $p \in \Phi_j$ for every $j \in D$, then $p \in G((\Phi_i)_{i \in N})$.*

Definition 5. *Given a hierarchy on N and induced by $<$, N can be divided into subgroups $M_1, \cdots, M_n$, where $\emptyset \neq M_i \subseteq N$ for every $i \in N$, $\bigcup_{i=1}^n M_i = N$ and M_i is inductively defined as follows:*

- $M_1 = \{i \in N : \nexists j \in N.i < j\}$
- $M_{k+1} = \{i \in N \backslash (\bigcup_{i=1}^k M_i) : \nexists j \in N \backslash (\bigcup_{i=1}^k M_i).i < j\}$

We finally come to the following result, which displays that a proposition is not rejected by every of the superiors is sufficient and necessary to make the subgroup composed of the immediate inferiors a decisive set on this proposition.

Theorem 2. *Given a hierarchy on N, and let $M_1, \cdots, M_n$ be the subgroups of each level, for every $k \in \{1, \cdots, n\}$ and $p \in X$, M_k is decisive on p for the quasi-lexicographic rule F if and only if $\neg p \notin \Phi_i$ for every $i \in \bigcup_{h=0}^{k-1} M_h$.*

This follows that non-oligarchs can have the power to make collective decisions on some proposition if and only if the proposition is not rejected by the oligarchs, which displays that with abstentions, oligarchic aggregation is no longer a single level determination but can also be a multiple-level democracy.

---

[3] If there is no infinite ascending sequence $i_1 < i_2 < i_3 < \cdots$, where $i_n \in N$, which is automatically satisfied by every hierarchy over N

[4] Refer to [10] for its precise definition.

5   Conclusion

In this paper, we have investigated the aggregation rules for judgment aggregation without requiring the completeness at both individual and collective levels. Different from the perspective in [7] which presents the first extension of Sen's liberal paradox, we focus on dealing with two questions: How does the hierarchical group generate the collective judgments? How can the non-oligarchs have the power to make the collective decision in an oligarchic environment? We have replied them by proposing a quasi-lexicographic rule for hierarchical groups, which is inspired by [11]. This judgment aggregation rule turns out to satisfy the desirable conditions and reveal that if certain issues are not rejected by the oligarchs, then non-oligarchs have the decisive power in making group decisions on these issues. This seems positive news to the result in [10], since this rule demonstrates that with abstentions, even an oligarchic aggregation procedure can also realize a multiple-level democracy. To some extent, this also explains why hierarchical (oligarchic) systems exist widely in the real world.

Acknowledgments. We are grateful to Laurent Perrussel and Meiyun Guo for valuable suggestions, and special thanks are due to three anonymous referees for their insightful comments. This research was partially supported by the Australian Research Council through Discovery Project DP0988750.

References

1. List, C., Pettit, P.: Aggregating sets of judgments: An impossibility result. Economics and Philosophy 18(1), 89–110 (2002)
2. List, C., Pettit, P.: Aggregating sets of judgments: Two impossibility results compared. Synthese 140(1-2), 207–235 (2004)
3. Pauly, M., Van Hees, M.: Logical constraints on judgement aggregation. Journal of Philosophical Logic 35(6), 569–585 (2006)
4. Dietrich, F.: Judgment aggregation (im)possibility theorems. Journal of Economic Theory 126(1), 286–298 (2006)
5. Mongin, P.: Factoring out the impossibility of logical aggregation. Journal of Economic Theory 141(1), 100–113 (2008)
6. Dietrich, F., List, C.: Propositionwise judgment aggregation: the general case. Social Choice and Welfare 40(4), 1067–1095 (2013)
7. Dietrich, F., List, C.: A liberal paradox for judgment aggregation. Social Choice and Welfare 31(1), 59–78 (2008)
8. Gärdenfors, P.: A representation theorem for voting with logical consequences. Economics and Philosophy 22(2), 181–190 (2006)
9. Dietrich, F.: A generalised model of judgment aggregation. Social Choice and Welfare 28(4), 529–565 (2007)
10. Dietrich, F., List, C.: Judgment aggregation without full rationality. Social Choice and Welfare 31(1), 15–39 (2008)
11. Andréka, H., Ryan, M., Schobbens, P.Y.: Operators and laws for combining preference relations. Journal of Logic and Computation 12(1), 13–53 (2002)

A Note on Bayesian Games

Yang Liu

Department of Philosophy, Columbia University
708 Philosohy Hall, New York, NY 10027, USA
yl2435@columbia.edu

Abstract. This paper tackles some conceptual problems in the epistemic foundations of classical game theory. Focus is placed on the discussions of the asymmetry of different epistemic standpoints in modeling a game and the thesis of no subjective probability for self-action in games.

1 Introduction

The theory of games is traditionally viewed as the study of strategic decision making of mutually dependent agents. A foundational approach to the subject aims at analyzing the basic concepts of the mathematical structures which serve as models for the games in question and making explicit the fundamental assumptions employed in these models. Among the basic concepts and assumptions, the notion of *epistemic mutual expectations* of different parties of a game plays an important role in comprehending the unfolding of a game. That is, in order for a player to act rationally in a game situation it is necessary that the player takes into account as to how other players reason about her, for the latter will influence their respective choices of actions, which in turn will take effect on her overall gain from the game. This line of thought has often been coined with the *epistemic approach* to the analysis of games. Yet, illuminating as this approach might be, great conceptual difficulties may emerge when attempts are made to lay out precisely how epistemic interactions are taking place in a given game situation. This has become a pressing issue in the foundations of game theory, which has sometimes been met with drastically opposing viewpoints. In this short note we revisit some problems concerning the epistemic status of the "solution concepts" in game theory, where focus will be placed on the discussion of the asymmetry of different epistemic standpoints in modeling a game. This will then lead to the analysis of the so-called no probability for self-action thesis, which, as we shall see in §3, posits a challenge to the epistemic configurations of current game-theoretic models. The arguments below are built on earlier works of Gaifman [3], Levi [5] (esp. [6]), Savage [7], Spohn [8], among others.

2 Different Points of View

One recurring issue in the foundations of games is the so-called problem of asymmetric epistemic viewpoints. That is, many of the core assumptions that are the

D. Grossi, O. Roy, and H. Huang (Eds.): LORI 2013, LNCS 8196, pp. 326–331, 2013.

basis of different "solution concepts" in classical game theory, such as the common knowledge of Bayesian rationality on the behaviors of the participants in a game and their prior probabilistic judgments on the incomplete information about the underlying game etc., are formed from the perspective of an external observer. However, from the viewpoint of an individual player, other players and their mutual beliefs and behaviors are just as much parts of the setting of the game as anything else, hence it seems that no principle of rationality mandates that one should, or even could, take the stance of an external observer and accept the recommendations prescribed from "the above and beyond," for the implementation of which has exceeded the individual player's epistemic capacity. Let us refer to the asymmetry between different perspectives in the analysis of a game, namely the point of view of an individual player and that of an external observer, as the first-person/third-person distinction in the epistemic standpoints of games. Then the aforementioned problem points to a conceptual gap in game-theoretic modeling created by the epistemic limitations of the participating parties.

Aumann in [1] made considerable attempts to fill this gap, where he reformulated the asymmetry of different epistemic viewpoints discussed above in terms of the tension between the "Bayesian" and the "game-theoretic" views of the world maintaining that the two accounts can be coherently integrated through the notion of subjective correlated equilibrium. He claims that in the framework of classical game theory the choices made by Bayesian rational players form a correlated equilibrium (cf. *ibid.* p.7). The cornerstone of Aumann's construction is the use of "all-inclusive" states of the world, an assumption that has been widely adopted in game theoretic literature.[1] More precisely, a state is assumed to be a *complete description* of the world (a possible world), which includes not only the information about the actions the players may adopt and their mutual beliefs about each other's actions and beliefs (call information of this type information at the *theoretic level*); it contains also substantial *meta-theoretic* information which includes players' prior probabilities judgments over *all* the states, their criteria for rational decision making, and their information structures/partitions over the states. The slogan is "conditional on one particular state, everybody knows everything!" In a given game situation the players are uncertain as to which state is the true state of the world, they however are, it is said, *informed of* a set of states which constitutes their information sets. It is hence easily seen that the first-person/third-person distinction made above is

---

[1] Aumann and Brandenburger in [2] adopted a more refined approach to the problem where each player's belief about other players' actions and beliefs are explicitly represented respectively through the notions of *conjectures* and *theories*. Both concepts are constituent components of the player's *types*, an idea originated by Harsanyi [4], which essentially plays the same role as all-inclusive states with perhaps less informative contents about the physical world. But, at any rate, each type profile (a state) includes a description of actions of all players and it is further assumed that there is a common prior defined over all states. This implies that the players have prior probabilistic judgments over their own actions. We will raise some concerns in regard with this features of the model in the next section.

really a nonstarter in any given state, for the latter encodes, as a basic assumption of the model, both theoretic information (obtainable by each individual player) and meta-theoretic information (accessible usually only to the theorist). Then, given the assumption of all-inclusive states, the problem of asymmetry of epistemic viewpoints seems to have disappeared.

However, the immediate epistemological question to ask here is: How can anyone come across a state of this kind in the first place? So it seems that our initial task of explicating the intriguing question about the epistemic asymmetry between different viewpoints in traditional game-theoretic analyses has been shifted to uncovering the status of all-inclusive states. In what follows, we discuss one of the problems facing the concept of all-inclusive states, namely the so-called no probabilities for acts thesis. It shows that there is no coherent way for the players to assign subjective probabilities to their own actions, the latter however is an unavoidable consequence of the assumption of all-inclusive states.

3   Probability and Act

The thesis that directly undermines the employment of all-inclusive states is this: "Any adequate quantitative decision model must not explicitly or implicitly contain any subjective probabilities for acts".[2] In the game-theoretic context, the principle says that *no subjective probabilities could be assigned by the players to their own acts in a meaningful way.*

The doctrine of no probabilities for self-actions is already hinted in Savage's discussion on the "small world" semantics where he mentioned in passing that probabilities for acts play no role in individual decision making. To illustrate, let us use the example Savage provided. Suppose that Jones is torn between either buying a sedan, or buying a convertible, or canceling the plan of purchase altogether and keeping the money. Now consider that, in a simple decision situation, the execution of an action may be solely determined by the relative ranking of the consequences to which the actions lead. Thus, if Jones prefers the convertible the most, then he shall just proceed to buy a convertible. "Chance and uncertainty are considered to have nothing to do with the situation." One might object that if Jones likes equally, say, the sedan, he might come to make a decision on his future action of purchasing either a convertible or a sedan by tossing a coin or by utilizing some internal randomizing mechanism (whim, impulse, etc.) to help him to decide, then in this case there seems to be a sense in which one could say that Jones will take an action with certain probability. However the formulation of the objection itself make it clear that the derived probability is essentially about the randomizer in use, which can hardly qualify for the agent's genuine subjective probability assignment for acts themselves.

In a more complicated situation where more careful deliberation is required, the agent's actions are further evaluated in terms of their respective consequences

---

[2]   Here we quote the version that is explicitly stated in [8]. Levi's well-known thesis that "deliberation crows out prediction" is largely in agreement with this principle, see [6].

under different circumstances, and, according to Savage, only the latter are subject to probabilistic evaluations. Say that Jones has finally reached a decision to buy a convertible because he realized that it's very likely that he will be taking a vacation in Monterey next month, in which case the utility of driving a convertible by the seaside in warm spring breeze will be maximally materialized. Hence, as seen in both situations, the choices of actions are never deliberated by a decision maker in isolation, they are always placed in a *context* in which they are to be evaluated, and the context itself will provide the relevant details (including the circumstances in which an act may take place and the consequences it results in) under which the values of acts are measured and decided. These considerations led Savage to his *belief-act-consequence* model, where acts are treated as functions mapping from (act-independent) states to consequences, but *not directly as the subjects of uncertainty* (see [7] §5.5).

Thus, according to this contextual interpretation of acts, when we say, for instance, that "it is more likely that I will stay home reading this afternoon than go playing tennis" we are not directly making any probabilistic judgements over our future actions (not even a comparative probabilistic judgment over the two possible acts). Rather, in deliberating these future actions, we are presenting to ourselves the circumstances under which the acts are to take place. That is to say, in our example, what actually come across in my mind might be things like what would be the chance that the wind will stop blowing so wildly, or how likely my partner could make to the tennis court on time, or how possible the book I am about to read would interest me, etc.; and, on the current view, only the latter are considered to be the subjects of probabilistic or value judgements.

Perhaps, part of the difficulties in incorporating the concept of all-inclusive states in a game-theoretic model is originated from an unquantified reading of Bayesianism. "According to the Bayesian view, subjective probabilities should be assignable to *every prospect*, including that of players choosing certain strategies in certain games" ([1] p.2, my emphasis). Here, it is not entire clear as to which subjective Bayesian theory is under consideration, yet it has been made explicit that each player is assumed to "conform to the Savage theory." However, as we have discussed a few lines above, any assignment of the players' subjective probabilities to their own acts/strategies actually falls outside the scope of the decision model put forward by Savage.

Yet even if we grant a more general reading on Bayesianism in the respect that subjective probabilities are manifested in their betting interpretation it will soon be clear that no non-trivial probabilities can be assigned to one's own future acts. To see this, suppose that Jones is faced with choices of either going to an Italian restaurant or to a French restaurant for dinner. In an attempt to elicit his subjective probabilities assigned on the two possible future actions, Jones is offered a bet with the condition that he wins X if he goes to an Italian restaurant and nothing if he chooses French (here it is assumed that the monetary reward of X is not too significant to take effect on Jones' choice of restaurant but, at the same time, it is not too small to be easily ignored by Jones either). Then, if Jones' subjective probability for his going to an Italian restaurant is p, then he shall be willing to pay a fee of pX to accept the bet in exchange of a reward of X on the event that he indeed is having Italian food for dinner. So far the example accords well with the standard betting interpretation of subjective probabilities, but story will turn once we notice that Jones' willingness to accept the bet of his going to an Italian restaurant at a cost of $pX > 0$ implies that he will be going to an Italian restaurant *for sure*! For, otherwise, it would be extremely unwise for him to *knowingly* pay a fee of pX but actually go to a French restaurant while gaining nothing from the bet he paid for, which can easily be avoided by simply rejecting the bet. And this is true for any $0 < p \leq 1$. Further, if $p = 0$ then Jones will be going to a French restaurant *for sure*. It follows that the betting rate upon which Jones is willing to pay a fair price for his acts collapses into 1 or 0. In another word, personal probabilities tend to be "gappy" when it comes to the agents' own actions.[3]

On a further note, there has been suggestions that there is a straightforward sense in which one can meaningfully assign subjective probabilities to one's own actions by referring to past experiences on choices in similar situations. According to this view, the reason why I believe that it's more likely that I will stay at home reading than playing tennis is perhaps because this is how it had always been resolved in the past for such a windy and chilly weekday afternoon when my partner is too occupied with work anyway. This line of thought however misses the point on at least two fronts. First, by referring to "similar situations" one has decomposed the deliberation of one's own future actions into scenarios in which these actions are taking place. As agued above, it is these contexts, rather than the isolated actions themselves, that are subject to genuine probabilistic analysis. More importantly, it seems that, in order for this frequentist view to be credited as a basic philosophical account for personal probabilities over *self*-actions (as opposed to an operational process attributed to another agent) it is necessary that the players are treated as if they have no freedom to choose their future actions: the acts that an agent will most likely to adopt are the ones that have been most frequently repeated. This reduces the process of deliberation for future actions to mere recollecting of past occurrences, then to fulfill an action is just to answer to the statistics. This however undermines the very presumption of the entire enterprise of decision sciences where great efforts are being made to assist decision makers to maximize their eventual gains *as a result* of deliberate and, in many cases, highly sophisticated planning/strategizing for their future actions over which they are expected to have full control.

Acknowledgments. Thanks are due to Haim Gaifman, Isaac Levi, Rohit Parikh, and Rush Stewart for fruitful exchanges; to two anonymous reviewers for helpful comments.

References

1. Aumann, R.J.: Correlated equilibrium as an expression of bayesian rationality. Econometrica: Journal of the Econometric Society, 1–18 (1987)

---

[3] See [8] p.115 and [5] p.32 for further discussions on this point.

2. Aumann, R.J., Brandenburger, A.: Epistemic conditions for nash equilibrium. Econometrica 63(5), 1161–1180 (1995)
3. Gaifman, H.: Self-reference and the acyclicity of rational choice. Annals of Pure and Applied Logic 96(1-3), 117–140 (1999)
4. Harsanyi, J.: Games with incomplete information played by "Bayesian" players, I-III. Management Science 14(3), 159–182, 320–334, 486–502 (1967-1968)
5. Levi, I.: Rationality, prediction, and autonomous choice. In: The Covenant of Reason: Rationality and the Commitments of Thought, pp. 19–39. Cambridge University Press (1997, 1989)
6. Levi, I.: Prediction, deliberation and correlated equilibrium. In: The Covenant of Reason: Rationality and the Commitments of Thought, ch. 5. Cambridge University Press (1997, 1996)
7. Savage, L.J.: The Foundations of Statistics, 2nd revised edn. Dover Publications, Inc. (1972)
8. Spohn, W.: Where Luce and Krantz do really generalize Savage's decision model. Erkenntnis 11(1), 113–134 (1977)

A Logic for Extensive Games with Short Sight

Chanjuan Liu[1], Fenrong Liu[2], and Kaile Su[1,3]

[1] School of Electronics Engineering and Computer Science, Peking University,
Beijing, China
[2] Department of Philosophy, Tsinghua University, Beijing, China
[3] Institute for Integrated and Intelligent Systems, Griffith University, Brisbane,
Australia
{chanjuan.pkucs,kailepku}@gmail.com, fenrong@tsinghua.edu.cn

1 Introduction

To characterize the structures and reason about strategies of extensive games, much work has been done to provide the logical systems for such games. These logic systems focus on various perspectives of extensive games: (Harrenstein *et al.*, 2003) concentrated on describing equilibrium concepts and strategic reasoning. (van Benthem, 2002) used dynamic logic to describe games as well as strategies.

The assumption of common knowledge on game structures in traditional extensive games is sometimes too strong and unrealistic. For instance, in a game like chess, the actual game space is exponential in the size of the game configuration, and may have a computation path too long to be effectively handled by most existing computers. So we often seek sub-optimal solutions by considering only limited information or bounded steps foreseeable by a player that has relatively small amount of computation resources. Grossi and Turrini proposed the concept of *games with short sight* (Grossi and Turrini, 2012), in which players can only see part of the game tree. However, there is no work on the logical reasoning of the strategies in this game model.

Inspired by the previous logics for extensive games, this paper is devoted to the logical analysis of game-theoretical notions of the solutions concepts in games with short sight. The closely related work is (Harrenstein *et al.*, 2003), in which a logic was proposed for strategic reasoning and equilibrium concepts. In this work, however, we present a new logical system called LS for games with short sight. This logic introduces new additional modalities $[\lhd], [(\sigma_i)], [\mathring{\sigma}^s]$ to capture interesting features such as restricted sight and limited steps.

2 Preliminaries

In this section, we recall the definition of finite games in extensive form with perfect information and games with short sight proposed by (Grossi and Turrini, 2012).

Definition 1. (*Extensive game(with perfect information)*) *A finite extensive game (with perfect information) is a tuple* $G=(N, V, A, t, \Sigma_i, \succeq_i)$, *where* (V, A) *is a tree with* V, *a set of nodes or vertices including a root* v_0, *and* $A \subseteq V^2$ *a set of*

D. Grossi, O. Roy, and H. Huang (Eds.): LORI 2013, LNCS 8196, pp. 332–336, 2013.
© Springer-Verlag Berlin Heidelberg 2013

arcs. N is a non-empty set of the players, and $\succeq_i$ represents preference relation for each player i, which is a partial order over V. For any two nodes v and v', if $(v, v') \in A$, we call v' a successor *of v, thus A is also regarded as the successor relation. Leaves are the nodes that have no successors, denoted by Z. t is turn function assigning a member of N to each non-terminal node. Σ_i is a non-empty set of strategies. A strategy of player i is a function $\sigma_i : \{v \in V \backslash Z | t(v) = i\} \to V$ which assigns a successor of v to each non-terminal node when $t(v) = i$.*

As usual, $\sigma = (\sigma_i)_{i \in N}$ represents a strategy profile which is a combination of strategies of all players and Σ represents the set of all strategy profiles. For any $M \subseteq N$, σ_{-M} denotes the collection of strategies in σ excluding those for players in M. We define an outcome function $O : \Sigma \to Z$ assigning leaf nodes to strategy profiles, i.e., $O(\sigma)$ is the outcome if the strategy profile σ is followed by all players. $O(\sigma_{-M})$ is the set of outcomes players in M can enforce provided that the other players strictly follow σ. $O(\sigma'_i, \sigma_{-i})$ is the outcome if player i uses strategy σ' while all other players employ σ.

Preference relation here is different from the conventional ones: In the literature the notion of preference is assumed to be a linear order over leaves, while in this paper it is a partial order over all nodes in V. We assume that players may not be able to precisely determine entire computation paths leading to leave nodes, and allow them to make estimations or even conjecture a preference between non-terminal nodes. This assumption also provides technical convenience for discussing games with short sight later.

In games with short sight, players' available information is limited in the sense that they are not able to see the nodes in some branches of the game tree or have no access to some of the terminal nodes.

The following definition makes the notion of *short sight* mathematically precise.

Definition 2. *(sight function). Let $G = (N, V, A, t, \Sigma_i, \succeq_i)$ be an extensive game. A short sight function for G is a function $s : V \backslash Z \to 2^{V|v} \backslash \emptyset$, associating to each non-terminal node v a finite subset of all the available nodes at v, and satisfying:*

$v' \in s(v)$ implies that $v'' \in s(v)$ for every $v'' \lhd v'$ with $v'' \in V|_v$, i.e. players' sight is closed under prefixes.($\lhd$ is the transitive closure of successor relation A.)

Intuitively, function s associates any choice point with vertices that each player can see.

Definition 3. *(Extensive game with short sight). An extensive game with short sight (Egss) is a tuple $S = (G, s)$ where G is a finite extensive game and s a sight function for G.*

Each game with short sight yields a family of finite extensive games, one for each non-terminal node $v \in V \backslash Z$:

Definition 4. *(sight-filtrated extensive game) Let S be an Egss given by (G, s) with $G=(N, V, A, t, \Sigma_i, \succeq_i)$. Given any non-terminal node v, a tuple $S\lceil_v$ is a finite extensive game by sight-filtration: $S\lceil_v = (N\lceil_v, V\lceil_v, A\lceil_v, t\lceil_v, \Sigma_i\lceil_v), \succeq_i\lceil_v)$ where*

- $N\lceil_v = N$;
- $V\lceil_v = s(v)$, which is the set of nodes within the sight from node v. The terminal nodes in $V\lceil_v$ are the nodes in $V\lceil_v$ of maximal distance, denoted by $Z\lceil_v$;
- $A\lceil_v = A \cap (V\lceil_v)^2$;
- $t\lceil_v = V\lceil_v \backslash Z\lceil_v \rightarrow N$ so that $t\lceil_v(v') = t(v')$;
- $\Sigma_i\lceil_v$ is the set of strategies for each player available at v and restricted to $s(v)$. It consists of elements $\sigma_i\lceil_v$ such that $\sigma_i\lceil_v(v') = \sigma_i(v')$ for each $v' \in V\lceil_v$ with $t\lceil_v(v') = i$;
- $\succeq_i\lceil_v = \succeq_i \cap (V\lceil_v)^2$.

Accordingly, we define the outcome function $O\lceil_v \colon \Sigma\lceil_v \rightarrow Z\lceil_v$ assigning leaf nodes of $S\lceil_v$ to strategy profiles.

3   A Logic of Extensive Games with Short Sight

In this section we present a modal logic LS (Logic of Extensive Games with Short sight). This logic supports reasoning about strategies in extensive games with short sight.

3.1 $\mathcal{LS}$: Syntax and Semantics

A language for general extensive games is proposed in (Harrenstein *et al.*, 2003), in which a strategy profile is taken as a modal operator, corresponding to an accessibility relation connecting a non-terminal node to leaf nodes. This language makes strategic reasoning simple, since one only needs to consider the outcome of this strategy without getting confused with all the actions at every choice point. To characterize what players can see in extensive games with short sight, we extend their language mainly by adding the modality $[\lhd]$. Let P be the set of propositional variables, and Σ be the set of strategy profiles. The language $\mathcal{LS}$ is given by the following BNF:

$$\varphi ::= p|\ \neg\varphi|\ \varphi_0 \wedge \varphi_1|\ \langle\leq_i\rangle\varphi|\ \langle\mathring{\sigma}\rangle\varphi|\ \langle\mathring{\sigma}_{-i}\rangle\varphi|\ \langle\lhd\rangle\varphi|\ \langle\mathring{\sigma}^s\rangle\varphi|\ \langle\mathring{\sigma}^s_{-i}\rangle\varphi$$

where $p \in P$, $\sigma \in \Sigma$. As usual, The dual of $\langle.\rangle\varphi$ is $[.]\varphi$. We begin with a brief explanation of the intuition behind the logic.

- The label $\leq_i$ denotes player i's preference relation.
- The label $\mathring{\sigma}$ stands for the outcomes of strategy profiles. $(v, v') \in R_{\mathring{\sigma}}$ iff v' is the terminal node reached from v by following σ.
- $(v, v') \in R_{\mathring{\sigma}_{-i}}$ iff v' is one of the leaf nodes extending v that player i can enforce provided that the other players strictly follow their strategies in σ.
- The label $\lhd$ is sight function for the current player, and $\langle\lhd\rangle\varphi$ means "φ holds in some node within the player $t(v)$'s sight at the present node v."
- $\langle\mathring{\sigma}^s\rangle\varphi$ means "φ holds in some state v' in $S\lceil_v$, which is the terminal node of $S\lceil_v$ that is reachable from the starting point v when all players adopt the strategy profile σ, i.e., $v' = O\lceil_v(\sigma\lceil_v)$." The interpretation for $\langle\mathring{\sigma}^s_{-i}\rangle\varphi$ is similar.

Let $S = (N, V, A, t, \Sigma_i, \succeq_i, s)$ be an Egss. The tuple of $(V, R_{\leq_i}, R_{\mathring{\sigma}}, R_{\mathring{\sigma}_{-i}}, R_{\vartriangleleft},$ $R_{\mathring{\sigma}^s}, R_{\mathring{\sigma}^s_{-i}})$ is defined as the frame F_S for $\mathcal{LS}$, where for each player i, strategy profile σ, nodes v, v', the accessibility relations are given as follows.

$$vR_{\leq_i}v' \quad \text{iff} \quad v' \succeq_i v$$
$$vR_{\mathring{\sigma}}v' \quad \text{iff} \quad v' = O|_v(\sigma|_v)$$
$$vR_{\mathring{\sigma}_{-i}}v' \quad \text{iff} \quad v' \in O|_v(\sigma_{-i}|_v)$$
$$vR_{\vartriangleleft}v' \quad \text{iff} \quad v' \in s_{t(v)}(v)$$
$$vR_{\mathring{\sigma}^s}v' \quad \text{iff} \quad v' = O\lceil_v(\sigma\lceil_v)$$
$$vR_{\mathring{\sigma}^s_{-i}}v' \quad \text{iff} \quad v' \in O\lceil_v(\sigma_{-i}\lceil_v)$$

A model M for $\mathcal{LS}$ is a pair (F, π) where F is a frame for $\mathcal{L}$ and π a function assigning to each proposition p in P a subset of V, i.e., $\pi : P \to 2^V$. The interpretation for $\mathcal{LS}$ formulas in model M are defined as follows:

$$M, v \models p \qquad \text{iff} \quad v \in \pi(p).$$
$$M, v \models \neg\varphi \qquad \text{iff} \quad \text{not } M, v \models \varphi.$$
$$M, v \models \varphi \wedge \psi \qquad \text{iff} \quad M, v \models \varphi \text{ and } M, v \models \psi.$$
$$M, v \models \langle\leq_i\rangle\varphi \qquad \text{iff} \quad M, u \models \varphi \text{ for some } u \in V \text{ with } vR_{\leq_i}u.$$
$$M, v \models \langle\mathring{\sigma}\rangle\varphi \qquad \text{iff} \quad M, u \models \varphi \text{ for some } u \in V \text{ with } vR_{\mathring{\sigma}}u.$$
$$M, v \models \langle\mathring{\sigma}_{-i}\rangle\varphi \qquad \text{iff} \quad M, u \models \varphi \text{ for some } u \in V \text{ with } vR_{\mathring{\sigma}_{-i}}u.$$
$$M, v \models \langle\vartriangleleft\rangle\varphi \qquad \text{iff} \quad M, u \models \varphi \text{ for some } u \in V \text{ with } vR_{\vartriangleleft}u.$$
$$M, v \models \langle\mathring{\sigma}^s\rangle\varphi \qquad \text{iff} \quad M, u \models \varphi \text{ for some } u \in V \text{ with } vR_{\mathring{\sigma}^s}u.$$
$$M, v \models \langle\mathring{\sigma}^s_{-i}\rangle\varphi \qquad \text{iff} \quad M, u \models \varphi \text{ for some } u \in V \text{ with } vR_{\mathring{\sigma}^s_{-i}}u.$$

The validities of a formula φ in models and frames are the same as the standard definitions (Blackburn *et al.*, 2001).

3.2   Axiom System

First, we have the following standard axioms.

(A_0) *Taut*, any classical tautology.

(A_1) K axiom for all modalities $[\leq_i], [\mathring{\sigma}], [\mathring{\sigma}_{-i}], [\vartriangleleft], [\mathring{\sigma}^s], [\mathring{\sigma}^s_{-i}]$.

Table 1 lists the other axioms of LS. The first column (N) is the *name* of the axiom. The second column denotes the *modalities* that each axiom is applied to. The third column shows the formula *schema*. The fourth column describes the *property* of the corresponding accessibility relation R.

K is used in all variants of the standard modal logic. T and 4 determine the preference of players to be *reflexive* and *transitive*. The sight of a player is reflexive. D ensures that a node reachable by a strategy profile σ from a node v is *determined*. I says that every outcome of strategy σ is *included* in the sets of outcomes by letting i free, and the other players following σ. M guarantees the final outcome vertices to be *terminated*. Y shows the *visibility* of all the nodes that can be reached from the current node v in sight-filtrated game $S\lceil_v$. D and I are the same as that for $[\mathring{\sigma}]$ and $[\mathring{\sigma}_{-i}]$.

The inference rules for LS are Modus Ponens (*MP*) and Necessitation (*Nec*).

Theorem 1. (*Soundness and Completeness Theorem*) *Logic of Extensive Games with Short sight LS is sound and complete w.r.t. all $\mathcal{LS}$-models.*

Table 1. Valid principles of LS

N	Modality	Schema	Property
T	$[\leq_i]$ $[\lhd]$	$[\leq_i]\varphi \to \varphi$ $[\lhd]\varphi \to \varphi$	reflexivity
4	$[\leq_i]$	$[\leq_i]\varphi \to [\leq_i][\leq_i]\varphi$	transtivity
D	$[\hat{\sigma}]$ $[\hat{\sigma}^s]$	$[\hat{\sigma}]\varphi \leftrightarrow \langle\hat{\sigma}\rangle\varphi$ $[.]\varphi \leftrightarrow \langle.\rangle\varphi$	determinism
I	$([\hat{\sigma}], [\hat{\sigma}_{-i}])$ $([\hat{\sigma}^s], [\hat{\sigma}^s_{-i}])$	$[\hat{\sigma}_{-i}]\varphi \to [\hat{\sigma}]\varphi$ $[\hat{\sigma}_{-i}]\varphi \to [\hat{\sigma}]\varphi$	inclusiveness
M	$[\hat{\sigma}]$ $[\hat{\sigma}_{-i}]$	$[\hat{\sigma}]([\hat{\sigma}']\varphi \leftrightarrow \varphi)$ $[\hat{\sigma}_{-i}]([\hat{\sigma}'_{-i}]\varphi \leftrightarrow \varphi)$	terminating
Y	$([\lhd], [\hat{\sigma}^s])$ $([\lhd], [\hat{\sigma}^s_{-i}])$	$[\lhd]\varphi \to [\hat{\sigma}^s]\varphi$ $[\lhd]\varphi \to [\hat{\sigma}^s_{-i}]\varphi$	visibility

Due to the length limit, we omit the proof here.

In the future, we would like to look into the model checking problem, especially, comparing the complexity of the problem in the standard game model and that in games with short sight.

Acknowledgement. This work is partially supported by Tsinghua University project (NO.2012WHYX003) and 973 Program (NO.2010CB328103). We would like to thank Johan van Benthem for his helpful comments.

References

Blackburn *et al.*, 2001. Blackburn, P., de Rijke, M., Venema, Y.: Modal logic. Cambridge University Press (2001)

Grossi and Turrini, 2012. Grossi, D., Turrini, P.: Short sight in extensive games. In: AAMAS, pp. 805–812 (2012)

Harrenstein *et al.*, 2003. Harrenstein, P., van der Hoek, W., Meyer, J.-J.C., Witteveen, C.: A modal characterization of nash equilibrium. Fundamenta Informaticae 57(2-4), 281–321 (2003)

van Benthem, 2002. van Benthem, J.: Extensive games as process models. Journal of Logic, Language and Information 11(3), 289–313 (2002)

Aggregated Beliefs and Informational Cascades

Rasmus K. Rendsvig

Dept. of Media, Cognition and Communication, University of Copenhagen
rendsvig@gmail.com

In the 1992 paper [1] Bikchandani et al. show how it may be rational for Bayesian agents in a sequential decision making scenario to ignore their private information and conform to the choices made by previous agents. If this occurs, an agent ignoring her private information is said to be *in a cascade*.

To illustrate, consider the following example: a set of agents must decide which of two restaurants to choose, one lying on the left side of the street, one on the right, with one being the better – L or R. Initially, agents have no information about which; each agent i has prior probabilities $Pr_i(L) = Pr_i(R)$. Every agent has two choices: either to go to the restaurant on the left, l_i, or to go the one on the right, r_i. All agents prefer to go to the better restaurant, and are punished for making the wrong choice, specified by pay-offs $u_i(l_i, L) = u_i(r_i, R) = v_1 > 0$ and $u_i(l_i, R) = u_i(r_i, L) = v_2 < 0$ with $v_1 + v_2 = 0$.

Before choosing, every agent receives a *private signal* indicating that either the restaurant on the left (L_i) or the one on the right (R_i) is the better one. The signals are assumed to be equally informative and positively correlated with the true state, in the sense that $Pr(L_i|L) = Pr(R_i|R) = q > .5$ and $Pr(L_i|R) = Pr(R_i|L) = 1 - q$. Given this setup, rational agents will follow their private signal, the majority choosing the better restaurant.

If agents are assumed to *choose sequentially* and *observe the choice of those choosing before them*, a cascade may result, possibly leading the majority to pick the worse option. The argument for this [1] rests on higher-ordering reasoning *not represented in the Bayesian framework*, and goes as follows. Given either L_1 or R_1, agent 1 will choose as her signal indicates, hereby revealing her signal to all subsequent agents. Agent 2 therefore as two pieces of information: his own signal together with that deduced from the choice of 1. If 2 receives the same signal as 1, he will make the same choice; given two opposing signal, assume he will invoke a self-biased tie-breaking rule, and go by his own signal. In both cases, 2's choice will also reveal his private signal to all subsequent agents. Assume that 1 and 2 received signals L_1, L_2. Then *no matter which signal 3 receives, she will choose* l_3: agent 3 will have three pieces of information, either L_1, L_2, L_3 or L_1, L_2, R_3. In either case, when conditionalizing on these, the posterior probability of L being the true state will be higher than that of R. So 3 will choose l_3, and thereby be in a cascade. Further, agent 4 will also be in a cascade: as 3 chooses l_3 no matter what, *her choice does not reveal her private signal*, why also 4 has three pieces of information, either L_1, L_2, L_4 or L_1, L_2, R_4. 4 is thus in the same epistemic situation as 3, and will choose l_4. As 4 is in a cascade, his choice will not reveal his private signal, *and the situation thus repeats for all subsequent agents*.

Notice that *cascades may not be truth conducive*: there is a $Pr(L_1|R) \cdot Pr(L_2|R)$ risk that *all* agents will choose the wrong restaurant – e.g., if signals are correct with probability $\frac{2}{3}$, all agents choose wrong with probability $\frac{1}{9}$.

D. Grossi, O. Roy, and H. Huang (Eds.): LORI 2013, LNCS 8196, pp. 337–341, 2013.
© Springer-Verlag Berlin Heidelberg 2013

Aim and Methodology. We construct a formal model that completely represents the reasoning made by agents in the sequential setup, for any input string of private signals. The type of model constructed is a dynamic epistemic logic variant of a state machine, in lack of terms called a *system*. A system operates by having for each state (Kripke model) some set of *transition rules* which as a function of the current state pick the next update to be invoked, hereby specifying the ensuing state. It is initiated from some *initial state* and terminates when an *end condition* is met.

The informational cascades system ($\mathcal{IC}$) constructed captures the following four elements of each agent's turn: *i)* earlier agents' actions are observed from which *ii)* their private signals (beliefs) are deduced and combined with *iii)* the private signal (belief) of the current agent after which *iv)* the chosen action is executed, observed by all.

$\mathcal{IC}$ diverges from the model of [1] in a number of aspects: it is *not probabilistic*, but *qualitative*; related, information aggregation is not done by Bayesian conditionalization, but by the *aggregation of perceived beliefs*; a *finite* set of agents is used; and *no pay-off structure nor rationality is assumed*, agency instead captured by transition rules.

An advantage of $\mathcal{IC}$ over the model from [1] is that $\mathcal{IC}$ fully specifies the intended scenario formally: all steps are defined for any string of private signals and all higher-order reasoning is represented. $\mathcal{IC}$ is thus a complete model for informational cascades.

In the present, only the system $\mathcal{IC}$ and results are presented, together with novel machinery required to define a system. A detailed walk-through of a cascading run with arguments for modeling choices, the presupposed definitions and references may be found in [2], the extended version of the present abstract.

Transition Rules and Systems

Before commencing with definitions, let us fix notation for presupposed machinery. Assume a finite set of agents $\mathcal{A} = \{1, 2, ..., n\}$, atoms $P \in \Phi$, a set Prop_Φ given by $\varphi ::= P \mid \varphi \mid \neg\varphi \mid \varphi \wedge \psi \mid B_i\varphi \mid K_i\varphi$ and definitions of *pointed epistemic plausibility models* (EPMs) $(\mathbf{S}, s_0) = (S, \leq_i, \|\cdot\|, s_0)_{i \in \mathcal{A}}$, *propositions* $\|\varphi\|_{\mathbf{S}} \subseteq S$ over EPMs for $\varphi \in \mathsf{Prop}_\Phi$, *pointed action plausibility models with postconditions* (APMs) $(\mathbf{E}, \sigma_0) = (\Sigma, \preceq_i , pre, post, \sigma_0)_{i \in \mathcal{A}}$, *doxastic programs* $\Gamma \subseteq \Sigma$ over APMs, *action priority update product* $\mathbf{S} \otimes \mathbf{E}$ (an anti-lexicographic belief revision operation on EPMs and APMs), and *dynamic modalities* $[\Gamma]$ with associated propositions $\|[\Gamma]\varphi\|_{\mathbf{S}}$.

Transition Rules. A *transition rule* $\mathcal{T}$ is an expression $\varphi \rightsquigarrow [X]\psi$ where $\varphi, \psi \in \mathsf{Prop}_\Phi$. Transition rules are *prescriptive* and read "if φ, then the next update must be such that after it, ψ".

Solutions. A set of transition rules dictates the choice for the next APM by finding the transition rule(s)'s *solution*. A *solution* to $\mathcal{T} = \varphi \rightsquigarrow [X]\psi$ over pointed EPM $(\mathbf{S}, s)$ is a doxastic program Γ such that $\mathbf{S}, s \models \varphi \rightarrow [\Gamma]\psi$. Γ is a solution to the set $\mathbb{T} = \{\mathcal{T}_1, ..., \mathcal{T}_n\}$ with $\mathcal{T}_k = \varphi_k \rightsquigarrow [X]\psi_k$ over $(\mathbf{S}, s)$ if $\mathbf{S}, s \models \bigwedge_1^n(\varphi_k \rightarrow [\Gamma_k]\psi_k)$, i.e. if Γ is a solution to all $\mathcal{T}_i$ over $(\mathbf{S}, s)$ *simultaneously*.[1] Finally, a set of doxastic programs $\mathbb{G}$ is a solution to $\mathbb{T}$ over $\mathbf{S}$ iff for every t of $\mathbf{S}$, there is a $\Gamma \in \mathbb{G}$ such that Γ is a solution to $\mathbb{T}$ over $(\mathbf{S}, t)$.

---

[1] Note the analogy with numerical equations; for both $2 + x = 5$ and $\{2 + x = 5, 4 + x = 7\}$, $x = 3$ is the (unique) solution.

Next APM Choice. If $\mathbb{G}$ is a solution to $\mathbb{T}$ over $\mathbf{S}$, then given any state from $\mathbf{S}$, the transition rules in $\mathbb{T}$ will specify one (or more) programs from $\mathbb{G}$ as the *next APM choice*, denoted $next(\mathbf{S})_{\mathbb{T},\mathbb{G}}$, subscripts omitted. A *deterministic* choice will be made if $\mathbb{G}$ is selected suitably, i.e. if it contains a *unique* Γ for each s. In the ensuing, solution spaces will be chosen thusly.

System. A *system* is a tuple $\mathcal{S} = \langle \mathbf{S_0}, \mathbb{T}, \mathbb{G}, end \rangle$ where $\mathbf{S_0}$ is an EPM, called the *initial state*, $\mathbb{T}$ is a set of sets $\mathbb{T}(\mathbf{S})$ each a set of transition rules (those of EPM $\mathbf{S}$),[2] $\mathbb{G}$ is a set of sets $\mathbb{G}(\mathbb{T}(\mathbf{S}))$, each a set of doxastic programs (the solution space for $\mathbb{T}(\mathbf{S})$), and $end \in \mathsf{Prop}_\Phi$ is called the *end condition*.

A system provides for each EPM from some chosen set, a set of transition rules with associated solution space, and is run using next APM choice. The first next APM choice is made when the actual state of $\mathbf{S_0}$ is specified. A system runs until either the end condition is met, or until it constructs an EPM for which no transition rules are specified or no solution is available. Care must be taken to avoid the latter possibilities.

An Informational Cascades System Based on Aggregated Beliefs

Atoms. Let Φ consist of two "types" of atomic doxastic propositions; $\{L\}$ with $\neg L =:$ R, representing respectively that the restaurant on the left or the one on the right is better, and $\{\alpha_i L, \alpha_i R\}_{i \in A}$ with $\alpha_i L \cap \alpha_i R = \emptyset$, representing i's restaurant choice.[3] $\alpha_i R$ is *not* short for $\neg \alpha_i L$ as i may not yet have made *any* choice.

Aggregated Beliefs. To accumulate information, a notion of *perceived aggregated beliefs* is used. Introduce an operator $A_{i|G}$, representing the beliefs of agent i when aggregating information from her beliefs about the beliefs of agents from group G. $A_{i|G}$ is defined using simple majority 'voting' with a self-bias tie-breaking rule:

$$\mathbf{S}, s \models A_{i|G}\varphi \text{ iff } \alpha + |\{j \in G : \mathbf{S}, s \models B_i B_j \varphi\}| > \beta + |\{j \in G : \mathbf{S}, s \models B_i B_j \neg\varphi\}|$$

with tie-breaking parameters α, β given by $\alpha = \frac{1}{2}$ if $s \in (B_i\varphi)_\mathbf{S}$, else $\alpha = 0$, and $\beta = \frac{1}{2}$ if $s \in (B_i\neg\varphi)_\mathbf{S}$, else $\beta = 0$. This definition leaves agent i's aggregated beliefs undetermined iff both i is agnostic whether φ and there is no strict majority on the matter.

Overview of $\mathcal{IC}$. As mentioned, each agent's turn consists of four steps. In $\mathcal{IC}$ defined below, these consist of: i) EPM $\mathbf{S_i}$, the initial state of i's turn, ii) APM $\mathbf{I_{i-1}}$, invoking the *interpretation* of agent $i-1$'s executed action, supplying i with information about $i-1$'s beliefs, iii) APM $\mathbf{P_i}$, the *private signal* of i, forming her private beliefs about L/R, and iv) either l_i or r_i, the action i finally executes. The initial state of $i+1$ is then given by $\mathbf{S_{i+1}} := ((\mathbf{S_i} \otimes \mathbf{I_{i-1}}) \otimes \mathbf{P_i}) \otimes next((\mathbf{S_i} \otimes \mathbf{I_{i-1}}) \otimes \mathbf{P_i})$, with $next((\mathbf{S_i} \otimes \mathbf{I_{i-1}}) \otimes \mathbf{P_i}) \in \{l_i, r_i\}$.

Three sets of transition rules are used to run the system: the first is a singleton, always invoking interpretation of the previous agent's action. The second is also a singleton, invoking a private signal specified by a vector defined together with the system. Third, a set of two rules which specify the choice of the agent as a function of her aggregated beliefs, hereby specifying the used agent type.

---

[2] It is assumed that *model names matter*: though $\mathbf{S} = \mathbf{S} \otimes \Gamma$, we allow that $\mathbb{T}(\mathbf{S}) \neq \mathbb{T}(\mathbf{S} \otimes \Gamma)$.

[3] $\alpha_i L$ and $\alpha_i R$ are *post-factual action descriptions*, not the actions themselves, as these are captured using APMs, see point 3. in the definition of the system $\mathcal{IC}$ below.

The System $\mathcal{IC}$. Define the system $\mathcal{IC} = \langle \mathbf{S_1}, \mathbb{T}, \mathbb{G}, end \rangle$ as follows: let the initial state be $\mathbf{S_1}$ (Fig. 1) and set $end := \alpha_m L \vee \alpha_m R$ with $m = max(\mathcal{A})$. That is, the system initiates with all agents uninformed about whether L or R, and terminates when the last agent has chosen at which restaurant to dine.

Fig. 1. The EPM $\mathbf{S_1}$ representing the initial uncertainty about the better restaurant. All agents know one restaurant is better, but does neither know nor believe which one. Labels L and R indicate truth of the atom, e.g. $s_0 \in \|L\|_{\mathbf{S_1}}$. For all $P \in \{\alpha_i L, \alpha_i R\}_{i \in \mathcal{A}}$, $\|P\|_{\mathbf{S_1}} = \emptyset$ as no agent has chosen.

Set $\mathbf{S_{n+1}} := (((\mathbf{S_n} \otimes \mathbf{I_{n-1}})) \otimes \mathbf{P_n}) \otimes next(((\mathbf{S_n} \otimes \mathbf{I_{n-1}})) \otimes \mathbf{P_n})$, and give $\mathbb{T}$ and $\mathbb{G}$ by

1. $\mathbb{T}(\mathbf{S_n}) = \{\mathcal{I}_{n-1} = \top \rightsquigarrow [X]\top\}$ with $\mathbb{G}(\{\mathcal{I}_{n-1}\}) = \{\mathbf{I_{n-1}}\}$, where $\mathbf{I_{n-1}}$ is the one state interpretation APM with preconditions

$$pre(i_{n-1}) := \alpha_{n-1}L \rightarrow A_{n-1|\mathcal{A}}L \wedge \alpha_{n-1}R \rightarrow A_{n-1|\mathcal{A}}R$$

with special case $\mathbf{I_0}$ having $pre(i_0) = post(i_0) = \top$.

2. $\mathbb{T}(\mathbf{S_n} \otimes \mathbf{I_{n-1}}) = \{\mathcal{P}_n = \top \rightsquigarrow [X]\top\}$ with $\mathbb{G}(\{\mathcal{P}_n\}) = \{(\mathbf{P_n}, x_n)\}$, where $\mathbf{P_n}$ is the private signal APM (Fig. 2) indexed for n, with x_n the actual state as given by a *private signal vector* $\mathbf{P} = (x_1, x_2, ..., x_m)$ with $x_k \in \{\sigma_L, \sigma_R\}$, determining whether n receives a signal that L (σ_L) or that R (σ_R).

$\langle L; \top \rangle$ $\overset{i}{\overbrace{}}$ $\langle R; \top \rangle$

$\sigma_L \leftarrow - - - \rightarrow \tau_L$

$\mathcal{A} \setminus \{i\} \uparrow \qquad\qquad \uparrow$

$\langle L; \top \rangle$ $\sigma_R \leftarrow - - - \rightarrow \tau_R$ $\langle R; \top \rangle$

$\underset{i}{\underbrace{}}$

Fig. 2. APM $\mathbf{P_i}$: i receives private signal while others remain uninformed about *which*. State labels $\langle \varphi; \psi \rangle$ specify pre- and postconditions. Transitive and reflexive arrows are not drawn.

3. $\mathbb{T}(((\mathbf{S_n} \otimes \mathbf{I_{n-1}})) \otimes (\mathbf{P_n}, x_n)) = \{\mathcal{A}_L, \mathcal{A}_R\}$ (Fig. 3) with $\mathbb{G}(\{\mathcal{A}_L, \mathcal{A}_R\}) = \{l_n, r_n\}$, the singleton doxastic programs over the APM in Fig. 3, all indexed for n.

Aggregator rules:

$\mathcal{A}_L = A_{i|G}L \rightsquigarrow [X]\alpha_i L$

$\mathcal{A}_R = A_{i|G}R \rightsquigarrow [X]\alpha_i R$

$\mathcal{A} \circlearrowleft_{l_i} \qquad\qquad \circlearrowright^{\mathcal{A}}_{r_i}$

$\langle \top; \alpha_i L \rangle \qquad \langle \top; \alpha_i R \rangle$

Fig. 3. Aggregator transition rules specifying an agent type who bases decisions on aggregated beliefs, and the APM $\mathbf{A_i}$ over which the two possible actions for agent i is given; i may choose to go to either the restaurant on the left (l_i) or the one on the right (r_i).

Given the cumbersome definition of $\mathcal{IC}$, it is worth verifying that the system in fact runs appropriately. By induction it may be shown that for every agent $i \leq m$, the system will produce state $\mathbf{S_{i+1}}$ satisfying $\alpha_i L \vee \alpha_i R$, yielding the following proposition.

Proposition 1. *The system $\mathcal{IC}$ runs until $end := \alpha_m L \vee \alpha_m R$ is satisfied at $\mathbf{S_{m+1}}$, irrespectively of which initial state or which signal vector $\mathbf{P}$ is used for input.*

In a Cascade. With $\mathcal{IC}$ defined, it is possible to precisely define the notion of *being in a cascade*: agent i is *in a cascade* iff

i) $next((\mathbf{S_i} \otimes \mathbf{I_{i-1}})) \otimes (\mathbf{P_i}, x_i)) = l_i$ for both $x_i \in \{\sigma_L, \sigma_R\}$, or
ii) $next((\mathbf{S_i} \otimes \mathbf{I_{i-1}})) \otimes (\mathbf{P_i}, x_i)) = r_i$ for both $x_i \in \{\sigma_L, \sigma_R\}$.

The definition captures that i acts in accordance with an established majority, *irrespective of her own signal*.[4]

The following lemma captures a crucial property regarding the higher-order reasoning occurring in cascades, namely that *the choice of an agent in a cascade provides no information about their private beliefs* (hence neither about their private signal).

Lemma 1. $\mathbf{S_{n+1}} \otimes \mathbf{I_n} \models B_{n+1} B_n L \vee B_{n+1} B_n R$ *iff n is not in a cascade.*

To state the main result, notation for the agents in cascade who ignored which signals is handy. Let $\mathbf{P_i}$ be the private signals for agents $j < i$, i.e. the initial segment of $\mathbf{P}$ of length $i - 1$. Let $C_{Li} = \{j < i : j$ is in cascade and $x_j = \sigma_L\}$ and $C_{Ri} = \{j < i : j$ is in cascade and $x_j = \sigma_R\}$. We may then state the main result.

Theorem 1. *Agent i is in cascade iff two more agents have received private signal of one type than have received signals of the other type, not counting signals of agents in a cascade. Precisely: i is in cascade of type i) iff*

$$|\{\sigma_L \in \mathbf{P_i}\}| - |C_{Li}| \geq (|\{\sigma_R \in \mathbf{P_i}\}| - |C_{Ri}|) + 2,$$

and agent i is in cascade of type ii) iff

$$(|\{\sigma_L \in \mathbf{P_i}\}| - |C_{Li}|) + 2 \leq |\{\sigma_R \in \mathbf{P_i}\}| - |C_{Ri}|.$$

The theorem provides necessary and sufficient conditions on the private signal string for an agent to be in a cascade. The sufficient conditions are identical to those from [1], see p. 1005-06, here shown for a model which explicitly represents all higher-order reasoning and agent decision making.

Corollary 1. *Cascades in $\mathcal{IC}$ are irreversible: if i is in a cascade of type i) resp. type ii), then for all $k > i$, k will be in a cascade of type i) resp. type ii).*

The corollary captures the quintessential effect of cascades, namely that they propagate through the remaining group.

These results show that the system $\mathcal{IC}$ functions as the informal reasoning supposed in [1]. For proofs, further conclusions, discussion, venues for future research and relevant references, the reader is referred to [2].

References

1. Bikhchandani, S., Hirshleifer, D., Welch, I.: A Theory of Fads, Fashion, Custom, and Cultural Change as Informational Cascades. Journal of Political Economy 100(5), 992–1026 (1992)
2. Rendsvig, R.K.: Aggregated Beliefs and Information Cascades (extended) (2013), http://vince-inc.com/rendsvig/papers/IC1.pdf

---

[4] The definition thus closely mirrors that from the original paper: "An informational cascade occurs if an individual's action does not depend on his private signal." [1, p. 1000]

Dynamic Attitudes, Fixed Points
and Minimal Change

Ben Rodenhäuser

ILLC, University of Amsterdam

Abstract. According to the principle of minimal change, an agent should
not change her belief state more than is strictly required to accomodate
new information. We propose a novel approach to the issue by considering a notion of optimality of belief revision policies that is sensitive to
the *target of revision*.

1 Introduction

The concept of *minimal change* is crucial for many theories of belief change (cf.,
eg., [12], [11], [9]). The thesis usually associated with the concept is that in order
to revise with a proposition P, one should transform one's belief state in such
a way as to ensure that one afterwards believes P, but in doing, so should keep
the "difference" to the original belief state *as small as possible*. This idea can
be motivated by appeal to two principles, the *principle of conservatism*, and
the *principle of informational economy*: one is justified to keep prior doxastic
commitments one is not forced to give up (*priniple of conservatism*, cf. [12])); but
dropping commitments one is justified to keep is a disproportionate response,
overly costly at the least (*principle of informational economy*, cf., e.g., [5]). Hence
it is rational to maintain all commitments one is not forced to give up (*principle
of minimal change*). Granting this, however, one needs to ask: minimal *for what
purpose*? Boutilier called his favourite belief revision method "natural revision,"
because it seemed to him to appropriately formalize a notion of minimal change,
or "conservatism" of belief change.[1] Darwiche and Pearl, on the other hand,
have criticized Boutilier's method, arguing, essentially, that it produces beliefs
that are not "robust" enough under further revisions (cf. [10], [6]).

This type of dissent leads me to propose, in this abstract, a formalization
of the problem of minimal change that is sensitive to the *target of revision*: if
truth in the most plausible states of some given order is the target, Boutilier's
policy should indeed count as "optimal". If, on the other hand, the belief state
resulting from revision is required to satisfy additional constraints (as Darwiche
and Pearl argued), other policies need to be taken into account so as to *meet those
constraints*. To implement this idea, I make use of the framework for modeling
dynamic doxastic attitudes developed in recent joint work ([3], [4]), a framework
which essentially generalizes the usual notion of a "belief revision policy" to bring
a much wider class of belief change operators into view, and study systematically
their properties.

---

[1] Cf. [7]. In this abstract, "natural revision" will be called *minimal upgrade*.

D. Grossi, O. Roy, and H. Huang (Eds.): LORI 2013, LNCS 8196, pp. 342–346, 2013.

2   Background

This section introduces, very briefly, the fundamental notions we use.[2]

Fix a finite non-empty set W, called *the set of all (possible) worlds*. A *proposition* is a subset of W, i.e., a set of possible worlds. A *plausibility order* is a pair $\mathcal{S} = (S, \leq)$, where $S \subseteq W$ is a finite set of possible worlds, and $\leq \subseteq S \times S$ is a total preorder on S, i.e., a transitive and connected (and thus reflexive) relation.[3] The fact that $w \leq v$ indicates that world w is *at least as plausible as* world v (from the perspective of our (implicit) agent). The *best P-worlds* (or *most plausible P-worlds*) in a plausibility order $\mathcal{S}$, denoted with $\text{best}_{\mathcal{S}} P$, are given by the proposition $\text{best}_{\mathcal{S}} P := \{w \in P \cap S \mid \forall v \in P : w \leq v\}$. The *best worlds* (or *most plausible worlds*) in $\mathcal{S}$ are given by $\text{best}\,\mathcal{S} := \text{best}_{\mathcal{S}} S$. Note that $\text{best}\,\mathcal{S} = \varnothing$ iff $S = \varnothing$.

A *doxastic proposition* $\mathcal{P}$ is a function $\mathcal{P} : \mathcal{S} \longmapsto \mathcal{P}(\mathcal{S})$ that assigns a proposition $\mathcal{P}(\mathcal{S}) \subseteq S$ to each plausibility order $\mathcal{S}$. We call $\mathcal{P}(\mathcal{S})$ the *proposition denoted by* $\mathcal{P}$ *in* $\mathcal{S}$. A *(doxastic) propositional attitude* is an indexed family of doxastic propositions $A := \{AP\}_{P \subseteq W}$, indexed by arbitrary propositions $P \subseteq W$, satisfying $A_{\mathcal{S}} P = A_{\mathcal{S}}(P \cap S)$. This requirement expresses, essentially, that whether AP is satisfied in $\mathcal{S}$ should not depend on P-worlds that are not part of S. A propositional attitude is *introspective* iff for any order $\mathcal{S}$ and proposition P, $A_{\mathcal{S}} P \in \{S, \varnothing\}$. We write $\mathcal{S} \models AP$ iff $A_{\mathcal{S}} P = S$.

Examples of propositional attitudes include: *(irrevocable) knowledge* K, defined by $\mathcal{S} \models KP$ iff $S \subseteq P$; *(simple) belief* B, defined by $\mathcal{S} \models BP$ iff $\text{best}\,\mathcal{S} \subseteq P$; *strong belief* Sb, defined by $\mathcal{S} \models SbP$ iff $\mathcal{S} \models BP$ and $\forall x, y \in S :$ if $x \in P, y \notin P$, then $x < y$.

A *(doxastic) upgrade* u is a function $u : \mathcal{S} \longmapsto \mathcal{S}^u$ that takes a given plausibility order $\mathcal{S} = (S, \leq)$ to a plausibility order $\mathcal{S}^u := (S^u, \leq^u)$, satisfying $S^u \subseteq S$.

Examples of upgrades include: the *update* $!P$, mapping, for each proposition P, each plausibility order $\mathcal{S}$ to the conditionalization of the order on P; the *lexicographic upgrade* $\Uparrow P$, making all P-worlds strictly better than all non-P-worlds, while keeping the order within each zone the same (cf. [13]); the *positive lexicographic upgrade* $\Uparrow^+ P$, defined exactly as lexicographic upgrade $\Uparrow P$, except that, for any proposition P and order $\mathcal{S}$ such that $P \cap S = \varnothing$, the result of applying $\Uparrow^+ P$ to $\mathcal{S}$ is the empty plausibility order; the *minimal upgrade* $\uparrow P$, making the best P-worlds the best worlds overall, keeping the order otherwise the same (cf. [8]); the *positive minimal upgrade* $\uparrow^+ P$ defined exactly as minimal upgrade $\uparrow P$, except that, for any proposition P and order $\mathcal{S}$ such that $P \cap S = \varnothing$, the result of applying $\uparrow^+ P$ to $\mathcal{S}$ is the empty plausibility order.

A *dynamic (doxastic) attitude* τ is a family of upgrades $\{\tau P\}_{P \subseteq W}$, satisfying

- *C1:* $\tau P \cdot \tau P = \tau P$.
- *C2:* $\mathcal{S}^{\tau P} = \mathcal{S}^{\tau(P \cap S)}$.
- *C3:* If $P = \varnothing$ or $P = W$, then $\tau P = \varnothing$ or $\tau P = id$.

---

[2] For motivation and explanations, the reader is refered to [2], [3].
[3] A preorder $\mathcal{S}$ on W is *total* iff for any two worlds $w, v \in W$: either $w \leq v$ or $v \leq w$.

For motivation of these requirements, we refer the reader to [3]. Examples of dynamic attitudes include: *infallible trust* !, mapping each proposition P to the update $!P$; *(positive) strong trust* $\Uparrow^{(+)}$, mapping each proposition P to the (positive) lexicographic upgrade $\Uparrow^{(+)}P$; *(positive) minimal trust* $\uparrow^{(+)}$, mapping each proposition P to the (positive) minimal upgrade $\uparrow^{(+)}P$.

Intuitively, an upgrade τP is *redundant*, or *uninformative*, in an order $\mathcal{S}$ if the order remains unchanged when applying τP, i.e., if $\mathcal{S}^{\tau P} = \mathcal{S}$. Now "redundancy of τ" can itself be seen as a *propositional* attitude, that we will denote by $\bar{\tau}$.

Formally, the *fixed point* $\bar{\tau}$ of a *dynamic attitude* τ is the propositional attitude given by $\mathcal{S} \models \bar{\tau}P$ iff $\mathcal{S}^{\tau P} = \mathcal{S}$. Observe that fixed points of dynamic attitudes are in general introspective. Fixed points capture the *target* of a given dynamic attitude τ, i.e., the propositional attitude *realized* by τ. Jumping ahead, this notion will allow us to study our main question: whether a given dynamic attitude (representing a belief revision policy) realizes its target in a minimal way.

As shown in [[3]], our examples of dynamic attitudes and our examples of propositional attitudes match along the notion of a fixed point:

- The fixed point of ! is K.
- The fixed point of $\Uparrow$ is $K^{\neg} \vee Sb$.[4]
- The fixed point of $\Uparrow^+$ is Sb.
- The fixed point of $\uparrow$ is $K^{\neg} \vee B$.
- The fixed point of $\uparrow^+$ is B.

Notice that $\Uparrow^+$ and $\uparrow^+$, rather than $\Uparrow$ and $\uparrow$, characterize the propositional attitudes K and B. In view of this observation, we focus on the former pair of dynamic attitudes.

3   Similarity, Optimality and Canonicity

Similarity. Given any two plausibility orders $\mathcal{S}$ and $\mathcal{S}'$, we can compare the two by the extent to which they agree on the relative plausibility of given pairs of worlds. Formally, $\mathcal{S}$ and $\mathcal{S}'$ *agree on* $(w, v) \in W^2$ iff $(w, v) \in \mathcal{S}$ iff $(w, v) \in \mathcal{S}'$. We denote by $agree(\mathcal{S}, \mathcal{S}')$ (the *agreement set of $\mathcal{S}$ and $\mathcal{S}'$*) the set of pairs (w, v) such that $\mathcal{S}$ and $\mathcal{S}'$ agree on (w, v).

For any order $\mathcal{S}$, the *weak similarity order* $\preceq_{\mathcal{S}}$ is then defined in terms of inclusion of agreement sets. For any plausibility orders $\mathcal{S}', \mathcal{S}''$:

$$\mathcal{S}' \preceq_{\mathcal{S}} \mathcal{S}'' \text{ iff } S', S'' \subseteq S \text{ and } agree(\mathcal{S}, \mathcal{S}'') \subseteq agree(\mathcal{S}, \mathcal{S}').$$

If $\mathcal{S}' \preceq_{\mathcal{S}} \mathcal{S}''$, then we say that $\mathcal{S}'$ is *at least as similiar (or close) to $\mathcal{S}$ as $\mathcal{S}''$*. The *weak similarity order* $\preceq_{\mathcal{S}}$ is defined as usual.

---

[4] Given introspective propositional attitudes A and A', we define the propositional attitude $A \vee A'$ by means of $\mathcal{S} \models (A \vee A')P$ iff $\mathcal{S} \models AP$ or $\mathcal{S} \models A'P$; and we define $A^{\neg}$ by means of $\mathcal{S} \models A^{\neg}P$ iff $\mathcal{S} \models A(\neg P)$.

Optimality. Let τ be a dynamic attitude. We say that τ is *optimal* iff there is no order $\mathcal{S}$, proposition P and dynamic attitude σ such that

$$\overline{\sigma} = \overline{\tau} \text{ and } \mathcal{S}^{\sigma P} \prec_{\mathcal{S}} \mathcal{S}^{\tau P}.$$

An attitude τ is thus optimal if it creates its fixed point $\overline{\tau}$ *in a minimal way*: for no given order $\mathcal{S}$ and proposition P does there exist an attitude σ *with the same fixed point* $\overline{\tau}$ such that $\mathcal{S}^{\sigma P}$ is closer to $\mathcal{S}$ than $\mathcal{S}^{\tau P}$.

Given a propositional attitude A, we say that τ is *optimal for* A if τ is optimal and $\overline{\tau} = A$. This notion gives us a general measure of minimal change *relative to the target of revision*: the optimality of τ is evaluated taking into account the fixed point $\overline{\tau}$ of τ.

Not all dynamic attitudes are optimal for their fixed point. As an example, we mention $?K$ ("test for irrevocable knowledge") defined by

$$\mathcal{S}^{?KP} := \begin{cases} \mathcal{S} & \mathcal{S} \models KP, \\ \varnothing & \mathcal{S} \not\models KP. \end{cases}$$

It is easy to see that $?K$ is not optimal for its fixed point K (hint: compare it to infallible trust !). On the other hand, one can show the following:

Proposition 1. *For a propositional attitude A, the following are equivalent:*

1. *A is introspective.*
2. *There exists a dynamic attitude τ that is optimal for A.*

The main examples of dynamic attitudes we consider here are all optimal:

Proposition 2.

1. *Infallible trust ! is optimal for irrevocable knowledge K.*
2. *Positive strong trust $\Uparrow^+$ is optimal for strong belief Sb.*
3. *Minimal positive trust $\uparrow^+$ is optimal for simple belief B.*

Each of these dynamic attitudes thus provides a solution to the minimal change problem (relative, in each case, to the target of revision). These operations have been at the center of interest in recent research in Dynamic Epistemic Logic. Our result justifies formally the claim that this was not an arbitrary choice.

Canonicity. Every now and then, a dynamic attitude τ is *uniquely optimal* in the sense that τ is the *only* optimal dynamic attitude whose fixed point is $\overline{\tau}$. If this is the case, we will call τ "canonical". Formally, for a given dynamic attitude τ, we say that τ is *canonical* if τ is optimal and for any attitude σ: if $\overline{\sigma} = \overline{\tau}$ and σ is optimal, then $\sigma = \tau$. Given a propositional attitude A, τ is *canonical for A* if τ is canonical and $\overline{\tau} = A$.

Proposition 3.

1. *Infallible trust ! is canonical for K.*

2. *Strong positive trust* $\Uparrow^+$ *is canonical for Sb.*
3. *Positive minimal trust* $\uparrow^+$ *is not canonical for B.*

The argument for the third item relies on the observation that, starting from an order $\mathcal{S}$ in which P is not believed but $\text{best}_{\mathcal{S}}\,P$ contains more than one world, it is sufficient to promote *some* best P-world, making it best overall: making *all* P-worlds best overall is not necessary. For this reason, there are policies deviating from positive minimal trust that are also optimal for belief. So $\uparrow^+$ is not canonical. Notice that it follows that *no dynamic attitude is canonical for simple belief.* Our formalization of minimal change thus does not give rise to a unique solution of the minimal change problem for simple belief. Assuming our formalization, such a unique solution simply does not exist. What is the significance of this observation? One reaction is to be satisfied with the earlier result that $\uparrow^+$ is *optimal.* However, another promising route is to study what tweaks to our formal notions allow a canonicity result after all. Furthermore, a question of obvious interest is to study necessary and sufficient criteria for canonicity in our setting, i.e., characterize the attitudes that are canonical according to our definitions! These avenues are explored in the full version of the paper.

References

1. Alchourron, C.E., Gardenfors, P., Makinson, D.: On the logic of theory change. Journal of Symbolic Logic 50, 510–530 (1985)
2. Baltag, A., Smets, S.: A qualitative theory of dynamic interactive belief revision. Texts in Logic and Games 3 (2008)
3. Baltag, A., Rodenhäuser, B., Smets, S.: Doxastic attitudes as belief revision policies. In: Proceedings of the ESSLLI Workshop on Strategies for Learning, Belief Revision and Preference Change (2012)
4. Baltag, A., Rodenhäuser, B., Smets, S.: Dynamic attitudes as strategies for belief change (unpublished manuscript 2013)
5. Board, O.: Dynamic interactive epistemology (2002)
6. Booth, R., Chopra, S., Meyer, T.: Admissible and restrained revision. Journal of Artificial Intelligence Research (2006)
7. Boutilier, C.: Revision sequences and nested conditionals. In: Proceedings of the Thirteenth International Joint Conference on Artificial Intelligence (IJCAI 1993), pp. 519–525 (1993)
8. Boutilier, C.: Iterated revision and minimal change of conditional beliefs. Journal of Philosophical Logic 25(3), 263–305 (1996)
9. Costa, H.A., Levi, I.: Contraction. on the decision-theoretic origins of minimal change and entrenchment (2005)
10. Darwiche, A., Pearl, J.: On the logic of iterated belief revision. Artificial Intelligence 89(1-29) (1996)
11. Gärdenfors, P.: Knowledge in Flux: Modeling the Dynamics of Epistemic States. The MIT Press (1988)
12. Harman, G.: Change in View. MIT Press (1986)
13. Nayak, A.C.: Iterated belief change based on epistemic entrenchment. Erkenntnis 41, 353–390 (1994)

Logic of Evidence-based Knowledge

Chenwei Shi

Department of Philosophy, Tsinghua University, Beijing, China
shichenwei88@gmail.com

Abstract. This paper presents logics for reasoning about *sound enough evidence* and its relation to evidence and knowledge. The logic of sound enough evidence is based on van Benthem and Pacuit's evidence logic and Holliday's formalization of Nozick's tracking theory. And the newly defined knowledge based on *sound enough evidence* does not imply the evidence-based belief but respects the epistemic closure. The related philosophical issues are also discussed.

1 Introduction

Example 1 (The Witness of The Murder). One day a man was murdered in a room. The detective Sherlock took charge of this case. Later a man came to him and said that he had witnessed the whole murder and recognized the murderer (Killer) because the window of that room was opposite to that of his own room and at that time he was looking out of the window

Scenario 1 without any instruments;
Scenario 2 with a telescope

and happened to see the murder. With the help of polygraph, Sherlock was certain about the honesty of the witness. But he still could not make the judgement that the murderer was definitely the man identified by the witness. (To be continued)

The explanation for Sherlock's caution is simple: He can only be sure that the witness has the evidence to make him *believe* that Killer is the murderer but can not be sure whether the evidence the witness has is *sound enough* to make him *know* that Killer is the murderer. In the above explanation, three notions, "evidence", "belief" and "knowledge", play a crucial role. To understand it better, a further analysis of the relation between these concepts is needed. Furthermore, we would like to answer the following question: How does Sherlock make sure that the witness's evidence is sound enough?

In this paper we will follow the analysis of the relationship between evidence and belief in [4], in which evidence structure was introduced to standard doxastic models and a dynamic logic of evidence-based belief was proposed. We attempt to analyse the relationship between evidence and knowledge and propose a new way of defining knowledge based on evidence.

Then the main issue comes down to the problem how to find the right evidences which underlie an agent's knowledge. The observation is $E_K \subseteq E_B$, where E_B and E_K represent the set of propositions directly supported by the agent's evidences from which

D. Grossi, O. Roy, and H. Huang (Eds.): LORI 2013, LNCS 8196, pp. 347–351, 2013.

she can derive her belief and knowledge respectively. In other words, while the propositions in E_B are only the propositions we have evidence for, the propositions in E_K are the propositions we have sound enough evidence for. The definition of knowledge in [1] suggests that the relation between E_B and E_K should be $E_K = E_T \subseteq E_B$, where E_T denotes the set of true propositions in E_B. Different from [1], however, we claim that the relation should be $E_K \subseteq E_T \subseteq E_B$, for which we will argue philosophically and technically in this paper. Therefore it is pivotal to find a mechanism which can pick the propositions in E_K from the propositions in E_B to define knowledge.

This paper presents a way in which E_K can be picked out from E_B, inspired by Holliday's [2] formalization of Nozick's [3] tracking theory, and then defines knowledge based on evidences.

2   Logic of Evidence-based Knowledge(EKL)

We first continue the story in Example 1
To make clear whether the witness knows that the Killer is the murderer(we assume that in fact the Killer is the murderer), Sherlock took another suspect Keller who was also the acquaintance of the witness to the room where the murder happened and stimulated the crime scene including the time, position of the murderer, light and so on. And he asked Keller to stand at the position of murderer and the witness to stand in front of his window to identify who the guy at the position of murderer was

Scenario 1 *without any instruments;*
Scenario 2 *with a telescope.*

At last, the witness

Scenario 1 *did not recognize who he was;*
Scenario 2 *recognized who he was.*

Therefore Sherlock

Scenario 1 *still could not conclude that Killer is the murderer;*
Scenario 2 *could conclude that the Killer is the murderer.*

This section develops a logic of evidence-based knowledge, which can be used to explain the difference between the two scenarios in the examples.

2.1   Language of EKL

Definition 1 (Language of EKL). *Let* **At** *be a set of atomic sentence symbols, the language* $\mathcal{L}_{EK}$ *is defined by*

$$\varphi_0 := p \mid \neg\varphi_0 \mid \varphi_0 \wedge \varphi_0 \mid \Box\varphi_0$$
$$\varphi := \varphi_0 \mid \neg\varphi \mid \varphi \wedge \varphi \mid K^E\varphi_0 \mid \varphi \,\Box\!\!\rightarrow \varphi$$

where $p \in$ **At.** *Additional connectives are defined as usual.*

The intended reading of $\Box\varphi$ is "the agent has evidence that implies φ(the agent has "evidence for" φ). $\Box\!\!\rightarrow$ is the operator for counterfactual conditionals and $\varphi \,\Box\!\!\rightarrow\, \psi$ can be read as "If it were the case that φ, then it would be the case that ψ".The new operator K^E is for "evidence-based knowledge" and $K^E\varphi$ can be read "the agent knows φ based on her sound enough evidence"

In this logic, "sound evidence for" is a derived operator, defined as an abbreviation by putting $\Box_K\varphi := \varphi \wedge (\neg\varphi \,\Box\!\!\rightarrow\, \Box\neg\varphi) \wedge (\varphi \,\Box\!\!\rightarrow\, (\Box\varphi \wedge \neg\Box\neg\varphi))$. The interpretation of $\Box_K\varphi$ is that the agent has sound enough evidence for φ, while $\Box\varphi$ says that the agent has evidence for φ. For simplicity in this version of EKL and some philosophical reason[1], evidence for counterfactual conditionals is not taken into consideration. It is for this reason that only φ_0 formulas can appear inside "evidence for" operator. And we write the language without operator $\Box\!\!\rightarrow$ as $\mathcal{L}_0$.

2.2 Counterfactual Evidence Model

The definition of knowledge in [2] can be seen as a selection from one's belief: what kind of belief can be qualified as knowledge? And this selection is a counterfactual test of the agent's epistemic state to check whether the agent's epistemic state meets the requirement of knowledge. Recall our question at the beginning: what kind of evidences for the agent's belief can be qualified as evidences for the agent's knowledge? Then it is natural to think that this counterfactual test can be used again, not of the agent's epistemic state, but of the agent's evidential state.

To capture such test, it is necessary to introduce the relation $\leqslant$ into the evidence model in [4], by which the relevance or realisticity [2] between different possible worlds can be compared in the new model:

Definition 2 (Counterfactual evidence (CE) models). *A counterfactual evidence model is a tuple $\mathcal{M}$ of the form $\langle W, E, \leqslant, V\rangle$ where*

1. *W is a non-empty set;*
2. *$E \subseteq W \times \wp(W)$ is an evidence relation*
3. *$\leqslant$ assigns to each $w \in W$ a binary relation $\leqslant_w$ on some $W_w \subseteq W$:*
   (a) $\leqslant_w$ is reflexive and transitive;
   (b) for all $v \in W_w, w \leqslant_w v$;
4. *V assigns to each $p \in$ **At** a set $V(p) \subseteq W$.*

Besides, we also assume that $\leqslant_w$ is total and well-founded for the more perspicuous truth definitions. And Two constraints are imposed on the evidence sets:

- *For each $w \in W, \emptyset \notin E(w)$ (evidence per se is never contradictory);*
- *For each $w \in W, W \in E(w)$ (agents know their space).*

Truth of formulas in $\mathcal{L}_E$ is defined as follows:

---

[1] The counterfactual conditionals, relating to the relevance ordering of the possible worlds, seems totally different from the fact which can be known by us through empirical sense (evidence) and inference.

[2] cf. §2.2 of [2] for the theory of relevant alternatives

Definition 3 (**Truth conditions**). *Given a model* $M = \langle W, E, \leqslant \rangle$ *and state* $w \in W$, *truth of a formula* $\varphi \in \mathcal{L}_E$ *is defined inductively as follows:*

- $M, w \models p$ *iff* $w \in V(p)$ ($p \in At$)
- $M, w \models \neg\varphi$ *iff* $M, w \nvDash \varphi$
- $M, w \models \varphi \wedge \psi$ *iff* $M, w \models \varphi$ *and* $M, w \models \psi$
- $M, w \models \varphi \square\!\!\rightarrow \psi$ *iff* $\forall v \in Min_{\leqslant_w}([\![\varphi]\!]) : M, w \models \psi$
- $M, w \models \square\varphi$ *iff there is an* $X \in E(w)$ *and for all* $v \in X$, $M, v \models \varphi$ *and* $\varphi \in \mathcal{L}_0$
  $M, w \models \square_K\varphi$ *iff* $\exists X \subseteq [\![\varphi]\!] : X \in E(w)$ & $\forall Y \subseteq \overline{[\![\varphi]\!]} : Y \notin E(w)$, *and*
  $\qquad\qquad \forall v \in Min_{\leqslant_w}(\overline{[\![\varphi]\!]})(\exists X \subseteq \overline{[\![\varphi]\!]} : X \in E(v))$, *and*
- $\qquad\qquad \forall v \in Min_{\leqslant_w}([\![\varphi]\!])(\exists X \subseteq [\![\varphi]\!] : X \in E(v)$ & $\forall Y \subseteq \overline{[\![\varphi]\!]} : Y \notin E(v))$
  $\qquad\qquad \varphi \in \mathcal{L}_0$
- $M, w \models K^E\varphi$ *iff* $\{\psi \mid M, w \models \square_K\psi\} \models \varphi$

where $[\![\varphi]\!] = \{v \in W \mid M, v \models \varphi\}$ *and* $\overline{[\![\varphi]\!]} = \{v \in W \mid v \notin [\![\varphi]\!]\}$ *and* $Min_{\leqslant_w}(S) = \{v \in S \cap W_w \mid$ *there is no* $u \in S$ *such that* $u <_w v\}$.

To avoid some obvious counterexamples, we revise Nozick's tracking theory: the agent has sound enough evidence for φ if and if only in the actual state where φ is true, she has evidence for φ and has no evidence for $\neg\varphi$, at the same time, she would have evidence for $\neg\varphi$ in the most relevant counterfactual states, and would have evidence for φ and accept no evidence for $\neg\varphi$ in the most relevant possible states where φ was the case. We can also call such evidential state truth-tracking evidential state. Collecting all the propositions supported directly in the truth-tracking evidential state, we form our knowledge based on them (E_K) by inference and reasoning.

Now with the fact that $\square_K\varphi \rightarrow K^E\varphi$ is valid it can be explained why Sherlock made different judgements in different scenarios. The difference between the two scenarios can be shown in the CE models M_1 and M_2 (see Figure 2.2 and Figure 2.3) where Killer is the murderer (i) in world w, Keller is the murderer (e) in world v, Killer and Keller are both the murderer ($i \wedge e$) in world u and Killer and Keller are neither the murderer in world t: It is easy to verify that $M_2, w \models \square_K i$ but $M_1, w \models \neg\square_K i$. It follows immediately that Sherlock knows that Killer is the murderer in Scenario 2 but not in Scenario 1.

Fig. 1. CE model for Scenario 1: M_1

2.3   Axiomatization of the Evidence Logic

Although the axiomatization of EKL is left as an open question, the axiom system CES of EKL without the operator K^E (sound enough evidence logic) can be given as follows:

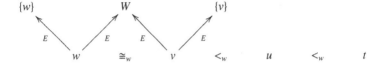

Fig. 2. CE model for Scenario 2: $\mathcal{M}_2$

taut: all propositional tautologies

$\square\!\!\rightarrow$-1 $\varphi \square\!\!\rightarrow \varphi$

$\square\!\!\rightarrow$-2 $((\varphi \square\!\!\rightarrow \psi) \wedge (\varphi \square\!\!\rightarrow (\psi \rightarrow \chi))) \rightarrow (\varphi \square\!\!\rightarrow \chi)$

$\square\!\!\rightarrow$-3 $(\varphi \square\!\!\rightarrow \psi) \rightarrow (((\varphi \wedge \psi) \square\!\!\rightarrow \chi) \leftrightarrow (\varphi \square\!\!\rightarrow \chi))$

$\square\!\!\rightarrow$-4 $\neg(\varphi \square\!\!\rightarrow \neg\psi) \rightarrow (((\varphi \wedge \psi \square\!\!\rightarrow \chi) \leftrightarrow (\varphi \square\!\!\rightarrow (\psi \rightarrow \chi)))$

$\square\!\!\rightarrow$-5 $\varphi \wedge (\varphi \square\!\!\rightarrow \psi) \rightarrow \psi$

$\top$**- and non $\bot$-evidence:** $\square\top \wedge \neg\square\bot$

$\square$**-monotonicity:** $\dfrac{\varphi \rightarrow \psi}{\square\varphi \rightarrow \square\psi}$

MP: Modus Ponens

LE: From $\varphi \leftrightarrow \psi$ infer $(\varphi \square\!\!\rightarrow \chi) \leftrightarrow (\psi \square\!\!\rightarrow \chi)$.

$N_\bigcirc$: Necessitation for $\bigcirc = \square, \varphi \square\!\!\rightarrow$

Theorem 1. *The sound enough evidence logic is sound and weakly complete for the class of counterfactual evidence models.*

3   Relationship between $\square$, $\square_K$, B and K^E

Given the truth condition of $\square$, $\square\!\!\rightarrow$, $\square_K$ and B^3, we can discuss the relationship between them like $\models \varphi \wedge (\neg\varphi \square\!\!\rightarrow \square\neg\varphi) \wedge (\varphi \square\!\!\rightarrow (\square\varphi \wedge \neg\square\neg\varphi)) \leftrightarrow \square_K\varphi$ and $\nvDash K^E\varphi \rightarrow B\varphi$.

More discussion can be found in the complete version.

Acknowledge. I thank Fenrong Liu for the helpful discussion and her instruction, Wesley H. Holliday, Johan van Benthem and Zhaoqing Xu for their valuable feedback, and finally the anonymous reviewers for their useful suggestions and comments.

This research is supported by Project of National Social Science Foundation of China (NO.13AZX018) and Tsinghua University Project (NO. 2012WHYX003).

References

1. Cornelisse, I.: Context dependence of epistemic operators in dynamic evidence logic. Master's thesis, University of Amsterdeam, ILLC (2011)
2. Holliday, W.: Knowing what follows: epistemic closure and epistemic logic. Ph.D. thesis, Stanford University (2012)
3. Nozick, R.: Philosophical Explanations. Harvard University Press (1981)
4. Van Benthem, J., Pacuit, E.: Dynamic logics of evidence-based beliefs. Studia Logica 99, 61–92 (2011)

---

^{3} Because the CE model is in fact a simple extension of the evidence model, the operators belief (B) can be defined in it as in the evidence model.

Backward Induction Is PTIME-complete

Jakub Szymanik*

Institute of Logic Language and Computation, University of Amsterdam
J.K.Szymanik@uva.nl

Abstract. We prove that the computational problem of finding backward induction outcome is PTIME-complete.

1 Introduction

Higher-order reasoning of the form 'I believe that Ann knows that Peter thinks ...' is an attractive topic for logical analysis. The logical investigations often go hand in hand with game theory. In this context, one of the common topics among researchers in logic and game theory has been backward induction (henceforth, BI), the process of reasoning backwards, from the end to determine a sequence of optimal actions. BI is a common method for determining sub-game perfect equilibria in the case of finite extensive-form games. BI can be understood as an inductive algorithm defined on a game tree – an algorithm that tells us which sequence of actions will be chosen by agents that want to maximize their own payoffs, assuming common knowledge of rationality.

Games have been also extensively used to design experimental paradigms aiming at studying social cognition, with a particular focus on higher-order social cognition. Often the experimental turn-based games can be modeled as extensive-form games and solved by applying BI. As it is hard to determine what the reasoning strategies used by participants in such games are, formal findings on backward induction have been used to better understand humans' strategic reasonings [1].

Recently, Van Benthem and Gheerbrant [2] have studied the logical definability of BI. They have observed that it can be defined in the first-order logic extended with the least fixed-point operator as well as in a variety of other dynamic epistemic formalisms. Obviously, from the least fixed-point definability result it follows that BI is in PTIME [3]. But is it also hard, and therefore complete for PTIME?

2 Preliminaries

Let us start by recalling that the reachability problem on alternating graphs is PTIME-complete [3].

---

* The research was supported by Veni Grant NWO-639-021-232. The author also wish to thank Rineke Verbrugge for many comments and suggestion as well as her Vici project NWO-277-80-001.

D. Grossi, O. Roy, and H. Huang (Eds.): LORI 2013, LNCS 8196, pp. 352–356, 2013.

Definition 1. *Let an alternating graph* $G = (V, E, A, s, t)$ *be a directed graph whose vertices,* V, *are labeled universal or existential.* $A \subseteq V$ *is the set of universal vertices.* $E \subseteq V \times V$ *is the edge relation.*

Definition 2. *Let* $G = (V, E, A, s, t)$ *be an alternating graph. We say that* t *is reachable from* s *iff* $P_a^G(s, t)$, *where* $P_a^G(x, y)$ *is the smallest relation on vertices of* G *satisfying:*

(1) $P_a^G(x, x)$.
(2) If x *is existential and* $P_a^G(z, y)$ *holds for some edge* (x, z) *then* $P_a^G(x, y)$.
(3) If x *is universal, there is at least one edge leaving* x, *and* $P_a^G(z, y)$ *holds for all edges* (x, z) *then* $P_a^G(x, y)$.

The idea here is that for t to be reachable from an existential node x there must exist a path from x to t, while the condition for a universal node y is stronger: t is reachable from y if and only if every path from y leads to t. One can think about alternating reachability in terms of a competitive game, where the player controlling existential vertices wants to get to t and the player controlling universal vertices is trying to prevent that. For example, in the alternating graph of Fig. 1, t is not reachable from s (i.e., there is no winning strategy for the existential player). To see it just imagine that the first player will move from s to v. Then the second player has only one choice leading to the dead-end. It means, that not every move of the first player controlling the universal node s is on the path to t.

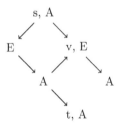

Fig. 1. t is not reachable from s

Now, we can define the alternating reachability problem, that is a class of alternating graphs in which t is reachable from s. One can think about that as a decision problem: given an alternating graph G and nodes s, t check whether t is reachable from s.

Definition 3. $REACH_a = \{G | P_a^G(s, t)\}$

The following computational complexity result will be crucial for us.

Theorem 1 ([4]). $REACH_a$ *is PTIME-complete via first-order reductions.*

Proof. The original proof of Immerman simulates directly an alternating Turing machine (ATM) to show that the problem is complete for ATM logarithmic space, known to be equal to P [5].

As the proof of Theorem 1 simulates ATM computation tree it follows that:

Corollary 1. $REACH_a$ *is PTIME-complete on trees.*

Note, that given a game tree T and an existential node s, the problem $REACH_a$ over T intuitively corresponds to the question: 'Is s a winning position for the first player in the zero-sum game T, i.e., can the first player force the game from node s towards node t against all possible counterstrategies of the second player?' (see [6], Problem A.11.1).

3   Backward Induction Problem

Now we are ready to define the computational decision problem corresponding to BI for extensive form, non zero-sum games. Intuitively: can the first player force the game from node s towards node t against all possible *rational* (= pay-off maximizing) counterstrategies of the second player? The difference here is that we consider only rational strategies as the non zero-sum games do not have to be strictly competitive.

Definition 4. *A   two-player   finite   extensive   form   game T = $(V, E, V_1, V_2, V_{end}, f_1, f_2, s, t)$, where V is the set of nodes, $E \subseteq V \times V$ is the edge relation (available moves). For $i = 1, 2$, $V_i \subseteq V$ is the set of nodes controlled by Player i, and $V_1 \cap V_2 = \emptyset$. V_{end} is the set of end nodes. Finally, $f_i : V_{end} \longrightarrow \mathbb{N}$ assigns pay-offs for Player i.*

Without loss of generality let us restrict attention to *generic games*:

Definition 5. *A game T is generic, if for each player, distinct end nodes have different pay-offs.*

Definition 6. *Let T be a two-player game. We define the backward induction accessibility relation on T. Let $P_{bi}^T(x, y)$ be the smallest relation on vertices of T such that:*

(1) $P_{bi}^T(x, x)$.
(2) Take $i = 1, 2$. Assume that $x \in V_i$ and $P_{bi}^T(z, y)$. If the following two conditions hold, then also $P_{bi}^T(x, y)$ holds:
   (a) $E(x, z)$;
   (b) there is no w, v such that $E(x, w)$, $P_{bi}^T(w, v)$, and $f_i(v) > f_i(y)$.

For example, in the tree of Fig. 3 t is not a backward induction solution for the game starting from s. Player 2 will rather start the game by going to the state w than v. And, t is not reachable from w.

We can again define the corresponding decision problem – whether in the game represented by tree T and starting in node s the first player can force the output t – as a class of game trees where s and t belong to the backward induction accessibility relation on T?

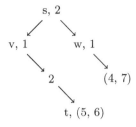

Fig. 2. t is not reachable from s

Definition 7. $\mathbb{BI} = \{T | P_{bi}^T(s,t)\}$

The problem $\mathbb{BI}$ intuitively corresponds to the question whether t is a sub-game perfect equilibrium in game T starting at node s [7].

4   Complexity of BI

The definability result of Van Benthem and Gheerbrant implies that it can be decided in polynomial time whether node t is a subgame perfect equilibrium of the game, i.e., the result of a gameplay following a BI strategy. First of all, note that it also follows that given an arbitrary finite extensive game with starting node s one can find a BI solution of the game in polynomial time. Simply, it is enough to run the polynomial decision algorithm for every node of the game. In this section we prove that computing backward induction relation is not only in PTIME but it is actually a PTIME-complete problem.

Theorem 2. $\mathbb{BI}$ *is PTIME-complete via first-order reductions.*

Proof. First of all, $\mathbb{BI}$ is in PTIME by providing FO(LFP) definition [3]. Now, it suffices to show PTIME-hardness. For that we will reduce the $REACH_a$ problem on trees (cf. Corollary 1) to the $\mathbb{BI}$ problem. We take any alternating tree $T = (V, E, A, s, t)$. Without loss of generality let us assume that s is existential. We construct a two player game, $T' = (V', E', V_1, V_2, V_{end}, f_1, f_2, s', t')$, where: $V = V'$, $t \in V_{end} = \{$end nodes of V$\}$, $E = E'$, $V_1 = V - A$, $V_2 = A$, $s = s$, $t = t$, and for every $v \in V'$, if $v \neq t$, then $f_1(t) > f_1(v)$ and $f_2(t) < f_2(v)$.

Now, we need to prove that $T \in REACH_a$ iff $T' \in \mathbb{BI}$. Assume, that $T \in REACH_a$. It means that whatever Player 2 does in the game T', Player 1 has a strategy to force outcome t. As t gives strictly the best pay-off for Player 1, then $P_{bi}^{T'}(s,t)$. Hence, $T' \in \mathbb{BI}$. For the other direction, assume for contradiction that $T \notin REACH_a$. This means that there is a node $v \neq t$ such that Player 2 can guarantee the game T' to end in v. From the pay-off construction for T', v is more attractive to Player 2 than t. Therefore, it is not the case that $P_{bi}^{T'}(s,t)$. Hence, $T' \notin \mathbb{BI}$.

What does this tell us about the complexity of backward induction? First of all, problems in PTIME are usually taken to be tractable [8], so relatively easy to solve, also for humans [9]. Furthermore, given assumptions on non-collapse, PTIME-completeness suggests that the problem of deciding whether a given node is a sub-game perfect equilibrium of the game is difficult to effectively parallelize (it lies outside NC^1) and solve in limited space (it lies outside LOGSPACE).

References

1. Szymanik, J., Meijering, B., Verbrugge, R.: Using intrinsic complexity of turn-taking games to predict participants' reaction times. In: Knauff, M., Pauen, M., Sebanz, N., Wachsmuth, I. (eds.) Proceedings of the 35th Annual Conference of the Cognitive Science Society, pp. 1426–1432. Cognitive Science Society, Austin (2013)
2. van Benthem, J., Gheerbrant, A.: Game solution, epistemic dynamics and fixed-point logics. Fundamenta Informaticae 100(1-4), 19–41 (2010)
3. Immerman, N.: Descriptive Complexity. Texts in Computer Science. Springer, New York (1998)
4. Immerman, N.: Number of quantifiers is better than number of tape cells. Journal of Computer and System Sciences 22(3), 384–406 (1981)
5. Chandra, A.K., Kozen, D.C., Stockmeyer, L.J.: Alternation. J. ACM 28(1), 114–133 (1981)
6. Greenlaw, R., Hoover, J.H., Ruzzo, W.L.: Limits to Parallel Computation: P-Completeness Theory. Oxford University Press, USA (1995)
7. Osborne, M.J., Rubinstein, A.: A Course in Game Theory. The MIT Press, Cambridge (1994)
8. Edmonds, J.: Paths, trees, and flowers. Canadian Journal of Mathematics 17, 449–467 (1965)
9. Frixione, M.: Tractable competence. Minds and Machines 11(3), 379–397 (2001)

---

[1] A problem is in NC if there exist constants c and k such that it can be solved in time $O(log^c n)$ using $O(n^k)$ parallel processors.

On Fuzzy Propositional Logic
with Different Negations

Shengli Zhang

Department of Computer Science, Xingyi Normal University for Nationalities,
Xingyi 562400, P.R. China
zs18203@163.com

Abstract. Pan pointed out that there are three kinds of different negations: contradictory negation, opposite negation and medium negation.
He also proposed the novel fuzzy set FScom with the above three kinds
of negations, followed by one improved fuzzy set IFScom that is established based on FScom. In this paper, a novel fuzzy propositional logic
calculus system that equipped with three kinds of negations is set up
to correspond to the FScom and IFScom. The soundness theorem and
completeness theorem of the system is proved.

1 Introduction

In the field of knowledge representation and reasoning, the negative knowledge
plays a particular and important role, which is as informative as positive knowledge. In recent years, this significant but interesting topic has attracted many
researchers to investigate. Wagner and Analyti et al [1,2] considered that there
exist(at least) two kinds of negations as follows: a weak negation expressing
non-truth and a strong negation expressing explicit falsity. Kaneiwa [3] proposed description logic with classical and strong negation. Ferré [4] introduced
an epistemic extension of the notion of negation in Logical Concept Analysis and
Natural Language. Pan [5] pointed out that there are three kinds of negations
in fuzzy knowledge and its negative relations, which are contradictory negation,
opposite negation and medium negation. He established a novel type of fuzzy set
called the fuzzy set with contradictory negation, opposite negation and medium
negation(FScom). Moreover, an improved fuzzy set, denoted by IFScom, was
proposed in [6] which overcomes the shortage of FScom in terms of sketching
fuzzy knowledge.

However, the fuzzy logic corresponding to fuzzy sets has only one kind of
negation cannot effectively distinguish and cope with the above-mentioned three
negations in fuzzy knowledge. Therefore, it is the necessity and natural demand
for efficiently tackling fuzzy knowledge to construct one new fuzzy logic calculus
with contradictory, opposite and medium negation(FPcom). At the same time,
we expect that this fuzzy logic system is able to provide a kind of logic tool for
fuzzy sets FScom and IFScom.

D. Grossi, O. Roy, and H. Huang (Eds.): LORI 2013, LNCS 8196, pp. 357–361, 2013.

2   Constructing Fuzzy Logic System FPcom

Definition 1.   Let S be a non-empty set, every element of S is called atomic formula, $\neg$, $\dashv$, $\sim$ and $\vee$ are connectives respectively, (and) are brackets, we stipulate that
(i) for any $A \in S$,A is one atomic fuzzy formula;
(ii) if A and B are atomic fuzzy formulas, then all of $\neg A$, $\dashv A$, $\sim A$ and $(A \vee B)$ are fuzzy formulas;
(iii) the set of all the fuzzy formulas if and only if generated by (i) and (ii), denoted by $F(S)$.

In fact, $F(S)$ is the free algebra of type $(\neg, \dashv, \sim$ $, \vee)$ generated by S [7]. We call the member of $F(S)$ medium fuzzy formula, or formula for short. Since there exists contradictory negation$\neg$, opposite negation $\dashv$ and medium negation $\sim$ in $F(S)$, $F(S)$is said to the fuzzy formula algebra with contradictory negation, opposite negation and medium negation, or fuzzy formula algebra for short.

Definition 2.   Let $F(S)$ be arbitrary fuzzy formula algebra, Γ, Δ be a set of formulae in $F(S)$ and A, A_i, B and C be medium fuzzy formulas in $F(S)$, where $i = 1, 2, 3, \cdots, n$. The following formal inferences are called inference rules of $F(S)$:
($\in$) $A_1, A_2 \ldots, A_n \vdash A_i$;
(τ) If $\Gamma \vdash \Delta \vdash A$, then $\Gamma \vdash A$; if $\vdash A$, then $\Gamma \vdash A$;
($\vee_1$) $A \vdash A \vee B$; if $\Gamma, A \vdash C$ and $\Gamma, B \vdash C$, then $\Gamma, A \vee B \vdash C$ (where Γ may be an empty set);
($\vee_2$) $A \vee B \vdash B \vee A$;
($\wedge$) $A, B \vdash A \wedge B$; $A \wedge B \vdash A, B$;
($\neg$) if $\Gamma, \neg A \vdash B, \neg B$, then $\Gamma \vdash A$;
($\rightarrow_-$) $A \rightarrow B, A \vdash B$; $\sim A \rightarrow B, A \vdash B$;
($\rightarrow_\wedge$) $A \rightarrow B, A \rightarrow C \vdash A \rightarrow (B \wedge C)$;
($\dashv_\vee$) $\dashv(A \vee B) \vdash \dashv A, \dashv B$; $\dashv A, \dashv B \vdash \dashv(A \vee B)$;
($\sim_\vee$) $\sim (A \vee B) \vdash (\sim A \wedge \sim B) \vee (\sim A \wedge \dashv B) \vee (\dashv A \wedge \sim B)$; $(\sim A \wedge \sim B) \vee (\sim A \wedge \dashv B) \vee (\dashv A \wedge \sim B) \vdash \sim (A \vee B)$;
($\rightarrow_+$) If $\Gamma, A \vdash B$ and $\Gamma, \sim A \vdash B$, then $\Gamma \vdash A \rightarrow B$;
($\sim\sim$) $A \rightarrow A \vdash \sim\sim A$; ($Y$) $\neg\dashv A, \neg \sim A \vdash A$;
($Y_\sim$) $\sim A \vdash \neg A, \neg\dashv A$; ($\dashv\dashv$) $A \vdash \dashv\dashv A$; $\dashv\dashv A \vdash A$.
where the formal expression "$\nabla, * \vdash \Diamond$" means from ∇ and $*$,$\Diamond$ follows; "$\nabla \vdash *, \Diamond$" means $*$ and $\Diamond$ follows by ∇; $P \rightarrow Q$ is defined as $\dashv P \vee Q$; $P \wedge Q$ is the abbreviation of the formula $\dashv(\dashv P \vee \dashv Q)$.

The system consisting of $F(S)$ and the above inference rules is said to Fuzzy Proposition logic with Contradictory negation, Opposite negation and Medium negation, denoted by FPcom.

Definition 3. In the fuzzy propositional logic FPcom, the relation among $\neg A, \dashv A$and $\sim A$ is as follows
$$\neg A = \dashv A \vee \sim A$$

Like other logic calculus system, the one can obtain many significant conclusions through developing mathematically FPcom step by step according to the above inference rules. Here, we only give some more significant conclusions as follows.

Proposition 1. (i) $A, \neg A \vdash B$; (ii) $A, \dashv A \vdash B$; (iii) $A, \sim A \vdash B$; (iv) $\dashv A, \sim A \vdash B$.

Proposition 2. (i) $\neg\neg A \vdash A$; (ii) $A \vdash \neg\neg A$; (iii) If $\Gamma, A \vdash B, \neg B$, then $\Gamma \vdash \neg A$.

Proposition 3. (i) $\neg A, \neg\dashv A \vdash\sim A$; (ii) $\dashv A \vdash\dashv \neg A, \neg \sim A$; (iii) $A \vdash \neg\dashv A, \neg \sim A$, where "$\triangle \vdash\dashv *$" means "$\triangle \vdash *$" and "$* \vdash \triangle$". The symbol "$\vdash\dashv$" below means the same.

Proposition 4. (i) If $\Gamma, \dashv A \vdash B, \neg B$ and $\Gamma, \sim A \vdash C, \neg C$, then $\Gamma \vdash A$; (ii) If $\Gamma, A \vdash B, \neg B$ and $\Gamma, \dashv A \vdash C, \neg C$, then $\Gamma \vdash\sim A$; (iii) If $\Gamma, A \vdash B, \neg B$ and $\Gamma, \sim A \vdash C, \neg C$, then $\Gamma \vdash \dashv A$.

Proposition 5. (i) If $A \vdash B, \sim A \vdash\sim B, \dashv A \vdash \dashv B$, then $B \vdash A, \sim B \vdash\sim A, \dashv B \vdash \dashv A$; (ii) If $A \vdash\dashv B, \dashv A \vdash\dashv \dashv B$, then $\sim B \vdash\dashv\sim A$.

Proposition 6. (i) $\vdash \neg\dashv \sim A$; (ii) $\vdash \neg\dashv\neg A$; (iii) $\dashv \sim A \vdash B$; (iv) $\dashv\neg A \vdash B$.

Proposition 7 (substitutive theorem). In fuzzy propositional calculus system FPcom, if $A \vdash\dashv B$ and $\dashv A \vdash\dashv \dashv B$ and $\sim A \vdash\dashv\sim B$, thus we have $f(A) \vdash\dashv f(B)$ and $\dashv f(A) \vdash\dashv \dashv f(B)$ and $\sim f(A) \vdash\dashv\sim f(B)$, for arbitrary medium fuzzy formula $f(p)$.

Definition 4. In fuzzy propositional calculus system FPcom, if we have $f(A) \vdash\dashv f(B)$ for any medium fuzzy formula $f(p)$, then A and B are said to be provability-equivalent, denoted by $A \Leftrightarrow B$.

3   λ-truth Value and λ-satisfiability of the Medium Fuzzy Formula

Definition 5. Let S be a non-empty set. For any formula A in $F(S)$, we stipulate λ-evaluation $v_\lambda : F(S) \to [0, 1-\lambda)\cup(1-\lambda, \lambda)\cup(\lambda, 1]$, where $\lambda \in (0.5, 1)$, as follows.

1) if A is an atomic formula, $v_\lambda(A)$ takes a unique value in $[0, 1 - \lambda) \cup (1 - \lambda, \lambda) \cup (\lambda, 1]$;

2) $v_\lambda(\dashv A) = 1 - v_\lambda(A)$;

3)

$$
v_\lambda(\sim A) = \begin{cases}
\frac{2\lambda-1}{1-\lambda}v_\lambda(A) + 1 - \lambda & \text{when } v_\lambda(A) \in (0, 1 - \lambda), \\
\frac{2-2\lambda}{1-2\lambda}(1 - \lambda - v_\lambda(A)) + \lambda & \text{when } v_\lambda(A) \in (1 - \lambda, \frac{1}{2}), \\
\frac{2-2\lambda}{1-2\lambda}(v_\lambda(A) - \lambda) + \lambda & \text{when } v_\lambda(A) \in [\frac{1}{2}, \lambda), \\
\lambda - \frac{2\lambda-1}{1-\lambda}(v_\lambda(A) - \lambda) & \text{when } v_\lambda(A) \in (\lambda, 1), \\
\frac{1}{2} & \text{when } v_\lambda(A) = 0 \text{ or } 1;
\end{cases}
$$

4) $v_\lambda(A \vee B) = \max(v_\lambda(A), v_\lambda(B))$;

5) $v_\lambda(\neg A) = v_\lambda(\dashv A\vee \sim A) = \max(v_\lambda(\dashv A), v_\lambda(\sim A))$.

Theorem 1. Let $F(S)$ be one fuzzy formula algebra, $\lambda \in (\frac{1}{2}, 1)$, A be any formula in $F(S)$, then we have

1) $v_\lambda(A) > \lambda$ or $v_\lambda(A) < 1 - \lambda$, if and only if $1 - \lambda < v_\lambda(\sim A) < \lambda$.

2) $1 - \lambda < v_\lambda(A) < \lambda$ if and only if $\lambda < v_\lambda(\sim A) \leq 1$.

Definition 6. Let A be arbitrary medium fuzzy formula in $F(S)$, Γ be a collection of formulae in $F(S)$, $\lambda \in (\frac{1}{2}, 1)$. If $v_\lambda(A) > \lambda$ holds for every λ-evaluation v_λ for $F(S)$, then A is said to be λ-permanent valid, denoted by $\models A$; If $v_\lambda(A) < \lambda$ follows for every λ-evaluation v_λ for $F(S)$, then A is said to be λ-permanent false; for every λ-evaluation v_λ for $F(S)$, if $v_\lambda(\Gamma) > \lambda$ holds, one can get $v_\lambda(A) > \lambda$, thus the formal inference $\Gamma \vdash A$ is said to be λ-valid inference, written as $\Gamma \models A$.

Definition 7. Let A be arbitrary medium fuzzy formula in $F(S)$, $\lambda \in (\frac{1}{2}, 1)$. If there exists λ-evaluation v_λ for $F(S)$, such that $v_\lambda(A) > \lambda$, we call A λ-satisfiable; if there exists λ-evaluation v_λ for $F(S)$, such that $v_\lambda(A) < \lambda$, we call A λ-trick.

We have some vital consequences for fuzzy logic system FPcom as follows.

Lemma 1. Let A be arbitrary medium fuzzy formula in $F(S)$, $\lambda \in (\frac{1}{2}, 1)$. Then $v_\lambda(\neg A) < \lambda$ holds iff $v_\lambda(\daleth A) < v_\lambda(\sim A) < \lambda$ follows for every λ-evaluation v_λ for $F(S)$.

Lemma 2. Let A be arbitrary medium fuzzy formula in $F(S)$, $\lambda \in (\frac{1}{2}, 1)$. Thus, $v_\lambda(A) > \lambda$ follows iff $v_\lambda(\daleth A) < v_\lambda(\sim A) < \lambda$ holds for every λ-evaluation v_λ for $F(S)$.

From the lemmas 1 and 2, it is immediate to get the result as follows.

Theorem 2. Let A be arbitrary medium fuzzy formula in $F(S)$, $\lambda \in (\frac{1}{2}, 1)$. For every λ-evaluation v_λ for $F(S)$, then we have

(1) $v_\lambda(A) > \lambda$ iff $v_\lambda(\neg A) < \lambda$; (2) $v_\lambda(A) < \lambda$ iff $v_\lambda(\neg A) > \lambda$.

4   Soundness and Completeness of Fuzzy Logic System FPcom

Lemma 3. Let $F(S)$ be arbitrary fuzzy formula algebra, Γ, Δ be a collection of formulae in $F(S)$ and A, A_i, B, C be medium fuzzy formulas in $F(S)$, where $i = 1, 2, 3, \cdots, n$. Then all of the formal inference rules $\Gamma \vdash A$ are λ-permanent valid $\Gamma \models A$, namely,

(1) $A_1, A_2 \ldots, A_n \models A_i$; (2) If $\Gamma \models \Delta \models A$, then $\Gamma \models A$; if $\models A$, then $\Gamma \models A$; (3) $A \models A \vee B$; if $\Gamma, A \models C$ and $\Gamma, B \models C$, then $\Gamma, A \vee B \models C$ (where Γ may be an empty set); (4) $A \vee B \models B \vee A$; (5) $A, B \models A \wedge B$; $A \wedge B \models A, B$; (6) If $\Gamma, \neg A \models B, \neg B$, then $\Gamma \models A$; (7) $A \rightarrow B, A \models B$; $\sim A \rightarrow B, A \models B$; (8) $A \rightarrow B, A \rightarrow C \models A \rightarrow (B \wedge C)$; (9) $\daleth(A \vee B) \models \daleth A, \daleth B$; $\daleth A, \daleth B \models \daleth(A \vee B)$; (10) $\sim (A \vee B) \models (\sim A \wedge \sim B) \vee (\sim A \wedge \daleth B) \vee (\daleth A \wedge \sim B)$; $(\sim A \wedge \sim B) \vee (\sim A \wedge \daleth B) \vee (\daleth A \wedge \sim B) \models \sim (A \vee B)$; (11) If $\Gamma, A \models B$ and $\Gamma, \sim A \models B$, then $\Gamma \models A \rightarrow B$; (12) $A \rightarrow A \models \sim\sim A$; (13) $\neg\daleth A, \neg \sim A \models A$; (14) $\sim A \models \neg A, \neg\daleth A$; (15) $A \models \daleth\daleth A$; $\daleth\daleth A \models A$.

Theorem 4 (Soundness of FPcom). Let $F(S)$ be arbitrary fuzzy formula algebra, Γ be a collection of formulae in $F(S)$ and A be one medium fuzzy formula in $F(S)$. For fuzzy logic system FPcom, we have

(i) If $\Gamma \vdash A$, then $\Gamma \models A$. (ii) If $\vdash A$, then $\models A$.

Lemma 4. Let A be any medium fuzzy formula in fuzzy formula algebra $F(S), \lambda \in (0.5, 1)$. For any λ-evaluation v_λ for $F(S)$, we can see any formula B in $F(S)$ is λ-permanent valid whenever $\lrcorner \sim A$ is λ-permanent valid.

Lemma 5. Let A be any medium fuzzy formula in fuzzy formula algebra $F(S), p_1, p_2, \cdots, p_n$ be an enumeration of all the pairwise different atomic fuzzy formulas in A. For every λ-evaluation v_λ for $F(S)$, $\lambda \in (0.5, 1)$, we stipulate $(i = 1, 2, 3, \cdots, n)$

$$v_\lambda(A_i) = \begin{cases} p_i & \text{if } v_\lambda(p_i) > \lambda, \\ \lrcorner p_i & \text{if } v_\lambda(p_i) < 1 - \lambda, \\ \sim p_i & \text{if } 1 - \lambda < v_\lambda(p_i) < \lambda, \end{cases}$$

Then we have:
(1) If $v_\lambda(A) > \lambda$, then $A_1, A_2, \cdots, A_n \vdash A$;
(2) If $v_\lambda(A) < 1 - \lambda$, then $A_1, A_2, \cdots, A_n \vdash \lrcorner A$;
(3) If $1 - \lambda < v_\lambda(A) < \lambda$, then $A_1, A_2, \cdots, A_n \vdash \sim A$.

Theorem 5 (Completeness of FPcom). Let Γ be a collection of formulae in $F(S)$ and A be one medium fuzzy formula in $F(S)$. For every λ-evaluation v_λ for $F(S)$, $\lambda \in (0.51)$, we have
(1) If $\models A$, then $\vdash A$; (2) If $\Gamma \models A$, then $\Gamma \vdash A$.

Acknowledgments. This research is sponsored by the Science and Technology Department of Guizhou Province, P.R.China, Grant No. 2324[2012] Contract J of the Science and Technology Department of Guizhou Province.

References

1. Wagner, G.: Web rules need two kinds of negation. In: Bry, F., Henze, N., Małuszyński, J. (eds.) PPSWR 2003. LNCS, vol. 2901, pp. 33–50. Springer, Heidelberg (2003)
2. Analyti, A., Antoniou, G., Damasio, C.V., Wagner, G.: Negation and Negative Information in the W3C Resource Description Framework. Annals of Mathematics, Computing and Teleinformatics (AMCT) 1(2), 25–34 (2004)
3. Kaneiwa, K.: Negations in description logic-contraries, contradictories, and subcontraries. In: Proceedings of the 13th International Conference on Conceptual Structures (ICCS 2005), pp. 66–79 (2005)
4. Ferré, S.: Negation, Opposition, and Possibility in Logical Concept Analysis. In: Missaoui, R., Schmidt, J. (eds.) ICFCA 2006. LNCS (LNAI), vol. 3874, pp. 130–145. Springer, Heidelberg (2006)
5. Pan, Z.H.: Three kinds of fuzzy knowledge and their base of set. Chinese Journal of Computers 35(7), 1421–1428 (2012) (in Chinese)
6. Zhang, S.L., Pan, Z.H.: An improved set description of negative knowledge processing in fuzzy knowledge. Journal of Shangdong University (Natural Science) 46(5), 103–109 (2011) (in Chinese)
7. Wang, G.J.: Non-classical mathematical logics and approximate reasoning. Science Press, Beijing (2008) (in Chinese)

Author Index